Biopolymers for Biomedical and Biotechnological Applications

Biopolymers for Biomedical and Biotechnological Applications

Edited by

Bernd H. A. Rehm
M. Fata Moradali

WILEY-VCH

Editors

Dr. Bernd H. A. Rehm
Centre for Cell Factories and Biopolymers
Griffith Institute for Drug Discovery
Menzies Health Institute Queensland
Griffith University
Don Young Road, Nathan, QLD 4111
Australia

Dr. M. Fata Moradali
Department of Oral Immunology and Infectious Diseases
University of Louisville
United States

Library of Congress Card No.:
applied for

British Library Cataloguing-in-Publication Data
A catalogue record for this book is available from the British Library.

Bibliographic information published by the Deutsche Nationalbibliothek
The Deutsche Nationalbibliothek lists this publication in the Deutsche Nationalbibliografie; detailed bibliographic data are available on the Internet at <http://dnb.d-nb.de>.

Print ISBN: 978-3-527-34530-4
ePDF ISBN: 978-3-527-81828-0
ePub ISBN: 978-3-527-81830-3
oBook ISBN: 978-3-527-81831-0

Cover Design Adam-Design, Weinheim, Germany
Typesetting SPi Global, Chennai, India
Printing and Binding
CPI Group (UK) Ltd, Croydon CR0 4YY
Printed on acid-free paper

10 9 8 7 6 5 4 3 2 1

Contents

1

Advances in Biocompatibility: A Prerequisite for Biomedical Application of Biopolymers

Matthew R. Jorgensen, Helin Räägel, and Thor S. Rollins

Nelson Laboratories, LLC, 6280 S Redwood Rd, Salt Lake City, UT 84123, USA

1.1 Introduction

Biocompatibility is a concept that, in one form or another, has existed since the dawn of medicine. At the base of Vesuvius in ancient Rome was the house of a surgeon, home to an impressive collection of medical instruments that were preserved by ash when the mountain exploded. Without a doubt, patrons of the ancient surgeon subjected themselves to these devices with the expectation and trust that they would be getting better – not worse – due to the treatment they received. While biocompatibility has not always been explicitly defined through history, the safety of a tool in a doctor's hand is central to the mission of the doctor. Following the industrial revolution, instruments have become mass-produced and marketed as effective tools for the practice of medicine, making doctors rely on the diligence of the manufacturer to ensure patient safety. Concurrently, our knowledge of toxicology has expanded through experience, and medical journals have become widely available to share clinical experiences. These platforms have been and are currently successfully used to notify doctors and also the public about medical instruments thought to be safe, but which actually did more harm than good, and discuss options for mitigating the risks associated with the use of these devices.

To protect patients from being harmed by medical devices, which for one reason or another might be unsafe due to negligence on the part of the device manufacturer, medical device safety has become regulated. These regulations require medical device manufacturers making a device or product to demonstrate that what they are producing performs appropriately when used as intended. Past experience and modern toxicology have identified what sorts of health risks are associated with the use of a given medical device. The most modern and comprehensive overview of biocompatibility is the suite of documents that make up the international standard ISO 10993; the first document in the series, *ISO 10993-1,* provides the high-level framework for evaluation of biocompatibility as a whole, while the other documents in the series explore specific topics in more detail.

The modern concept and definition of biocompatibility is the ability of a medical device (or material) to "perform with an appropriate host response" when

Biopolymers for Biomedical and Biotechnological Applications, First Edition.
Edited by Bernd H. A. Rehm and M. Fata Moradali.

used as intended. This means that the device or material should not cause an unacceptable biological risk when used, taking into account the nature of use in terms of contact site and duration, as well as the potential benefit of using the device. ISO 10993-1, Annex A, lists several key biological risks associated with specific types and durations of patient contact. As the contact duration goes up, and the devices or materials become more invasive, the types of potential risks multiply. For example, a device that is used on an intact skin is not very invasive, and therefore the associated risks are minimal; the skin is an organ effective at protecting the body from our natural environment that is often replete with biological risks. In contrast, consider a neurological stent; this invasive device is in permanent contact with brain tissues. For such a device, risks range from immediate toxicity to thrombosis to more chronic systemic toxicities like cancer. Therefore, even the more modern concept of biocompatibility encompasses the broader idea well captured by the oft-repeated phrase in medicine "First, do no harm," which certainly applies to the materials used with the intention of healing.

1.2 Biocompatibility Evaluation of Biopolymeric Materials and Devices

Biopolymers represent a special subset of materials useful in medicine, being derived or produced by living organisms or synthesized from basic biological building blocks. Compared with synthetic polymers, the advantages from the perspective of biocompatibility are clear: because these materials are made by living systems, from building blocks ubiquitous to life, it would seem like the potential for adverse biological reactions would be reduced. For implants, like biocomposite bone anchors used by Arthrex® in hip arthroscopy procedures (Figure 1.1), if the goal is to mimic the tissue being replaced, using a material made from

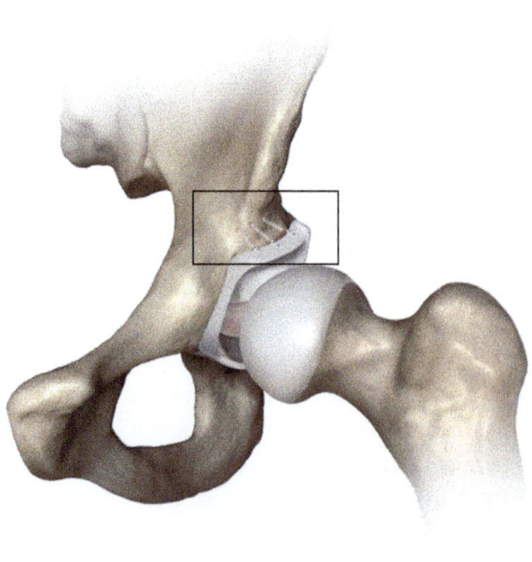

Figure 1.1 BioComposite Knotless SutureTak® anchor used in hip arthroscopy procedures. *Source:* Courtesy of Arthrex®.

natural building blocks is logical. The scope and range of biopolymers has been discussed in detail within this text and elsewhere in literature [1–3]. Briefly, they include polysaccharides (such as chitin, hyaluronic acid, and cellulose), polyesters (such as polylactic acid [PLA]), proteins (such as silk, collagen, and casein), and others like latex rubber and shellac. As varied as the possible biopolymers are their individual chemical properties; therefore, broad grouping of biopolymers for biocompatibility is not possible. Rather, these materials should be considered without special allowance, in terms of their intended use and durability in the body.

The biocompatibility evaluation process, in general, begins by determining what potential biological risks the use of the material would present. Once risks are determined, a plan to evaluate those risks should be developed. Often, the risk identification process begins by answering the following questions:

1. **What is the intended use of the device (or material)?**
   a. What tissues or fluids will it contact in the body (either directly or indirectly)?
   b. How long is the cumulative amount of time it may contact the body?
   c. Who will be exposed to the device (infants, pediatrics, adults)?
2. **What is known about the device materials and their fate in the body?**
   a. What processing, packaging, and sterilization are the materials exposed to?
   b. Are the materials known to degrade over time?
   c. What previous clinical experience is there with the device (or materials)?

Annex A in ISO 10993-1 contains a chart of biological risks for consideration, stratified by contact duration (limited ≤24 hours, prolonged >24 hours to 30 days, long term >30 days) and contact type. These risks can provide a starting point for understanding the risks presented by a device for both the device manufacturer and those who would in the end approve the device for use. To illustrate how Annex A is used, two commonly used biopolymeric devices are put through the thought process as examples:

- Device 1: A chitin-based hemostatic agent for acute treatment during massive hemorrhage in an open wound
- Device 2: A polycaprolactone (PCL) implant for infants, designed to degrade and resorb over a period of two to three years

How the description of Device 1 and Device 2 translates into a classification and set of biological risks is shown in Table 1.1.

The risks identified by ISO 10993-1, Annex A (outlined for the two devices in Table 1.1), are not necessarily all-inclusive or exhaustive. The spirit of the document is to provide a starting point and basis for a biological evaluation; if other potential biological or toxicological risks are known through clinical experience, those would also need to be addressed. For instance, if a medical instrument is known or has been shown to chip during a surgical procedure, leaving fragments of the device possibly permanently in the patient, this should be addressed in the biocompatibility assessment.

Table 1.1 Example classification and associated risks for two representative devices.

	Hemostatic	**Implant**
Contact tissues	Bleeding wound	Muscle and bone
Contact duration	Expected to be less than 24 h, but could extend beyond	Device resorbs over 2–3 yr
Target patient population	Adults	Infants
Classification per Annex A	Category: surface medical device Contact: breached or compromised skin Contact duration: prolonged	Category: implant medical device Contact: tissue/bone Contact duration: permanent
Biological risks to be addressed (per ISO 10993-1, Annex A)	• Cytotoxicity • Sensitization • Irritation • Material-mediated pyrogenicity • Acute systemic toxicity • Subacute toxicity • Implantation effects	• Cytotoxicity • Sensitization • Irritation • Material-mediated pyrogenicity • Acute systemic toxicity • Subacute toxicity • Subchronic toxicity • Chronic toxicity • Implantation effects • Genotoxicity • Carcinogenicity • Degradation

It should also be recognized that the risks identified by Annex A are not highlighted in the standard as an explicit "checklist for testing." Fortunately, the latest ISO 10993-1 released in 2018 more clearly defines this statement within the document. Based on the updated verbiage in the standard, each of the biological risks (or endpoints) can be evaluated using a risk-based approach, taking into consideration chemical and material information, existing endpoint-specific data, or a written rationale why testing or further data is not needed to address a particular risk. In any case, the biocompatibility of a device or material must be spelled out, addressing directly each of the specific risk identified, mitigating concern through testing results or written evaluation in a biological risk assessment.

1.3 Using a Risk-Based Approach to Biocompatibility

After the specific biological risks for a particular implementation of a device are identified, the strategy for how the biological safety will be proven must be decided. In the past, the expectation was that because devices are typically made by competitors in unique environments, and with proprietary processing, categorically calling a material "biocompatible" was not possible, and testing should be executed anew for each device coming to market. The list of biological risks was pretty much a shopping list, more or less blindly ordered and executed. Since that time, there has been a dramatic shift toward a more thoughtful scientific approach to the evaluation of biocompatibility.

The shift from check-listing tests to a risk-based approach has been motivated by several factors:

- Consideration of animal welfare, with a charge to reduce animal testing as much as possible
- A broader and better consolidated body of data on materials and toxicology
- Better analytical chemistry tools to evaluate manufacturing residuals, material leachables, and degradation products

Knowing that the key is to protect patient safety by proving biocompatibility of a device to the skeptical reviewer while at the same time avoiding as much unnecessary testing as possible is the heart of evaluating biocompatibility using a risk-based approach. There is an art to a biocompatibility evaluation, balancing commonsense measures to ensure safety with currently available data on one hand and the expectations of regulatory bodies across the spectrum on the other. Understanding the role the material information has and how this broadly impacts the testing strategy (along with the cost and time burden of testing) is central to the strategy.

In the best case, material information and written assessment alone can be sufficient to mitigate and address all of the biological risks associated with a device. To be convincing, however, a great deal of detail is needed. Often, the question of biocompatibility is not about the bulk material itself at all, but rather about the processing of that material that takes place both upstream and downstream. Consider a polycaprolactone (PCL) implant, manufactured using 3D printing from a powder starting material. To the manufacturer, the name PCL along with its assigned chemical abstracts service (CAS) number defines the material. But there are many ways to synthesize PCL [4] that may influence its safety profile in terms of impurities that (while not obvious from bulk properties) will affect toxicology. Consider the PCL pipeline upstream from the device manufacturer.

1. Preparation of the monomer (either ε-caprolactone or 6-hydroxycaproic acid) at raw chemical supplier:
   a. ε-Caprolactone and 6-hydroxycaproic acid may be produced naturally by oxidation of cyclohexanol by microorganisms and then harvested and purified (all steps removing or introducing impurities to varying degrees).
   b. ε-Caprolactone can also be produced industrially through a reaction of cyclohexanone with peracetic acid.
2. The monomer is purified, packaged, sold, and shipped to the maker of the polymer without knowledge that the monomer will end up in a medical device:
   a. Purity and performance metrics are based on bulk properties (not toxicological endpoints).
3. The monomer is polymerized by another manufacturer:
   a. Polymerization occurs using a variety of different possible techniques, using different activators and/or catalysts, several of which are complex

organometallic complexes of questionable safety (see, for example, those contained in Ref. [4]).

4. The polymer is powdered and purified by the manufacturer using a proprietary cryogenic process.

Most (or all) of the details of the upstream process are unknown to the medical device manufacturer, yet they can impact device safety. It could matter, from a toxicological perspective, if the PCL in a device is manufactured using lithium diisopropylamide or *tert*-butoxy potassium as a catalyst. If the device manufacturer were to ask their polymer supplier what catalyst or what monomer is used and the method of manufacture, the information is likely considered intellectual property, and medical device manufacturers are typically not big enough customers of polymer manufacturers to be able to make demands. Therefore, in these cases, it is up to the device manufacturer to prove the biocompatibility of their materials acknowledging that very little is known about the impurity profile of their device.

Knowing what you do not know and how that gap in knowledge might be interpreted by a regulator or a patient receiving the device is key in developing a testing strategy for biocompatibility (Figure 1.2). Regulators have been witness to all sorts of mischief on the part of manufacturers, and patients have been injured by devices made from misunderstood materials, elevating further the concern for each device that is in the process of clearance for market. For biopolymer devices, it is typically not known what trace chemicals may be in the material. Another gap in knowledge is often how different processing steps influence the degradation rate of resorbable biopolymers. To answer those questions, we turn to chemistry.

1.3.1 Chemistry of Biopolymers and Risk

Based on their physicochemical properties, various biopolymers so far used in the medical industry can loosely be placed into three categories: polysaccharides, proteins, and polyesters. Some examples of common biopolymers are shown in Table 1.2.

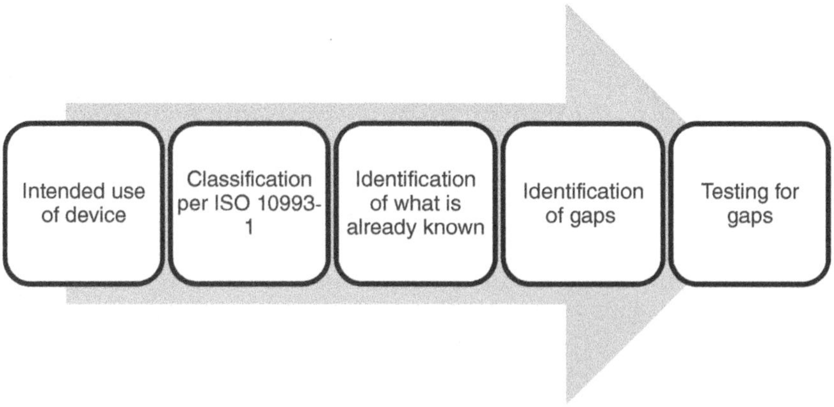

Figure 1.2 Thought process for using ISO 10993-1 for biological evaluation of medical devices.

Table 1.2 Examples of common biopolymers.

Classification	Example biopolymer	Notes on production	Risks
Polysaccharides	Hyaluronic acid, HA (polymer of D-glucuronic acid and *N*-acetylglucosamine)	Primarily produced using bacteria including *Streptococcus* [5–7]	Production by pathogenic bacteria coproduces myriad other potentially toxic biological products that must be removed during subsequent purification steps
	Cellulose (polymer of D-glucose)	From plant products, cellulose is dissolved from other plant materials in an alkali process, followed by purification. Produced bacterially using *Acetobacter xylinum* [8, 9]	Industrial purification steps can introduce impurities. Bacterial production coproduces myriad other potentially toxic biological products that must be removed during subsequent purification steps
Proteins	Silk Primarily fibroin, a repeating amino acid sequence of (Gly-Ser-Gly-Ala-Gly-Ala)	Primarily from the mulberry silkworm *Bombyx mori* [10]	Industrial post-processing and purification steps can introduce impurities
Polyesters	Polylactic acid	Primarily ring-opening polymerization of lactide (cyclic lactic acid dimer) [11]	Crude lactic acid contains many impurities (acids, alcohols, metals)

While the chemistry of biopolymers and the source of these materials' building blocks are very diverse, there is a commonality among them when it comes to potential patient risk: there is always concern over side products and manufacturing residuals. While it is accepted that biopolymers have an inherent advantage from being similar chemically to substances naturally found in the body, they also have the same disadvantage facing all medical device materials from being processed. For that reason, the chemical evaluation strategy used for medical devices made from biopolymers is very similar to what is used for devices made from fully synthetic materials. The heart of the strategy is acknowledging that the manufacturer of the device does not know what they do not know, and the only way to safeguard against unpleasant surprises is to screen for everything that might reasonably be in or on the device.

1.3.2 Chemistry Screening of Biopolymers

It is important to start the design of a chemistry testing strategy with the end goal in mind. In the case of chemistry for biocompatibility, the end goal is to be able to screen for unexpected contaminants with enough sensitivity and with enough

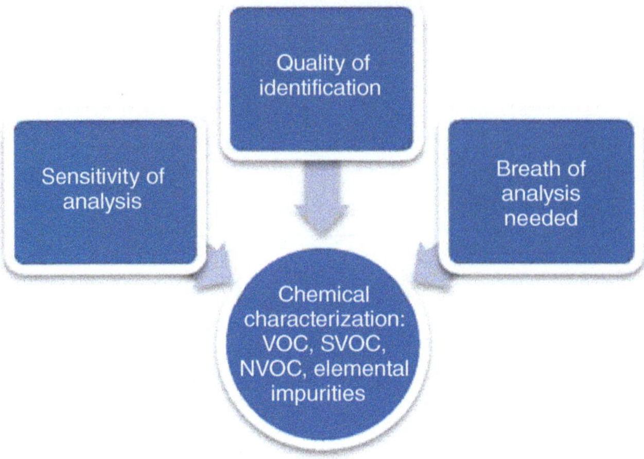

Figure 1.3 Important aspects for setting up a chemical characterization study.

accuracy that toxicological conclusions can be made based on the data produced (Figure 1.3). Determining the proper sensitivity can be a matter of debate but should be low enough so that any chemicals that are present – but not reported because they are below the sensitivity – are known to not be toxicologically concerning. In other words, a threshold of toxicological concern (TTC) is needed.

The TTC concept was developed to define an acceptable intake for any unstudied/understudied chemical that, if below the TTC, would pose a negligible risk of carcinogenicity, systemic toxicity, and reproductive toxicity. The concept was developed for chemicals present in the human diet and is accepted by the US Food and Drug Administration (FDA), International Conference on Harmonization (ICH), and the European Medicines Agency (EMA) for the evaluation of impurities in pharmaceuticals. It has also been used for assessing contaminants in consumer products and environmental contaminants. The methods upon which the TTC is based are generally considered very conservative since they involve data for the most sensitive species and most sensitive site induction (several "worst-case" assumptions). The TTC concept provides an estimate of safe exposures values for any compound not on the TTC exclusion list (i.e. metals, nitrosamines, and polycyclic aromatic hydrocarbons). The most conservative TTC value has been set at 1.5 μg/d and is assigned for greater than 10 years to a lifetime of exposure. A TTC of 120 μg/d has been proposed for genotoxic exposures limited to one month or less [12]. Exceeding the TTC is not necessarily associated with an increased risk given the conservative assumptions employed in the derivation of the TTC value [13–17]. When adequate evidence exists that a constituent is non-carcinogenic, a non-carcinogenic TTC value may be used to address the constituent (e.g. Cramer classification) [18, 19].

The TTC concept for medical devices was formalized in ISO 21726 published in February 2019. This brief international standard outlines the appropriate strategy for using the Cramer class and TTC. When adequate toxicological data is not available in the literature, the Cramer classification should be used for non-cancer

Table 1.3 Recommended TTC values from ISO 21726.

Medical device contact category	Limited (<24 h)	Prolonged (24 h to 30 d)	Long term[a] (>30 d)		
Duration of body contact	≤1 mo		>1–12 mo	>1–10 yr	>10 yr to lifetime
TTC for any one compound (μg/d)	120		20	10	1.5[b]

a) Considered permanent according to ISO 10993-1.
b) This value incorporates a 10^{-5} cancer risk for a 60 kg adult.

effects; for cancer-based effects, the ICH M7 TTC values should be used based on the contact duration of the device. Cramer classification stratifies compounds into three groups (I, II, and III, with III being the highest risk); the acceptable daily exposures are 1800 μg/d for class I, 540 μg/d for class II, and 90 μg/d for class III compounds. The TTC values from ISO 21726 for carcinogenic endpoints depend on contact duration and are shown in Table 1.3.

In addition to the sensitivity, the breadth of the analysis is critical. ISO 10993-12, ISO 10993-17, and ISO 10993-18 provide guidance on the sample preparation and scope of analysis to give the required breadth. The device should be extracted in multiple solvents covering a range of polarities to be representative of the range of matrices that are found in the body. Extraction conditions should be selected to appropriately exaggerate the amount of chemicals found. For example, extraction of the device at 50 °C for 72 hours is prescribed by ISO 10993-12 and is the most commonly used extraction condition. Typical extraction solvents are purified water, isopropyl alcohol, and hexane. Following extraction, the extracts must be analyzed for volatile organic compounds (VOCs), semi-volatile organic compounds (SVOCs), non-volatile organic compounds (NVOCs), and metals using a suite of techniques that are both qualitative and quantitative; these are almost always chromatography with mass spectroscopy (MS) for organic compounds and inductively coupled plasma for metals.

VOCs are typically analyzed for only in aqueous extracts, as semipolar and nonpolar solvents are often VOCs themselves. Two main techniques are available for VOCs: headspace gas chromatography with mass spectroscopy (HS-GC/MS) and purge and trap GC/MS. HS-GC/MS measures the volatiles present in the gas above a water sample in a closed vial; the vial might be slightly heated to encourage volatiles to enter the gas phase above the liquid. The gas is directed through a gas chromatograph, which separates molecules in the gaseous mixture by polarity. Different molecular polarities are retained in the instrument for different amounts of time; how long a molecule remains in the instrument is referred to as the retention time. After separation, the molecules are identified using mass spectroscopy. Briefly, mass spectroscopy works by fragmenting molecules into electrically charged pieces and then measuring the weight of those pieces very precisely. With knowledge of both the retention time and mass fragmentation patterns, VOCs can almost always be positively identified by comparison with large public or commercial databases. Purge and trap measurements differ from headspace only in the way compounds are sampled; first volatile organics are purged from the water by bubbling inert gas through the liquid and trapped in an adsorbent tube. VOCs are released from the tube into the GC/MS for analysis as with HS-GC/MS.

SVOC measurement methods provide the single broadest source of information regarding the content of extracts and are amenable to both aqueous and non-aqueous extraction matrices. The term SVOC is ill defined in the medical device community but generally is considered to be those compounds most well suited for analysis by direct injection GC/MS. The distinction of this definition is important, as there are many molecules amenable to direct injection GC/MS that are considered to be NVOCs by every other definition. The methods used for SVOCs by GC/MS are mostly characterized by the details of their sample preparation and rigor of data analysis; instrumental details of the GC/MS remain largely harmonized. Water extracts are prepared for analysis by first doing a solvent exchange to a solvent compatible with GC/MS. Typically this is accomplished by repeatedly shaking the extract with methylene chloride under acidic, neutral, and basic conditions. The methylene chloride can then be concentrated and directly injected into the instrument. Organic solvents do not need a solvent exchange and are typically concentrated and then directly injected.

NVOCs not amenable for analysis by GC/MS are most clearly those compounds that have such a high molecular weight or polarity that they are not capable of vaporization without decomposition. For these compounds, liquid chromatography with mass spectroscopy (LC/MS) must be used. Unlike GC/MS analyses, which have more or less standardized instrument parameters, LC methods are highly variable. Because of this variability, large public databases are of limited utility, and effective interpretation of data relies much more on the level of expertise of the analyst and internal experience of the analyzing lab. LC techniques coupled with advanced mass spectroscopy tools providing high-resolution accurate mass (HRAM) such as quantitative time of flight (qTOF) or Orbitrap can be a significant advantage, as these more sensitive methods can greatly narrow down the number of possible compounds in the identification process.

One of the key variables in chemical analysis for toxicological risk assessment and biocompatibility is the degree of certainty in the identification and quantification of compounds. Quality of identification can range from a fully automated comparison to a public database, without peer review of the results to fully confident identification. Fully automated identification can lead to scenarios where compounds with very low match scores are reported as compounds for which they are almost certainly not. On the other end of the identification spectrum is a fully validated identification where the compound in question has been injected using a standard on the same instrument and under the same conditions and under expert review. Of course, in practice, results can be a mix. It is not possible to inject standards for every compound that might occur from a biomaterial. With respect to quantification, results can vary based on the amount of evidence that is present to support the accuracy and precision of the presented results. On one end of the spectrum, results can be fully validated with calibration curves and precision and accuracy measurements. On the other end, results may be estimates based only on the concentration of an internal standard. Because patient safety may hinge on the result, often toxicologists want something more than a blind estimate of concentration of the compound is on the edge of being considered safe.

Chemistry results must be evaluated and assessed through the lens of toxicology to understand the possible systemic risks associated with the findings and the route of exposure of the device per ISO 10993-17. This assessment should complement the results of traditional biocompatibility tests performed on biopolymeric device materials.

1.4 Specific Biological Endpoint Evaluations

For most biological endpoints per ISO 10993-1, a biopolymer would be tested very similarly to any other polymer. The main concern with a biopolymer is the degradation profile and the impact of the degradation on the test system. The testing system that needs the most consideration for the individual degradation profile of a material is in cytotoxicity, systemic toxicity, implantation, and material/chemical characterization.

1.4.1 Cytotoxicity

In general, cytotoxicity tests are a broad range of assays that look for the impact of a substance on individual cells grown under *in vitro* conditions. The test can be performed on different cell lines and can look at (qualitatively) or assess (quantitatively) different cellular endpoints. The various internationally accepted cytotoxicity assays are summarized in part 5 of the ISO 10993 series (i.e. ISO 10993-5). All the tests usually run using the L929 mouse fibroblast cell line. Although it is possible to use other cell lines for testing, the L929 cell line is the one that has historically been used and is therefore recommended for comparison. Additionally, despite the availability of many different versions of cytotoxicity tests, the standard testing for biocompatibility of medical devices consist of either MEM elution, MTT/XTT assays, or neutral red uptake assay. Each assay has different cytotoxicity evaluation endpoints and sensitivity, so comparing results from one assay to the other has proven to be difficult.

The cytotoxicity test is a very sensitive test and is the most likely test to cause trouble with any medical device, but specifically with biopolymers. This trouble comes from the fact that some biopolymers lack the mechanical properties and stability in the extraction fluid that is used to prepare a sample for the cytotoxicity test. This lack of stability may be caused a high concentration of ions in the extraction fluid that could result in a cytotoxic response in the assay. Crosslinking can be used in the attempt to improve the results, but this can also cause potential cytotoxicity as these crosslinking agents themselves can be cytotoxic (e.g. glutaraldehyde).

Therefore, the best approach for assessing cytotoxicity of biopolymers is a risk-based approach. As mentioned before, the cytotoxicity test is historically the most sensitive test available and is thus often used as a screening test for materials, process residuals, and the final device configuration. In the ANSI/AAMI/ISO 10993-5 Guidance section 10, it states "Any cytotoxic effect can be of concern. However, it is primarily an indication of potential for *in vivo* toxicity and the device cannot necessarily be determined to be unsuitable for a given clinical application based solely on cytotoxicity data." When elevated cytotoxicity results are seen, a risk assessment should be performed to identify the source of observed cytotoxicity. Then on, the risk assessment should evaluate the toxic potential of the material or compound to determine the clinical impact. The investigation should include a review of the procedures to determine the effectiveness of the test system, additional testing to evaluate clinical risk of the results, and then a clinical risk assessment of the toxicity using additional animal testing along with chemical analysis and toxicological assessment of the detected compounds.

Based upon examination of the biopolymer, its history of use in medical industry, inherent surface properties of the device material, surface area in contact with the user, use and contact type, duration of contact, and the route of exposure, this cytotoxicity failure may not be clinically relevant, and subsequently it can be concluded that adverse effects in patients are unlikely to develop.

1.4.2 Systemic Toxicity (Acute, Subacute, Subchronic, and Chronic)

Systemic toxicity is a potential adverse generalized response including organ or organ system effects that can result from the absorption, distribution, and metabolism of leachates from the device or its materials to parts of the body that are not in direct contact with the device or material. The type of test recommended per ISO 10993 is dependent on the duration of exposure to the patient:

- Acute toxicity is defined as an adverse systemic effect occurring at any time within 72 hours after single, multiple, or continuous exposures of a test sample for 24 hours.
- Subacute toxicity is defined as an adverse effect occurring after multiple or continuous exposure between 24 hours and 28 days. The term subacute might be somewhat misleading since generally "sub" is understood as less, and subacute would, based on this logic, be considered as less than acute. Since this term is confusing, it is best to consider subacute toxicity as any adverse effects occurring within a short-term repeated exposure during a systemic toxicity study. This is generally done with time intervals between 14 and 28 days for intraperitoneal injection studies; intravenous studies are generally defined as treatment durations or exposure of more than 24 hours but less than 14 days.
- Subchronic toxicity is any adverse effect occurring after the repeated or continuous administration of an extract of a material or device for (typically) 90 days in rodents or in other species for duration of exposure that does not exceed 10% of the life span of the test animal. Subchronic intravenous studies are generally defined as treatment durations of 14–28 days for rodents and non-rodents, respectively.

Table 1.4 Standard device extraction ratios used for biocompatibility (per ISO 10993-12).

Thickness (mm)	Extraction ratio
<0.5	6 cm^2/ml
0.5–1.0	3 cm^2/ml
>1.0	3 cm^2/ml
>1.0 (elastomeric devices)	1.25 cm^2/ml
Irregular solid devices	0.2 g/ml
Irregular porous devices	0.1 g/ml

- Chronic toxicity is any adverse effect occurring after the repeated or continuous administration of a test sample for a major part of the test animal's life span; these are usually studies with duration of 6–12 months.

The main consideration point for systemic toxicity and biopolymers is regarding the dose. The standard biocompatibility test is performed on the basis of surface area or mass to volume; these ratios are spelled out in Table 1.4.

As Table 1.4 points out, the more surface area or mass a device has, the more extraction volume is added to the device during sample preparation. This approach works well for solid, stable materials such as metals and hard plastics but can be challenging with materials such as biopolymers, especially if they are produced with a porous microarchitecture or are biodegradable.

Another giant gap in the approach that uses surface area or mass for calculating the extraction volume is that it does not take into consideration the actual dose that a single patient will be exposed to. Typically, each biological test requires a certain minimal volume of fluid to run, and because of this limitation the sample amount needed for the testing is directly portioned to the logistics demanded by the test itself and not on the actual clinical use of the device. For example, let us say during a surgical procedure, a patient will only receive one PLA screw that is 0.5 g in weight. For the biocompatibility assessment of the screw, a standard subacute study was run. For testing, up to 112 screws were included in order to conform to the required sample volumes that were repeatedly dosed to the test animals, resulting in an exposure that is in actuality multiple times the clinical mass to body weight dose. This leads to a vast overestimation of the exposure risks of the biopolymer.

A better way to design the different systemic toxicity studies of biopolymers is based on dose per body weight of the patient. The standard weights per patient population are described in Table 1.5. In this case, one would determine the appropriate worst-case target population for the medical device or material and determine a dose per kg of body weight based on that criterion. Subsequently, the testing would be done with a sample size that would expose the specific animal to a safety-factor-corrected dose that represents the appropriate clinical dose.

An example of a test design according to the clinical dose approach would be as follows: a surgical procedure where up to two screw PLA screws (each weighing

Table 1.5 Standard body weight parameters.

Population	Standard body weight used (kg)
Adult man	70
Adult woman	58
Children	10
Neonates (<1 yr)	3.5

Table 1.6 Example of specific population doses for 1 g PLA screw.

Population	Gram of screw per kg body weight	With 10X safety factor
Adult man	0.01	0.14
Adult woman	0.02	0.17
Children	0.10	1.00
Neonates (<1 yr)	0.29	2.86

0.5 g) will be implanted into a patient, the worst-case exposure per patient will be 1 g of PLA, and the specific clinical prescribed doses are outlined in Table 1.6.

In a rat subchronic study, if the worst-case target population is adult women and the test rat weighs 500 g, the dose would be calculated as follows:

Desired ratio with safety factor = 0.17 g of screw per body weight

$$\frac{0.17\text{ g of screw}}{\text{kg body weight}} \times \frac{1\text{ kg}}{1000\text{ g}} \times \frac{500\text{ g}}{1\text{ Rat}} = \frac{0.085\text{ g of Screw}}{\text{Rat}}$$

This approach would ensure an accurate exposure dose to the animal and would present a more clinically relevant evaluation for the risks of systemic toxicity for the device.

1.4.3 Implantation

The most difficult and complex test design for many biopolymers revolves around implantation risks. It is important not to walk into an implant study with haste and without careful planning. Indeed, in this case, failing to plan could lead to a failing test. It is important that the study is planned in sufficient detail such that all relevant information can be extracted from the study, as the implant test is usually the longest test in the biocompatibility suite, and therefore, it is imperative to have the design right up front.

The main issue with testing a biopolymer in an implant test is the absorption profile. Physical characteristics (such as form, absorption rate, metabolism characteristics, density, and surface hardness) can all influence the tissue response to the test material. Also, the choice of control articles should be matched as closely as reasonably possible to the test sample physical characteristics. This is recommended in order to allow comparison of the specific tissue reaction(s) with that of a similar material whose clinical acceptability and biocompatibility characteristics have been established to determine acceptance criteria for the test.

Another key consideration for the implant test for a biopolymer is with the implantation time points. ISO 10993-6 states: "For absorbable materials, the test period shall be related to the estimated degradation time of the test product at a clinically relevant implantation site. When determining the time points for sample evaluation, an estimation of the degradation time shall be made." Usually, in practice we try to estimate the absorption profile based on the specific metabolism rate and method of the material and the implant system. After this, we set three time periods: one where we first see degradation (usually between two and four weeks), second when half the sample is degraded, and third when we see a "steady state" in the sample material. A steady state is defined as a point in time where the body is no longer interacting with the material and no additional changes are happening. For example, *in vivo* implantation tests with a PLLA density scaffold demonstrated fast degradation in the first three weeks, after which the degradation rate progressively decreased [20]. This milestone is reached when the body has either encapsulated or otherwise dealt with the foreign material or when full degradation of the material has occurred.

As mentioned above, an appropriate control is the basis for the acceptance criteria of the test itself, making it an essential component for a relevant and applicable test system. The implantation test is set up so that the evaluation is conducted by comparing the result of the test site histopathology with the control site. Thus, if the chosen control article is a hard piece of metal or plastic that would not induce interaction with the surrounding tissues, then the comparison with the implant site of the biopolymer would probably not be favorable, leading to a higher tissue reactivity and making it look like the test material is non-biocompatible. However, if an appropriate control is used, then the histopathological comparison of the test and control article sites can be made with confidence, and a correct understanding of the implantation risk of the material can be drawn.

1.5 Conclusion

Biopolymers occupy a unique and advantageous space as a medical device material. Devices made from these naturally occurring or biomimetic substances have the distinct advantage that the material itself is akin to those tissues the device contacts. From a bulk perspective, there is no concern regarding the material as a foreign body. Biopolymers also have environmental and manufacturing advantages as they are often produced not from petroleum derivatives but by living systems.

In contrast to the major advantages presented by biopolymers within the context of biocompatibility, there are a couple of key concerns that must be addressed. The natural origin of these materials does not mean that they are free from manufacturing residuals. Contact with solvents through manufacturing and purification steps can introduce contamination, as can contact with storage and primary packaging materials. Chemical analysis screening for these compounds can be complicated by the complex organic nature of the device material. Additionally, many biopolymers are degradable or resorbable by the body. While this is, in principle, a positive therapeutic effect, it can be difficult to prove that the safety of the device does not change over the degradation lifetime.

The pallet of materials afforded by biopolymers allows an even broader spectrum of medical devices with huge potential to help mankind. The biocompatibility principles discussed in this chapter can be applied to biopolymers to address concerns with regard to their safety. Use of thoughtful risk-based testing strategies can conservatively mitigate risk, allowing more of these devices to reach full maturity in development and arrive on the market.

References

1 Rebelo, R., Fernandes, M., and Fangueiro, R. (2017). Biopolymers in medical implants: a brief review. *Procedia Engineering* 200: 236–243.

2 Yadav, P., Yadav, H., Shah, V.G. et al. (2015). Biomedical biopolymers, their origin and evolution in biomedical sciences: a systematic review. *Journal of Clinical and Diagnostic Research: JCDR* 9 (9): ZE21.

3 Sahana, T. and Rekha, P. (2018). Biopolymers: applications in wound healing and skin tissue engineering. *Molecular Biology Reports* 45: 2857–2867.

4 Labet, M. and Thielemans, W. (2009). Synthesis of polycaprolactone: a review. *Chemical Society Reviews* 38 (12): 3484–3504. https://doi.org/10.1039/B820162P.

5 Chien, L.J. and Lee, C.K. (2007). Enhanced hyaluronic acid production in *Bacillus subtilis* by coexpressing bacterial hemoglobin. *Biotechnology Progress* 23 (5): 1017–1022.

6 Chien, L.J. and Lee, C.K. (2007). Hyaluronic acid production by recombinant *Lactococcus lactis*. *Applied Microbiology and Biotechnology* 77 (2): 339–346.

7 Liu, L., Wang, M., Du, G., and Chen, J. (2008). Enhanced hyaluronic acid production of *Streptococcus zooepidemicus* by an intermittent alkaline-stress strategy. *Letters in Applied Microbiology* 46 (3): 383–388.

8 Helenius, G., Bäckdahl, H., Bodin, A. et al. (2006). In vivo biocompatibility of bacterial cellulose. *Journal of Biomedical Materials Research Part A* 76A (2): 431–438.

9 Hoenich, N.A. (2007). Cellulose for medical applications: past, present, and future. *BioResources* 1 (2): 270–280.

10 Rockwood, D.N., Preda, R.C., Yücel, T. et al. (2011). Materials fabrication from *Bombyx mori* silk fibroin. *Nature Protocols* 6 (10): 1612–1631.

11 Auras, R.A., Lim, L.-T., Selke, S.E., and Tsuji, H. (2011). *Poly(lactic Acid): Synthesis, Structures, Properties, Processing, and Applications*. Wiley.

12 ISO/TS 21726:2019 (2016). Biological evaluation of medical devices – Application of the threshold of toxicological concern (TTC) for assessing biocompatibility of medical device constituents. Standardization IOf.

13 Kroes, R., Kleiner, J., and Renwick, A. (2005). The threshold of toxicological concern concept in risk assessment. *Toxicological Sciences* 86 (2): 226–230.

14 Munro, I., Renwick, A., and Danielewska-Nikiel, B. (2008). The threshold of toxicological concern (TTC) in risk assessment. *Toxicology Letters* 180 (2): 151–156.

15 Hennes, E. (2012). An overview of values for the threshold of toxicological concern. *Toxicology Letters* 211 (3): 296–303.

16 Guideline IHT (2014). Assessment and control of DNA reactive (mutagenic) impurities in pharmaceuticals to limit potential carcinogenic risk. *International Conference on Harmonization of Technical Requirements for Registration of Pharmaceuticals for Human Use (ICH)*, Geneva. Citeseer (31 May 2017). http://academy.gmp-compliance.org/guidemgr/files/M7_R1_ADDENDUM_STEP_4_2017_0331.PDF

17 European Food Safety Authority and World Health Organization (2016). Review of the Threshold of Toxicological Concern (TTC) Approach and Development of New TTC Decision Tree. EFSA Supporting publication 2016:EN-1006. https://efsa.onlinelibrary.wiley.com/doi/pdf/10.2903/sp.efsa.2016.EN-1006

18 Cramer, G., Ford, R., and Hall, R. (1976). Estimation of toxic hazard – a decision tree approach. *Food and Cosmetics Toxicology* 16 (3): 255–276.

19 Munro, I.C., Ford, R.A., Kennepohl, E., and Sprenger, J. (1996). Correlation of structural class with no-observed-effect levels: a proposal for establishing a threshold of concern. *Food and Chemical Toxicology* 34 (9): 829–867.

20 Weng, W., Song, S., Cao, L. et al. (2014). A comparative study of bioartificial bone tissue poly-L-lactic acid/polycaprolactone and PLLA scaffolds applied in bone regeneration. *Journal of Nanomaterials* 2014: 236.

2

Advanced Microbial Polysaccharides

Filomena Freitas[1], Cristiana A.V. Torres[1], Diana Araújo[1], Inês Farinha[1,2], João R. Pereira[1], Patrícia Concórdio-Reis[1], and Maria A.M. Reis[1]

[1] UCIBIO-REQUIMTE, Chemistry Department, Faculty of Sciences and Technology, Universidade Nova de Lisboa, Campus da Caparica, Caparica 2829-516, Portugal
[2] 73100 Lda., Rua Ivone Silva nº 6 4º piso, 1050-124 Lisboa, Portugal

2.1 Introduction

Microbial polysaccharides are high molecular weight (Mw) carbohydrate polymers produced by microorganisms, namely, bacteria, fungi, yeast, and microalgae [1–3]. They include intracellular polysaccharides that are accumulated in the cytoplasm of the cells as carbon and energy reserves (e.g. glycogen), cell-wall polysaccharides that contribute to the cells' structural stability (e.g. chitin), and extracellular polysaccharides that are secreted by the cells, forming either a capsule that remains associated with the cell surface (capsular polysaccharides [CPS]) or a slime that is loosely bound to the cell surface (exopolysaccharides [EPS]) [2, 4]. Of the last type, CPS are mostly associated with the pathogenicity of bacteria and virulence-promoting factors [5], while EPS have been proposed to provide protection against environmental stress, cell adherence to surfaces, and carbon or water storage reserves [6].

Polysaccharides can be used into two main areas of application: (i) as structuring agents, based on their ability to form polymeric structures, such as films, gels, emulsions, microparticles, and nanoparticles, and (ii) as biological active materials/compounds that can be used for the development of novel pharmaceutical drugs or replace some of the currently used products [7, 8]. Other applications include their use as sources of high-value monomers, such as rare sugars (e.g. fucose, rhamnose, ribose, glucuronic acid, etc.), to generate oligosaccharides (e.g. galactooligosaccharides, fucooligosaccharides) that can be used in nutraceuticals [9].

This chapter starts with a brief overview on microbial polysaccharide diversity in terms of functional properties and their main areas of application (Section 2.2), followed by a more detailed analysis of the currently more relevant and emerging areas (Sections 2.3–2.7).

Biopolymers for Biomedical and Biotechnological Applications, First Edition.
Edited by Bernd H. A. Rehm and M. Fata Moradali.

2.2 Functional Properties and Applications of Microbial Polysaccharides

The main components of microbial polysaccharides are carbohydrates. Glucose, galactose, and mannose are the most common monomers, but other neutral sugars such as rhamnose, arabinose, and fucose and some uronic acids and amino sugars are also frequently found in such biopolymers. In addition to carbohydrates, microbial polysaccharides may also contain several organic acyl substituents in their molecular chains, such as ester-linked groups and pyruvate ketals [4, 10, 11]. Given this diversity of polysaccharides' components that may include different sugar monomers as well as several noncarbohydrate groups, there is a wide range of possible molecular structures for these macromolecules. Consequently, microbial polysaccharides can display distinct physical and chemical properties.

The possible multiple combinations of monomeric units in polysaccharide molecules, along with the stereospecificity of glycosidic linkages (α or β anomers), lead to very complex chemical structures ranging from linear homopolysaccharides to highly branched heteropolysaccharides. Molecular mass distribution, chemical composition, and structure, namely, the presence of ionizable groups that confer the polysaccharides a polyelectrolyte behavior, greatly affect their properties, as well as the nature and number of intra-/intermolecular interactions. Moreover, the properties of polysaccharides may be altered using mixtures with other components, for example, by blending with other polymers or adding salts and crosslinking agents.

Some examples of microbial polysaccharides that have been developed at the industrial level and are currently commercialized include xanthan, a heteropolysaccharide composed of glucose, mannose, glucuronic acid, pyruvate, and acetate, produced by *Xanthomonas* sp., which is used mainly in food, pharmaceutical, and personal care products and in oil recovery [12]; dextran, a water-soluble glucan secreted by lactic acid bacteria of the genera *Leuconostoc, Streptococcus, Weissella, Pediococcus,* and *Lactobacillus,* which is used in food, cosmetic, and medical applications [13]; hyaluronic acid (HA), a linear polymer of glucuronic acid and *N*-acetylglucosamine units, produced by *Streptococcus zooepidemicus* that is used in cosmetics, pharmaceuticals, and medicine [14]; pullulan, a linear α-glucan secreted by the black yeast-like *Aureobasidium pullulans* that is used as a thickening agent and edible coating [6, 15]; and scleroglucan, a water-soluble β-glucan secreted by the plant pathogen *Sclerotium* sp. that is used in applications such as enhanced oil recovery, food, cosmetic, and pharmaceutical [16]. Additionally, there are a growing number of recent reports on newly discovered polysaccharides, with novel molecular structures, obtained from species belonging to different taxonomic groups, which display numerous biological activities (Tables 2.1 and 2.3) and may turn into novel applications.

Biopolymers of microbial origin have been studied lately due to their improved properties and easy production when compared with other natural polymers [4, 78]. Microorganisms usually have higher growth rates than algae and plants, and their production processes can easily be manipulated to improve yields and

Table 2.1 Examples of polysaccharides produced by bacteria and fungi displaying biological activity.

Microbial source	Polysaccharide	Biological activity	References
Bacteria			
Acetobacter xylinum NCIM2526	Fructan	Antioxidant; anti-inflammatory	[17]
Bacillus licheniformis T4	Fructo-fucan	Anti-cytotoxic; antiviral	[18]
Bacillus tequilensis PS21	Heteropolysaccharide (xylose, glucose, ribose, rhamnose, galactose)	Antioxidant	[19]
Enterobacter A47	Heteropolysaccharide (fucose, glucose, galactose, glucuronic acid, acetate, succinate, pyruvate)	Antioxidant	[20]
Enterobacter cloacae Z0206	Heteropolysaccharide (fucose, glucose, galactose, glucuronic acid)	Antioxidant; antidiabetic; hypolipidemic	[21]
Enterococcus faecium K1	Heteropolysaccharide (mannose, glucose, galactose)	Hypocholesterolemic; antibiofilm; antioxidant	[22]
Lactobacillus sp.Ca6	α-(1,6)-Glucan	Antioxidant; antibacterial; wound healing	[23]
Lactobacillus casei SB27	Heteropolysaccharide (galactose, glucose)	Antitumor	[24]
Lactobacillus gasseri FR4	Heteropolysaccharide (glucose, mannose, galactose, rhamnose, fucose)	Antioxidant; antimicrobial	[25]
Lactobacillus kefiranofaciens DN1	Heteropolysaccharide (mannose, arabinose, glucose, galactose, rhamnose)	Antibacterial	[26]
Lactobacillus plantarum BR2	Glucomannan	Antioxidant; antidiabetic; hypocholesterolemic	[27]
Pediococcus parvulus 2.6	β-(1,3)-Glucan	Anti-inflammatory; probiotic	[28]
Pseudoalteromonas sp. S-5	Heteropolysaccharide (mannose, glucose, galactose)	Anticancer	[29]
Fungi			
Antrodia cinnamomea	Heteropolysaccharide (fucose, glucosamine, galactose, glucose, mannose, sulfate)	Anticancer	[30]
Aspergillus sp. Y16	Galactomannan	Antioxidant	[31]
Candida utilis	Glucomannan	Antiarthritis; antioxidant	[32]
Diaporthe sp.	β-Glucan	Antitumor	[33]
Fusarium equiseti ANP2	Glucomannan	Antioxidant	[34]
Fusarium solani SD5	Rhamnogalactan	Anti-inflammatory; anti-allergic	[35]
Trichoderma kanganensis	Heteropolysaccharide (mannose, galactose, glucose, glucuronic acid)	Anticancer; antioxidant	[36]

productivity [79]. Moreover, the production process is not climate or seasonal dependent and can rely on the use of low-cost by-products or wastes as raw materials [80]. Microbial polysaccharides have unique features and properties that make them suitable to a wide range of applications. More specifically, these biopolymers have been extensively used in food, pharmaceutical, medical, and cosmetic products due to their unique performances as thickening, stabilizing, and binding agents. Some of these biopolymers (e.g. bacterial alginate, gellan, FucoPol) can also have intermolecular interactions that could result in polymeric matrices, allowing the physical manipulation of polysaccharides into structured materials such as gels (e.g. hydrogels) or films that could be used in biomedical applications [4, 81]. Adding to this property, the ability of polysaccharides to interact with different inorganic materials represents an important feature for the encapsulation of bioactive substances (e.g. pharmaceuticals for their controlled release) and for the incorporation of nanostructures (such as carbon nanotubes or metallic nanoparticles [MNPs]) to produce enhanced biomaterials (with synergetic conductive or magnetic properties, respectively) [82].

2.3 Commercially Relevant Microbial Polysaccharides: Established Uses and Novel/Prospective Applications

The great diversity of microbial polysaccharide composition and functional properties enables their application in several industrial fields (e.g. medical, food products, pharmaceutical, biomedicine). Although only few are commercialized, among them are pullulan, scleroglucan, xanthan gum, dextran, levan, gellan gum, and hyaluronic acid.

2.3.1 Pullulan

Pullulan is a fungal polymer produced by *A. pullulans*. It is a neutral, linear glucose homopolysaccharide composed of α-(1,6) maltotriose units [83, 84]. Pullulan presents a unique linkage pattern showing remarkable physical properties, such as high solubility, adhesiveness, forming fibers, and thin biodegradable film capacity, which are transparent and impermeable to oxygen [15, 85]. Therefore, pullulan offers a variety of potential industrial and medical applications. For a long time, pullulan membranes/films have been used as coating and packaging material in the food industry, but nowadays they are also being used in dietary capsule formulations. Pullulan-based oral care products are also being commercialized. Due to the easy decomposition, it is used as coating in the paper industry. Modified pullulan is used as raw material in pharmaceutical applications, namely, nanoparticles, bioimaging, plasma expander, tissue engineering, etc. [84–86].

The cancer therapy and bioremediation are emerging markets for pullulan, due to its bioactivity with some cytotoxic molecules and the adsorption capacity for some heavy metals [84, 85]. Moreover, currently several research groups are studying novel pullulan composites blended with other biodegradable materials (e.g. pullulan/dextran, pullulan/rice starch gel, and pullulan/cellulose) in order to obtain desired properties, such as thermal stability, high tensile strength, and emulsion stability [84]. Hydrogels of pullulan have been studied for three-dimensional (3D) printing of scaffolds [87]. Hayashibara Co., Ltd. (Japan), is one of the companies that commercialized pullulan to be applied mainly as a thickening agent and edible coating.

2.3.2 Scleroglucan

Scleroglucan is a water-soluble homopolymer of β-glucans produced by filamentous fungi, especially of the genus *Sclerotium* as part of the adhesion mechanism to plant tissues [16, 88]. Scleroglucan was first commercialized in the 1970s, being currently available under different trademarks (e.g. Clearogel, Polytetran, Polytran FS, Actigum) [88, 89].

Scleroglucan is a thermostable biopolymer and, due to its nonionic nature, is stable over a wide range of pH (2.5–12). Scleroglucan solution exhibits shear thinning behavior; it acts as foam stabilizer and has a good emulsifying capacity. Further, it also exhibits biological activity. The interesting physicochemical and biological properties enable the use of scleroglucan on several industrial applications. Initially, scleroglucan was used in the oil industry [90, 91]. Nowadays it is used as thickener in paintings and in pesticides [16]. In the biomedical field it is used in edible films, tablets, and granulates, showing to be a good matrix for the controlled release of active substances. Pharmaceutical applications include the use in tablet coatings, ophthalmic solutions, injectable antibiotic suspensions, and calamine lotion [92]. In the food industry scleroglucan is used for the stabilization of dressings and ice creams. Numerous patents describe quality improvement of frozen or heat-treated edibles. In Europe it is mainly used in cosmetic products for the skin, such as body washes, shampoos, conditioners, and eyeliners [16, 93, 94].

2.3.3 Xanthan Gum

Xanthan gum is a heteropolysaccharide isolated from *Xanthomonas campestris* composed of a 1,4-β-D-glucose backbone having a trisaccharide side chain (a glucuronic acid residue between two mannose residues) attached to alternate glucose residues [95, 96]. It forms highly viscous aqueous solutions, even at low concentrations, with shear thinning behavior, and is stable in terms of temperature, salts, and a wide range of pH. It is the most widely accepted commercial bacterial polysaccharide due to its exceptional rheological properties [7, 95, 97]. Xanthan's major commercial producers comprise CP Kelco, Merck, Pfizer, Rhône-Poulenc, Sanofi-Elf, and Jungbunzlauer. It is used in several industries, such as foods, food packaging, personal care products, cosmetics, drug delivery systems, water treatments, and drilling fluids [12, 98].

Recent trends on the use of xanthan-based polysaccharide are focused on the use of formulations for various tissue engineering applications. Moreover, the shear thinning and gelling properties of xanthan are interesting in the area of 3D bioprinting of the tissue scaffolds and/or tissue models for future tissue engineering applications [12, 98]. The use of xanthan in the form of luminescent composites, namely, nanocrystals, for biomedical applications shows good potential but is barely explored [98]. Xanthan-based polymers have also been studied for solid polymeric electrolytes [99, 100].

2.3.4 Dextrans

Dextrans are homopolysaccharides composed of D-glucose units. Species belonging to the genera *Leuconostoc, Weissella, Lactobacillus, Pediococcus,* and *Streptococcus* produced dextran [101, 102]. Dextrans have interesting physicochemical properties, such as thickening capacity, emulsifier or stabilizing ability, and high solubility in water. Additionally, its flexible structure is an important characteristic that enables the dextran use as functional hydrocolloid. Dextran is used as thickening agent in cosmetics and foods (e.g. bakery products and confectionery) to improve texture and moisture and prevent sugar crystallization. It is also used in pharmaceutical/biomedicine applications and formerly was applied for drug delivery and blood plasma polymer expander [7, 103]. The potential of dextran-based polymers is being evaluated to form nanofibers for controlled drug release [104], as anticancer therapeutics [105, 106], and as hydrogels for wound healing [107]. Dextran is commercialized by some companies, such as Oregon Green and ATTO-TEC.

2.3.5 Curdlan

Curdlan is a glucan produced by *Alcaligenes, Rhizobium,* and *Bacillus* species. It is alkaline-soluble and water-insoluble gel-forming polymer, which limits its industrial applications [108–110]. Hence, curdlan has traditionally been used as a stabilizer, texturizer, and thickener in the food industry [111, 112].

Further, some studies showed interest in converting curdlan into more soluble oligosaccharides, exhibiting notable biomedical functions (e.g. antitumor and immunological activities) [113, 114]. Besides, the oligosaccharides may also have an efficient role in the improvement for sustainable agriculture since it can efficiently activate the plant innate immune defense system [115]. Research on curdlan and its derivatives are being performed to develop environmentally friendly alternatives to oil-based plastics [111].

2.3.6 Gellan Gum

Industrially, gellan gum is produced by the bacterium *Sphingomonas paucimobilis* ATCC 31461 [116]. It is a heteropolysaccharide composed of L-rhamnose, D-glucose, D-glucuronic acid, and D-glucose monomers and side chains of acetyl and glyceryl substituents. Gellan gum have the ability to form gels, whose properties are dependent on the acyl group content: low acyl form produces rigid, nonelastic, brittle, and thermostable gels, while the high acyl gellan produces soft, elastic, non-brittle, thermoreversible gels [117, 118]. Gellan gum is mainly used in the food industry, namely, as a thickener, stabilizer, and binder agent. It is used in desserts and drinking jellies, as well as in low-calorie (sugar-free) jams, fruit preparations, yogurt, sauces, nonfat salad dressings, and films [119, 120]. Gellan gum is commercialized as Kelcogel, a gelling agent in the food industry, and is marketed by CP Kelco.

Moreover, gellan is described as a multifunctional additive for various pharmaceutical products, especially for controlled release forms, including oral, ophthalmic, nasal, and other formulations [121, 122]. It is already used in ophthalmic formulations (e.g. Timoptol). Recent reports suggest that gellan-based materials can also be used in tissue engineering, regenerative engineering, or gene transfer technology [118, 123, 124].

2.3.7 Levan

Levan is a homofructan composed of fructose residues and can be secreted by several microorganisms (e.g. *Acetobacter* sp., *Halomonas* sp., *Zymomonas* sp., *Lactobacillus* sp.) or produced by plants [7, 125, 126]. Levan does not swell in water, and it has a very low intrinsic viscosity value. It is water and oil soluble, is compatible with salts and surfactants, and has emulsifying capacity, biological activity, and adhesive ability [127, 128]. Its functional properties turn it suitable for use in food (e.g. prebiotic agent), cosmetics (e.g. dermal filler), and pharmaceuticals. The levan low viscosity promotes its use in pharmaceuticals to produce capsules or coatings and in several therapeutic applications. Levan nanoparticles have potential to delivery peptides and protein drugs. Further, levan is used in the green synthesis of silver and gold nanoparticles [126, 129]. It is produced by Montana Polysaccharides Corp. in the United States.

2.3.8 Hyaluronic Acid

Hyaluronic acid (HA) is a linear polymer composed of repeating disaccharide units of glucuronic acid and *N*-acetylglucosamine [130, 131]. It is produced by *S. zooepidemicus*; however the concern about the pathogenicity of *Streptococcus* has driven the efforts toward transforming generally recognized as safe (GRAS) nonproducers (e.g. *Lactococcus lactis, Bacillus subtilis*) into HA producers [130]. HA shows a great swelling capacity, biocompatibility, non-immunogenicity, biodegradability, and viscoelasticity. Its physicochemical and biological properties render HA potential for applications in cosmetics (e.g. dermal filler), pharmaceuticals, and medicine (e.g. osteoarthritis treatment, tissue engineering) [132–134]. Ongoing research on HA and its numerous modifications/blends shows the development of materials with improved properties for drug delivery and tissue engineering technologies [135, 136].

2.4 Hydrogels Based on Microbial Polysaccharides

Among various physical structures such as films, fibers, and beads, microbial polysaccharides are biopolymers suitable to fabricate a half liquid-like and half solid-like material, known as hydrogel [137–139]. Hydrogel is a 3D crosslinked polymeric network capable of absorbing and retaining large amounts of water, commonly used in a wide range of applications in the most diversified fields [140, 141].

Over recent decades, several polysaccharides including alginate, gellan gum, cellulose, dextran, hyaluronic acid, xanthan, and chitin/chitosan, either alone or in blends, all attainable through microbial production, have been used for the design and fabrication of hydrogels.

According to their preparation method and physical structure, hydrogels can be produced by physical or chemical crosslinking (Figure 2.1) [138, 141]. Physically crosslinked hydrogels are reversible under specific conditions, and polymer chains are weakly stabilized by secondary forces such as ionic interactions, hydrogen bonding, or hydrophobic interactions. Despite their shape instability due to the reversibility of the formed bonds, physical hydrogels are generally harder than the chemical hydrogels [138]. A well-known example of hydrogels formed by ionic interaction is the crosslinking of alginate using divalent cations as Ca^{2+} [143]. On the other hand, chemically crosslinked hydrogels are irreversible and stable, with strong covalent bonds involving reactions of polymeric backbone with a crosslinking agent. Chemical hydrogels can be produced by different techniques such as radiation and graft copolymerization or in the presence of a crosslinking agent. In the case of polysaccharides, the most common technique is the use of a crosslinking agent involving active reaction sites as –OH groups on its backbone [141].

Depending on the types of monomers involved, hydrogels can be classified as homopolymer hydrogels, if composed by one single monomer unit; copolymer hydrogels, if constituted by two or more monomeric units, one of which must be hydrophilic; and interpenetrating polymeric network (IPN) hydrogels when two independent crosslinked networks intermesh each other in the presence of crosslinker. Therefore, hydrogels can be semi-IPN if one of the components is a non-crosslinked polymer [140].

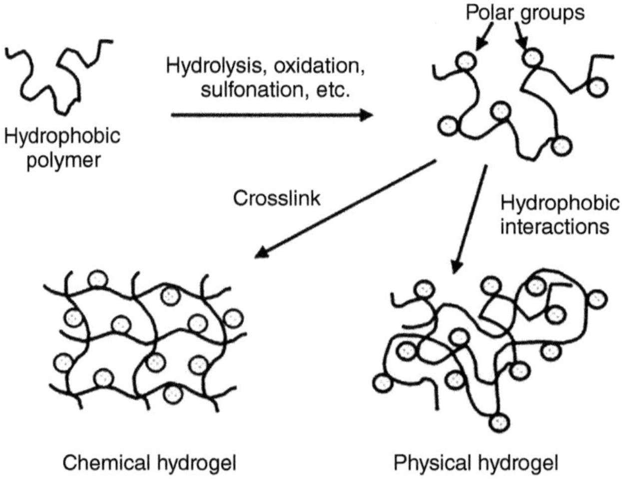

Figure 2.1 Schematic representation of the fabrication of chemically and physically crosslinked hydrogels. *Source:* Hoffman 2012 [142]. Reprinted with permission of Elsevier.

Additionally, polysaccharide hydrogels can also be categorized on the basis of ionic charges as non-anionic (neutral), ionic (cationic or anionic), and ampholytic hydrogels. Such classification refers to the overall charge, namely, no charge groups are present in neutral hydrogels, and cationic and anionic hydrogels are characterized by the presence of positively and negatively charged groups, respectively. In the presence of both negatively and positively charged groups, ampholytic hydrogels are produced [141, 144]. Ionic and ampholytic hydrogels are also known as polyelectrolytes.

Considering their final application, hydrogels can be designed to be stimulus sensitive, responding distinctively toward the external condition such as temperature, pH, ionic strength, and magnetic or electric field [140, 145]. In fact, the ability to respond to external stimuli makes them usually called "smart" or "stimuli-sensitive" hydrogels. Moreover, exhibiting "smart" characteristics is an advantage to be useful in biomedical applications such as controlled drug delivery [146, 147] or agricultural applications [148].

An example of naturally thermoresponsive microbial polysaccharide is gellan gum. As earlier mentioned, gellan is an anionic extracellular bacterial polysaccharide with the ability to fabricate thermoreversible gels that can have distinct mechanical properties depending on their composition. Therefore, while acetylated form of gellan produces soft and elastic gels, with deacetylated gellan hard and brittle gels are produced [144, 149]. Generally, gellan has an upper critical solution temperature (UCST), which means that at a high temperature a polymer solution is obtained and the gel is produced upon cooling the solution. In particular, the temperature of gelation for gellan is within a range from 35 to 42 °C, varying with molecular weight, processing conditions, and the presence of cations [150]. Although the most common application of gellan gels is in food industry as food additive and as thickener or gelling agent [151], their potential to be applied in some biomedical applications including drug delivery and tissue engineering approaches has been investigated [149, 152]. Due to their properties, gellan hydrogels are suitable to be used as injectable system for long-term cartilage regeneration, as reported by Gong et al. [153].

The fabrication of hydrogels based on microbial polysaccharides is emerging mainly due to their less toxicity, biocompatibility, and biodegradability properties ally to acceptable mechanical strength [141]. One of the main polysaccharides studied to be used in hydrogel design and production is chitosan. Microbial chitosan is a semicrystalline cationic polysaccharide obtained by deacetylation of chitin present in yeasts and fungi cell wall. Chitosan-based hydrogels have been prepared either by physically or chemically crosslinked methods to develop materials suitable to be applied in biomedical field as drug delivery systems or wound healing dressing [154, 155]. It is well known that chitosan has low water solubility and can be maintained in solution under acid conditions. Consequently, the neutralization of a chitosan solution to a pH above 6.2 (amine pK_a) displays the gelation phenomenon [156]. This behavior allows the utilization of chitosan to produce pH-sensitive hydrogels [154]. In fact, chitosan pH-responsive hydrogels can be transformed into thermosensitive by the incorporation of polyol- or sugar-phosphate salts such as glycerophosphate [156]. Contrary to gellan hydrogels, chitosan-based thermoreversible hydrogels have a low critical solution

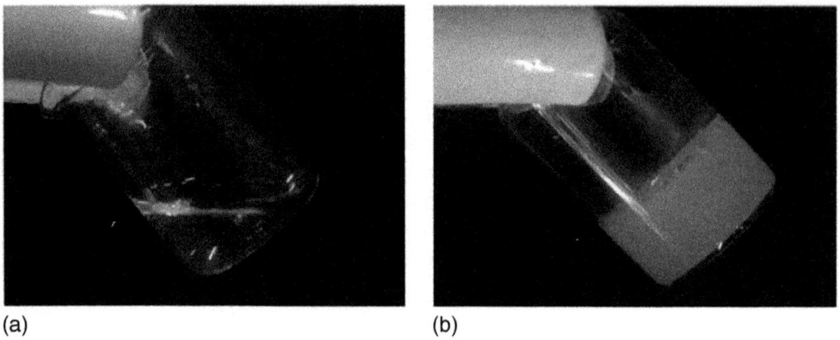

Figure 2.2 Macroscopic aspect of injectable chitosan-based hydrogels at (a) 4 °C and (b) 37 °C. *Source:* From Cao et al. 2015 [158].

temperature (LCST), which means that at room temperature a polymer solution presents low viscosity and above an LCST a gel is obtained [154]. Chenite et al. reported the production of an injectable hydrogel by physical mixture of glycerophosphate and chitosan for tissue engineering application. Neutralization of ammonium groups of chitosan by the phosphates enables the hydrophobic and hydrogen bonding between chitosan chains at high temperatures. By that way, mixture remains liquid at room temperature and forms a gel at 37 °C [157]. Similar hydrogels were fabricated by Cao et al. for the treatment of chronic rhinosinusitis (Figure 2.2) [158].

Another microbial polysaccharide used to fabricate hydrogels is dextran. As previously mentioned, dextran is a water-soluble bacterial EPS [159]. Similar to other polysaccharides, dextran has hydroxyl groups that allow derivatization and, consequently, chemical and physical crosslinking. Various authors have reported the synthesis of dextran-based hydrogels mainly for drug delivery and tissue engineering applications [144, 159]. The utilization of poly(ethylene glycol)-grafted dextran and α-cyclodextrins for the fabrication of thermoresponsive hydrogels was reported by Huh et al. [160]. Results showed that polyethylene glycol (PEG) grafts form inclusion complexes with α-cyclodextrin molecules through hydrophobic interactions and thermoreversible gelation occurs through supramolecular assembly and dissociation. Similarly, Pescosolido et al. [161] have synthesized biodegradable dextran-based hydrogels via UV polymerization of hydroxyethyl-methacrylate-derivatized dextran (Dex-HEMA) and hyaluronic acid for bioprinting applications (Figure 2.3).

Xanthan gum is other polyanionic polysaccharide often used as an attractive material for the fabrication of hydrogels. Generally, xanthan is combined with other polysaccharides to improve gelation characteristics since by itself xanthan is only able to produce transient weak gels. Recently, a hydrogel made from xanthan and galactomannan from the seeds of the Brazilian native tree *Schizolobium parahyba* was reported by Koop et al. [162]. The binary hydrogel was loaded with curcumin for topical and cutaneous wound applications. The results revealed that such hydrogels allowed prolonged exposure to the skin without any irritation and the possibility to treat skin diseases such as psoriasis. In other study, xanthan was chemically crosslinked with starch to perform

Figure 2.3 Top view of 3D printed hyaluronic acid (6% w/v) and Dex-HEMA (10% w/v) hydrogel. The scale bar represents 25 mm. *Source:* Reprinted with permission from Pescosolido et al. [161]. Copyright 2011, American Chemical Society.

hydrogels for drug delivery applications [163]. Starch–xanthan gum hydrogels exhibited selective permeability depending on drug charges and revealed to be a promising material for controlled release of several drugs.

As described, a wide and diverse range of polysaccharides have been used as attractive materials to design and fabrication of hydrogels [139, 141]. Among others, remarkable properties of microbial polysaccharides make them promising materials for different biomedical applications including tissue engineering, drug delivery, and cell therapies.

2.5 Bionanocomposites Based on Microbial Polysaccharides

MNPs are valuable nanostructures with proven applicability in areas such as molecular diagnostics and biomedicine [164, 165]. Their unique physical properties can be tailored based on the size and composition of the inorganic material that can include noble metals (e.g. gold, silver), magnetic elements (e.g. iron, cobalt), or semiconductors (e.g. carbon nanotubes) [164]. The encapsulation of such MNPs in an inorganic (e.g. ceramic) or organic (e.g. biopolymeric) matrix generates multiphase materials (nanocomposites), wherein the synergetic effect between the components adds novel features to this material. Over the past years, the interest in the study and development of nanocomposites has grown considerably due to their valuable physical properties and countless applications that range from packaging to biomedicine [166].

Nanocomposites can have different types of matrices: ceramic, metallic, or polymeric. The properties and functionalities of these matrices can be enhanced by using diverse nanostructures such as ceramics, carbon nanotubes, metal nanoparticles, or even active biological substances [166]. Giving this, nowadays it is possible to produce nanocomposites for all sorts of applications. One good example is the production of a biocompatible hydrogel with conductive properties to be used in biomedicine. By using a polymer-based matrix (e.g. chitosan) incorporated with a conductor inorganic material (e.g. carbon nanotubes), it is possible to produce a nanocomposite with potential use in the design of electrochemical biosensors [82, 167]. Another great example is the nanocomposites produced with magnetic nanoparticles. These MNPs have their potentialities well established in the medical field. Usually, these applications take advantage of three unique features inherent to magnetic nanoparticles. These properties are the field-induced mobility (for the development of drug delivery systems), their ability to modify magnetic relaxation times of surrounding molecules (for magnetic resonance imaging [MRI] applications), and their capacity to heat under an alternative magnetic field (for hyperthermia applications) [168, 169]. The encapsulation of magnetic nanoparticles in a polymeric matrix (e.g. polyacrylamide) creates a thermoresponsive hydrogel. If this nanocomposite also contains a pharmaceutical drug, it could be used as a controlled release system. By applying an alternative magnetic field, the nanoparticles will increase their temperature changing the hydrogel structure to a more flexible state, allowing the easier delivery of the pharmaceutical substance [82]. In the previous examples, it has been established the importance of using polymeric matrices to produce nanomaterials with enhanced properties, as well as the potentialities of using these nanocomposites for medical applications.

More recently, it has been studied the use of biological polymers in the production of nanocomposites (bionanocomposites). The wide diversity of available natural polymers with distinct structures and properties has driven the interest toward the development of novel biopolymer nanocomposites with unique or improved functionalities. Unlike synthetic polymers, biopolymers have inherent favorable interaction with living systems [78]. Moreover, due to their chemical and structural diversity, biopolymers can provide excellent matrices for incorporation of different active substances (e.g. MNPs, hydrophilic and hydrophobic drugs), being more sustainable and having limited environmental impact due to their inherent biodegradability [80].

As previously mentioned, microbial polysaccharides have unique properties suitable to a wide range of applications. Some of these biopolymers are also known to have interactions with biological systems. These properties are usually associated with the composition of the polysaccharide. For instance, FucoPol is a biopolymer with potential antitumor and anti-inflammatory properties due to its high fucose content. Fucose is a rare sugar with reported anticarcinogenic, antiaging, and anti-inflammatory properties [4]. For these reasons, microbial polysaccharides can be used in the development of bionanocomposites not only as a matrix material due to their ability to form structured materials (e.g. hydrogels and films) but also as an active substance. This duality presented by the polysaccharides is an important feature for the development of bionanomaterials especially for biomedical applications [168, 170].

Lately, the use of bionanocomposites with polysaccharides in several fields ranging from the degradation of pollutants to the development of hyperthermia agents and targeted delivery systems has been studied (Figure 2.4). For these types of applications, the biodegradability, the biocompatibility, and the biological response are important properties that can be easily attainable by using polysaccharides (especially from microbial origin) [172]. In Table 2.2, the use

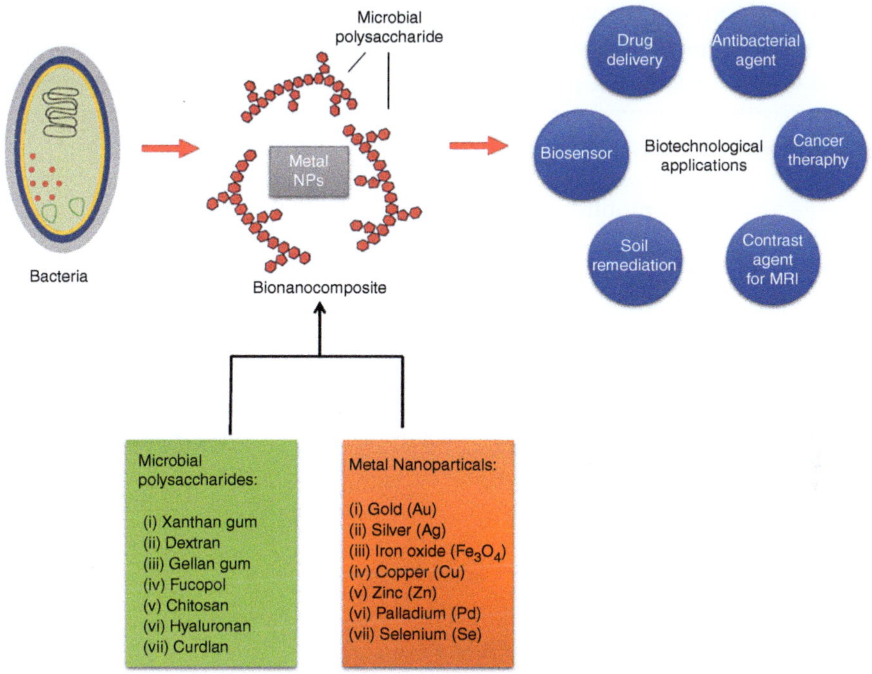

Figure 2.4 Microbial polysaccharide-based nanocomposites with metal nanoparticles for biotechnological applications. *Source:* Adapted from Manivasagan and Oh 2016 [81] and Escárcega-González et al. 2018 [171].

of distinct polysaccharides (all attainable through microbial production) and different MNPs for the development of nanocomposites for several applications has been shown. It is curious that the use of Au and Ag nanoparticles is usually associated with the production of nanocomposites with anticancer and antibacterial properties, respectively. Meanwhile, Fe_3O_4 nanoparticles are frequently related to the development of nanomaterials for drug delivery systems, hyperthermia, and contrast agents for MRI. Given this, the potentialities of the use of bionanocomposites with microbial polysaccharides especially in biomedicine remain well established.

Microbial polysaccharides show variability, versatility, stability, biocompatibility, and biodegradability. Adding to this, their important feature to act as a thickening, stabilizing, or binding agent makes them suitable to a wide range of applications, from food industry to pharmaceutical, medical, and cosmetic products. Moreover, since some of these biopolymers can form structured materials (such as gels and films), they could be used as matrices for the development of bionanocomposites with incorporated nanoparticles (Au^0, Ag^0, Fe_3O_4) for distinct applications in the biomedical field (anticancer, antibacterial, or hyperthermia agent, respectively). Polysaccharide-based nanomaterials are an excellent source for nanotechnological applications in food, pharmaceutical, biomedicine, and cosmetic industries.

Table 2.2 Bionanocomposites containing polysaccharides and their applications.

Polysaccharide	Nanoparticles	Possible applications	References
Hyaluronan	Fe_3O_4	Contrast agents for MRI; drug delivery systems	[169]
Dextran	Fe_3O_4	Cellular MRI and fluorescence imaging; drug delivery systems	[169]
	Au^0	Anticancer agent	[171]
	Ag^0	Sensor for cysteine detection; antibacterial and antifungal agent	[169]
Chitosan	Fe_3O_4	Drug delivery systems	[169]
	$Cu^0/Fe^0/CdS$	Degradation of Congo red and heavy metals in water (e.g. Cr(VI))	[170]
Xanthan gum	Fe_3O_4	Contrast agents for MRI; hyperthermia agent	[168]
	Au^0	Drug delivery systems; anticancer therapy	[171]
	Ag^0	Antibacterial and catalytic agent	[173]
	Pd/Fe	Soil remediation	[171]
Gellan Gum	Au^0	Drug delivery systems; anticancer agent	
	Ag^0	Antibacterial and topical treatment	[171]
	Fe_3O_4	Drug delivery systems; anticancer agent	

MRI, magnetic resonance imaging.

2.6 Bioactive Polysaccharides from Microalgae: An Emerging Area

Microalgae (including cyanobacteria) can use CO_2 as carbon source and incorporate it in complex organic molecules, accounting for nearly half of CO_2 capture by photosynthetic organisms annually. Their growth requirements are quite simple: a seawater medium supplemented with a source of nitrogen (apart from nitrogen-fixing microorganisms), phosphate, iron, magnesium, calcium, and other minor salts [174]. As photosynthetic organisms, no carbon source is added to the cultivation medium; thus less wastewater is generated at the end of the process. Furthermore, algae cultivation is possible with non-potable water and using the sunlight, which contributes to a more ecological and economical process [68].

Currently, there are more than 30 000 known species of microalgae [175] that are present in different evolutionary lines and have contrasting ecological requirements, meaning this group has an enormous metabolic diversity and great potential for biotechnology. In fact, microalgae production is an emergent market with an expected yearly growth of 10% [6]. Besides the use of biomass as feed in aquaculture and livestock production [176, 177], microalgae are the source of high-value products, such as natural pigments (e.g. β-carotene and astaxanthin), polyunsaturated fatty acids, proteins, and antioxidants, that are commercialized mainly for the nutraceutical [178] and skin care [179] industries. Moreover, microalgae are increasingly investigated as a new sustainable and environmentally friendly alternative to fossil fuel resources, since they accumulate high amounts of lipids and carbohydrates that can be used as feedstock for biofuel production [180, 181]. However, almost all commercial microalgae products are obtained from the biomass, and only recently their relevance as producers of valuable polysaccharides has started to be considered.

2.6.1 Polysaccharide-Producing Microalgae

Microalgae are a large group of photosynthetic unicellular or multicellular organisms, which includes both prokaryotic (Cyanophyta, i.e. cyanobacteria/blue-green algae) and eukaryotic organisms, belonging to the phyla Chlorophyta (green algae), Rhodophyta (red algae), Chrysophyta (diatoms), and Pyrrophyta (dinoflagellates). Polysaccharide-producing microalgae are found in all microalgae phyllo (Table 2.3), and their EPS are characterized by complex chemical structures generally with a high diversity of sugar monomers in the same macromolecule, including rare sugars such as fucose, rhamnose, and ribose, which are known to confer the biopolymers' biological activity [6]. Examples include the EPS secreted by *Arthrospira platensis* [43], *Porphyridium marinum* [40], and *Rhodella* sp. [46]. Fructose was also found in EPS secreted by *Dunaliella salina* [72]. Of notice is the presence of methyl-derivate sugars (e.g. *Dictyosphaerium chlorelloides*) [71] and uronic acids, mainly glucuronic and galacturonic acid [182–184]. As can be seen in Table 2.3, sulfate groups are also found within algal EPS (sEPS), which further contributes to their unique properties [6].

There are few studies where the glycosidic bonds and polysaccharide structure were evaluated. Examples include (i) spirulan, the polysaccharide produced by *A. platensis*, which is composed of two disaccharide repeating units (→3)-α-L-Rha-(1→2)-α-L-Aco-(1→, where Aco (acofriose) was a sulfated 3-*O*-methyl-Rha, and *O*-hexuronosyl-Rha (aldobiuronic acid) [185]), and (ii) nostoflan, a polysaccharide from *Nostoc flagelliforme* composed of →4)-β-D Glc-(1→4)-D-Xyl-(1 and →4)-[β-D-GlcA-(1→6)-]-β-D-Glc-(1→4)-D-Gal-(1→ [66].

It should be noted that the chemical composition, type of linkage, sulfate content, and position might be significantly different depending on the species, the cultivation and extraction conditions, and the analytical methods employed [182, 183, 186, 187]. As so, their biological activity can be significantly different. For example, the molecular weight of the EPS produced by *Porphyridium cruentum* influenced the immunomodulatory activity, with the polymer with lower molecular weight (Mw) having the strongest immunoenhancing effect [188].

2.6.2 Biological Activity and Potential Applications

Due to the structural and chemical diversity of microalgal EPS, they have been the subject of recent research on their biological activity and potential application in the fields of biomedicine, pharmaceuticals, cosmeceuticals, and therapeutics [189]. Reported bioactive properties are presented in Table 2.3 and

Table 2.3 Polysaccharide-producing microalgae and polysaccharide characterization.

Organism	Sugar composition	Other	Mw (Da)	Bioactivity and applications	References
Rhodophyta					
Porphyridium sp.	Xyl, Gal, Glc, GlcA	Sulf	2.4×10^5 to 1.8×10^6	Anti-inflammatory, antioxidant, hypocholesterolemic, biolubricant	[37–41]
Porphyridium purpureum	Gal, Xyl, Glc, GlcA, Fuc	Sulf	n.a.	Antiviral, antimicrosporidian activity	[42, 43]
Porphyridium cruentum	Gal, Glc, Ara, Man, Fuc, Xyl, Rha	Protein, sulf, UA	n.a.	Antibacterial, antiviral, antiglycemic	[44, 45]
Rhodella sp.	Xyl, Gal, Glc, Rha, Ara, GlcA	Sulf	n.a.	n.a.	[46]
Rhodella maculate	Gal, Xyl, GlcA, Rha, Ara, Glc	Protein, sulf	n.a.	n.a.	[43]
Rhodella reticulata	n.a.	n.a.	n.a.	Antioxidant	[47]
Rhodella violacea	Gal, Xyl, Glc, GlcA, Rha, Ara	Protein, sulf	n.a.	n.a.	[43]
Pyrrophyta (dinoflagellates)					
Cochlodinium polykrikoides	Man, Gal, Glc	UA, sulf	n.a.	Antiviral	[48]
Gymnodinium sp.	Gal	Sulf, lactic acid	1.3×10^6	Antitumor activity	[49, 50]
Gyrodinium impudicum	Gal	Sulf, UA	1.9×10^7	Immunomodulatory and antitumor activity	[51–53]
Cyanophyta (cyanobacteria)					
Aphanothece halophytica	Glc, Fuc, Man, Ara, GlcA	—	2.0×10^6	Adjuvant activity, antiviral, anticancer	[54–57]
Anabaena augstmalis	Glc, Gal, Man, Xyl, Fuc, Rha, GalN, GlcN, GalA, GlcA	Sulf	n.a.	n.a.	[58]
Arthrospira platensis	Rha, Gal, Glc, Fuc, Xyl	Sulf	n.a.	Antiviral, antibacterial, antioxidant, anticoagulant, skin repair	[42, 59–63]
Microcoleus vaginatus	Glc, Gal, Ara, Xyl, Man, Rha, Fuc, GalA, GluA	Protein	n.a.		[64]
Nostoc sp.	Glc, Gal, Xyl, Rha, Man, GalA, GlcA	Protein	n.a.	n.a.	[65]

(*continued*)

Table 2.3 (Continued)

Organism	Sugar composition	Other	Mw (Da)	Bioactivity and applications	References
Nostoc flagelliforme	Glc, Xyl, GlcA, Gal	—	n.a.	Antiviral	[66, 67]
Phormidium autumnale	Rha, Rib, Man, Glc, Fuc, Gal, Ara, GalA, GlcA	Sulf	n.a.	n.a.	[58]
Synechocystis aquatilis	Fuc, Glc, Rha, Xyl, Man, GlcN, GalA, GlcA	Sulf	n.a.	n.a.	[58]
Chlorophyta					
Chlamydomonas reinhardtii	GalA, Rib, Ara, Xyl, Glu, Gal, Rha	Pyr	2.3×10^5	Antioxidant	[68]
Chlorella stigmatophora	Glc, GlcA, Xyl, Rib/Fuc	Sulf, UA	22×10^3	Anti-inflammatory, immunomodulatory (immunosuppressant)	[69]
Cyanobacterium aponinum	GalA, Fuc, 3-OMe-GalA, Glc, Ara, Gal, Man, Rha, 4-OMe-GlcA	—	1.06×10^6	Immunomodulatory	[70]
Dictyosphaerium chlorelloides	Gal, Me-Gal, Rha, Man, Me-hexose, Glc, Me-Glc, Xyl, Me-Xyl, Ara	Protein	9.6×10^5	Antiproliferative immunostimulation of pro- and anti-inflammatory cytokines	[71]
Dunaliella salina	Glc, Gal, Fru, Xyl	—	n.a.	n.a.	[72]
Dunaliella tertiolecta	Glc	—	n.a.	n.a.	[73]
Graesiella sp.	Fuc, Gal, Ara, Glc, Man, Xyl, Rib, Rha	Sulf, UA, protein	n.a.	Antioxidant, antiproliferative	[74]
Haematococcus pluvialis	Rib, Ara, Man, Glc	Acetyl and amino groups	23×10^6	Antiaging, immunomodulatory	[75]
Parachlorella kessleri	Ara, Rha, Xyl, Man, Gal	—	65×10^3	Antitumor, immunomodulatory	[76]
Chromophyta (diatoms)					
Navicula directa	Fuc, Xyl, Gal, Man, Rha, Glc, GlcA	Protein, sulf	2.2×10^5	Antiviral	[77]
Phaeodactylum tricornutum	Glc, GlcA, Man	Sulf, UA	27 to 449×10^3	Anti-inflammatory, immunomodulatory (immunostimulatory)	[69]

n.a., not available; Ara, arabinose; Fuc, fucose; Fru, fructose; Gal, galactose; GalA, galacturonic acid; GalN, galactosamine; Glc, glucose; GlcA, glucuronic acid; GlcN, glucosamine; Man, mannose; Me, methyl derivatives; Pyr, pyruvate; Rha, rhamnose; Rib, ribose; Sulf, sulfate; UA, uronic acids; Xyl, xylose.

include anti-inflammatory, immunomodulatory, antioxidant, antiviral, antibacterial, antitumor, and antihyperlipidemic activity, as well as anticoagulant and/or antithrombotic properties [183, 184, 189, 190].

2.6.2.1 Antiviral Activity

Generally, polysaccharides rich in sulfate groups have antiviral capacity. These molecules seemed to be able to block the binding of the virions to the host cell surfaces [77, 183, 191] and inhibit reverse transcriptase in human immunodeficiency virus (HIV), interfering with the replication and the production of new viral particles [183, 191, 192]. For example, sulfur-containing EPS from *Arthrospira* presented antiviral activity against numerous viruses, including human cytomegalovirus (HCMV), antihuman immunodeficiency virus (anti-HIV), herpes simplex virus type 1 (HSV-1), human herpes virus type 6 (HHV-6), measles virus, mumps virus, influenza A virus, ectromelia virus (ECTV), and vaccinia virus (VACV) [42]. *P. cruentum* secreted an EPS that had a higher antiviral activity against vesicular stomatitis virus than the chemical compounds ribavirin, brivudine, cidofovir, and ganciclovir [45]. Naviculan, a sulfated polysaccharide from diatom *Navicula directa*, inhibited enveloped viruses HSV-1, HSV-2, and influenza A. This polysaccharide inhibited host cell binding and virus penetration and had an inhibitory effect on cell–cell fusion between CD4-expressing and HIV virus gp160-expressing cells that were used as a model of HIV infection [77]. The antiviral activity of the polysaccharides produced by red microalgae was related to their anticancer properties, as these sulfated polysaccharides were able to inhibit murine leukemia retrovirus and cell transformation by murine sarcoma virus by interfering with early steps of replication, virus absorption into the host cells, and after provirus integration into the host genome [193]. Moreover, the sEPS from *Gyrodinium impudicum* inhibited or slowed the effect of encephalomyocarditis virus in infected HeLa cells, without any cytotoxic effects to the host cells [52]. The antiviral efficiency is also related to the molecule size, content in sulfate, monosaccharide composition, and linkage diversity [191].

2.6.2.2 Immunomodulatory, Anti-inflammatory, and Anticancer Activities

The immune system comprises immune organs, immune cells, and immune substances and is the most important system in pathogen defense and tumor control [194]. Microalgal polysaccharides with immunomodulatory activity (Table 2.3) were able to promote macrophages response; induce mature dendritic cells (DCs); affect the functions of B cells, T cells, and NK cells; and interfere with migration and adhesion of leukocytes. Antitumor activity of microalgal EPS has been evaluated *in vitro* with cancer cell lines or *in vivo* using animal models. Anticancer activity can be direct via inducing apoptosis of cancer cells (i.e. depolarization of mitochondrial membrane) or cell cycle arrest and/or prevent metastasis or indirect by enhancing the innate immune system that leads to anticancer efficiency (i.e. nitric oxide [NO] pathway) [194, 195]. Khan et al. [195] have reviewed the mechanisms behind polysaccharide's anticancer potential.

Stimulation of Macrophage Response Several studies report that microalgal EPS can led to the activation of macrophages. Macrophage activation is achieved by (i) improving macrophage proliferation, (ii) enhancing phagocytic activity, (iii) increasing NO and reactive oxygen species (ROS) production, and (iv) stimulating or regulating cytokines and chemokines release [188]. *In vitro*, EPS produced by *P. cruentum* promoted the proliferation of peritoneal macrophages and increased the NO production. Furthermore, the EPS enhanced lymphocyte proliferation in tumor-bearing mice, which suggests that the mechanism for the antitumor activity against implanted S180 tumor might be related to the immunomodulatory activity of this polysaccharide [188]. Sulfated EPS produced by the harmful *G. impudicum* induced the activation of murine peritoneal macrophages that had a cytotoxic effect on tumor cells, which was mainly due to the production of NO [51]. The EPS from *D. chlorelloides* exhibit a significant immunomodulatory effect by stimulation of anti- and pro-inflammatory cytokines [71].

Effect of Polysaccharides in T, B, D, and NK Cells DCs are the most efficient cells in delivering antigens to T cells and expressing several co-stimulatory molecules that led to their activation. When activated, T cells are target specific and highly efficient in tumor treatment [194]. PCEPS, an EPS produced by *Parachlorella kessleri*, inhibited cell growth of both murine and human colon carcinoma cells. Moreover, PCEPS stimulated the growth of splenocytes and bone marrow cells, increasing specific subpopulations of the cells, namely, $CD19^+$ B cells, $33D1^+$ DCs and $CD68^+$ macrophages, and $CD8^+$ cytotoxic T cells. In a murine colon carcinoma peritoneal dissemination model with syngeneic mice, the polysaccharide attenuated tumor growth. These results suggest that PCEPS has both direct and indirect tumoricidal activity, respectively, via inhibiting cell growth and stimulating the host immune responses [76].

Antiproliferative and Direct Anticancer Potential Other microalgal EPS have shown direct antitumor and antiproliferative effect. *Gymnodinium* sp. EPS (GA3) promoted the apoptosis of human myeloid leukemia K562 cells by the inhibition of topoisomerases I and II [50]. Calcium spirulan of *Arthrospira* was reported to prevent pulmonary metastasis by inhibiting adhesion and proliferation of tumor cells [196]. EPSAH, an EPS produced by *Aphanothece halophytica*, induced apoptosis on the HeLa human cervical cancer cell line by targeting the protein regulator Grp78 that induces the loss of mitochondrial membrane potential and the p53–survivin pathway, resulting in caspase-3 activation and causing apoptosis [55].

Anti-inflammatory Activity EPS produced by microalgae species such as *Porphyridium*, *Cyanobacterium aponinum*, *Phaeodactylum tricornutum*, and *Chlorella stigmatophora* have been reported to have anti-inflammatory properties associated with their immunomodulatory activity. Both polysaccharides from *P. tricornutum* and *C. stigmatophora* had anti-inflammatory activity in the carrageenan-induced paw edema test. *P. tricornutum* enhanced phagocytic activity both *in vitro* and *in vivo*, while *C. stigmatophora* showed immunosuppressant effects [69]. Sulfated polysaccharides from red microalgae *Porphyridium* inhibited the migration and adhesion of polymorphonuclear leukocytes *in vitro*, which seems to be related to the *in vivo* anti-inflammatory effect of the polysaccharide that also inhibited cutaneous erythema in human subjects, as leukocyte migration to the site of injury contributes to the production of NO and cytokine release, resulting in the inhibition of over-inflammation [39]. Furthermore, EPS from *C. aponinum*, a cyanobacteria from Blue Lagoon ecosystem (Iceland), affected human DC maturation *in vitro*, leading to the secretion of immunosuppressive cytokine IL-10 and consequent stimulation of T-cell differentiation with a decreased Th17 but increased Treg phenotype. These results suggested that EPS from *C. aponinum* might be related to the beneficial effects on psoriasis observed in patients bathing in the Blue Lagoon [70].

2.6.2.3 Anticoagulant and Antithrombotic Activity

Coagulation factors are essential elements in blood coagulation as they stop blood flow through an injured vessel wall. Anticoagulants interfere with the coagulation factors and are used for therapeutic treatments, for example, in hemophilic patients [197]. The most commonly used anticoagulant/antithrombotic compound, heparin, can cause several unwanted side effects [198]; thus many studies focused on the anticoagulant properties of bioactive molecules, including algal polysaccharides. The anticoagulant activity of macroalgal polysaccharides such as carrageenan and fucoidan has already been reported [197, 198]. However, little is known for microalgal polysaccharides, even though the anticoagulant activity seems to be directly related to high sulfate content of the polysaccharides, which is often a characteristic of microalgal EPS [191]. Despite the high content in sulfate, *Cochlodinium polykrikoides* did not have anticoagulant activity [48], which indicates that other factors might be involved in the anticoagulant effect of polysaccharides, namely, the sugar composition, branching pattern, position, and distribution of sulfate groups [187, 191]. Nevertheless, calcium spirulan had a significant anticoagulant effect based on specific enhancement of thrombin inhibition by heparin cofactor II [61].

2.6.2.4 Antioxidant Activity

Free radicals such as ROS can cause oxidative stress in living organisms, and, in humans, oxidative stress is associated with the occurrence of cancer, diabetes, neurodegenerative diseases, and cardiovascular problems [198, 199]. The central nervous system (CNS) is particularly vulnerable to oxidative stress due to its high oxygen requirements and lipid content. Oxidative stress in the CNS is related to neuronal death and the progression of several neurodegenerative diseases such as multiple sclerosis, Alzheimer's diseases, and Parkinson's diseases [200]. Antioxidants can enhance the activity of antioxidant enzymes, scavenge free radicals, and inhibit lipid oxidation [41]. Crude EPS from *Rhodella reticulata* had antioxidant activity, and a ROS scavenging capacity was much higher than the standard antioxidant α-tocopherol [47]. The molecular weight of the polysaccharide might influence the antioxidant properties of these molecules. For example, only the low molecular weight EPS from *P. cruentum* inhibited oxidative damage in mouse cells and tissues [41]. Furthermore, the sulfate and uronic acid content, structure, and conformation of the polysaccharide might impact their antioxidant properties [191]. Due to their valuable antioxidant properties, microalgal polysaccharides have a great potential to be used as neuroprotective agents; however, further investigation in the CNS is needed.

2.6.2.5 Other Biological Properties

Microalgal polysaccharides might be useful to treat microbial infections since they can block the adhesion to the host cells and inhibit pathogen growth. In fact, sulfated polysaccharides from microalgae sources inhibited the adhesion of human pathogen *Helicobacter pylori* and the fish pathogens *Vibrio campbellii, Vibrio ordalii, Streptococcus saprophyticus,* and *Aeromonas veronii* [201]. Antimicrobial activity has also been reported for polysaccharides secreted from *P. cruentum* [45] and *A. platensis* [60]. EPS from *Porphyridium purpureum* had antimicrosporidian activity, inhibiting the growth of the honeybee parasite *Nosema ceranae* [43].

Furthermore, sulfated polysaccharides from macroalgae and marine animals were successful in the inhibition of the enzyme responsible for cholesterol absorption in the intestine, human pancreatic cholesterol esterase. This effect seems to be related to high sulfate content and molecular weight and the presence of 3-sulfate in the monosaccharide molecule [191]. Thus, microalgal polysaccharides might have an important role in cholesterol regulation. Despite the lack of studies (Table 2.3), both *Porphyridium* sp. biomass and sEPS had hypocholesterolemic effect, decreasing the accumulation of hepatic cholesterol and lowering the plasma triglycerides and very-low-density lipoprotein (VLDL) cholesterol levels in rats fed with high cholesterol diets [38]. Moreover, EPS from *P. cruentum* had antiglycemic properties and reduced the blood glucose level in diabetic rodents [44].

Sulfated polysaccharides might have a promising future as biolubricants due to their rheological properties. In fact, the sEPS from *Porphyridium* sp. had a better lubricant effect than the most used lubricant hyaluronic acid, with better results in reducing friction and wear under different simulated joint efforts. These results are related to the great rheological stability over a range of temperatures, pH, and salinities [37]. Furthermore, this sEPS was successfully patented to be a compound in joint-lubricating products to treat degenerative joint disorders caused by arthritis, due to the promising results demonstrated by injecting the sEPS in joints of rabbits' knees [202].

2.6.3 Commercialization Prospects

The exploitation of microalgal polysaccharides holds great potential of development, representing a sustainable and environmentally friendly strategy to capture CO_2 while producing high-value products with interesting application in therapeutics, cosmetics, and regenerative medicine. However, microalgal polysaccharides had difficulties to enter the market, mainly due to the existence of cheaper alternatives such as macroalgal (e.g. carrageenan, fucoidan, agar) and plant (e.g. guar gum and xanthan gum) products [203]. Nevertheless, there are already some successful examples of the application of microalgal polysaccharides mainly associated with the cosmetic area, such as the sEPS from the red algae *Porphyridium* sp., marketed by Frutarom under the trade name Alguard™ [6]. Alguard has been used in cosmetics as an antiaging, anti-irritant, UVB damage protection, anti-inflammatory, lip balms, healing, and smoothing creams [191]. Also, Algenist™ is a cosmetic product capable of improving health and skin appearance, consisting of a mixture of microalgal polysaccharides named alguronic acid [204]. Furthermore, polysaccharides from microalgae *Haematococcus pluvialis* and *P. cruentum* showed antiaging effects, envisaging their use as active ingredients in cosmetic and pharmaceutical formulations [75, 205].

2.7 Applications of Chitinous Polymers

2.7.1 Chitin, Chitosan, and Chitinous Polysaccharides

Chitin is the second most abundant polysaccharide available in nature after cellulose. Composed of *N*-acetyl-D-glucosamine and *N*-glucosamine units (Figure 2.5), this polysaccharide can be found on both animal and microbial sources. The chitin extracted from microorganisms is advantageous over the chitin from animal sources, due to several factors: the microbial cultures achieved high growth rates and high cell densities, there is no seasonality of the raw material, the composition of polysaccharide is more stable and controllable, and the microbial chitin is allergen-free, which is very important in case of pharma/biomedical applications [207].

Microbial chitin is a cell-wall component of several yeast and fungi strains, conferring rigidity and protection to the cells [208, 209]. Depending of the fungal

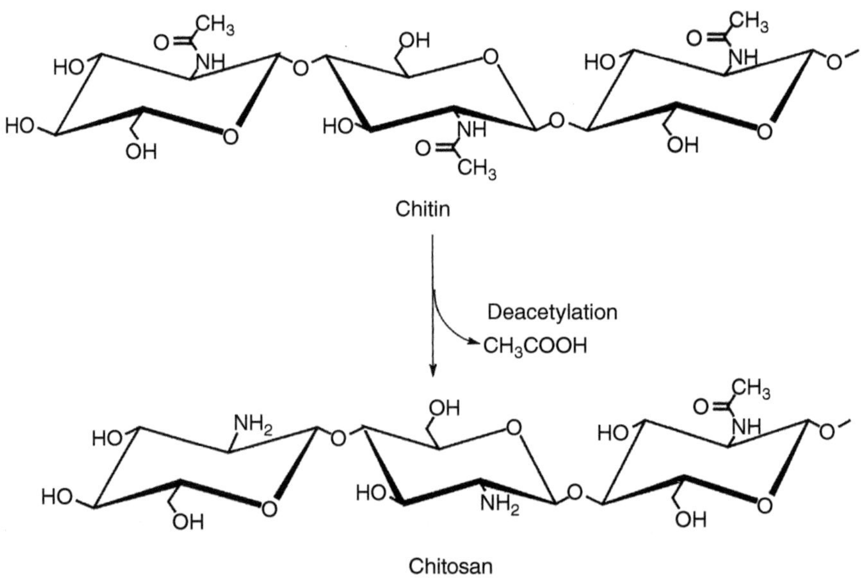

Figure 2.5 Deacetylation reaction of chitin to chitosan. *Source:* Croisier and Jérôme 2013 [206]. Reproduced with permission of Elsevier.

source, chitin can achieve a cell-wall content up to 60% on a dry basis [207]. It is also possible to find chitin polysaccharides in some algae species [210–212].

Fungal chitin can also be found covalently linked to β-glucans to form chitin complexes such as chitin–glucan complex (CGC) [208, 213]. Among the microbial cultures that can be used as chitin/chitinous polysaccharide sources are filamentous fungi (e.g. *Aspergillus niger, Aspergillus fumigatus, Rhizopus oryzae, Mucor rouxii*) [213–217], mushrooms (e.g. *Ganoderma lucidum, Agaricus bisporus, Schizophyllum commune*) [218–221], and yeasts (e.g. *Saccharomyces cerevisiae, Komagataella* (*Pichia*) *pastoris, Candida albicans*) [216, 222–224]. The chitinous polysaccharide extraction from fungal cell walls can be performed with mechanical and/or nonmechanical methods. Some examples of mechanical methods are high-pressure homogenization, bead mill, or ultrasound, while the nonmechanical methods include the alkaline hydrolysis (with NaOH or KOH as solvents) or enzymatic treatments [225].

Chitosan is the most known chitin derivative, obtained from the deacetylation reaction of chitin, in high alkali and temperature conditions or by enzymatic hydrolysis (Figure 2.5) [226–228]. Depending on the alkaline treatment performed for fungal cell-wall lysis, the extraction procedure can also result in the extraction of chitosan or chitosan–glucan complex (ChGC) [215, 221, 229, 230].

2.7.2 Properties of Chitinous Polysaccharides

Chitin is a crystalline and hydrophobic polysaccharide with a molecular weight of 1 to 2×10^5 Da [226, 227, 231, 232]. Chitinous polysaccharides, such as CGC, also have a similar molecular weight (4 to 5×10^5 Da), but since they are covalently linked to β-glucans, they present an amorphous nature [222]. Chitin and chitinous polysaccharides are insoluble in water and in most of organic solvents. The most common solvent systems used for these biomaterials are *N*,*N*-dimethylacetamide/LiCl, NaOH/urea, LiOH or NaOH at low temperatures, or, more recently, ionic liquids [227, 231, 233]. In opposition, chitosan is soluble in diluted organic acids such as acetic, formic, or hydrochloric acid solutions [227, 231]. Chitosan is also characterized for its high deacetylation degree (DD), usually above 50% and a molecular weight of 2×10^4 to 2×10^5 Da [216, 234].

Chitin, chitosan, and their complexes are also known by their biological activity. Behind being biodegradable and biocompatible, these polysaccharides also have antimicrobial, anti-inflammatory, antioxidant, and antitumoral activity [235–237]. Moreover, fungal chitosan also has emulsifying properties with vegetable oils [214].

2.7.3 Applications of Chitinous Polysaccharides

In application development of microbial chitinous polysaccharides, several polymer structures can be constructed from nanofibers or nanoparticles for food formulations or biomedical devices, hydrogels for drug delivery systems or films, and scaffolds/composites or tablets for pharmaceutical formulations [155, 215, 238, 239].

Table 2.4 Microbial chitinous polysaccharide properties and applications.

Polysaccharide	Source	Properties	Applications	References
Chitin	*Saccharomyces cerevisiae*	Antifungal	Disease resistance against *Botrytis cinerea* infection in tomato fruit	[224]
Chitin	*Rhizopus oryzae*	Anti-hepatotoxic effect, antioxidant	Pharma formulation to prevent the liver damage against paracetamol and increase the body's defense system	[240]
Chitosan	*S. cerevisiae, Rhizopus arhizus*	Metal ion sorption	Wastewater treatment	[241]
Chitosan	*Aspergillus niger*	Biocompatibility and swelling capacity	Hydrogel support for controlled release of the histidine analogous, betahistine	[215]
Chitin–glucan complex	*A. niger*	Anticholesterol	Fiber with potential therapeutic utility for coronary heart diseases, obesity, and diabetes	[242–245]
Chitin–glucan complex	*A. niger*	Heavy metal and nucleotide absorption capacity	Food additive	[246]
Chitin–glucan complex	*A. niger*	Water absorption capacity, antiaging effect	Cosmetic formulation for antiaging products	[247, 248]
Chitin/chitosan–glucan complex	*Schizophyllum commune*	Antibacterial, noncytotoxic, wound healing capacity	Micro and non-woven fiber for wound healing	[229]

The diversity of polymer structures and their physicochemical and biological properties have attracted an increasing interest on these polysaccharides for several and diverse areas, such as biomedical devices, pharmaceutical industry, food products, cosmetic formulations, agriculture, and textiles or even on wastewater treatment (Table 2.4) [234, 235, 241].

2.7.3.1 Biomedical Applications

Chitin, chitosan, and chitinous polysaccharides have been subject of increasing interest on the biomedical field due to their biocompatibility, biodegradability, and nontoxicity [237]. Moreover, these polysaccharides also have biological activity including anticancer and antibacterial activity [234]. They are also immunostimulating, antithrombogenic, hemostatic, and wound healing [234, 235]. These properties make these biopolymers interesting materials for drug delivery systems, tissue engineering, or wound healing products.

Chitin and chitosan can be used as drug/vitamin carriers and excipients in films, hydrogels, or powders for drug delivery systems or ophthalmologic formulations [228, 234]. For example, chitosan hydrogels were already used as support for the controlled release of betahistine (a histidine analog) [215] (Table 2.4). Chitosan can also be used in contact lenses for eyes due to its optical clarity, mechanical stability, and wettability [235].

Chitin-based materials are promising materials for tissue engineering of nerves, blood vessels, or scaffolds for regenerative medicine. These structures have high porosity and structural integrity that promote cell adhesion and encourage cell function as important features for regenerative materials [228, 235]. These polysaccharides also contribute to the activation of immune system, which was already demonstrated in some *in vitro* and *in vivo* trials [249].

Chitinous polysaccharides are also commonly used as wound healing fibers [229] (Table 2.4). Some chitin and chitosan products are already commercially available, for example, in form of sponges, nonwoven dressings for burns or skin abrasions, and several other injuries, as well as gels for internal wounds [228].

2.7.3.2 Pharmaceutical Applications

Similar to biomedical applications, chitinous polysaccharides are also interesting materials for the pharmaceutical industry, due to its antimicrobial, antitumor, antioxidant, anticholesterol and hemostatic effect, and wound healing [250]. Chitin extracted from *R. oryzae* was shown to have anti-hepatotoxic effect, preventing the liver damage against paracetamol and contributes to the increasing of body's defense system [240]. *A. niger* CGC was shown to prevent the early development of atherosclerosis in *in vivo* studies [251]. In pharma applications, these biomaterials can be used as carriers of encapsulated drugs for drug delivery devices, vaccine adjuvants, or an excipient powder that can be used as binder, disintegrator, and/or a lubricant agent [234, 235, 252].

2.7.3.3 Food Applications

Chitinous polysaccharides are also used as dietary fibers in some food products, since they have the capacity of lowering the cholesterol levels by blocking the absorption of fat and cholesterols in organism and are a glucose-lowering agent [234, 250]. In this way, these fibers facilitate the weight and fat loss and have a control over nutrition [234]. For example, there are several reports showing that the addition of fungal CGC as a fiber in food formulations has a beneficial effect in the development of obesity and diabetes and in the treatment of heart diseases [242, 243, 245] (Table 2.4).

Chitosan is also known to be beneficial for food preservation as a coating for food packaging, acting as an inhibitor of microbial growth [207].

2.7.3.4 Other Applications

Chitosan and CGC are used as a component of some cosmetic products such as hand and body creams, hair care products, or toothpaste. For example, chitosan protects hair from mechanical damage and exhibits an anti-electrostatic effect on hair. Other chitosan benefits on cosmetic formulations include the water resistance and sun irradiation protection of chitosan emulsions, activation of fibroblasts, collagen deposition improvement, antiseptic properties, and skin protection from microbial infections [207, 235]. Similar to chitosan, *A. niger*

CGC is also already used as a natural scaffold for skin moisturization and rejuvenation [247, 248].

In textile industry, chitin fibers are also used despite their low tensile strength. In this way, chitin is usually used as a coating material for cellulosic, nylons, cotton, and wool fibers. The use of these modified fibers includes the production of wound dressings, medical textiles, sanitary absorbents, underwear, and waterproof textiles [207].

Chitinous polysaccharides are antimicrobial and have metal ion absorption capacity, features that make these biomaterials also valuable to agricultural products. These polysaccharides are considered valuable alternatives to synthetic chemical agents (bactericides), due to their antibacterial activity. There are several reports showing the role of chitinous polysaccharides in some fruit and vegetable resistance against fungal infections [207, 217, 224] (Table 2.4). Moreover, it was also reported that chitin can retain nutrients in the soil [234]. Another application of chitinous polysaccharides in agriculture is as a metal and mineral absorbent for the clarification of wines [253].

The metal absorption capacity of chitinous polysaccharides also enables their use in wastewater treatment and bioremediation. For example, chitosan and CGC have the ability to remove copper, nickel, zinc, or cobalt ions from wastewater [241, 254].

2.8 Microbial Polysaccharides: A World of Opportunities

Microbial polysaccharides are characterized by a range of valuable properties that render them suitable for broad areas of applications, including high-value market niches like cosmetics, pharmaceuticals, and biomedicine. Such biopolymers often possess novel or improved properties compared with synthetic polysaccharides, as well as those obtained from other natural sources (e.g. plants, algae, and animals). Besides their already established uses mostly as rheology modifiers, texturing agents, and bioemulsifiers, many microbial polysaccharides are currently arising as relevant bioactive materials displaying antioxidant, antitumor, immune enhancer, antiaging, and cholesterol lowering, which render them potentially suitable for the development of novel pharmaceuticals, nutraceuticals, and cosmetic products. Moreover, there is intense research related to their use in biopolymeric structures (e.g. hydrogels, bionanocomposites, etc.) possessing advantageous properties compared with other commercial materials. There are a growing number of reports on the isolation of new microbial sources that synthesize polysaccharides with distinctive properties, which may translate into the development of novel biomaterials or active drugs. Therefore, microorganisms represent a still underexplored source of products that, given the current research interest in their study, will surely translate into high-value novel products, particularly for the pharmaceutical and nutraceutical industries.

Acknowledgments

The authors acknowledge the Unidade de Ciências Biomoleculares Aplicadas (UCIBIO), which is financed by national funds from FCT/MCTES (UID/Multi/04378/2019). Diana Araújo and Patrícia Concórdio-Reis acknowledge FCT/MCTES for fellowships SFRH/BD/140829/2018 and SFRH/BD/131947/2017, respectively.

References

1 Moscovici, M. (2015). Present and future medical applications of microbial exopolysaccharides. *Frontiers in Microbiology* 6: 1012.
2 Rehm, B.H.A. (2009). *Microbial Production of Biopolymers and Polymer Precursors: Applications and Perspectives*. Caister Academic Press.
3 Rühmann, B., Schmid, J., and Sieber, V. (2015). Methods to identify the unexplored diversity of microbial exopolysaccharides. *Frontiers in Microbiology* 6: 565.
4 Freitas, F., Alves, V.D., and Reis, M.A.M. (2011). Advances in bacterial exopolysaccharides: from production to biotechnological applications. *Trends in Biotechnology* 29 (8): 388–398.
5 Cress, B.F., Englaender, J.A., He, W. et al. (2014). Masquerading microbial pathogens: capsular polysaccharides mimic host-tissue molecules. *FEMS Microbiology Reviews* 38: 660–697.
6 Freitas, F., Torres, C.A.V., and Reis, M.A.M. (2017). Engineering aspects of microbial exopolysaccharide production. *Bioresource Technology* 245: 1674–1683.
7 Yildiz, H. and Karatas, N. (2018). Microbial exopolysaccharides: resources and bioactive properties. *Process Biochemistry* 72: 41–46.
8 Yu, Y., Shen, M., Song, Q., and Xi, J. (2018). Biological activities and pharmaceutical applications of polysaccharide from natural resources: a review. *Carbohydrate Polymers* 183: 91–101.
9 Roca, C., Alves, V.D., Freitas, F., and Reis, M.A.M. (2015). Exopolysaccharides enriched in rare sugars: bacterial sources, production and applications. *Frontiers in Microbiology* 6: 288.
10 Freitas, F., Alves, V.D., and Reis, M.A.M. (2014). Bacterial polysaccharides: production and cosmetic applications. In: *Polysaccharides: Bioactivity and Biotechnology*. Part III. Bioactive Polysaccharides (eds. J.-M. Merillon and K.G. Ramawat), 1–24. Germany: Springer-Verlag GmbH.
11 Rehm, B.H.A. (2010). Bacterial polymers: biosynthesis, modifications and applications. *Nature Reviews in Microbiology* 8: 578–592.
12 Kumar, A., Rao, K.M., and Soo, S. (2018). Application of xanthan gum as polysaccharide in tissue engineering: a review. *Carbohydrate Polymers* 180: 128–144.

13 Patel, A. and Prajapati, J.B. (2013). Food and health applications of exopolysaccharides produced by lactic acid bacteria. *Advances in Dairy Research* 1: 2.

14 Amado, I.R., Vásquez, J.A., Pastrana, L., and Teixeira, J.A. (2016). Cheese whey: a cost-effective alternative for hyaluronic acid production by *Streptococcus zooepidemicus*. *Food Chemistry* 198: 54–61.

15 Prajapati, V.D., Jani, G.K., and Khanda, S.M. (2013). Pullulan: an exopolysaccharide and its various applications. *Carbohydrate Polymers* 95: 540–549.

16 Schmid, J., Meyer, V., and Sieber, V. (2011). Scleroglucan: biosynthesis, production and application of a versatile hydrocolloid. *Applied Microbiology and Biotechnology* 91: 937–947.

17 Srikanth, R., Siddartha, G., Reddy, C.H.S.S.S. et al. (2015). Antioxidant and anti-inflammatory levan produced from *Acetobacter xylinum* NCIM2526 and its statistical optimization. *Carbohydrate Polymers* 123: 8–16.

18 Gugliandolo, C., Spanò, A., Lentini, V. et al. (2013). Antiviral and immunomodulatory effects of a novel bacterial exopolysaccharide of shallow marine vent origin. *Journal of Applied Microbiology* 116: 1028–1034.

19 Luang-In, V., Saengha, W., and Deeseenthum, S. (2018). Characterization and bioactivities of a novel exopolysaccharide produced from lactose by *Bacillus tequilensis* PS21 isolated from Thai milk kefir. *Microbiology and Biotechnology Letters* 46 (1): 9–17.

20 Lourenço, S.C., Torres, C.A.V., Nunes, D. et al. (2017). Using a bacterial fucose-rich polysaccharide as encapsulation material of bioactive compounds. *International Journal of Biological Macromolecules* 104: 1099–1106.

21 Huang, M., Wang, F., Zhou, X. et al. (2015). Hypoglycemic and hypolipidemic properties of polysaccharides from *Enterobacter cloacae* Z0206 in KKAy mice. *Carbohydrate Polymers* 117: 91–98.

22 Bhat, B. and Bajaj, B.K. (2018). Hypocholesterolemic and bioactive potential of exopolysaccharide from a probiotic *Enterococcus faecium* K1 isolated from *kalarei*. *Bioresource Technology* 254: 264–267.

23 Trabelsi, I., Ktari, N., Slima, S.B. et al. (2017). Evaluation of dermal wound healing activity and in vitro antibacterial and antioxidant activities of a new exopolysaccharide produced by *Lactobacillus* sp.Ca6. *International Journal of Biological Macromolecules* 103: 194–201.

24 Di, W., Zhang, L., Wang, S. et al. (2017). Physicochemical characterization and antitumour activity of exopolysaccharides produced by *Lactobacillus casei* SB27 from yak milk. *Carbohydrate Polymers* 171: 307–315.

25 Rani, R.P., Anandharaj, M., and Ravindran, A.D. (2018). Characterization of a novel exopolysaccharide produced by *Lactobacillus gasseri* FR4 and demonstration of its in vitro biological properties. *International Journal of Biological Macromolecules* 109: 772–783.

26 Jeong, D., Kim, D.H., Kang, I.B. et al. (2017). Characterization and antibacterial activity of a novel exopolysaccharide produced by *Lactobacillus kefiranofaciens* DN1 isolated from kefir. *Food Control* 78: 436–442.

27 Sasikumar, K., Vaikkath, D.K., Devendra, L., and Nampoothiri, K.M. (2017). An exopolysaccharide (EPS) from a *Lactobacillus plantarum* BR2 with

potential benefits for making functional foods. *Bioresource Technology* 241: 1152–1156.
28 Pérez-Ramos, A., Mohedano, M.L., Pardo, M.Á., and López, P. (2018). β-Glucan producing *Pediococcus parvulus* 2.6: test of probiotic and immunomodulatory properties in zebrafish models. *Frontiers in Microbiology* 9: 1684.
29 Chen, G., Qiana, W., Li, J. et al. (2015). Exopolysaccharide of Antarctic bacterium *Pseudoalteromonas* sp. S-5 induces apoptosis in K562 cells. *Carbohydrate Polymers* 121: 107–114.
30 Cheng, J.J., Chang, C.C., Chao, C.H., and Lu, M.K. (2012). Characterization of fungal sulfated polysaccharides and their synergistic anticancer effects with doxorubicin. *Carbohydrate Polymers* 90: 134–139.
31 Chen, Y., Mao, W., Tao, H. et al. (2011). Structural characterization and antioxidant properties of an exopolysaccharide produced by the mangrove endophytic fungus *Aspergillus* sp. Y16. *Bioresource Technology* 102: 8179–8184.
32 Bauerová, K., Mihalová, D., Drábiková, K. et al. (2012). Effects of glucomannan isolated from *Candida utilis* on adjuvant arthritis in Lewis rats. *Current Topics in Nutraceutical Research* 10 (1): 13–30.
33 Orlandelli, R.C., da Silva, M.L.C., Vasconcelos, A.F.D. et al. (2017). β-(1→3,1→6)-D-Glucans produced by *Diaporthe* sp. endophytes: purification, chemical characterization and antiproliferative activity against MCF-7 and HepG2-C3A cells. *International Journal of Biological Macromolecules* 94: 431–437.
34 Prathyusha, A.M.V.N., Sheela, G.M., and Bramhachari, P.V. (2018). Chemical characterization and antioxidant properties of exopolysaccharides from mangrove filamentous fungi *Fusarium equiseti* ANP2. *Biotechnology Reports* 19: e00277.
35 Mahapatra, S. and Banerjee, D. (2012). Structural elucidation and bioactivity of a novel exopolysaccharide from endophytic *Fusarium solani* SD5. *Carbohydrate Polymers* 90: 683–689.
36 Lu, Y., Xu, L., Cong, Y. et al. (2019). Structural characteristics and anticancer/antioxidant activities of a novel polysaccharide from *Trichoderma kanganensis*. *Carbohydrate Polymers* 205: 63–71.
37 Arad, S.M., Rapoport, L., Moshkovich, A. et al. (2006). Superior biolubricant from a species of red microalga. *Langmuir* 22 (17): 7313–7317.
38 Dvir, I., Stark, A.H., Chayoth, R. et al. (2009). Hypocholesterolemic effects of nutraceuticals produced from the red microalga *Porphyridium* sp. in rats. *Nutrients* 1 (2): 156–167.
39 Matsui, S.M., Muizzudin, N., Arad, S.M., and Marenus, K. (2003). Sulfated polysaccharides from red microalgae anti-inflammatory properties *in vitro* and *in vivo*. *Applied Biochemistry and Biotechnology* 104: 13–22.
40 Soanen, N., Da Silva, E., Gardarin, C. et al. (2016). Improvement of exopolysaccharide production by *Porphyridium marinum*. *Bioresource Technology* 213: 231–238.
41 Sun, L., Wang, C., Shi, Q., and Ma, C. (2009). Preparation of different molecular weight polysaccharides from *Porphyridium cruentum* and their

antioxidant activities. *International Journal of Biological Macromolecules* 45 (1): 42–47.

42 Radonic, A., Thulke, S., Achenbach, J. et al. (2010). Anionic polysaccharides from phototrophic microorganisms exhibit antiviral activities to vaccinia virus. *Journal of Antivirals and Antiretrovirals* 2 (4): 51–55.

43 Roussel, M., Villay, A., Delbac, F. et al. (2015). Antimicrosporidian activity of sulfated polysaccharides from algae and their potential to control honeybee nosemosis. *Carbohydrate Polymers* 133: 213–220.

44 Huang, J., Liu, L., Yu, Y. et al. (2006). Reduction in the blood glucose level of exopolysaccharide of *Porphyridium cruentum* in experimental diabetic mice. *Journal of Fujian Normal University* 22 (3): 77–80.

45 Raposo, M.F.J., Morais, A.M.M.B., and Morais, R.M.S.C. (2014). Influence of sulfate on the composition and antibacterial and antiviral properties of the exopolysaccharide from *Porphyridium cruentum*. *Life Sciences* 101 (1–2): 56–63.

46 Villay, A., Laroche, C., Roriz, D. et al. (2013). Optimisation of culture parameters for exopolysaccharides production by the microalga *Rhodella violacea*. *Bioresource Technology* 146: 732–735.

47 Chen, B., You, W., Huang, J. et al. (2010). Isolation and antioxidant property of the extracellular polysaccharide from *Rhodella reticulata*. *World Journal of Microbiology and Biotechnology* 26 (5): 833–840.

48 Hasui, M., Matsuda, M., Okutani, K., and Shigeta, S. (1995). In vitro antiviral activities of sulfated polysaccharides from a marine microalga (*Cochlodinium polykrikoides*) against human immunodeficiency virus and other enveloped viruses. *International Journal of Biological Macromolecules* 17 (5): 293–297.

49 Hasui, M., Matsuda, M., Yoshimatsu, S., and Okutani, K. (1995). Production of a lactate-associated galactan sulfate by a dinoflagellate *Gymnodinium* A3. *Fisheries Science* 61 (2): 321–326.

50 Umemura, K., Yanase, K., Suzuki, M. et al. (2003). Inhibition of DNA topoisomerases I and II, and growth inhibition of human cancer cell lines by a marine microalgal polysaccharide. *Biochemical Pharmacology* 66 (3): 481–487.

51 Bae, S.-Y., Yim, J.H., Lee, H.K., and Pyo, S. (2006). Activation of murine peritoneal macrophages by sulfated exopolysaccharide from marine microalga *Gyrodinium impudicum* (strain KG03): involvement of the NF-κB and JNK pathway. *International Immunopharmacology* 6 (3): 473–484.

52 Yim, J.H., Kim, S.J., Ahn, S.H. et al. (2004). Antiviral effects of sulfated exopolysaccharide from the marine microalga *Gyrodinium impudicum* strain KG03. *Marine Biotechnology* 6 (1): 17–25.

53 Yim, J.H., Kim, S.J., Ahn, S.H., and Lee, H.K. (2007). Characterization of a novel bioflocculant, p-KG03, from a marine dinoflagellate, *Gyrodinium impudicum* KG03. *Bioresource Technology* 98 (2): 361–367.

54 Li, P., Liu, Z., and Xu, R. (2001). Chemical characterization of the released polysaccharide from the cyanobacterium *Aphanothece halophytica* GR02. *Journal of Applied Phycology* 13 (1): 71–77.

55 Ou, Y., Xu, S., Zhu, D., and Yang, X. (2014). Molecular mechanisms of exopolysaccharide from *Aphanothece halophytica* (EPSAH) induced apoptosis in HeLa cells. *PLoS One* 9 (1): e87223.

56 Zheng, W., Chen, C., Cheng, Q. et al. (2006). Oral administration of exopolysaccharide from *Aphanothece halophytica* (Chroococcales) significantly inhibits influenza virus (H1N1)-induced pneumonia in mice. *International Immunopharmacology* 6 (7): 1093–1999.

57 Zhu, L., Zhang, F., Yang, L.-J. et al. (2016). EPSAH, an exopolysaccharide from *Aphanothece halophytica* GR02, improves both cellular and humoral immunity as a novel polysaccharide adjuvant. *Chinese Journal of Natural Medicines* 14 (7): 541–548.

58 Di Pippo, F., Ellwood, N.T.W., Gismondi, A. et al. (2013). Characterization of exopolysaccharides produced by seven biofilm-forming cyanobacterial strains for biotechnological applications. *Journal of Applied Phycology* 25 (6): 1697–1708.

59 Ahmed, M., Moerdijk-Poortvliet, T.C.W., Wijnholds, A. et al. (2014). Isolation, characterization and localization of extracellular polymeric substances from the cyanobacterium *Arthrospira platensis* strain MMG-9. *European Journal of Phycology* 49 (2): 143–150.

60 Challouf, R., Trabelsi, L., Ben Dhieb, R. et al. (2011). Evaluation of cytotoxicity and biological activities in extracellular polysaccharides released by cyanobacterium *Arthrospira platensis. Brazilian Archives of Biology and Technology* 54 (4): 831–838.

61 Majdoub, H., Mansour, M.B., Chaubet, F. et al. (2009). Anticoagulant activity of a sulfated polysaccharide from the green alga *Arthrospira platensis. Biochimica et Biophysica Acta (BBA) - General Subjects* 1790 (10): 1377–1381.

62 Reichert, M., Bergmann, S.M., Hwang, J. et al. (2017). Antiviral activity of exopolysaccharides from *Arthrospira platensis* against koi herpesvirus. *Journal of Fish Diseases* 40 (10): 1441–1450.

63 Wang, B., Liu, Q., Huang, Y. et al. (2018). Extraction of polysaccharide from spirulina and evaluation of its activities. *Evidence-Based Complementary and Alternative Medicine* 2018: 1–8.

64 Ge, H., Zhang, J., Zhou, X. et al. (2014). Effects of light intensity on components and topographical structures of extracellular polymeric substances from *Microcoleus vaginatus* (Cyanophyceae). *Phycologia* 53 (2): 167–173.

65 Ge, H., Xia, L., Zhou, X. et al. (2014). Effects of light intensity on components and topographical structures of extracellular polysaccharides from the cyanobacteria *Nostoc* sp. *Journal of Microbiology* 52 (2): 179–183.

66 Ahmadi, A., Zorofchian Moghadamtousi, S., Abubakar, S., and Zandi, K. (2015). Antiviral potential of algae polysaccharides isolated from marine sources: a review. *BioMed Research International* 2015: 1–10.

67 Kanekiyo, K., Hayashi, K., Takenaka, H. et al. (2007). Anti-herpes simplex virus target of an acidic polysaccharide, nostoflan, from the edible blue-green alga *Nostoc flagelliforme. Biological and Pharmaceutical Bulletin* 30 (8): 1573–1575.

68 Bafana, A. (2013). Characterization and optimization of production of exopolysaccharide from *Chlamydomonas reinhardtii*. *Carbohydrate Polymers* 95 (2): 746–752.
69 Guzmán, S., Gato, A., Lamela, M. et al. (2003). Anti-inflammatory and immunomodulatory activities of polysaccharide from *Chlorella stigmatophora* and *Phaeodactylum tricornutum*: anti-inflammatory and immunomodulatory activities. *Phytotherapy Research* 17 (6): 665–670.
70 Gudmundsdottir, A.B., Omarsdottir, S., Brynjolfsdottir, A. et al. (2015). Exopolysaccharides from *Cyanobacterium aponinum* from the Blue Lagoon in Iceland increase IL-10 secretion by human dendritic cells and their ability to reduce the IL-17^+RORγt^+/IL-10^+FoxP3^+ ratio in CD4^+ T cells. *Immunology Letters* 163 (2): 157–162.
71 Halaj, M., Paulovičová, E., Paulovičová, L. et al. (2018). Biopolymer of *Dictyosphaerium chlorelloides* – chemical characterization and biological effects. *International Journal of Biological Macromolecules* 113: 1248–1257.
72 Mishra, A. and Jha, B. (2009). Isolation and characterization of extracellular polymeric substances from micro-algae *Dunaliella salina* under salt stress. *Bioresource Technology* 100 (13): 3382–3386.
73 Goo, B.G., Baek, G., Jin Choi, D. et al. (2013). Characterization of a renewable extracellular polysaccharide from defatted microalgae *Dunaliella tertiolecta*. *Bioresource Technology* 129: 343–350.
74 Trabelsi, L., Chaieb, O., Mnari, A. et al. (2016). Partial characterization and antioxidant and antiproliferative activities of the aqueous extracellular polysaccharides from the thermophilic microalgae *Graesiella* sp. *BMC Complementary and Alternative Medicine* 16 (1): 210.
75 Liu, X., Zhang, M., Liu, H. et al. (2018). Preliminary characterization of the structure and immunostimulatory and anti-aging properties of the polysaccharide fraction of *Haematococcus pluvialis*. *RSC Advances* 8 (17): 9243–9252.
76 Ishiguro, S., Uppalapati, D., Goldsmith, Z. et al. (2017). Exopolysaccharides extracted from *Parachlorella kessleri* inhibit colon carcinoma growth in mice via stimulation of host antitumor immune responses. *PLoS One* 12 (4): e0175064.
77 Lee, J.-B., Hayashi, K., Hirata, M. et al. (2006). Antiviral sulfated polysaccharide from *Navicula directa*, a diatom collected from deep-sea water in Toyama Bay. *Biological and Pharmaceutical Bulletin* 29 (10): 2135–2139.
78 Torres, C.A.V., Marques, R., Antunes, S. et al. (2011). Kinetics of production and characterization of the fucose containing exopolysaccharide from *Enterobacter* A47. *Journal of Biotechnology* 156: 261–267.
79 Alves, V.D., Freitas, F., Torres, C.A.V. et al. (2010). Rheological and morphological characterization of the culture broth during exopolysaccharide production by *Enterobacter* sp. *Carbohydrate Polymers* 81: 758–764.
80 Kumar, S.A., Mody, K., and Jha, B. (2007). Bacterial exopolysaccharides – a perception. *Journal of Basic Microbiology* 47: 103–117.
81 Manivasagan, P. and Oh, J. (2016). Marine polysaccharide-based nanomaterials as a novel source of nanobiotechnological applications. *International Journal of Biological Macromolecules* 82: 315–327.

82 Aimé, C. and Coradin, T. (2012). Nanocomposites from biopolymer hydrogels: blueprints for white biotechnology and green materials chemistry. *Journal of Polymer Science Part B: Polymer Physics* 50: 669–680.

83 Prasongsuk, S., Loytakul, P., Ali, I. et al. (2018). The current status of *Aureobasidium pullulans* in biotechnology. *Folia Microbiologica* 63 (2): 129–140.

84 Tabasum, S., Noreen, A., Maqsood, M.F. et al. (2018). A review on versatile applications of blends and composites of pullulan with natural and synthetic polymers. *International Journal of Biological Macromolecules* 120: 603–632.

85 Cheng, K., Demirci, A., and Catchmark, J.M. (2011). Pullulan: biosynthesis, production, and applications. *Applied Microbiology and Biotechnology* 92: 29–44.

86 Singh, R.S., Saini, G.K., and Kennedy, J.F. (2008). Pullulan: microbial sources, production and applications. *Carbohydrate Polymers* 73 (4): 515–531.

87 Giustina, G.D., Gandin, A., Brigo, L. et al. (2019). Polysaccharide hydrogels for multiscale 3D printing of pullulan scaffolds. *Materials and Design* 165: 107566.

88 Taskin, M., Erdal, S., and Canli, O. (2010). Utilization of waste loquat (*Eriobotrya japonica* Lindley) kernels as substrate for scleroglucan production by locally isolated *Sclerotium rolfsii*. *Food Science and Biotechnology* 19: 1069–1075.

89 Castillo, N.A., Valdez, A.L., and Fariña, J.I. (2015). Microbial production of scleroglucan and downstream processing. *Frontiers in Microbiology* 6: 1106.

90 Davison, P. and Mentzer, E. (1982). Polymer flooding in North-sea reservoirs. *Society of Petroleum Engineers Journal* 22 (3): 353–362.

91 Pirri R., Gadioux J., Rivenq R. (1995) Scleroglucan gel applied in the oil industry. EP 1995/0484217A1.

92 Survase, S.A., Saudagar, P.S., Bajaj, I.B., and Singhal, R.S. (2007). Scleroglucan: fermentative production, downstream processing and applications. *Food Technology and Biotechnology* 45 (2): 107–118.

93 Asjadi, S.E., Nerderpel, Q.A., Cotiuga, I.M. et al. (2018). Biopolymer scleroglucan as an emulsion stabilizer. *Colloids and Surfaces A: Physicochemical and Engineering Aspects* 546: 326–333.

94 Zhang, Y., Kong, H., Fang, Y. et al. (2013). Schizophyllan: a review on its structure, properties, bioactivities and recent developments. *Bioactive Carbohydrates and Dietary Fibre* 1 (1): 53–71.

95 Imeson, A. (ed.) (2010). *Food Stabilisers, Thickening and Gelling Agents*. Wiley Blackwell: United Kingdom.

96 Katzbauer, B. (1998). Properties and applications of xanthan gum. *Polymer Degradation and Stability* 59 (1–3): 81–84.

97 Rottava, I., Batesini, G., Silva, M.F. et al. (2009). Xanthan gum production and rheological behaviour using different strain of *Xanthomonas* sp. *Carbohydrate Polymers* 77 (1): 65–71.

98 Petri, D.F.S. (2015). Xanthan gum: a versatile biopolymer for biomedical and technological applications. *Journal of Applied Polymer Science* 132 (23): 420–435.

99 Caldeira, I., Lüdtke, A., Tavares, F. et al. (2018). Ecologically friendly xanthan gum-PVA matrix for solid polymeric electrolytes. *Ionics* 24: 413–420.

100 Tavares, F.C., Dörr, D.S., Pawlicka, A., and Avellaneda, C.O. (2018). Microbial origin xanthan gum-based solid polymer electrolytes. *Journal of Applied Polymer Science* https://doi.org/10.1002/app.46229.

101 Naessens, M., Cerdobbel, A., Soetaert, W., and Vandamme, E.J. (2005). *Leuconostoc* dextransucrase and dextran: production, properties and applications. *Journal of Chemical Technology and Biotechnology* 80: 845–860.

102 Zhou, Q., Feng, F., Yang, Y. et al. (2018). Characterization of a dextran produced by *Leuconostoc pseudomesenteroides* XG5 from homemade wine. *International Journal of Biological Macromolecules* 107: 2234–2241.

103 Kamoun, E.A., Kenawy, E.S., and Chen, X. (2017). A review on polymeric hydrogel membranes for wound dressing applications: PVA-based hydrogel dressings. *Journal of Advanced Research* 8: 217–233.

104 Maslakci, N.N., Ulusoy, S., Uygun, E. et al. (2017). Ibuprofen and acetylsalicylic acid loaded electrospun PVP-dextran nanofiber mats for biomedical applications. *Polymer Bulletin* 74: 3283–3299.

105 Rai, S., Kureel, A.K., Dutta, P.K., and Mehrotra, G.K. (2018). Phenolic compounds based conjugates from dextran aldehyde and BSA: preparation, characterization and evaluation of their anti-cancer efficacy for therapeutic applications. *International Journal of Biological Macromolecules* 110: 425–436.

106 Singh, A., Dutta, P.K., Kumar, H. et al. (2018). Synthesis of chitin-glucan-aldehyde-quercetin conjugate and evaluation of anticancer and antioxidant activities. *Carbohydrate Polymers* 193: 99–107.

107 Alibolandi, M., Mohammadi, M., Taghdisi, S.M. et al. (2017). Synthesis and preparation of biodegradable hybrid dextran hydrogel incorporated with biodegradable curcumin nanomicelles for full thickness wound healing. *International Journal of Pharmaceutics* 532 (1): 466–477.

108 Cao, L., Tian, H., Wu, M. et al. (2018). Determination of curdlan oligosaccharides with high performance anion exchange chromatography with pulsed amperometric detection. *Journal of Analytical Methods in Chemistry* 2018: 3980814.

109 Liu, Y., Gu, Q., Ofosu, F.K., and Yu, X. (2015). Isolation and characterization of curdlan produced by *Agrobacterium* HX1126 using -lactose as substrate. *International Journal of Biological Macromolecules* 81: 498–503.

110 Yang, M., Zhu, Y., Li, Y. et al. (2016). Production and optimization of curdlan produced by *Pseudomonas* sp. QL212. *International Journal of Biological Macromolecules* 89: 25–34.

111 Chien, C.Y., Enomoto-Rogers, Y., Takemura, A., and Iwata, T. (2017). Synthesis and characterization of regioselectively substituted curdlan hetero esters via an unexpected acyl migration. *Carbohydrate Polymers* 155: 440–447.

112 Miwa, M., Nakao, Y., and Nara, K. (1994). Food applications of Curdlan. In: *Food Hydrocolloids, Structures, Properties, and Functions* (eds. K. Nishinari and E. Doi), 119–124. New York: Plenum Press.

113 Hida, T.H., Ishibashi, K., Miura, N.N. et al. (2009). Citokine induction by a linear 1,3-glucan, curdlan-oligo, in mouse leukocytes in vitro. *Inflammation Research* 58 (1): 9–14.

114 Miyanishi, N., Iwamoto, Y., Watanabe, E., and Oda, T. (2003). Induction of TNF-α production from human peripheral blood monocytes with β-1,3-glucanoligomer prepared from laminarin with β-1,3-glucanase from *Bacillus clausii* NM-1. *Journal of Bioscience and Bioengineering* 95 (2): 192–195.

115 Fu, Y., Zhao, X., and Du, Y. (2011). Research progresses on plant resistance induced by β-glucan oligomer. *Chinese Journal of Biological Control* 27 (2): 260–275.

116 Bajaj, I.B., Survase, S.A., Saudagar, P.S., and Singhal, R.S. (2007). Gellan gum: fermentative production, downstream processing and applications. *Food Technology and Biotechnology* 45 (4): 341–354.

117 Freitas, F., Alves, V.D., Coelhoso, I., and Reis, M.A.M. (2013). Production and food applications of microbial biopolymers. In: *Engineering Aspects of Food Biotechnology*, . Part I: Use of Biotechnology in the Development of Food Processes and Products (eds. J.A. Teixeira and A.A. Vicente), 61–83. Boca Raton, FL: CRC Press, Taylor & Francis Group.

118 Zia, K.M., Tabasum, S., Khan, M.F. et al. (2018). Recent trends on gellan gum blends with natural and synthetic polymers: a review. *International Journal of Biological Macromolecules* 109: 1068–1087.

119 Mariod, A. and Fadul, H. (2013). Review: gelatin, source, extraction and industrial applications. *Acta Scientarum Plonorum Technologia Alimentaria* 12 (2): 135–147.

120 Saha, D. and Bhattacharya, S. (2010). Hydrocolloids as thickening and gelling agent in food: a critical review. *Journal of Food Science and Technology* 47 (6): 587–597.

121 Pacelli, S., Paolicelli, P., Avitabile, M. et al. (2018). Design of a tunable nanocomposite double network hydrogel based on gellan gum for drug delivery applications. *European Polymer Journal* 104: 184–193.

122 Paolicelli, P., Petralito, S., Varani, G. et al. (2018). Effect of glycerol on the physical and mechanical properties of thin gellan gum films for oral drug delivery. *International Journal of Pharmaceutics* 547 (1–2): 226–234.

123 Hill, L.J., Moakes, R.J.A., Vareechon, C. et al. (2018). Sustained release of decorin to the surface of the eye enables scarless corneal regeneration. *Regenerative Medicine* 3: 23.

124 Osmalek, T., Froelich, A., and Tasarek, S. (2014). Application of gellan gum in pharmacy and medicine. *International Journal of Pharmaceutics* 466 (1–2): 328–340.

125 Öner, E.T., Hernàndez, L., and Combie, J. (2016). Review of levan polysaccharide: from a century of past experiences to future prospects. *Biotechnology Advances* 34 (5): 827–844.

126 Srikanth, R., Reddy, C.H.S.S.S., Siddartha, G. et al. (2015). Review on production, characterization and applications of microbial levan. *Carbohydrate Polymers* 120: 102–114.

127 Silbir, S., Dagbagli, S., Yegin, S. et al. (2014). Levan production by *Zymomonas mobilis* in batch and continuous fermentation. *Carbohydrate Polymers* 99: 454–461.

128 Zhang, T., Li, R., Qian, H. et al. (2014). Biosynthesis of levan by levansucrase from *Bacillus methylotrophicus* SK 21.002. *Carbohydrate Polymers* 101: 975–981.

129 González-Garcinuño, A., Tabernero, A., Dominguez, A. et al. (2018). Levan and levansucrase: polymer enzyme, microorganisms and biomedical applications. *Biocatalysis and Biotransformation* 36: 233–244.

130 Jia, Y., Zhu, J., Chen, X. et al. (2013). Metabolic engineering of *Bacillus subtilis* for the efficient biosynthesis of uniform hyaluronic acid with controlled molecular weights. *Bioresource Technology* 132: 427–431.

131 Vázquez, J.A., Montemayor, M.I., Fraguas, J., and Murado, M.A. (2009). High production of hyaluronic and lactic acids by *Streptococcus zooepidemicus* in fed-batch culture using commercial and marine peptones from fishing by-products. *Biochemical Engineering Journal* 44 (2–3): 125–130.

132 Allemann, I.B. and Baumann, L. (2008). Hyaluronic acid gel (Juvéderm) preparations in the treatment of facial wrinkles and folds. *Clinical Interventions in Aging* 3 (4): 629–634.

133 Benedini, L.J. and Santana, M.H. (2013). Effects of soy peptone on the inoculum preparation of *Streptococcus zooepidemicus* for production of hyaluronic acid. *Bioresource Technology* 130: 798–800.

134 Berkó, S., Maroda, M., Bodnár, M. et al. (2013). Advantages of cross-linked versus linear hyaluronic acid for semisolid skin delivery systems. *European Polymer Journal* 49 (9): 2511–2517.

135 Rezaeeyiazdi, M., Colombani, T., Memic, A., and Bencherif, S.A. (2018). Injectable hyaluronic acid-*co*-gelatin cryogels for tissue-engineering applications. *Materials* 11 (8): 1374.

136 Tiwari, S., Patil, R., and Bahadur, P. (2019). Polysaccharide based scaffolds for soft tissue engineering applications. *Polymers* 11 (1): 1.

137 Ahmad, N.H., Mustafa, S., and Che Man, Y.B. (2015). Microbial polysaccharides and their modification approaches: a review. *International Journal of Food Properties* 18 (2): 332–347.

138 Barbucci, R., Spera, V., Armenia, E., and Quagliariello, V. (2017). Microarchitecture of water confined in hydrogels. In: *Hydrogels: Design, Synthesis and Application in Drug Delivery Systems and Regenerative Medicine* (eds. T.R.R. Singh, G. Laverty and R. Donnelly), 1–31. CRC Press, Taylor & Francis Group.

139 Leone, G. and Barbucci, R. (2009). Polysaccharide based hydrogels for biomedical applications. In: *Hydrogels* (ed. R. Barbucci), 25–41. Milano: Springer.

140 Ahmed, E.M. (2015). Hydrogel: preparation, characterization and applications: a review. *Journal of Advanced Research* 6: 105–121.

141 Bathia, J.K., Kaith, B.S., and Kalia, S. (2013). Polysaccharide hydrogels: synthesis, characterization and applications. In: *Polysaccharide Based Graft Copolymers* (eds. S. Kalia and M.W. Sabaa), 271–290. Berlin, Heidelberg: Springer-Verlag.

142 Hoffman, A.S. (2012). Hydrogels for biomedical applications. *Advanced Drug Delivery Reviews* 64: 18–23.

143 Gacesa, P. (1988). Alginates. *Carbohydrate Polymers* 8: 161–182.

144 Oliveira, J.T. and Reis, R.L. (2008). Hydrogels from polysaccharide-based materials: fundamentals and applications in regenerative medicine. In: *Natural-Based Polymers for Biomedical Applications* (eds. R.L. Reis, N.M. Neves, J.F. Mano, et al.), 485–514. Woodhead Publishing.

145 Usta, U. and Asmatulu, R. (2015). Hydrogels in various biomedical applications. In: *Polymer Science: Research Advances, Practical Applications and Educational Aspects* (eds. A. Méndez-Vilas and A. Solano), 248–257. Formatex Research Center.

146 Deen, G.R. and Loh, X.J. (2018). Stimuli-responsive cationic hydrogels in drug delivery applications. *Gels* 4 (1): 1–13.

147 Soppimath, K.S., Aminabhavi, T.M., Dave, A.M. et al. (2002). Stimulus-responsive "smart" hydrogels as novel drug delivery systems. *Drug Development and Industrial Pharmacy* 28 (8): 957–974.

148 Rudzinski, W.E., Dave, A.M., Vaishnav, U.H. et al. (2002). Hydrogels as controlled release devices in agriculture. *Designed Monomers and Polymers* 5 (1): 39–65.

149 Ferris, C.J., Gilmore, K.J., Wallace, G.G., and Panhuis, M. (2013). Modified gellan gum hydrogels for tissue engineering applications. *Soft Matter* 9: 3705–3711.

150 Graham, S., Marina, P.F., and Blencowe, A. (2019). Thermoresponsive polysaccharides and their thermoreversible physical hydrogel networks. *Carbohydrate Polymers* 207: 143–159.

151 Giavasis, I., Harvey, L.M., and McNeil, B. (2000). Gellan gum. *Critical Reviews in Biotechnology* 20 (3): 177–211.

152 Matricardi, P., Cencetti, C., Ria, R. et al. (2009). Preparation and characterization of novel gellan gum hydrogels suitable for modified drug release. *Molecules* 14 (9): 3376–3391.

153 Gong, Y., Wang, C., Lai, R.C. et al. (2009). An improved injectable polysaccharide hydrogel: modified gellan gum for long-term cartilage regeneration in vitro. *Journal of Materials Chemistry* 19: 1968–1977.

154 Bhattarai, N., Gunn, J., and Zhang, M. (2010). Chitosan-based hydrogels for controlled localized drug delivery. *Advanced Drug Delivery Reviews* 62 (1): 83–89.

155 Jayakumar, R., Prabaharan, M., Kumar, P.T.S. et al. (2011). Biomaterials based on chitin and chitosan in wound dressing applications. *Biotechnology Advances* 29 (3): 322–337.

156 Tahrir, F.G., Ganji, F., and Ahooyi, T.M. (2015). Injectable thermosensitive chitosan/glycerophosphate-based hydrogels for tissue engineering and drug delivery applications: a review. *Recent Patents on Drug Delivery & Formulation* 9 (2): 107–120.

157 Chenite, A., Chaput, C., Wang, D. et al. (2000). Novel injectable neutral solutions of chitosan form biodegradable gels in situ. *Biomaterials* 21 (21): 2155–2161.

158 Cao, C., Chunhong, Y., Hu, Z., and Zhou, S. (2015). Potential application of injectable chitosan hydrogel treated with siRNA in chronic rhinosinusitis therapy. *Molecular Medicine Reports* 12 (5): 6688–6694.

159 Van Tomme, S.R. and Hennink, W.E. (2007). Biodegradable dextran hydrogels for protein delivery applications. *Expert Review of Medical Devices* 4 (2): 147–164.

160 Huh, K.M., Ooya, T., Lee, W.K. et al. (2001). Supramolecular-structured hydrogels showing a reversible phase transition by inclusion complexation between poly(ethylene glycol) grafted dextran and α-cyclodextrin. *Macromolecules* 34 (25): 8657–8662.

161 Pescosolido, L., Schuurman, W., Malda, J. et al. (2011). Hyaluronic acid and dextran-based semi-IPN hydrogels as biomaterials for bioprinting. *Biomacromolecules* 12 (5): 1831–1838.

162 Koop, H.S., Freitas, R.A., Souza, M.M. et al. (2015). Topical curcumin-loaded hydrogels obtained using galactomannan from *Schizolobium parahyba* and xanthan. *Carbohydrate Polymers* 116: 229–236.

163 Shalviri, A., Liu, Q., Abdekhodaie, M.J., and Wu, X.Y. (2010). Novel modified starch-xanthan gum hydrogels for controlled drug delivery: synthesis and characterization. *Carbohydrate Polymers* 79 (4): 898–907.

164 Larguinho, M., Canto, R., Cordeiro, M. et al. (2015). Gold nanoprobe-based non-crosslinking hybridization for molecular diagnostics. *Expert Review of Molecular Diagnostics* 15 (10): 1355–1368.

165 Laurent, S., Forge, D., Port, M. et al. (2008). Magnetic iron oxide nanoparticles: synthesis, stabilization, vectorization, physicochemical characterizations, and biological applications. *Chemistry Reviews* 108: 2064–2110.

166 Camargo, P.H.C., Satyanarayana, K.G., and Wypych, F. (2009). Nanocomposites: synthesis, structure, properties and new application opportunities. *Materials Research* 12: 1–39.

167 Wang, S.-F., Shen, L., Zhang, W.-D., and Tong, Y.-J. (2005). Preparation and mechanical properties of chitosan/carbon nanotubes composites. *Biomacromolecules* 6: 3061–3072.

168 Dias, A.M.G.C., Hussain, A., Marcos, A.S., and Roque, A.C.A. (2011). A biotechnological perspective on the application of iron oxide magnetic colloids modified with polysaccharides. *Biotechnology Advances* 29: 142–155.

169 Swierczewska, M., Han, H.S., Kim, K. et al. (2016). Polysaccharide-based nanoparticles for theranostic nanomedicine. *Advanced Drug Delivery Reviews* 99: 70–84.

170 Zhao, X., Lv, L., Pan, B. et al. (2011). Polymer-supported nanocomposites for environmental application: a review. *Chemical Engineering Journal* 170: 381–394.

171 Escárcega-González, C.E., Garza-Cervantes, J.A., Vázquez-Rodríguez, A., and Morones-Ramírez, J.R. (2018). Bacterial exopolysaccharides as reducing and/or stabilizing agents during synthesis of metal nanoparticles with biomedical applications. *International Journal of Polymer Science* 2018: 1–15.

172 Özcan, E. and Öner, E.T. (2015). Microbial production of extracellular polysaccharides from biomass sources. In: *Polysaccharides* (eds. K.G. Ramawat and J.-M. Mérillon), 161–184. Switzerland: Springer International Publishing.

173 Hussain, A., Zia, K.M., Tabasum, S. et al. (2017). Blends and composites of exopolysaccharides; properties and applications: a review. *International Journal of Biological Macromolecules* 94: 10–27.

174 Rossi, F. and De Philippis, R. (2016). Exocellular polysaccharides in microalgae and cyanobacteria: chemical features, role and enzymes and genes involved in their biosynthesis. In: *The Physiology of Microalgae. Developments in Applied Phycology* (eds. M. Borowitzka, J. Beardall and J. Raven), 565–590. Springer International Publishing.

175 El-Hack, M.E.A., Abdelnour, S., Alagawany, M. et al. (2019). Microalgae in modern cancer therapy: current knowledge. *Biomedicine & Pharmacotherapy* 111: 42–50.

176 Roy, S.S. and Pal, R. (2015). Microalgae in aquaculture: a review with special references to nutritional value and fish dietetics. *Proceedings of the Zoological Society* 68 (1): 1–8.

177 Spolaore, P., Joannis-Cassan, C., Duran, E., and Isambert, A. (2006). Commercial applications of microalgae. *Journal of Bioscience and Bioengineering* 101 (2): 87–96.

178 Nicoletti, M. (2016). Microalgae nutraceuticals. *Food* 5 (3): 54.

179 Stolz, P. and Obermayer, P. (2005). Manufacturing microalgae for skin care. *Cosmetics & Toiletries* 120 (3): 99–106.

180 Chen, C.-Y., Zhao, X.-Q., Yen, H.-W. et al. (2013). Microalgae-based carbohydrates for biofuel production. *Biochemical Engineering Journal* 78: 1–10.

181 Markou, G. and Nerantzis, E. (2013). Microalgae for high-value compounds and biofuels production: a review with focus on cultivation under stress conditions. *Biotechnology Advances* 31 (8): 1532–1542.

182 Delattre, C., Pierre, G., Laroche, C., and Michaud, P. (2016). Production, extraction and characterization of microalgal and cyanobacterial exopolysaccharides. *Biotechnology Advances* 34 (7): 1159–1179.

183 Raposo, M.J., de Morais, A., and Morais, R. (2015). Marine polysaccharides from algae with potential biomedical applications. *Marine Drugs* 13 (5): 2967–3028.

184 Singh, S., Kant, C., Yadav, R.K. et al. (2019). Chapter 17: Cyanobacterial exopolysaccharides: composition, biosynthesis, and biotechnological applications. In: *Cyanobacteria* (eds. A.K. Mishra, D.K. Tiwari and A.N. Rai), 347–358. Academic Press.

185 Lee, J.-B., Hayashi, T., Hayashi, K., and Sankawa, U. (2000). Structural analysis of calcium spirulan (Ca–SP)-derived oligosaccharides using electrospray ionization mass spectrometry. *Journal of Natural Products* 63 (1): 136–138.

186 Kumar, D., Kaštánek, P., and Adhikary, S.P. (2018). Exopolysaccharides from cyanobacteria and microalgae and their commercial application. *Current Science* 115 (2): 9.

187 Phélippé, M., Gonçalves, O., Thouand, G. et al. (2019). Characterization of the polysaccharides chemical diversity of the cyanobacteria *Arthrospira platensis*. *Algal Research* 38: 101426.

188 Sun, L., Wang, L., and Zhou, Y. (2012). Immunomodulation and antitumor activities of different-molecular-weight polysaccharides from *Porphyridium cruentum*. *Carbohydrate Polymers* 87 (2): 1206–1210.

189 Xiao, R. and Zheng, Y. (2016). Overview of microalgal extracellular polymeric substances (EPS) and their applications. *Biotechnology Advances* 34 (7): 1225–1244.

190 Liu, L., Pohnert, G., and Wei, D. (2016). Extracellular metabolites from industrial microalgae and their biotechnological potential. *Marine Drugs* 14 (10): 191.

191 Raposo, M.F.J., Morais, A.M.M.B., and Morais, R.M.S.C. (2014). Bioactivity and applications of polysaccharides from marine microalgae. In: *Polysaccharides* (eds. K.G. Ramawat and J.-M. Mérillon), 1–38. Springer International Publishing.

192 Amaro, H.M., Guedes, A.C., and Malcata, F.X. (2011). Antimicrobial activities of microalgae: an invited review. In: *Science Against Microbial Pathogens: Communicating Current Research and Technological Advances*, vol. 3 (ed. A. Méndez-Vilas), 1272–1284. Formatex Research Center.

193 Talyshinsky, M.M., Souprun, Y.Y., and Huleihel, M.M. (2002). Anti-viral activity of red microalgal polysaccharides against retroviruses. *Cancer Cell International* 2: 8.

194 Li, J., Shen, B., Nie, S., and Chen, K. (2019). Marine microbial polysaccharides: promising immunomodulatory and anticancer potential. In: *Marine Polysaccharides: Advances and Multifaceted Applications* (eds. S. Ahmed and A. Soundararajan), 13–28. Pan Stanford Publishing Pte. Ltd.

195 Khan, T., Date, A., Chawda, H., and Patel, K. (2019). Polysaccharides as potential anticancer agents – a review of their progress. *Carbohydrate Polymers* 210: 412–428.

196 Morais, M.G., Stillings, C., Dersch, R. et al. (2010). Preparation of nanofibers containing the microalga Spirulina (*Arthrospira*). *Bioresource Technology* 101 (8): 2872–2876.

197 Wijesekara, I., Pangestuti, R., and Kim, S.-K. (2011). Biological activities and potential health benefits of sulfated polysaccharides derived from marine algae. *Carbohydrate Polymers* 84 (1): 14–21.

198 Ngo, D.-H. and Kim, S.-K. (2013). Sulfated polysaccharides as bioactive agents from marine algae. *International Journal of Biological Macromolecules* 62: 70–75.

199 Deng, R. and Chow, T.-J. (2010). Hypolipidemic, antioxidant, and antiinflammatory activities of microalgae spirulina: hypolipidemic, antioxidant, and antiinflammatory activities of microalgae spirulina. *Cardiovascular Therapeutics* 28 (4): 33–45.

200 Pangestuti, R. and Kim, S.-K. (2011). Neuroprotective effects of marine algae. *Marine Drugs* 9 (5): 803–818.

201 Guzmán-Murillo, M.A. and Ascencio, F. (2000). Anti-adhesive activity of sulfated exopolysaccharides of microalgae on attachment of red sore disease-associated bacteria and *Helicobacter pylori* to tissue culture cells. *Letters in Applied Microbiology* 30 (6): 473–478.

202 Arad, S.M. and Atar, D. (2007). Viscosupplementation with algal polysaccharides in the treatment of arthritis. WIPO patent WO/2007/066340, filed 07 December 2006 and issued 14 June 2007.

203 Borowitzka, M.A. (2013). High-value products from microalgae – their development and commercialization. *Journal of Applied Phycology* 25 (3): 743–756.

204 Ariede, M.B., Candido, T.M., Jacome, A.L.M. et al. (2017). Cosmetic attributes of algae – a review. *Algal Research* 25: 483–487.

205 Bayona, K.C.D. (2012). Activity of sulfated polysaccharides from microalgae *Porphyridium cruentum* over degenerative mechanisms of the skin. *International Journal of Science and Advanced Technology* 2 (8): 85–92.

206 Croisier, F. and Jérôme, C. (2013). Chitosan-based biomaterials for tissue engineering. *European Polymer Journal* 49: 780–792.

207 Synowiecki, J. and Al-Khateeb, N.A. (2003). Production, properties, and some new applications of chitin and its derivatives. *Critical Reviews in Food Science and Nutrition* 43 (2): 145–171.

208 Arroyo, J., Farkaš, V., Sanz, A.B., and Cabib, E. (2016). Strengthening the fungal cell wall through chitin–glucan cross-links: effects on morphogenesis and cell integrity. *Cellular Microbiology* 18 (9): 1239–1250.

209 Gow, N.A.R., Latge, J.-P., and Munro, C.A. (2017). The fungal cell wall: structure, biosynthesis, and function. *Microbiology Spectrum* 5 (3) https://doi.org/10.1128/microbiolspec.FUNK-0035-2016.

210 Kapaun, E. and Reisser, W. (1995). A chitin-like glycan in the cell wall of a *Chlorella* sp. (Chlorococcales, Chlorophyceae). *Planta* 197: 577–582.

211 Ogawa, Y., Kimura, S., Wada, M., and Kuga, S. (2010). Crystal analysis and high-resolution imaging of microfibrillar α-chitin from *Phaeocystis*. *Journal of Structural Biology* 171: 111–116.

212 Ozkan, A. and Rorrer, G.L. (2017). Effects of light intensity on the selectivity of lipid and chitin nanofiber production during photobioreactor cultivation of the marine diatom *Cyclotella* sp. *Algal Research* 25: 216–227.

213 Kang, X., Kirui, A., Muszyński, A. et al. (2018). Molecular architecture of fungal cell walls revealed by solid-state NMR. *Nature Communications* 9: 2747.

214 Abdel-Gawad, K.M., Hifney, A.F., Fawzy, M.A., and Gomaa, M. (2017). Technology optimization of chitosan production from *Aspergillus niger* biomass and its functional activities. *Food Hydrocolloids* 63: 593–601.

215 Muñoz, G., Valencia, C., Valderruten, N. et al. (2015). Extraction of chitosan from *Aspergillus niger* mycelium and synthesis of hydrogels for controlled release of betahistine. *Reactive and Functional Polymers* 91–92: 1–10.

216 Pochanavanich, P. and Suntornsuk, W. (2002). Fungal chitosan production and its characterization. *Letters in Applied Microbiology* 35: 17–21.

217 Wu, T., Zivanovic, S., Draughon, F.A. et al. (2005). Physicochemical properties and bioactivity of fungal chitin and chitosan. *Journal of Agricultural and Food Chemistry* 53: 3888–3894.

218 Álvarez, S.P.O., Cadavid, D.A.R., Sierra, D.M.E. et al. (2014). Comparison of extraction methods of chitin from *Ganoderma lucidum* mushroom obtained in submerged culture. *BioMed Research International* 2014: 169071.

219 Singh, A. and Dutta, P.K. (2017). Extraction of chitin-glucan complex from *Agaricus bisporus*: characterization and antibacterial activity. *Journal of Polymer Materials* 34 (1): 1–9.

220 Smirnou, D., Krcmar, M., and Prochazkova, E. (2011). Chitin-glucan complex production by *Schizophyllum commune* submerged cultivation. *Polish Journal of Microbiology* 60 (3): 223–228.

221 Wu, T., Zivanovic, S., Draughon, F.A., and Sams, C.E. (2004). Chitin and chitosans – value-added products from mushroom waste. *Journal of Agricultural and Food Chemistry* 52: 7905–7910.

222 Farinha, I., Duarte, P., Pimentel, A. et al. (2015). Chitin–glucan complex production by *Komagataella pastoris*: downstream optimization and product characterization. *Carbohydrate Polymers* 130: 455–464.

223 Holan, Z., Pokorný, V., Beran, K. et al. (1981). The glucan-chitin complex in *Saccharomyces cerevisiae. Archives of Microbiology* 130: 312–318.

224 Sun, C., Fu, D., Jin, L. et al. (2018). Chitin isolated from yeast cell wall induces the resistance of tomato fruit to *Botrytis cinerea. Carbohydrate Polymers* 199: 341–352.

225 Liu, D., Ding, L., Sun, J. et al. (2016). Yeast cell disruption strategies for recovery of intracellular bio-active compounds – a review. *Innovative Food Science and Emerging Technologies* 36: 181–192.

226 Hamed, I., Özogul, F., and Regenstein, J.M. (2016). Industrial applications of crustacean by-products (chitin, chitosan, and chitooligosaccharides): a review. *Trends in Food Science and Technology* 48: 40–50.

227 Rinaudo, M. (2006). Chitin and chitosan: properties and applications. *Progress in Polymer Science* 31: 603–632.

228 Singh, R., Shitiz, K., and Singh, A. (2017). Chitin and chitosan: biopolymers for wound management. *International Wound Journal* 14 (6): 1276–1289.

229 Abdel-Mohsen, A.M., Jancar, J., Massoud, D. et al. (2016). Novel chitin/chitosan-glucan wound dressing: isolation, characterization, antibacterial activity and wound healing properties. *International Journal of Pharmaceutics* 510: 86–99.

230 Muzzarelli, R.A.A., Boudrant, J., Meyer, D. et al. (2012). Current views on fungal chitin/chitosan, human chitinases, food preservation, glucans, pectins and inulin: a tribute to Henri Braconnot, precursor of the carbohydrate polymers science, on the chitin bicentennial. *Carbohydrate Polymers* 87: 995–1012.

231 Pillai, C.K.S., Paul, W., and Sharma, C.P. (2009). Chitin and chitosan polymers: chemistry, solubility and fiber formation. *Progress in Polymer Science* 34: 641–678.

232 Teng, W.L., Khor, E., Tan, T.K. et al. (2001). Concurrent production of chitin from shrimp shells and fungi. *Carbohydrate Research* 332: 305–316.

233 Silva, S.S., Mano, J.F., and Reis, R.L. (2017). Ionic liquids in the processing and chemical modification of chitin and chitosan for biomedical applications. *Green Chemistry* 19: 1208–1220.

234 Philibert, T., Lee, B.H., and Fabien, N. (2017). Current status and new perspectives on chitin and chitosan as functional biopolymers. *Applied Biochemistry and Biotechnology* 181 (4): 1314–1337.

235 Elieh-Ali-Komi, D. and Hamblin, M.R. (2016). Chitin and chitosan: production and application of versatile biomedical nanomaterials. *International Journal of Advanced Research (Indore)* 4 (3): 411–427.

236 Giavasis, I. (2014). Bioactive fungal polysaccharides as potential functional ingredients in food and nutraceuticals. *Current Opinion in Biotechnology* 26: 162–173.

237 Park, B.K. and Kim, M.-M. (2010). Applications of chitin and its derivatives in biological medicine. *International Journal of Molecular Sciences* 11: 5152–5164.

238 Ifuku, S., Nomura, R., Morimoto, M., and Saimoto, H. (2011). Preparation of chitin nanofibers from mushrooms. *Materials* 4: 1417–1425.

239 Wang, W., Li, C., Du, G. et al. (2017). Characteristics and rheological properties of polysaccharide nanoparticles from edible mushrooms (*Flammulina velutipes*). *Journal of Food Science* 82 (3): 687–693.

240 Ramanathan, A. and Kittusamy, R. (2011). Antihepatotoxic effect of isolated chitin from *Rhizopus oryzae* against paracetamol-induced hepatotoxicity. *Bangladesh Journal of Pharmacology* 6: 64–67.

241 Gerente, C., Lee, V.K.C., Le Cloirec, P., and McKay, G. (2007). Application of chitosan for the removal of metals from wastewaters by adsorption – mechanisms and models review. *Critical Reviews in Environmental Science and Technology* 37: 41–127.

242 Bays, H.E., Evans, J.L., Maki, K.C. et al. (2013). Chitin-glucan fiber effects on oxidized low-density lipoprotein: a randomized controlled trial. *European Journal of Clinical Nutrition* 67: 2–7.

243 Bottin, J.H., Swann, J.R., Cropp, E. et al. (2016). Mycoprotein reduces energy intake and postprandial insulin release without altering glucagon-like peptide-1 and peptide tyrosine-tyrosine concentrations in healthy overweight and obese adults: a randomised-controlled trial. *British Journal of Nutrition* 116: 360–374.

244 Marzorati, M., Maquet, V., and Possemiers, S. (2017). Fate of chitin-glucan in the human gastrointestinal tract as studied in a dynamic gut simulator (SHIME®). *Journal of Functional Foods* 30: 313–320.

245 Neyrinck, A.M., Possemiers, S., Verstraete, W. et al. (2012). Dietary modulation of clostridial cluster XIVa gut bacteria (*Roseburia* spp.) by chitin-glucan fiber improves host metabolic alterations induced by high-fat diet in mice. *Journal of Nutritional Biochemistry* 23: 51–59.

246 Kulev, D. and Negrutsa, I. (2015). Chitin-glucan complex – food additive with sorbent properties. *Journal of Hygienic Engineering and Design* 11: 53–56.

247 Gautier, S., Bruyere, J.-M., and Maquet, V. (2010) Fine-granulometry fungal extract chitine-glucane. US Patent 2010/0003292, filed 20 November 2007.

248 Gautier, S., Xhauflaire-Uhoda, E., Gonry, P., and Piérard, G.E. (2008). Chitin-glucan, a natural cell scaffold for skin moisturization and rejuvenation. *International Journal of Cosmetic Science* 30: 459–469.

249 Dubey, L.K., Moeller, J.B., Schlosser, A. et al. (2014). Induction of innate immunity by *Aspergillus fumigatus* cell wall polysaccharides is enhanced by the composite presentation of chitin and beta-glucan. *Immunobiology* 219 (3): 179–188.

250 Freitas, F., Roca, C., and Reis, M.A.M. (2015). Fungi as sources of polysaccharides for pharmaceutical and biomedical applications. In: *Handbook of Polymers for Pharmaceutical Technologies, Volume 3, Biodegradable Polymers* (eds. V.K. Thakur and M.K. Thakur), 61–104. New Jersey and Massachusetts: Wiley-Scrivener.

251 Berecochea-Lopez, A., Decordé, K., Ventura, E. et al. (2009). Fungal chitin-glucan from *Aspergillus niger* efficiently reduces aortic fatty streak accumulation in the high-fat fed hamster, an animal model of nutritionally induced atherosclerosis. *Journal of Agricultural and Food Chemistry* 57: 1093–1098.

252 Freitas, F., Roca, C., Cruz, F., et al. (2013) Natural biocomposite powder prepared from *Pichia pastoris* biomass, method of preparation and its use as excipient. International patent WO2013/140222, filed 15 March 2013.

253 Bornet, A. and Teissedre, P.L. (2008). Chitosan, chitin-glucan and chitin effects on minerals (iron, lead, cadmium) and organic (ochratoxin A) contaminants in wines. *European Food Research and Technology* 226: 681–689.

254 Skorik, Y.A., Pestov, A.V., and Yatluk, Y.G. (2010). Evaluation of various chitin-glucan derivatives from *Aspergillus niger* as transition metal adsorbents. *Bioresource Technology* 101: 1769–1775.

3

Microbial Cell Factories for Biomanufacturing of Polysaccharides

M. Fata Moradali[1] and Bernd H.A. Rehm[2]

[1] *University of Florida, College of Dentistry, Department of Oral Biology, 1395 Centre Drive, Room D4-38A, Gainesville, FL 32610-0424, USA*
[2] *Menzies Health Institute Queensland (MHIQ), Griffith University (Gold Coast Campus), Griffith Institute for Drug Discovery (GRIDD), Centre for Cell Factories and Biopolymers (CCFB), Brisbane, QLD 4111, Australia*

3.1 Introduction

Microorganisms represent a vast source of diverse biopolymers in nature. They possess the metabolic capacity to divert metabolites and small molecules such as sugar derivatives toward the synthesis of diverse polymers including polysaccharides. Polysaccharides are composed of sugars and/or sugar acids that have specific biological roles such as in the formation of a cell wall, as storage materials such as glycogen, the capsule of bacteria, and the biofilm matrix. They largely contribute to the protection of bacterial cells against environmental stresses such as desiccation, toxic molecules/antibiotics, and recognition by the immune system of the host [1, 2]. Microbes have evolved with relevant genetic background for producing various enzymes responsible for polymerization and modification of polysaccharides into desired materials. In the current applied microbiology and biotechnology area, polysaccharide-producing microorganisms, in particular bacteria, have been considered as primary cell factories. The application of natural polysaccharides is increasingly extended into areas such as food industry, formulation of cosmetics, medicine, pharmaceutical industry, electronics, and biosensing. Recent advances in genome sequencing, synthetic biology, systems biology, and technological advancements in bioengineering and biomanufacturing have opened up new avenues toward designing novel cell factories that produce innovative materials for advanced applications such as in tissue engineering, cell therapy, controlled drug release, and diagnostics.

3.2 Prominent Microbial Polysaccharides and Their Properties and Applications

In this part, we review the diversity of prominent polysaccharides produced by microbes. These are diverse polymeric carbohydrates that can be classified

Biopolymers for Biomedical and Biotechnological Applications, First Edition.
Edited by Bernd H. A. Rehm and M. Fata Moradali.

either as homopolymers if they are composed of one type of sugar units or as heteropolymers if different types of sugar comonomers constitute their structure. Further, the diversity of polysaccharides is determined by the type of glycosidic linkage, unbranching or the degree of branching, the length of polymer (or molecular weight), the type of sugars in side chains, and the degree of substitutions such as acetyl, pyruvyl, and succinyl groups. All these chemical features impact physical behavior of polymers, intra- and intermolecular interactions, molecular dynamics, flexibility, and gross conformation and also determine their properties including solubility, rheological characteristics, viscoelastic properties, interaction with cations, ionic strength, crosslinking, gelation, water retention, strength, extension of the polymer, stability, and eventually their applications. For instance, under normal conditions, polysaccharides such as alginate, hyaluronate, carrageenan, gellan, curdlan, and pectin have excellent gelation properties in aqueous solutions, while xanthan, acetan, rhamsan, succinoglycan, glucomannan, and galactomannan are good thickeners. Besides abovementioned properties, processing microbial polysaccharides into high-value biomaterials for biomedical and pharmaceutical purposes is attributed to their biocompatibility and no toxicity. Here, we review chemistry and physicochemical properties of the foremost microbial polysaccharides that have been extensively studied and are used in various applications.

3.2.1 Xanthan and Acetan

Xanthan is a repeating heteropolymer that consists of β-(1-4)-linked glucose units in the backbone with two mannose units and one glucuronic acid as side chain (Figure 3.1). It is produced by the bacterium *Xanthomonas campestris* [3, 4]. Xanthan displays unique pseudoplastic rheological and viscosifying properties in solutions even at low concentrations, which is stable in a wide range of pH and temperature. Commercial products of xanthan gum produced by *X. campestris* have been extensively applied in food and cosmetic industries [3, 5].

Acetan, produced by the bacterium *Komagataeibacter xylinus* (previously known as *Acetobacter xylinum*), is an anionic heteropolysaccharide that shares its cellulosic backblock structure (β-(1-4)-linked glucose units) with xanthan, but with a longer charged side chain consisting of α-L-rhamnose-(1-6)-β-D-glucose-(1-6)-α-D-glucose-(1-4)-β-D-glucuronate-(1-2)-α-D-mannose where the mannose residue is attached to the linear cellulosic backblock via an α-(1-3) glycosidic linkage (Figure 3.1). Acetan does not contain any pyruvic acid substituents and similar to xanthan is acetylated [6, 7].

3.2.2 Succinoglycan and Galactoglucan

Succinoglycan is an acidic branched heteropolymer consisting of octasaccharide repeating units of β-linked glucose and galactose in the ratio of 7 : 1 with several modifications such as the addition of succinate, pyruvate, and acetate [8] (Figure 3.1). Galactoglucan is composed of disaccharide repeating units of

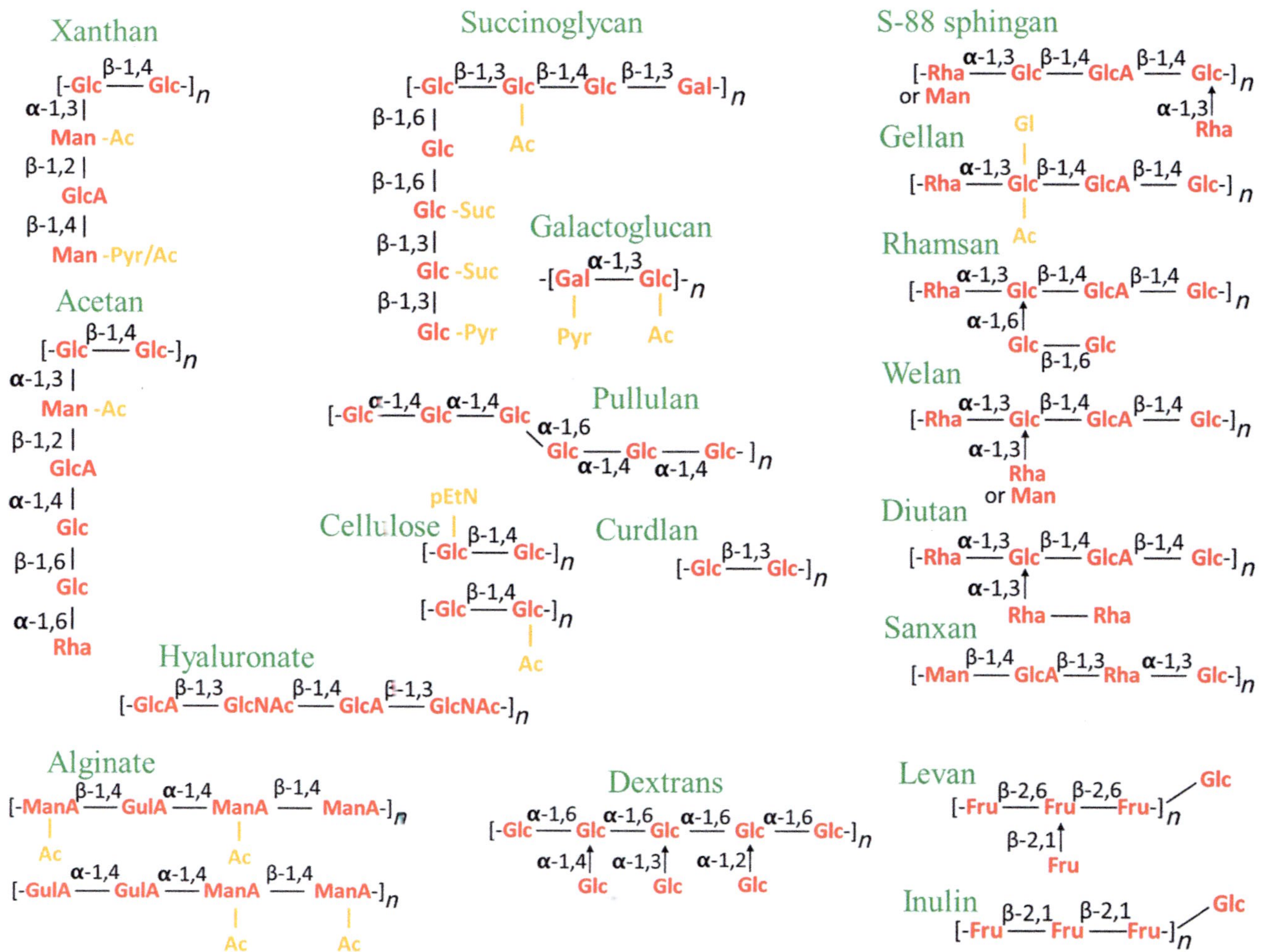

Figure 3.1 Chemical structure of prominent microbial polysaccharides. Glc, glucose; GlcA, glucuronic acid; Rha, rhamnose; Fru, fructose; Gal, galactose; GlcNAc, *N*-acetylglucosamine; ManA, mannuronic acid; GulA, guluronic acid; Man, mannose; Ac, acetyl; Pyr, pyruvyl; Suc, succinyl; Gl, glyceryl; pEtN, phosphoethanolamine.

alternatively (1→3)-linked D-glucose and D-galactose residues in the ratio of 1 : 1, but highly pyruvylated and acetylated [9, 10] (Figure 3.1). These polysaccharides have been used as suspending and thickening agents, texture enhancer, emulsifier, plasticizer, stabilizer, or flocculating agent [11]. Succinoglycan and galactoglucan are produced by bacteria belonging to the genera *Rhizobium*, *Agrobacterium*, *Alcaligenes*, and *Pseudomonas* [10, 12]. Succinoglycan has been produced by fermentation of *Agrobacterium tumefaciens* and commercialized under the name "Rheozan® SH" (Solvay Novecare) as a hydrocolloid rheology modifier, thickener, and stabilizer in cosmetic products, which is highly compatible with acidic agents for making aqueous solutions highly viscose even in low quantity that is not affected by temperature.

3.2.3 Sphingan Polysaccharides

A group of heteropolymers such as S-88 sphingan, diutan (or S-657), gellan (S-60), welan (S-130), and rhamsan (S-194) mainly composed of four sugars in backbone linked as (→4)-α-L-rhamnose-(1→3)-β-D-glucose-(1→4)-β-D-glucuronic acid-(1→4)-β-D-glucose-(1→), but variable in the composition and linkage of the side chains [13, 14] (Figure 3.1). Gellan is acetylated without side chain, but modified with L-glyceryl substituents. In interaction with cations, gellan forms a strong gel [14, 15]. Welan contains acetyl group, and its side chain has an L-rhamnosyl or L-mannosyl group. Welan does not have gelling property, but it is a good viscosifier, which is stable to high temperatures, high salt, and pH [16, 17]. Rhamsan is a thickener and suspending agent that is acetylated with two D-glucosyl units in the side chain, and S-88 sphingan is acetylated and has one rhamnose in the side chain [16]. Diutan possesses acetyl groups and a two-rhamnose side chain that is linked to a glucose residue via the (1→4) glycosidic linkage (Figure 3.1). It is a thickening, suspending, and stabilizing agent in aqueous solutions and fluids [16–18].

Sphingan polysaccharides are produced by bacteria belonging to the genus *Sphingomonas* [19] for various industrial applications such as in food as thickener, stabilizers of emulsions, suspending solids, film, and gel formers [16, 18, 20]. Further, *Sphingomonas sanxanigenens* NX02 (DSM 19645) produces sanxan exopolysaccharide that was found different from sphingans that has excellent thickening, gelling, shear thinning, and emulsifying properties [21].

3.2.4 Pullulan

A fungal exopolysaccharide produced by yeast-like fungus *Aureobasidium pullulans* (known as black yeast). It is a linear α-D-glucan consisting of maltotriose repeating units interconnected by α-(1→6) linkages (Figure 3.1). Its linearity and linkage pattern confer unique properties such as high flexibility, solubility, adhesive properties, forming fibers, and strong and oxygen-impermeable films as well as compression moldings. Pullulan has a wide range of commercial applications in biomedical, food, pharmaceutical, cosmetics, and electronics industries [22, 23].

3.2.5 Cellulose and Curdlan

Similar to plant cellulose, the backbone of microbial cellulose is composed of β-(1-4)-linked glucose units (Figure 3.1). However, microbial cellulose is present in the form of ribbon cellulose and highly porous matrix composed of bundles of microfibrils, while cellulose chains in plant cell wall are formed into clusters of high-order structures called fibril lamella and mesh-like bulky structures [24–26]. Microbial celluloses have high purity, high crystallinity, and moldability, and due to abundant hydroxyl groups in their structure, microbial celluloses possess high water-holding capacity (up to 100 times its weight) and hydrophilicity [27]. Microbial cellulose is biodegradable and has high biological affinity and higher degree of polymerization (or higher molecular weight) that confer greater tensile strength than plant cellulose [28]. Some bacterial species such as *Pseudomonas fluorescens* produce acetylated cellulose [29]. Production of cellulose provides an advantage to enterobacterial pathogens such as *Escherichia coli* and *Salmonella enterica* [30, 31]. *E. coli* causing urinary tract infections produces phosphoethanolamine cellulose in the biofilm matrix, which forms mortar-like structures to stabilize proteinaceous curli fibers. In complex biofilms, these fibers mediate strong connections of pathogens to host cells and provide resistance in high shear stress conditions associated with the bladder and urinary tract niches [32, 33].

Microbial cellulose has been largely considered for various applications for industrial purposes such as in food and chemical industries as well as pharmaceutical and biomedical applications [28, 34]. Microbial cellulose is produced by various bacterial species belonging to *Gluconacetobacter* (formerly *Acetobacter*), *Agrobacterium, Aerobacter, Achromobacter, Azotobacter, Rhizobium, Sarcina, Salmonella* and *Escherichia*, most groups of algae, and the slime mold *Dictyostelium* [28, 35]. However, genetic elements of cellulose producers involved in cellulose biosynthesis may vary among the producers; hence, there are some variations in chemical structure of theses polymers such as the formation of mixed-linkage (1→3,1→4)-β-D-glucan and curdlan [31, 36, 37]. *Gluconacetobacter hansenii* (formerly known as *Gluconacetobacter xylinus*) is reported as the most efficient cellulose producer among bacterial species and has been extensively studied as a model cellulose producer [38].

Curdlan is a high molecular mass linear β-(1-3)-glucan homopolymer with no side chain (Figure 3.1). It is insoluble in water, but has a strong gelation capability in aqueous suspension when it is heated that forms an elastic gel. Gelation is caused by aggregation of the rodlike triple helices that is thermoirreversible if it is heated above ~80 °C, but it becomes a thermoreversible gel if it is heated up to 55 °C [39]. Due to unusual rheological properties, curdlan is largely considered as a thickening and gelling agent in food industry as well as a unique biomaterial for pharmaceutical and biomedical purposes [40–42]. Many soil bacteria of the family *Rhizobiaceae* can produce curdlan [39].

3.2.6 Alginates

These heteropolymers are composed of two uronic acid derivatives including β-D-mannuronic acid (M) and its C5 epimer α-L-guluronic acid (G) linking via

1,4-glycosidic bonds that present in variable consecutive blocks of M, G, and MG residues in alginate strands [43] (Figure 3.1). Alginates are traditionally extracted from seaweeds, but bacteria belonging to the genera *Pseudomonas* and *Azotobacter* can produce alginates. The properties of bacterial alginates are significantly different from algal alginates as they have a very high molecular weight, for example, ~3.9 MDa obtained from *Pseudomonas aeruginosa*, and acetyl groups in the structure [44]. Alginates possess intrinsic viscoelasticity that depends on the frequency of constituting blocks as flexibility decreases in the order MG block > MM block > GG block. The most important property of alginates is the formation of hydrogels and crosslinked polymeric scaffolds in the presence of divalent cations such as Mg^{2+}, Mn^{2+}, and Ca^{2+} [45]. Alginates produced by seaweeds have been harnessed for more than a century by human for a wide industrial application such as in agriculture, food, textile, cosmetic, and pharmaceutical/biomedical industries [43, 46–49].

3.2.7 Hyaluronic Acid or Hyaluronate

It is a linear negatively charged heteropolysaccharide composed of alternating β-D-glucuronic acid and β-D-*N*-acetylglucosamine residues linked via β-(1-4) and β-(1-3)-glycosidic bonds [50] (Figure 3.1). Both vertebrates and bacteria belonging to group A *Streptococcus*, *Pasteurella multocida*, and *Cryptococcus neoformans* produce hyaluronic acid [51–54]. Some pathogens such as *Streptococcus pyogenes* and *Bacillus cereus* G9241 produce a capsule of hyaluronate to hide their antigenic surface from opsonization because these capsules mimic the structure of hyaluronate found in human connective tissues [55, 56]. Hyaluronate has a unique viscoelastic and gelation properties and an extremely high water retention capacity. It is biocompatible and nutritionally neutral; for this reason, it has been widely applied in cosmetic products as well as biomedical advances such as tissue engineering and drug delivery [54, 57, 58].

3.2.8 Dextrans

These are a family of glucopolysaccharides mainly composed of α-(1→6)-linked glucose units that have variable amounts of α-(1→2)-, α-(1→3)-, or α-(1→4)-branched linkages and vary from slightly to highly branched [59, 60] (Figure 3.1). Their molecular weight varies in the range of 1–40 000 kDa [61, 62]. Due to variation of the composition, structure, branching, and molecular mass, dextrans display different properties. Hence, various dextrans can be found as water soluble, water insoluble, or less soluble. In this regard, unbranched dextrans have high water solubility, while branched dextrans with >43% branching through α-(1→3) linkages display water insolubility [63, 64]. However, commercial dextrans are often water soluble, with low degree of branching (0.5%), and they are stable under mild acidic and basic conditions [64]. Due to possessing abundant hydroxyl groups in the structure, dextrans have been found suitable for modifications and conjugations [64]. Furthermore, molecular weight of dextrans determines their applications. Dextrans are used in foods as viscosifying, gelling, texturing, and emulsifying agents, and numerous potential pharmaceutical and biomedical applications have been implicated for low molecular weight dextrans [65, 66]. Dextrans are often extracellularly synthesized from sucrose by enzymatic activity of dextransucrases produced by *Leuconostoc mesenteroides* strains, which are commercially important as the producers of dextransucrases and also bacteria belonging to the genera *Leuconostoc*, *Streptococcus*, *Weissella*, and *Pediococcus* produce dextrans [67–69].

There are other forms of microbial polysaccharides that are chemically related to dextrans. Of these, mutan polysaccharide synthesized by mutansucrase has mostly α-(1→3) linkages with less than 12% α-(1→6) linkage [70, 71], alternan possesses alternating α-(1→3) and α-(1→6) linkages with long-chain branching and has very high molecular weight [72], and reuteran has a large amount of α-(1→ 4) linkages (58%) besides α-(1→ 6) linkages (42%) produced by *Lactobacillus reuteri* 121 [73].

3.2.9 Levan and Inulin

These are polymers of D-fructose that are linked via β-(2→6)-glycosidic linkage with occasional β-(2→1) branching in levan and β-(2→1)-glycosidic linkage in the backbone of inulin, and both have one terminal glucose [74] (Figure 3.1). Levans are produced by certain plants and bacteria; however, microbial levans have greater molecular weight (2000–100 000 kDa) than levans from plant (2–33 kDa) [75–77]. Microbial inulins have greater molecular weight and 15% more branching compared with plant inulins [78]. Levans are synthesized extracellularly by enzymatic activity of levansucrase produced by both Gram-positive and Gram -negative bacteria, while inulinosucrases synthesizing inulins are only present in lactic acid bacteria [76, 79].

Levan does not swell in water and has an uncommonly low intrinsic viscosity and high water solubility as well as oil solubility, but it is insoluble in most of organic solvents. Due to having the large number of hydroxyl groups, levans can form adhesive bonds with other substrates and can be strong adhesives, while their branching structure underlies their cohesive strengths [76, 80]. Levans are applied in food and cosmetic industries as stabilizers, thickeners, emulsifiers, and water-soluble film formers. Levans have the potential for pharmaceutical and biomedical applications such as for drug delivery, encapsulating, and blood plasma extender, and they are known as antihyperglycemic and antidiabetic agents [81–85]. Inulin is highly water soluble and shows gelling properties at high level. Inulin is widely used in processed foods as a fat, sugar replacer, or a prebiotic, and it is considered as a significant component of the dietary fiber complex for promoting human health [86].

3.3 Biosynthesis Pathways of Bacterial Polysaccharides

Biosynthesis of a polysaccharide is a very complex process and requires the involvement of many genes and specific proteins to accomplish it. Further, regulatory mechanisms at transcriptional and posttranscriptional level play critical roles in polysaccharide production because these processes are intertwined with other physiological processes as well as external factors and stimuli.

Understanding biosynthesis pathway of polysaccharides is necessary for establishing an efficient cell factory and production of tailor-made products. Variations in the type of enzymes mediating various steps of polysaccharide biosynthesis including the provision of various precursors and regulatory mechanisms underpin the diversity of biosynthesis pathways of polysaccharides among bacteria. Furthermore, polysaccharides are different from each other in terms of composition, backbone structure, branching, molecular weight, polydispersity, and post-polymerization modifications such as addition or elimination of chemical substitutions. Here, the authors present a few examples to illustrate how polysaccharides are synthesized in bacteria:

3.3.1 Genetic Background Required for Biosynthesis of Polysaccharides in Bacteria

Production of polysaccharides is initiated by the expression and production of active enzymes including glycosyltransferases that are responsible for providing precursor and/or polymerization. In bacteria, these genes, which are functionally related, are commonly clustered in operonic organizations. Different gene clusters and various encoded proteins responsible for biosynthesis of various microbial polysaccharides have been summarized in a few review articles [79, 87]. Each operon is often downstream to a potent specific promoter that controls transcription of whole genes in a polycistronic gene expression manner. Furthermore, the presence of internal promoters upstream of specific genes in an operon is a common phenomenon. Interaction of RNA polymerase with a given promoter region can be controlled by other cellular factors and stimuli that will be discussed later.

On the other hand, production of some enzymes and proteins may originate from other part of the genome whose function can be shared by multiple cellular processes. By contrast, protein-coding genes in eukaryotic systems are commonly organized monocistronically (i.e. one promoter drives the expression of one gene). Hence, the study of biosynthesis pathways of polysaccharide in prokaryotic systems is simpler and more feasible compared with eukaryotic systems. For instance, alginates are anionic exopolysaccharides that are produced by both eukaryotic (i.e. seaweeds) and prokaryotic systems (*Pseudomonas* spp. and *Azotobacter* spp.). While alginates have been extracted from seaweeds for more than a century for a wide variety of commercial purposes, our knowledge about biosynthesis pathways of alginates is majorly established on alginate-producing bacteria, mainly *P. aeruginosa*. However, alginates originating from certain seaweed have been only regarded as safe products by the US Food and Drug Administration (FDA), and bacterial alginates have not been qualified yet for non-research purposes.

The constituting genes and their organization within operons responsible for biosynthesis of microbial polysaccharides may vary among different species and strains. For example, compared with alginate (*alg*) operon that has a highly conserved organization among *Pseudomonas* and *Azotobacter* species, genetic organization of cellulose (*bcs*) operon is diverse among cellulose-producing species

[31, 88]. This might be explained by wide distribution of *bcs* operon in a variety of species belonging to genera *Gluconacetobacter*, *Rhizobium*, *Burkholderia*, *Pseudomonas*, *Dickeya*, *Erwinia*, *Agrobacterium*, *Salmonella*, and *Escherichia*. Based on the gene order and gene content, at least three major types and several subtypes of *bcs* operon have been characterized [31]. On the other hand, the streptococcal hyaluronate (*has*) operons varies among hyaluronate producers as they are subjected to intragenome gene duplications and rearrangements [89].

3.3.2 Production of Active Precursor, Polymerization, and Polysaccharide Modifications

Initially, one or more enzymes mediate the biosynthesis of active precursor(s) in the form of nucleotide sugars/sugar acids such as UDP-glucose, UDP-GlcNAc, and GDP-mannuronic acid. These are the donors of sugar monomers to the reducing end of the growing carbohydrate or acceptor at the catalytic site of a polymerase. Biosynthesis of active precursor is an energy-consuming process that commonly originates in central metabolism where these molecules are also applied for maintaining primary cellular functions. At polymerization step, polysaccharides are formed by the formation of glyosidic bonds catalyzed by glycosyltransferases with the corresponding sugar nucleotides as substrates. Glycosyltransferases can be anchored to the cytoplasmic membrane such as cellulose and alginate polymerases that may interact with other protein subunits to constitute a multiprotein complex for accomplishing polymerization, translocation of a nascent polymer across cell membrane or cell wall, modification of polymer structure, and eventually secretion of mature polymer outside the cells. Protein complexes responsible for production of an acetylated polyMG alginate in *P. aeruginosa*, non-acetylated cellulose in *K. xylinus*, acetylated cellulose in *P. fluorescens* SBW25, poly-β-D-*N*-acetylglucosamine in *Staphylococcus epidermidis*, and xanthan in *X. campestris* consist of at least 10, 5, 10, 4, and 12 different membrane and periplasmic protein subunits, respectively. Some proteins within multiprotein complexes not only interact with others to create a scaffold to span the bacterial envelope and facilitate polymer translocation across membrane but also take part in modification of polymer chain such as alginate epimerase AlgG and alginate acetyltransferase AlgX [47, 90–93].

The mechanisms of the production of polysaccharides have been comprehensively reviewed, and four general mechanisms have been described in bacteria [79]. Here, we briefly describe these mechanisms:

(1) *The Wzx/Wzy-dependent mechanism*: Upon this mechanism, individual repeating units of sugars that are covalently linked to an undecaprenol diphosphate are assembled by several glycosyltransferases at the cytoplasmic membrane followed by being translocated across the cytoplasmic membrane by a flippase (e.g. Wzx protein) [94]. Polymerization of long chain is mediated by the Wzy polymerase in the periplasm, and additional protein subunits including polysaccharide co-polymerase (PCP) and outer membrane polysaccharide export (OPX) proteins mediate the export of polymerized repeat units (Figure 3.2). Heteropolymers with highly diverse

sugar compositions such as xanthan, succinoglycan, colanic acid [101, 102], sphingan polysaccharides (e.g. diutan, gellan, and welan) [18], many capsular polysaccharides, and lipopolysaccharide (LPS) O-antigens [103, 104] are produced via this mechanism.

(2) *The ATP-binding cassette (ABC) transporter-dependent mechanism*: This mechanism is similar to the Wzx/Wzy pathway, but polymerization of long chain occurs at the cytoplasmic face of the inner membrane, and a complex of an ABC transporter, PCP, and OPX proteins mediate polymer translocation and secretion, respectively (Figure 3.2). Furthermore, a conserved glycolipid consisting of phosphatidylglycerol and a poly-2-keto-3-deoxyoctulosonic acid (Kdo) linker are found at the reducing terminus of these polysaccharides synthesized via ABC transporter-dependent mechanism [97]. Depending on how many glycosyltransferases involved in initial stages of synthesis, both homopolymers and heteropolymers are produced via this pathway. Some capsular polysaccharides are synthesized via ABC transporter-dependent pathway, which are linked to the cell surface of many bacteria [105, 106].

(3) *The synthase-dependent mechanism*: In this mechanism, entire polysaccharide strand is polymerized by a single glycosyltransferase followed by being secreted across the bacterial envelope [88]. Usually the polymerase unit is a part of multiprotein complex spanning the bacterial envelope and responsible for polymerization, translocation, and secretion of the polymer (Figure 3.2). Furthermore, some protein subunits of the protein complex may contribute to enzymatic modification of the polysaccharide after polymerization event. Hence, a polysaccharide synthesized via this pathway is often polymerized initially as a homopolymer strand, while other enzymes catalyze adding functional groups such as acetyl group or enzymatically modify nascent strand for creating a heteropolymer structure. For instance, alginate is initially synthesized as a polymannuronate (polyM) strand by Alg8 glycosyltransferase, and during translocation across membrane and the periplasm, AlgG (alginate epimerase) catalyze conversion of mannuronate to guluronate residues to form a polyMG strand, while AlgX acetyltransferase mediates O-acetylation of polyMG alginate [43, 44, 92, 93] (Figure 3.2). Alginate, cellulose, and curdlan are produced via the synthase-dependent mechanism. Hyaluronic acid that is a heteropolymer is also synthesized via similar pathway, but only a single glycosyltransferase mediates both polymerization and secretion events by using different types of precursors [54].

(4) *The extracellular synthesis*: In this mechanism, a single glycosyltransferase is secreted to the cell surface, and polysaccharide polymerization occurs extracellularly [59, 107] (Figure 3.2). Dextran, mutan, alternan, reuteran, levan, and inulin are synthesized extracellularly by enzymatic activity of dextransucrase, mutansucrase, alternansucrase, reuteransucrase, levansucrase, and inulinosucrase, respectively.

3.3.3 Regulatory Pathways and Posttranslational Modifications

Biosynthesis of polysaccharides is under the control of multi-tiered regulatory mechanisms at translational and posttranslational levels. At translational level,

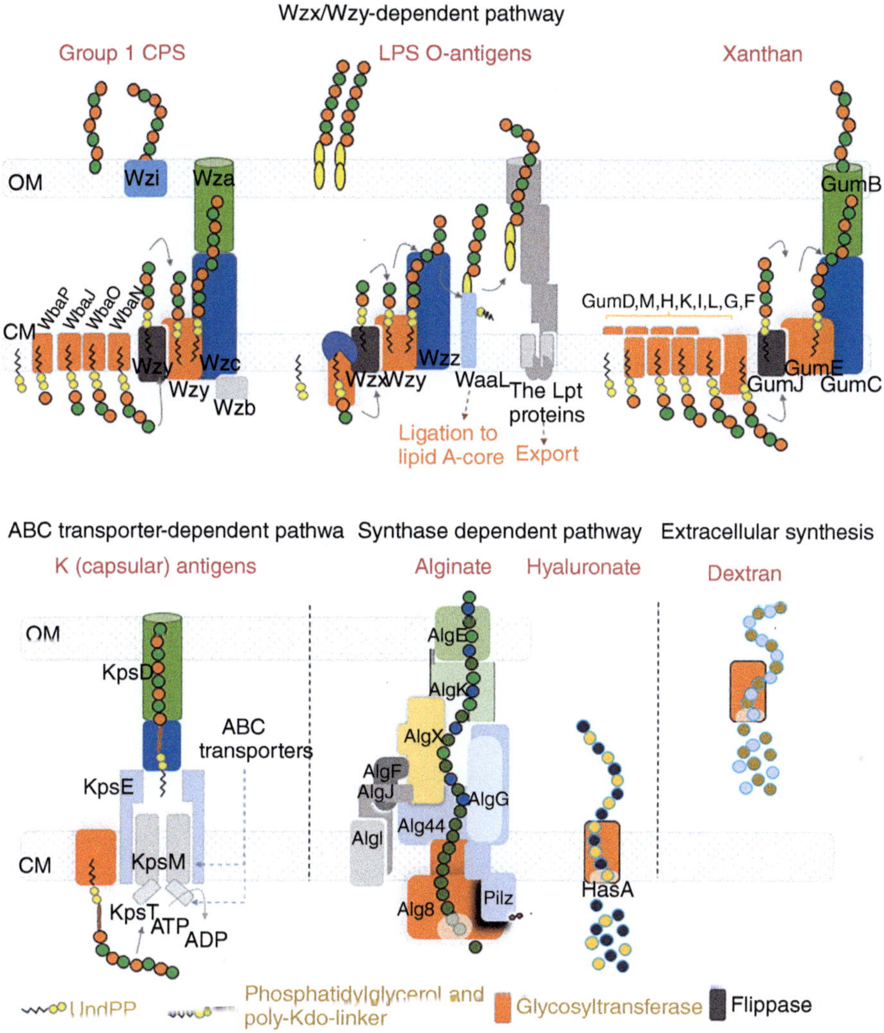

Figure 3.2 The mechanisms of the production of polysaccharides. Synthesis of group 1 CPS, LPS O-antigens, and K antigens is shown in *E. coli* strains [95–97]. The mechanism of xanthan synthesis is explained in *X. campestris* [98]. Alginate and hyaluronate biosynthesis are explained in *P. aeruginosa* and group A streptococci [99, 100]. CPS, capsular polysaccharide; LPS, lipopolysaccharide; UndPP, undecaprenyl pyrophosphate; kdo, 3-deoxy-D-manno-oct-2-ulosonic acid; CM, cytoplasmic membrane; OM, outer membrane.

transcription initiation factors such as sigma factors and master regulators mediate the activation of the promoters controlling the expression of functionally related genes required for the biosynthesis of polysaccharides. In principle, these factors are expressed or activated in response to specific stimuli such as metabolites and small molecules or conditions such as growth phase, temperature, and various physical and chemical stresses. For instance, in the biosynthesis of alginate and cellulose by *Pseudomonas* and *Salmonella*, AlgU (δ^{22}) and RpoS

(σ^{38}) sigma factors are the respective principal transcriptional regulators. These activate the RNA polymerase binding to promoter regions upstream to the *alg* and *bcs* operons, respectively. RpoS interacts with a small protein named Crl to accomplish its mission. Further, biosynthesis of polysaccharides can be controlled by alternative regulatory pathways as well as variable numbers of genomic elements and promoters in different species and strains. For instance, cellulose production in *S. enterica* is dependent on the master regulator CsgD but not in *E. coli* strains [108, 109].

In *P. aeruginosa*, transcription of alginate operon is principally depended on the *algD* promoter upstream to *algD* gene encoding a GDP-mannose 6-dehydrogenase to catalyze the formation of the precursor GDP-mannuronic acid. The activity of the *algD* promoter is principally under the control of AlgU sigma factor, but other sigma factors and DNA-binding proteins have been reported as other relevant regulatory factors [110]. Furthermore, two internal promoters have been identified upstream alginate-modifying genes *algG* and *algI* in *P. aeruginosa*, and it has been suggested that such internal promoters may regulate polysaccharide modification events [111]. On the other hand, in *Azotobacter vinelandii*, at least two promoters have been identified upstream the *algD* promoter in addition to several extra internal promoters upstream to *alg8* gene (alginate polymerase), *algG*, and *algA* genes and two promoters upstream *algC* [112–115]. Further, alginate-controlling sigma factor AlgU is sequestered by a membrane anchored anti-sigma factor MucA. In those strains unable to produce alginate (or non-mucoid strains), intact MucA protein sequesters AlgU at inner membrane in a stable complex, while upon destabilization of this complex in response to envelop stresses, AlgU is released to activate the transcription of alginate operon by binding to the *algD* promoter. AlgU itself is under the control of different regulatory proteins that negatively or positively regulate its activation. For instance, MucE, MucP, and AlgW were reported as positive regulators for AlgU activity at the envelope of *P. aeruginosa* [116, 117].

Small noncoding RNAs (sRNA) (25–500 nucleotides in length) are another large group of regulatory factors that control specific biological processes including production of polysaccharides at posttranslational level [118]. sRNA regulatory networks are very complex that suppress or activate polysaccharide production by binding to cognate mRNA or proteins in a concentration-dependent manner in response to specific stimuli [118, 119]. For instance, it is understood that the master regulator CsgD that positively regulate cellulose biosynthesis is negatively regulated by at least five sRNAs at the mRNA level [120, 121]. In contrast, alginate production in bacteria is positively regulated by Rsm/Csr-sRNAs that counteracts the repressor activity of the RsmA protein on the *algD* mRNA and induces the production of AlgD to catalyze the production of the alginate precursor GDP-mannuronic acid [122].

Regulation of sRNAs levels may link to the activity of another layer of regulatory networks established by two-component systems (TCS). TCS signaling pathways are stimulus–response coupling mechanisms that play major regulatory roles in both prokaryotic and eukaryotic systems [123]. The prototypical TCS consists of sensor kinases and response regulator proteins that mediate phosphorylation cascades and controls a wide variety of physiological functions at translational and posttranslational levels [124, 125]. The GacS/GacA TCS and their homologues control the expression of Rsm/Csr-sRNAs in many γ-proteobacteria. Regulation of alginate production via Rsm/Csr-sRNA signaling is positively regulated by the GacS/GacA TCS [126, 127].

As abovementioned, regulatory RNA-binding proteins such as RsmA and CsrA create other regulatory pathways for controlling the biosynthesis of polysaccharides in many bacteria. Upon binding to mRNA, these regulatory proteins control mRNA stability and translation process. RsmA binding to the *algD* mRNA and CrsA binding to the *pgaA* mRNA in the biosynthesis of alginate and poly-β-1,6-*N*-acetyl-D-glucosamine are two examples of direct effect of RNA-binding proteins on polysaccharide biosynthesis [122, 126, 128].

However, they may indirectly impact polysaccharide biosynthesis through controlling key regulatory proteins that alter physiological status of the cells. As an example, Hfq RNA-binding protein binds to the transcripts of enzymes that are involved in determining cellular concentration of cyclic 3′5′-diguanylic acid bis-(3′-5′)-cyclic dimeric guanosine monophosphate (c-di-GMP) second messenger, a key molecular switch that controls many biological processes including motility–sessility transition and exopolysaccharide production [129]. The second messenger c-di-GMP plays a key role in regulation of polysaccharide production at posttranscriptional level in a wide range of bacteria. c-di-GMP binds to its specific receptor called the PilZ domain and allosterically regulates the activity of specific proteins and protein complexes for a desired output via impacting protein–protein interactions or structural conformation [130–132]. In c-di-GMP signaling, two groups of proteins are main actors:

(1) c-di-GMP-metabolizing enzymes including diguanylate cyclases (DGC) (containing GGDEF motif) and phosphodiesterases (PDE) (containing EAL or HD-GYP motif). In response to external stimuli, DGC and PDE enzymes synthesize and degrade c-di-GMP; hence, they regulate c-di-GMP turnover in the cells because cellular concentration of c-di-GMP is a key determinant for either the suppression or activation of specific biological processes. The number of these enzymes in different bacterial species significantly varies; for example, at least 40 different c-di-GMP-metabolizing enzymes have been identified in *P. aeruginosa*. Many DGC and PDE enzymes are controlled by abovementioned regulatory pathways including the two-component signal transduction systems and sRNA regulatory pathways [31].
(2) c-di-GMP receptor/effector protein. c-di-GMP binding to a target protein is commonly mediated by the PilZ domain, a specific receptor that binds to c-di-GMP molecule. Based on highly conserved motifs involved in c-di-GMP binding (i.e. RxxxR and (D/N)x(S/A)xxG), proteins harboring the PilZ domain are readily identifiable via comparison of primary sequences. In the biosynthesis of alginate and cellulose, the PilZ domain and c-di-GMP binding are required for activation of polysaccharide polymerization. In these cases, the PilZ domains are found at the N-terminal and cytoplasmic part of the membrane-anchored protein Alg44 (alginate co-polymerase) and the C-terminal part of the membrane-anchored glycosyltransferase BcsA (cellulose synthase). Alg44 interacts with the membrane-anchored

glycosyltransferase Alg8 (alginate polymerase) to activate alginate polymerization upon the binding of c-di-GMP to Alg44. The mechanistic mechanisms for activation of cellulose and alginate polymerization upon c-di-GMP were described previously [130, 131, 133].

c-di-GMP-dependent regulation of the production of cellulose, alginate, Pel, and poly-β-1,6-*N*-acetylglucosamine has been studied in some bacteria. According to current working models, a pool of c-di-GMP is required for activation of polysaccharide production, and in the case of alginate synthesis in *P. aeruginosa*, it is provided by enzymatic activity of the membrane protein MucR (DGC/PDE) [134]. In *S. enterica* or *E. coli*, DGC enzymes AdrA and STM1987 exert similar function for activating cellulose production [135, 136]. The activity of these enzymes is linked to other regulatory pathways. For example, the production of AdrA is under the control of sRNA regulatory pathway at mRNA level as explained above [137].

Overall, besides various enzymes and the multiprotein complexes that mediate the biosynthesis of polysaccharides, regulatory pathways are very important to consider for engineering and optimizing cell factories. These regulatory components are ideal candidates to optimally control cellular processes and enzymatic pathways as well as overall biomanufacturing processes, for example, the enhancement of the productivity or the reduction of negative feedbacks imposed by unwanted by-products. We comprehensively discussed intricate regulatory mechanisms of alginate biosynthesis in a review article [110]. Furthermore, regulatory pathways of cellulose biosynthesis in *Arabidopsis* as a model for higher plants and in *Salmonella* and *E. coli* as model organisms were previously reviewed by others [108, 138–140]. In separate works, different groups reviewed and discussed various genetic and bioinformatic strategies for generating tailor-made transcriptional biosensors for optimizing prokaryotic and eukaryotic cell factories that can also be effectively applied for optimizing cell factories with regard to polysaccharide production and their productivity [141–144].

3.4 Strategies for Engineering Cell Factories

An efficient system for production of microbial products depends on wise selection of a cell factory. Engineering of the species with natural polysaccharide synthesis capacity is one of the primary approaches to increase the productivity. An ideal producer as well as produced polysaccharide must be qualified as "generally regarded as safe" (GRAS), a designation determined by US FDA that applies to substances accepted as safe for alimentary use by qualified experts. Generally, different countries have considered microbial polysaccharides as GRAS because they are nutritionally inert and nontoxic [145]. However, as a wide variety of infectious microbes are able to produce polysaccharides as part of extracellular polymeric substances (EPS), these products are limited to use even for food applications [145]. Many pathogenic species have been model organisms to understand biosynthesis pathways through genetic and metabolic engineering [2]. Sewalt et al. have reviewed the process of GRAS status for microbial strains to apply for industrial purposes [146]. Importantly, application of microbial products for therapeutic and biomedical purposes requires additional and specific levels of assessments and criteria, different from GRAS status for only alimentary applications [2]. Regardless of GRAS status of organisms, we present below some examples of possible strategies used by different groups to optimize cell factories and their productivity.

3.4.1 Enhancement of Productivity upon the Energetic State of the Cell and Metabolites

Synthesis of a polysaccharide is an energy-consuming process. The overall carbon and energy flux is a key determinant in the production of polysaccharides. Also, it is documented that the specific rate of production of polysaccharides is linked to the oxidation state of producer and both inversely correlate with cell growth or cellular mass; therefore, the integration of carbon and energy flux favors maximizing production rate than cell mass production [147, 148]. It has been suggested that understanding the major flux control points in biosynthesis of polysaccharides is a key step for increasing productivity [149]. For example, in the case of alginate, the synthesis of an active precursor from two molecules of glucose consumes three adenosine triphosphate (ATP) molecules, one guanosine diphosphate (GDP) molecule, two NADP, and eight NAD^+. However, when it is started from fructose, only three ATP molecules, one GDP molecule, and two NAD^+ are required. In the former scenario, a large portion of carbon flux is wasted in the form of six molecules of carbon dioxide due to complete oxidation of substrate that does not occur upon fructose utilization [147, 148]. Oxygen availability positively affects polysaccharide production such as alginate via affecting redox state of the cells, the carbon uptake rate, and enhancement of the tricarboxylic acid cycle (TCA cycle) and the pentose phosphate pathway [150]. These findings indicate that understanding of mass balances between utilized substrates, the glycolytic rate, the components of energy flux system, and production of polysaccharides is of paramount importance in biotechnological practices. As another example, the synthesis of the two precursors of hyaluronate requires three ATP molecules, two uridine-5′-triphosphate (UTP) molecules, two NAD^+ molecules, one molecule of acetyl-CoA, and one molecule of glutamine [151, 152]. The balance between molar concentration of hyaluronate precursors and the molar concentration of ATP and a proper balance in recycling NADH and NAD^+ were found critical for production of hyaluronate [151, 153]. In a metabolic engineering attempt for enhancing the production of hyaluronic acid, the increase of energy efficiency via removing excess NADH upon overexpression of NADH oxidase (NOX) and replacement of lactate production with acetate did result in increased biomass rather than hyaluronic acid increase [154]. Therefore, proper selection of cell factory and starting feedstock(s) can provide a balance in energetic state of the cell factory and metabolites for favoring production of polysaccharides.

3.4.2 Genetic and Metabolic Engineering of Cell Factories

Efficient provision and direction of an active precursor as a sugar donor molecule for initiating polymerization of polysaccharides is one of the primary factors for selecting a polysaccharide-producing cell factory. In principle, synthesis of an active precursor is a highly energy-consuming process that competes with central metabolism and other biological processes. In fact, a successful strategy must demonstrate enhancement of directing energy flux toward synthesis of active precursor and polymerization process rather than the increase of biomass production. A good example is the utilization of UDP-glucose and UDP-*N*-acetylglucosamine as the precursors for production of hyaluronic acid that compete with cell wall biosynthesis pathway [155].

Hence, it is reasonable to select a cell factory that is naturally competent in production of high quantity of an active precursor and polysaccharide.

Another strategy relies on appropriate application of high-energy substrates in medium that could lead to supplying high quantity of active precursors [156]. Upon this strategy, significant increase of an active precursor and enhancement of the production of the curdlan polysaccharide by *Agrobacterium* sp. ATCC 31749 was achieved via employing low polyphosphates ($Na_4P_2O_7$, $Na_5P_3O_{10}$, and $(NaPO_3)_6$) with high-energy phosphate bonds instead of KH_2PO_4 and K_2HPO_4 in the medium [157]. Therefore, amending metabolic pathways by minor genetic manipulation to redirect precursor provision toward a desired pathway can lead to enhancing productivity in metabolic engineering of cell factory [156]. Metabolic engineering is a promising strategy for manipulating biosynthetic pathways involved in the production of active precursors, and some aspects of metabolic engineering have been addressed in a review article [158].

However, biosynthesis of a sugar nucleotide molecule is often a complex multistep enzymatic pathway where an active sugar nucleotide molecule is generated from a carbon pool derived from central metabolism that is shared by other metabolic pathways. Liu and his colleagues successfully produced the P1 trisaccharide in a heterologous host via manipulation of a metabolic pathway that regenerated UDP-glucose and UDP-galactose independent from central metabolism. To this end, they performed *in trans* expression of a sucrose synthase, a UDP-glucose 4′-epimerase in addition to sucrose supplementation [159].

Another strategy for producing a desired active precursor and biopolymer is established on channeling and/or coupling the utilization of a specific precursor toward specific metabolic pathway(s). This is achievable by blocking unwanted pathways upon deletion of specific gene(s) and/or through *in trans* expression of exogenous genes that may bridge different metabolic pathways. A good example is large-scale production of sialyllactose (NeuAcα2-3Galβ1-4Glc) from exogenous *N*-acetylneuraminic acid (NeuAc) and lactose upon *in trans* expression of two recombinant synthases and NeuAc permease [160].

Further, redirecting metabolic pathway toward boosting production of an active precursor is achievable via deleting a gene that channels a precursor toward other pathways. This strategy was successfully applied for enhancing production of GDP-fucose from glucose utilization in *E. coli* via inactivation of *wcaJ* involved in colonic acid synthesis. Then, accumulated GDP-fucose was redirected toward coupling with lactose utilization and conversion by heterologous glycosyltransferases such as a fucosyltransferase. This could lead to the production of fucosylated *N*-acetyllactosamine oligosaccharides at several grams per liter of culture [161]. In a similar approach, carbohydrate moieties of gangliosides GM1and GM2 were largely produced by *E. coli* strains by up to 1.25 g/l. In this work, in order to redirect the metabolic pathways for boosting provision of active precursors, cell factories were metabolically engineered via deletion of β-galactosidase and sialic acid aldolase besides overproduction of several exogenous glycosyltransferases. In addition, applied glycerol culture was supplemented with exogenous lactose and sialic acid [162].

In another effort, large-scale production of polymeric derivatives of lactose was achieved via enhancing generation of active precursors in metabolically engineered *E. coli* strains. These strains were engineered via expressing the recombinant glycosyltransferases and notable production of lactose permease to increase lactose entry [160]. It is noteworthy to mention that the increase of the quantity of some substrates may pose a toxic or inhibitory effect on the cells, for example, those that are alternative sources of carbon and nitrogen such as amino sugars, e.g. glucosamine [163].

Also, a microbial coupling system has been considered as a strategy for increasing the efficiency of precursor provision. Sialyllactose trisaccharide was successfully produced from NeuAc and orotic acid upon combining four bacterial strains [164]. In this system, sequential recycling and secretion of nucleoside derivatives were exerted by the beginning and intermediate strains via coupling with orotic acid metabolism. Secreted recycled nucleoside derivatives were utilized by other metabolically compatible strains to generate and secret active cytidine monophosphate (CMP-NeuAc) that was converted to sialyllactose within the end producer upon coupling with lactose utilization. At the same time, the nucleoside derivatives leaving the reaction were secreted for being re-utilized and recycled by the beginners. However, although a microbial coupling system and the use of specialized strains seems effective for regenerating active precursors independent from central metabolism, the coupling of multiple microbial cultures might not be a cost-effective strategy from commercial perspective [158].

3.4.3 Strategies for Optimizing Physicochemical Properties of Polysaccharides

Physicochemical properties of polysaccharides are determined by polymerization degree or molecular weight, polydispersity, chemical nature of repeating units, chemical modifications, and branching. These are important determinants to the behavior of a polymer in solutions or in solid state, microscopic properties, viscoelasticity, molecular dynamics, water retention capacity, gelling, and intra- and intermolecular interactions. Indeed, such physicochemical properties determine the application of polysaccharides. For instance, polysaccharides with high molecular weight are highly desirable for biomedical applications [158] due to desired biocompatibility, high water retention capacity, excellent gelling properties, and a long half-life at physiological conditions. For instance, hyaluronate and alginates with high molecular mass (>10 kDa) have been largely considered for applications in cosmetics, pharmaceuticals, and biomedical products [2, 54, 165–168]. On the other hand, low molecular mass polysaccharides and their oligomers and derivatives (usually <10 kDa) such as shorter hyaluronate and alginates (oligo-G and D-mannuronate [called M2000]) display different biological effects and are considered as active pharmaceutical drugs desirable for therapeutic purposes such as promotion of angiogenesis, inhibition of tumor progression, or induction of pro-inflammatory mediators, anti-inflammatory substances, and anti-biofilm agents [54, 169–174]. However, for some polysaccharides such as carrageenan, low molecular weight species (<100 kDa) may cause adverse biological effects and are banned by FDA to apply in food products [175, 176].

Generally speaking, polymerization of all polymeric substances occurs via either a processive or non-processive (or distributive) mechanisms. In processive manner, a polymerase does not dissociate from the elongating polymer during multiple or even many thousands round of catalytic reaction that leads to the formation of long polymers such as many polysaccharides and DNA strand during replication [2, 177, 178]. In contrast, in non-processive or distributive mechanism, a polymerase dissociates from the growing polymer after limited numbers of catalytic reactions while it may reassociate with substrate/product for a new round of catalysis, leading to the formation of short stretches of a polymer such as during DNA repairing that requires polymerization of short DNA stretches [179, 180]. In contrast to processive polymerization of DNA and protein in which templates lead polymerization process, in a template-independent processive mechanism such as the polymerization of polysaccharides, other mechanisms can determine polymerization degree or polymer length that are discussed below:

(1) *Controlling processivity via substrate tethering mechanism*: This mechanism was explained by May et al. by using mycobacterial galactofuranosyltransferase GlfT2 that catalyzes polymerization of galactan polysaccharide (Figure 3.3). In an *in vitro* assay, GlfT2 and various synthetic substrates containing disaccharides of galactofuranose attached to lipids of various lengths were applied to explain substrate tethering mechanism [181]. In this mechanism, the lipid substituent with sufficient length, not solely the saccharide part, serves as a tether and is a key determinant of processivity of polymerization and polymer length (Figure 3.3). Lifetime of the enzyme–acceptor complex is prolonged upon binding of the lipid substituent to the lipid-binding secondary site of the glycosyltransferase that could facilitate the processivity of polymerization; therefore, stronger tethering such as by longer lipid substituents could result in production of longer polymers [181]. However, polymer length negatively correlated with association of enzyme to the growing polymer, i.e. as polymer length increases, the rate of dissociation would increase that favored polymerization termination [181].
(2) *Controlling processivity via coupling with modification events*: In this mechanism, the processivity of polysaccharide polymerization is linked to *in situ* enzymatic modifications such as epimerization and acetylation [44]

(Figure 3.3). This mechanism was well explained in the synthesis of alginate mediated by a multiprotein complex spanning the envelope of *P. aeruginosa* that is responsible for alginate polymerization, modification, and secretion [44, 90, 100, 133]. In our working model, interacting proteins Alg8 and Alg44, respectively known as an alginate polymerase and the c-di-GMP-binding protein, constitute alginate polymerizing unit at the cytoplasmic membrane that requires c-di-GMP binding for activation of alginate polymerization. Furthermore, these proteins interact with alginate-modifying enzymes AlgG (C5-mannuronan epimerase) and AlgX (acetyltransferase) in the periplasm [44, 133] (Figure 3.3). Alginate is initially polymerized as a polymannuronate chain by the cytoplasmic domain of Alg8 in a processive manner. During translocation of the polymer across the cytoplasmic membrane into the periplasm, AlgG catalyzes the conversion of mannuronate (M) residues to guluronate (G) residues. Concomitantly, AlgX catalyzes the addition of acetyl groups to the C2 and C3 hydroxyl groups of M residues. It was understood that the actual epimerization process interferes with the processivity of alginate polymerization, resulting in production of shorter chains of alginates, while acetylation process positively correlated with polymerization degree [44, 133]. This might be explained by the fact that AlgG-mediated epimerization shows a reaction mechanism similar to that of alginate lyase AlgL, which degrades alginate [92, 182, 183]. Hence, our finding may explain why algal alginates with abundant G blocks in chemical structure have much shorter length compared with bacterial alginates. Here, the stoichiometry of involved enzymes was found critical for not only determination of the composition but also for polymerization degree [44]. Because of the existence of at least two internal promoters upstream to the genes encoding alginate-modifying enzymes, changes in the stoichiometry of enzymes and protein subunits constituting the protein complex are postulated in response to external stimuli [111].

(3) *Controlling processivity via Wzz protein and its homologues*: In this mechanism a chain-length regulator protein mediates determination of polymer chain (Figure 3.3). This mechanism is explained for LPSs that are synthesized via the Wzy-dependent mechanism. LPSs exhibit a modal chain-length distribution that is determined with the length of LPS O-antigens. It has been understood that polymerization degree of LPS O-antigens is regulated by the chain-length regulator protein Wzz that is part of the machinery complex responsible for biosynthesis and secretion (Figure 3.3). In the absence of a Wzz protein, LPS was found with relatively shorter and randomly distributed chain lengths of the O-antigen [184, 185]. Current working models proposed for the mechanism of Wzz-mediated chain-length regulation have been debated, and Islam and Lam have summarized and assessed six proposed models [94]. Among them, a mechanism called the chain-feedback ruler has been proposed as the most logical model accounting for the most as yet provided evidences while being established upon hybridizing the ruler and chain-feedback models. In this model, at the cytoplasmic membrane, the bell-shaped quaternary structure of Wzz protein binds to the growing O-antigen chain to allow further catalytic activity by the Wzy polymerase and

formation of longer chains. As polymerization and growing chains continue, higher-order structure of the growing polymer in relation to the Wzz bell structure induces a mechanical feedback into the protein structure that leads to releasing the O-antigen polymer from Wzy. Subsequently, WaaL subunit receives freed O-antigen polymer for ligating to lipid A-core oligosaccharide to form a LPS molecule for secretion [94].

(4) *Controlling processivity via substrate concentration*: In this mechanism, the amount of substrate accessible for a polymerase determines the catalytic rate that not only affects the rate of polymerization and yield of final product but also determines the processivity of polymerization degree (Figure 3.3). In the case of hyaluronate with a heteropolymeric structure, the ratio of active precursors β-D-glucuronic acid and β-D-*N*-acetylglucosamine can determine molecular weight of the polymer. It was demonstrated that increasing the synthesis rate of UDP-*N*-acetylglucosamine via overproduction of enzymes in its cognate pathway let to increasing the molecular weight of hyaluronate chains to about 3.4 MDa, which is almost two times the molecular weight of the chains produced by wild type (~1.8 MDa). In contrast, overproduction of enzymes in the pathway of β-D-glucuronic acid synthesis decreased molecular weight of chains [186]. Therefore, we can assume that stereochemistry of UDP-*N*-acetylglucosamine may facilitate the processivity of polymerization. Substrate-dependent regulation of molecular weight have been also reported for controlling the length of other polysaccharides such as amylopectin and capsular polysaccharide [187, 188].

An efficient integration of carbon and energy flux is a key determinant for generating accessible activated sugar precursors that is under the control of different parameters in a culture medium such as dissolved oxygen, the degree of agitation, hydrodynamic conditions, pH, and the age of the culture. From biotechnological perspective, applying such controllable conditions must enhance carbon and energy flux toward the synthesis of high molecular mass biopolymer, not cell biomass and/or metabolic by-products. It is demonstrated that the amount of dissolved oxygen and stirring speed or sheer stress in the culture medium can affect molecular mass of produced polysaccharides such as alginate and hyaluronate. In these cases, increasing dissolve oxygen but mild to low agitation speed significantly increased molecular mass of both alginate and hyaluronate [189, 190]. Under anaerobic conditions, *Streptococcus zooepidemicus* produced much shorter hyaluronates (1.22 MDa) when compared with aerobic fermentation (2.19 MDa), while cell biomass and yield was found unchanged [189]. However, this approach might not be similarly effective for other types of polysaccharides and cell factories as other parameters might influence too [191]. For instance, both pH and dissolved oxygen were found determining molecular mass of pullulan [192].

(5) *Controlling processivity via the copy number of a polymerase/synthase*: In this mechanism, the processivity of polymerization depends on the copy numbers of polymerases and synthases (Figure 3.3). Here, the presence of the

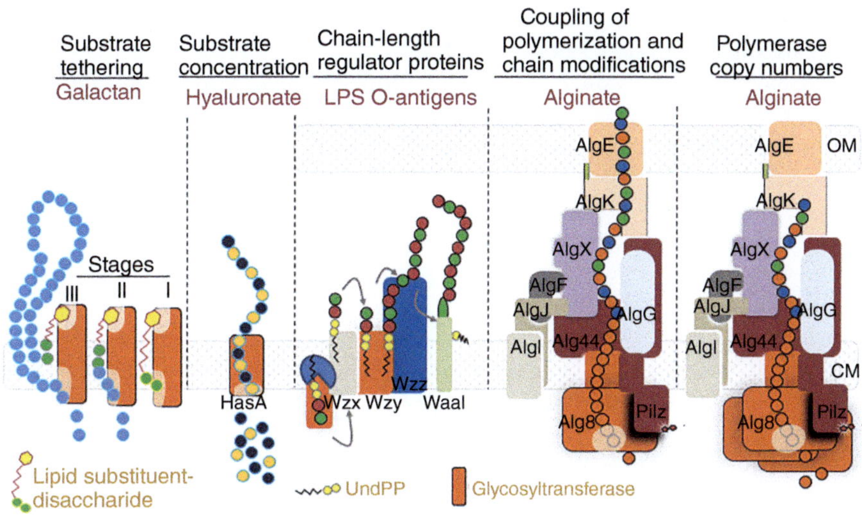

Figure 3.3 Various mechanisms for determination of polymerization degree or the length of polysaccharides in bacteria. CM, cytoplasmic membrane; OM, outer membrane.

number of different synthases competes for activated sugar precursors. Consequently, this reduces polymer chain lengths. In different studies, overproduction of hyaluronate synthase HasA or alginate polymerase Alg8 decreased the molecular weight, while yields were boosted [44, 186]. When alginate polymerase Alg8 was overproduced, the molecular weight of alginate was reduced (3.0 kDa) compared with that of produced from wild type (3.9 kDa), but yield was enhanced by almost 12 times [44] (Figure 3.3).

3.4.4 Recombinant Production of Polysaccharides and Tailor-Made Products

Microbial-derived contaminants such as LPS (also known as endotoxin), exotoxins, and other toxic molecules are common impurities associated with production of biopolymers that display immunogenicity and provoke immune responses and/or cause cytotoxicity in human and animal tissues. These unwanted impurities bring significant concerns that disqualify microbial products for medical and pharmaceutical applications. Development of nonpathogenic or avirulent cell factories upon engineering of a natural producer has been greatly considered for production of safe products. For instance, *P. aeruginosa* has been a primary model organism to study biosynthesis of alginate for many years. This species has a great natural capacity in biosynthesis of high molecular weight alginate, but it is an opportunistic human pathogen that produces a variety of toxic and immunogenic components. Recently, a highly attenuated *P. aeruginosa* PGN5 has been successfully engineered for commercial production of safe alginates [193]. On the other hand, many nonpathogenic *Pseudomonas* species harbor the *alg* operon and required genetic background for alginate production that can be novel candidates for developing avirulent and toxin-free cell factories. In this sense, exploring subgroups of natural producers lacking virulent genes in combination with the strategy of knocking out virulence factors such as a subgroup of *Pseudomonas syringae* that does not produce type III secretion system (T3SS) and relevant toxins [194] must be considered for generation of alginate-producing cell factories.

Recombinant production of polysaccharides in nonpathogenic, nonvirulent, and toxin-free cell factories is another promising strategy for producing safe products. However, recombinant production of polysaccharides is not an easy task as it requires sufficient knowledge about biosynthesis pathways, regulatory mechanisms, and recombinant protein production for the reconstruction of a polysaccharide biosynthesis pathway. Obviously, this would be a harder task when a pathway consists of multiple steps with membrane-anchored protein complex or it originates from Gram-negative bacteria whose envelope is tripartite (i.e. the cytoplasmic membrane, the periplasm, and the outer membrane).

From Gram-negative bacteria, *E. coli* strains are ideal cell factories as they have rapid growth kinetics, high productivity, and low susceptibility to culture conditions. They are well characterized, and multiple molecular tools are available for generating novel and optimized strains. Hence, various genetically modified strains are available with different genotypes that are applicable for specific purposes. An endotoxin-free *E. coli* ClearColi™ is a promising solution for elimination of endotoxin contamination issue [195]. On the other hand, a variety of plasmids are available for single or combinational transformations that enable the transfer of multiple genes into one cell for recombinant protein production. From Gram-positive species, some *Bacillus* species such as *B. subtilis*, *B. coagulans*, and *B. licheniformis* have been greatly considered as attractive cell factories, and they have GRAS status. They have high growth rates that reduce the time of fermentation cycle, with high capability of secreting proteins into the extracellular medium while they do not produce endotoxins associated with the cell wall, unlike Gram-negative species. Therefore, *Bacillus* species are highly considered as industrial workhorse over *E. coli* strains [196, 197]. As another example, *Bacillus megaterium* has found its way into industrial application for recombinant production of proteins and metabolites. As a Gram-positive species, *B. megaterium* does not produce endotoxins associated with the cell wall and grow on a variety of carbon sources and simple media [198]. A wide range of vectors have been generated for the *B. megaterium* system that are applicable for stable and high-yield protein production. Furthermore, *B. megaterium* is a good secretor, and due to lacking external alkaline proteases, proteolytic degradation is intrinsically minimized [199–201]. Further, a sporulation-negative mutant of *B. megaterium* was engineered for representing an improved industrial production strain for biotechnological processes [202]. Van Kranenburg et al. reviewed the genetic elements and engineering possibility of different species and strains belonging to the genera *Xanthomonas*, *Sphingomonas*, *Lactococcus*, and *Streptococcus* as potential cell factories for production of exopolysaccharides [203].

Recombinant production of cellulose and hyaluronate has been successfully achieved. While *G. hansenii* has been a model organism for understanding microbial cellulose biosynthesis, it has not been an ideal cell factory due to slow growth, susceptibility to culture conditions, low yield, and emerging mutants that are unable to produce cellulose [204]. Recombinant production of cellulose was achieved via expression of core genes *bcsABCD* and cognate upstream operon (*cmcax, ccp Ax*) in genetically modified *E. coli* strains including GM HMS174 (DE3) and in GM C41 (DE3) [204]. Compared with alginate operon, cellulose operon is curtailed and composed of four genes that are enough for complete cellulose production. However, upstream genes to the *bcs* operon were also required for enhancing cellulose production and proper localization of the Bcs protein complex in the cell membrane [204, 205]. Recombinant production of hyaluronate has been considered for many years because hyaluronate products from main resources, i.e. animal tissues and *Streptococcus* species (such as *S. zooepidemicus*) are likely contaminated with unwanted pollutants in addition to *Streptococcus* species, have pathogenic potentials that produce several endotoxins. In *Streptococcus* species, hyaluronate is synthesized by hyaluronate synthase (HasA) from two different active sugar donors including UDP-glucuronic acid and UDP-*N*-acetylglucosamine that are generated from glucose-6-phosphate via two pathways. Compared with alginate and cellulose operons, hyaluronate operon in *Streptococcus* is more curtailed and limited to three genes including *hasA*, *hasB*, and *hasC*. The genes *hasB* and *hasC* along with *pgm* (phosphoglucomutase) creates the first pathway to generate UDP-glucuronic acid, while *hasE* (phosphoglucoisomerase), *glmS* (amidotransferase), *glmM* (mutase), and *hasD* (acetyltransferase/pyrophosphorylase) constitute the second pathway to provide UDP-*N*-acetylglucosamine as part of the global cell metabolism [154]. Recombinant production of hyaluronate in *B. subtilis* enjoyed the facts that this species is a GRAS-qualified, safe, and nonpathogenic bacteria and hyaluronate is commonly produced by Gram-positive bacteria. The strategy leading to introduction of *B. subtilis* as an industrial workhorse for production of hyaluronate was based on heterologous expression of *hasA* originating from *Streptococcus equisimilis* in addition to overexpression of native genes of *B. subtilis* homologous to *hasB*, *hasC*, and *hasD* in *Streptococci* that catalyze the last steps of production of the precursors [206]. On the other hand, recombinant production of hyaluronate from Gram-negative bacteria such as *E. coli* is much more challenging due to the lack of key enzymes and lower level of gene expression from hyaluronate biosynthesis pathway [54]. Heterologous production of hyaluronate from *E. coli* strains JM109 was demonstrated by co-expressing hyaluronate synthase from *P. multocida* and UDP-glucose dehydrogenase from *E. coli* K5 strain. However, while hyaluronate yield from *E. coli* was a decent quantity as gram per liter level, still the productivity was about two to three times lower than the yield produced by *Streptococci* species [207]. Also, similar strategy was successfully applied for recombinant production of hyaluronate from *Agrobacterium* sp. ATCC 31749 (a commercial production strain for curdlan) and *Lactococcus lactis* [208, 209].

Recently plants have been considered as heterologous hosts for recombinant production of bacterial exopolysaccharides. Plants lack microbial-derived contaminants and pyrogens that enable an efficient purification and a large-scale production [210]. In 2014, Smith et al. provide the proof of the principle for recombinant production of bacterial polysaccharide in plants. To this end, they used the pCambia2301 vector and *A. tumefaciens*-mediated gene transfer to express the gene for the type 3 synthase (*cps3S*) from *Streptococcus pneumoniae* in *Nicotiana tabacum*. They successfully produced recombinant *S. pneumoniae* type 3 polysaccharide that was applicable in vaccine development [210].

Alginate production in *Pseudomonas* and *Azotobacter* species requires the direct involvement of at least 13 genes mediating various enzymatic steps from provision of precursor to alginate polymerization, modification, and secretion. In an effort to establish recombinant production of alginate via producing minimal protein requirement, the Rehm group assessed various easy-to-handle, nonpathogenic, and commercial *E. coli* strains (Rosetta 2, ClearColi® BL21 (DE3), SHuffle, and BL21 (DE3), BL21 Star (DE3)) and the Gram-positive *B. megaterium* YYBm1 [211]. Various commercial *E. coli* strains are available for providing specific cellular conditions required for recombinant protein production. *B. megaterium* was selected as a Gram-positive heterologous host for eliminating the requirement of alginate export that naturally occurs through the periplasm and the outer membrane in the Gram-negative species. The analysis using KEGG PATHWAY Database indicated that the production of fructose-6-phosphate derived from TCA cycle naturally occurs in *E. coli* and *B. megaterium* that is required for the production of GDP-mannuronic acid (or alginate precursor). Core genes from nonpathogenic *A. vinelandii* including precursor providers (i.e. *algA*, *algC*, and *algD*), alginate polymerase *alg8*, alginate co-polymerase, and c-di-GMP binding *alg44* and alginate exporter *algE* (also called *algJ* from *A. vinelandii*) were selected for recombinant production in heterologous hosts. *P. aeruginosa*'s *mucR* was also added to provide presumable c-di-GMP pool for activation of alginate polymerization. Two compatible plasmid systems were applied to transfer these genes into *E. coli* or *B. megaterium* strains. Upon assessment of enzymatic kinetics, the production of alginate precursor GDP-mannuronic acid was confirmed as being produced in the cytoplasm of the heterologous host *E. coli* Rosetta 2 and ClearColi BL21 (DE3). However, production of membrane-anchored alginate polymerase and co-polymerase (Alg8 and Alg44) was found challenging in the same strains using the same vector. However, *E. coli* SHuffle (pKNT25::*alg8*-6His::*alg44*) and BL21 Star (DE3) (pETDuet-1::*alg8*::*alg44*-12His) could support Alg8 and Alg44 production and proper localization in the cytoplasmic membrane [211]. However, production of alginate in heterologous host was found challenging, and more efforts are required to achieve this aim.

3.5 Conclusion and Future Perspective

In this chapter, we have highlighted the advances in understanding the biosynthesis of bacterial polysaccharides and their current and potential applications as biobased materials. Despite great advances in the design of cell factories for enhanced polysaccharide production as well as production of tailored biopolymers, challenges still remain. Biosynthesis of polysaccharides is a complex and energy-consuming process that requires the involvement of complex cellular pathways from provision of enzymes and proteins, channeling of carbon resource from central metabolism toward polysaccharide biosynthesis, and regulatory and signaling systems that control such cellular processes. Improved understanding of these complexities and reducing them through systems biology approaches will inform synthetic biology approaches for designing cell factories and efficient production of microbial polysaccharides toward development of biobased materials. When it comes to development of biomaterials for biomedical and pharmaceutical applications, rational engineering of novel GRAS-certified cell factories and polysaccharides remains challenging. Therefore, an efficient system for producing and harnessing microbial products depends on wise selection of a cell factory. In fact, development of novel endotoxin-free and nonpathogenic cell factories, improvement of purification techniques, processing, and safety assessment methods are urgent needs to promote the application of current microbial polysaccharides as high-value biomaterials for biomedical and biotechnological applications.

Acknowledgments

B.H.A. Rehm thanks past and present members of his research group at the Centre for Cell Factories and Biopolymers at Griffith University for contributions to our work on biopolymer synthesis.

References

1 Flemming, H.C., Wingender, J., Szewzyk, U. et al. (2016). Biofilms: an emergent form of bacterial life. *Nature Reviews Microbiology* 14: 563–575.

2 Moradali, M.F. and Rehm, B.H.A. (2020). Bacterial biopolymers: from pathogenesis to advanced materials. *Nature Reviews Microbiology* 18: 195–210, https://doi.org/10.1038/s41579-019-0313-3.

3 García-Ochoa, F., Santos, V.E., Casas, J.A., and Gómez, E. (2000). Xanthan gum: production, recovery, and properties. *Biotechnology Advances* 18: 549–579.

4 Jansson, P.E., Kenne, L., and Lindberg, B. (1975). Structure of extracellular polysaccharide from *Xanthomonas campestris*. *Carbohydrate Research* 45: 275–282.

5 Katzbauer, B. (1998). Properties and applications of xanthan gum. *Polymer Degradation and Stability* 59: 81–84.

6 Ojinnaka, C., Brownsey, G.J., Morris, E.R., and Morris, V.J. (1997). Effect of deacetylation on the synergistic interaction of acetan with locust bean gum or konjac mannan. *Carbohydrate Research* 305: 101–108.

7 Ojinnaka, C., Jay, A.J., Colquhoun, I.J. et al. (1996). Structure and conformation of acetan polysaccharide. *International Journal of Biological Macromolecules* 19: 149–156.

8 González, J.E., Semino, C.E., Wang, L.-X. et al. (1998). Biosynthetic control of molecular weight in the polymerization of the octasaccharide subunits of succinoglycan, a symbiotically important exopolysaccharide of *Rhizobium*

meliloti. *Proceedings of the National Academy of Sciences of the United States of America* 95: 13477.
9 Chandrasekaran, R., Lee, E.J., Thailambal, V.G., and Zevenhuizen, L.P.T.M. (1994). Molecular architecture of a galactoglucan from *Rhizobium meliloti*. *Carbohydrate Research* 261: 279–295.
10 Zevenhuizen, L.P.T.M. (1997). Succinoglycan and galactoglucan. *Carbohydrate Polymers* 33: 139–144.
11 Andhare, P., Delattre, C., Pierre, G. et al. (2017). Characterization and rheological behaviour analysis of the succinoglycan produced by *Rhizobium radiobacter* strain CAS from curd sample. *Food Hydrocolloids* 64: 1–8.
12 Harada, T. and Yoshimura, T. (1964). Production of a new acidic polysaccharide containing succinic acid by a soil bacterium. *Biochimica et Biophysica Acta, Mucoproteins and Mucopolysaccharides* 83: 374–376.
13 Banik, R.M., Kanari, B., and Upadhyay, S.N. (2000). Exopolysaccharide of the gellan family: prospects and potential. *World Journal of Microbiology and Biotechnology* 16: 407–414.
14 Campana, S., Ganter, J., Milas, M., and Rinaudo, M. (1992). On the solution properties of bacterial polysaccharides of the gellan family. *Carbohydrate Research* 231: 31–38.
15 Kuo, M.-S., Mort, A.J., and Dell, A. (1986). Identification and location of L-glycerate, an unusual acyl substituent in gellan gum. *Carbohydrate Research* 156: 173–187.
16 Kang, K.S. and Pettitt, D.J. (1993). Chapter 13: Xanthan, gellan, welan, and rhamsan. In: *Industrial Gums*, 3e (eds. R.L. Whistler and J.N. Bemiller), 341–397. London: Academic Press.
17 Lee, E.J. and Chandrasekaran, R. (1991). X-ray and computer modeling studies on gellan-related polymers: molecular structures of welan, S-657, and rhamsan. *Carbohydrate Research* 214: 11–24.
18 Coleman, R.J., Patel, Y.N., and Harding, N.E. (2008). Identification and organization of genes for diutan polysaccharide synthesis from *Sphingomonas* sp. ATCC 53159. *Journal of Industrial Microbiology and Biotechnology* 35: 263–274.
19 Pollock, T.J. (1993). Gellan-related polysaccharides and the genus *Sphingomonas*. *Microbiology* 139: 1939–1945.
20 Fialho, A.M., Moreira, L.M., Granja, A.T. et al. (2008). Occurrence, production, and applications of gellan: current state and perspectives. *Applied Microbiology and Biotechnology* 79: 889.
21 Huang, H., Wu, M., Yang, H. et al. (2016). Structural and physical properties of sanxan polysaccharide from *Sphingomonas sanxanigenens*. *Carbohydrate Polymers* 144: 410–418.
22 Leathers, T.D. (2003). Biotechnological production and applications of pullulan. *Applied Microbiology and Biotechnology* 62: 468–473.
23 Oğuzhan Yildiz, P. and Yangilar, F. (2013). Pullulan: production and usage in food industry. *African Journal of Food Science and Technology* (ISSN: 2141-5455) 4 (3): 57–63.

24 Chawla, P., Bajaj, I.B., Survase, S., and Singhal, R.S. (2009). Microbial cellulose: fermentative production and application. *Food Technology and Biotechnology* 47 (2): 107–124.

25 Dahman, Y. (2009). Nanostructured biomaterials and biocomposites from bacterial cellulose nanofibers. *Journal of Nanoscience and Nanotechnology* 9 (9): 5105–5122.

26 Esa, F., Tasirin, S.M., and Rahman, N.A. (2014). Overview of bacterial cellulose production and application. *Agriculture and Agricultural Science Procedia* 2: 113–119.

27 Klemm, D., Heublein, B., Fink, H.-P., and Bohn, A. (2005). Cellulose: fascinating biopolymer and sustainable raw material. *Angewandte Chemie International Edition* 44: 3358–3393.

28 Shoda, M. and Sugano, Y. (2005). Recent advances in bacterial cellulose production. *Biotechnology and Bioprocess Engineering* 10: 1.

29 Spiers, A.J., Bohannon, J., Gehrig, S.M., and Rainey, P.B. (2003). Biofilm formation at the air–liquid interface by the *Pseudomonas fluorescens* SBW25 wrinkly spreader requires an acetylated form of cellulose. *Molecular Microbiology* 50: 15–27.

30 Pontes, M.H., Lee, E.J., Choi, J., and Groisman, E.A. (2015). *Salmonella* promotes virulence by repressing cellulose production. *Proceedings of the National Academy of Sciences of the United States of America* 112: 5183–5188.

31 Römling, U. and Galperin, M.Y. (2015). Bacterial cellulose biosynthesis: diversity of operons, subunits, products, and functions. *Trends in Microbiology* 23: 545–557.

32 Hollenbeck, E.C., Antonoplis, A., Chai, C. et al. (2018). Phosphoethanolamine cellulose enhances curli-mediated adhesion of uropathogenic. *Proceedings of the National Academy of Sciences of the United States of America* 115: 10106–10111.

33 Thongsomboon, W., Serra, D.O., Possling, A. et al. (2018). Phosphoethanolamine cellulose: a naturally produced chemically modified cellulose. *Science* 359: 334–338.

34 Klemm, D., Schumann, D., Udhardt, U., and Marsch, S. (2001). Bacterial synthesized cellulose – artificial blood vessels for microsurgery. *Progress in Polymer Science* 26: 1561–1603.

35 Saxena, I.M. and Brown, J.R.M. (2005). Cellulose biosynthesis: current views and evolving concepts. *Annals of Botany* 96: 9–21.

36 Karnezis, T., Epa, V.C., Stone, B.A., and Stanisich, V.A. (2003). Topological characterization of an inner membrane (1→3)-β-D-glucan (curdlan) synthase from *Agrobacterium* sp. strain ATCC31749. *Glycobiology* 13: 693–706.

37 Pérez-Mendoza, D., Rodríguez-Carvajal, M.Á., Romero-Jiménez, L. et al. (2015). Novel mixed-linkage β-glucan activated by c-di-GMP in *Sinorhizobium meliloti. Proceedings of the National Academy of Sciences of the United States of America* 112: E757.

38 Bielecki, S., Krystynowicz, A., Turkiewicz, M., and Kalinowska, H. (2002). Bacterial cellulose. In: *Biopolymers (Polysaccharides I: Polysaccharides from*

Prokaryotes), vol. 5 (eds. J. Vandamme, S.D. Baets and S.A. Weinheim), 37–90. Wiley-VCH Verlag.

39 Mcintosh, M., Stone, B.A., and Stanisich, V.A. (2005). Curdlan and other bacterial (1→3)-β-D-glucans. *Applied Microbiology and Biotechnology* 68: 163–173.

40 Janeway, C.A. and Medzhitov, R. (2002). Innate immune recognition. *Annual Review of Immunology* 20: 197–216.

41 Kanke, M., Tanabe, E., Katayama, H. et al. (1995). Application of curdlan to controlled drug delivery. III. Drug release from sustained release suppositories in vitro. *Biological and Pharmaceutical Bulletin* 18 (8): 1154–1158.

42 Miwa, M., Nakao, Y., and Nara, K. (1993). Food applications of curdlan. In: *Food Hydrocolloids: Structures, Properties, and Functions* (eds. K. Nishinari and E. Doi), 119–124. Springer US: Boston, MA.

43 Rehm, B.H. and Valla, S. (1997). Bacterial alginates: biosynthesis and applications. *Applied Microbiology and Biotechnology* 48: 281–288.

44 Moradali, M.F., Donati, I., Sims, I.M. et al. (2015). Alginate polymerization and modification are linked in *Pseudomonas aeruginosa*. *mBio* 6: e00453–e00415.

45 Mørch, Ý.A., Donati, I., and Strand, B.L. (2006). Effect of Ca^{2+}, Ba^{2+}, and Sr^{2+} on alginate microbeads. *Biomacromolecules* 7: 1471–1480.

46 Helgerud, T., Gåserød, O., Fjæreide, T. et al. (2009). Alginates. In: *Food stabilizers, thickeners and gelling agents* (ed. A. Imeson), 50–56. Iowa: Willey-Blackwell.

47 Moradali, M.F., Ghods, S., and Rehm, B.H.A. (2018). Alginate biosynthesis and biotechnological production. In: *Alginates and Their Biomedical Applications* (eds. B.H.A. Rehm and M.F. Moradali), 1–25. Singapore: Springer.

48 Rehm, B. and Moradali, M.F. (2018). *Alginates and Their Biomedical Applications*. Singapore: Springer.

49 Skaugrud, O., Hagen, A., Borgersen, B., and Dornish, M. (1999). Biomedical and pharmaceutical applications of alginate and chitosan. *Biotechnology and Genetic Engineering Reviews* 16: 23–40.

50 Meyer, K. and Palmer, J.W. (1934). The polysaccharide of the vitreous humor. *Journal of Biological Chemistry* 107: 629–634.

51 DeAngelis, P.L., Jing, W., Drake, R.R., and Achyuthan, A.M. (1998). Identification and molecular cloning of a unique hyaluronan synthase from *Pasteurella multocida*. *Journal of Biological Chemistry* 273: 8454–8458.

52 Jong, A., Wu, C.-H., Chen, H.-M. et al. (2007). Identification and characterization of *CPS1* as a hyaluronic acid synthase contributing to the pathogenesis of *Cryptococcus neoformans* infection. *Eukaryotic Cell* 6: 1486.

53 DeAngelis, P.L., Papaconstantinou, J., and Weigel, P.H. (1993). Isolation of a *Streptococcus pyogenes* gene locus that directs hyaluronan biosynthesis in acapsular mutants and in heterologous bacteria. *Journal of Biological Chemistry* 268 (20): 14568–14571.

54 Sze, J.H., Brownlie, J.C., and Love, C.A. (2016). Biotechnological production of hyaluronic acid: a mini review. *3 Biotech* 6: 67.

55 Oh, S.Y., Budzik, J.M., Garufi, G., and Schneewind, O. (2011). Two capsular polysaccharides enable *Bacillus cereus* G9241 to cause anthrax-like disease. *Molecular Microbiology* 80: 455–470.

56 Wilkening, R.V. and Federle, M.J. (2017). Evolutionary constraints shaping *Streptococcus pyogenes*-host interactions. *Trends in Microbiology* 25: 562–572.

57 Cooper, C.A., Brown, K.K., Meletis, C.D., and Zabriskie, N. (2008). Inflammation and hyaluronic acid. *Alternative and Complementary Therapies* 14: 78–84.

58 Fagien, S. and Cassuto, D. (2012). Reconstituted injectable hyaluronic acid: expanded applications in facial aesthetics and additional thoughts on the mechanism of action in cosmetic medicine. *Plastic and Reconstructive Surgery* 130: 208–217.

59 Dols, M., Remaud-Simeon, M., Willemot, R.M. et al. (1998). Characterization of the different dextransucrase activities excreted in glucose, fructose, or sucrose medium by *Leuconostoc mesenteroides* NRRL B-1299. *Applied and Environmental Microbiology* 64: 1298.

60 Seymour, F.R. and Knapp, R.D. (1980). Structural analysis of dextrans, from strains of leuconostoc and related genera, that contain 3-*O*-α-D-glucosylated α-D-glucopyranosyl residues at the branch points, or in consecutive, linear positions. *Carbohydrate Research* 81: 105–129.

61 Polifka, J.E. and Habermann, J. (2015). 2.9 – Anticoagulants, thrombocyte aggregation inhibitors, fibrinolytics and volume replacement agents. In: *Drugs During Pregnancy and Lactation*, 3e (eds. C. Schaefer, P. Peters and R.K. Miller), 225–249. San Diego: Academic Press.

62 Spotti, M.J., Perduca, M.J., Piagentini, A. et al. (2013). Does dextran molecular weight affect the mechanical properties of whey protein/dextran conjugate gels? *Food Hydrocolloids* 32: 204–210.

63 Côté, G.L. and Robyt, J.F. (1983). The formation of α-D-(1→3) branch linkages by an exocellular glucansucrase from *Leuconostoc mesenteroides* NRRL B-742. *Carbohydrate Research* 119: 141–156.

64 Mehvar, R. (2000). Dextrans for targeted and sustained delivery of therapeutic and imaging agents. *Journal of Controlled Release* 69: 1–25.

65 Leemhuis, H., Pijning, T., Dobruchowska, J.M. et al. (2013). Glucansucrases: three-dimensional structures, reactions, mechanism, α-glucan analysis and their implications in biotechnology and food applications. *Journal of Biotechnology* 163: 250–272.

66 Naessens, M., Cerdobbel, A., Soetaert, W., and Vandamme, E. (2005). *Leuconostoc* dextransucrase and dextran: production, properties and applications. *Journal of Chemical Technology and Biotechnology* 80 (8): 845–860.

67 Ciszek-Lenda, M. (2011). Biological functions of exopolysaccharides from probiotic bacteria. *Central-European Journal of Immunology* 36: 51–55.

68 Kim, D. and Robyt, J.F. (1994). Production and selection of mutants of *Leuconostoc mesenteroides* constitutive for glucansucrases. *Enzyme and Microbial Technology* 16: 659–664.

69 Kothari, D., Das, D., Patel, S., and Goyal, A. (2015). Dextran and food application. In: *Polysaccharides* (eds. K. Ramawat and J.M. Mérillon), 735–752. Cham: Springer.

70 Ferretti, J.J., Gilpin, M.L., and Russell, R.R. (1987). Nucleotide sequence of a glucosyltransferase gene from *Streptococcus sobrinus* MFe28. *Journal of Bacteriology* 169: 4271.

71 Haag, A.P., Geesey, G.G., and Mittleman, M.W. (2006). Bacterially derived wood adhesive. *International Journal of Adhesion and Adhesives* 26: 177–183.

72 Cote, G., Leathers, T.D., Ahlgren, J.A. et al. (1997). Alternan and highly branched limit dextrans: Low-viscosity polysaccharides as potential new food ingredients. In: *Chemistry of novel foods* (eds. A.M. Spanier, M. Tamura, H. Okai and O. Mills), 95–110. Carol Stream, IL: Allured Publishing Corp.

73 Meng, X., Pijning, T., Dobruchowska, J.M. et al. (2016). Structural determinants of alternating ($\alpha 1 \rightarrow 4$) and ($\alpha 1 \rightarrow 6$) linkage specificity in reuteransucrase of *Lactobacillus reuteri*. *Scientific Reports* 6: 35261.

74 Mamay, Wahyuningrum, D., and Hertadi, R. (2015). Isolation and characterization of levan from moderate halophilic bacteria *Bacillus licheniformis* BK AG21. *Procedia Chemistry* 16: 292–298.

75 Keith, J., Wiley, B., Ball, D. et al. (1991). Continuous culture system for production of biopolymer levan using erwinia herbicola. *Biotechnology and Bioengineering* 38: 557–560.

76 Srikanth, R., Reddy, C.H., Siddartha, G. et al. (2015). Review on production, characterization and applications of microbial levan. *Carbohydrate Polymers* 120: 102–114.

77 Xu, W., Liu, Q., Bai, Y. et al. (2018). Physicochemical properties of a high molecular weight levan from *Brenneria* sp. EniD312. *International Journal of Biological Macromolecules* 109: 810–818.

78 Cho, S.S. and Samuel, P. (eds.) (2009). *Fiber Ingredients: Food Applications and Health Benefits*. Boca Raton, FL: CRC Press.

79 Schmid, J., Sieber, V., and Rehm, B. (2015). Bacterial exopolysaccharides: biosynthesis pathways and engineering strategies. *Frontiers in Microbiology* 6: 496.

80 Rhee, S.-K., Song, K.-B., Kim, C.-H. et al. (2005). Levan. In: *Biopolymers Online* (ed. A. Steinbüchel). Weinheim: Willey-WCH Verlag, doi:10.1002/3527600035.bpol5014.

81 Belghith, K.S., Dahech, I., Belghith, H., and Mejdoub, H. (2012). Microbial production of levansucrase for synthesis of fructooligosaccharides and levan. *International Journal of Biological Macromolecules* 50: 451–458.

82 Dahech, I., Belghith, K.S., Belghith, H., and Mejdoub, H. (2012). Partial purification of a *Bacillus licheniformis* levansucrase producing levan with antitumor activity. *International Journal of Biological Macromolecules* 51: 329–335.

83 Dahech, I., Belghith, K.S., Hamden, K. et al. (2011). Antidiabetic activity of levan polysaccharide in alloxan-induced diabetic rats. *International Journal of Biological Macromolecules* 49: 742–746.

84 Dawes, E.A., Ribbons, D.W., and Rees, D.A. (1966). Sucrose utilization by *Zymomonas mobilis*: formation of a levan. *Biochemical Journal* 98: 804–812.
85 Sezer, A.D., Kazak, H., Öner, E.T., and Akbuğa, J. (2011). Levan-based nanocarrier system for peptide and protein drug delivery: optimization and influence of experimental parameters on the nanoparticle characteristics. *Carbohydrate Polymers* 84: 358–363.
86 Shoaib, M., Shehzad, A., Omar, M. et al. (2016). Inulin: properties, health benefits and food applications. *Carbohydrate Polymers* 147: 444–454.
87 Rehm, B.H. (2010). Bacterial polymers: biosynthesis, modifications and applications. *Nature Reviews Microbiology* 8: 578–592.
88 Whitney, J.C. and Howell, P.L. (2013). Synthase-dependent exopolysaccharide secretion in Gram-negative bacteria. *Trends in Microbiology* 21: 63–72.
89 Blank, L.M., Hugenholtz, P., and Nielsen, L.K. (2008). Evolution of the hyaluronic acid synthesis (*has*) operon in *Streptococcus zooepidemicus* and other pathogenic streptococci. *Journal of Molecular Evolution* 67: 13–22.
90 Hay, I.D., Schmidt, O., Filitcheva, J., and Rehm, B.H. (2012). Identification of a periplasmic AlgK–AlgX–MucD multiprotein complex in *Pseudomonas aeruginosa* involved in biosynthesis and regulation of alginate. *Applied Microbiology and Biotechnology* 93: 215–227.
91 Hay, I.D., Ur Rehman, Z., Moradali, M.F. et al. (2013). Microbial alginate production, modification and its applications. *Microbial Biotechnology* 6: 637–650.
92 Jain, S., Franklin, M.J., Ertesvåg, H. et al. (2003). The dual roles of AlgG in C-5-epimerization and secretion of alginate polymers in *Pseudomonas aeruginosa*. *Molecular Microbiology* 47: 1123–1133.
93 Robles-Price, A., Wong, T.Y., Sletta, H. et al. (2004). AlgX is a periplasmic protein required for alginate biosynthesis in *Pseudomonas aeruginosa*. *Journal of Bacteriology* 186: 7369–7377.
94 Islam, S.T. and Lam, J.S. (2014). Synthesis of bacterial polysaccharides via the Wzx/Wzy-dependent pathway. *Canadian Journal of Microbiology* 60: 697–716.
95 Han, W., Wu, B., Li, L. et al. (2012). Defining function of lipopolysaccharide O-antigen ligase WaaL using chemoenzymatically synthesized substrates. *Journal of Biological Chemistry* 287: 5357–5365.
96 Nickerson, N.N., Mainprize, I.L., Hampton, L. et al. (2014). Trapped translocation intermediates establish the route for export of capsular polysaccharides across *Escherichia coli* outer membranes. *Proceedings of the National Academy of Sciences of the United States of America* 111: 8203–8208.
97 Willis, L.M. and Whitfield, C. (2013). Structure, biosynthesis, and function of bacterial capsular polysaccharides synthesized by ABC transporter-dependent pathways. *Carbohydrate Research* 378: 35–44.
98 Becker, A. (2015). Challenges and perspectives in combinatorial assembly of novel exopolysaccharide biosynthesis pathways. *Frontiers in Microbiology* 6: 687.
99 Ashbaugh, C.D., Albertí, S., and Wessels, M.R. (1998). Molecular analysis of the capsule gene region of group A *Streptococcus*: the *hasAB* genes are sufficient for capsule expression. *Journal of Bacteriology* 180: 4955–4959.

100 Rehman, Z.U., Wang, Y., Moradali, M.F. et al. (2013). Insights into the assembly of the alginate biosynthesis machinery in *Pseudomonas aeruginosa. Applied and Environmental Microbiology* 79: 3264–3272.

101 Garegg, P.J., Lindberg, B., Onn, T., and Sutherland, I.W. (1971). Comparative structural studies on the M-antigen from *Salmonella typhimurium, Escherichia coli* and *Aerobacter cloacae. Acta Chemica Scandinavica* 25: 2103–2108.

102 Grant, W.D., Sutherland, I.W., and Wilkinson, J.F. (1969). Exopolysaccharide colanic acid and its occurrence in the Enterobacteriaceae. *Journal of Bacteriology* 100: 1187–1193.

103 Raetz, C.R., Reynolds, C.M., Trent, M.S., and Bishop, R.E. (2007). Lipid A modification systems in gram-negative bacteria. *Annual Review of Biochemistry* 76: 295–329.

104 Whitfield, C. (2009). Structure and assembly of *Escherichia coli* capsules. *EcoSal Plus* 3 (2), doi:10.1128/ecosalplus.4.7.3.

105 Cuthbertson, L., Kos, V., and Whitfield, C. (2010). ABC transporters involved in export of cell surface glycoconjugates. *Microbiology and Molecular Biology Reviews* 74: 341.

106 Cuthbertson, L., Mainprize, I.L., Naismith, J.H., and Whitfield, C. (2009). Pivotal roles of the outer membrane polysaccharide export and polysaccharide copolymerase protein families in export of extracellular polysaccharides in gram-negative bacteria. *Microbiology and Molecular Biology Reviews* 73: 155.

107 Dunican, L.K. and Seeley, H.W. (1965). Extracellular polysaccharide synthesis by members of the genus *Lactobacillus*: conditions for formation and accumulation. *Journal of General Microbiology* 40: 297–308.

108 Da Re, S. and Ghigo, J.M. (2006). A CsgD-independent pathway for cellulose production and biofilm formation in *Escherichia coli. Journal of Bacteriology* 188: 3073–3087.

109 Grantcharova, N., Peters, V., Monteiro, C. et al. (2010). Bistable expression of CsgD in biofilm development of *Salmonella enterica* serovar *typhimurium. Journal of Bacteriology* 192: 456–466.

110 Hay, I.D., Wang, Y., Moradali, M.F. et al. (2014). Genetics and regulation of bacterial alginate production. *Environmental Microbiology* 16: 2997–3011.

111 Paletta, J.L. and Ohman, D.E. (2012). Evidence for two promoters internal to the alginate biosynthesis operon in *Pseudomonas aeruginosa. Current Microbiology* 65: 770–775.

112 Campos, M., Martínez-Salazar, J.M., Lloret, L. et al. (1996). Characterization of the gene coding for GDP-mannose dehydrogenase (*algD*) from *Azotobacter vinelandii. Journal of Bacteriology* 178: 1793–1799.

113 Gaona, G., Núñez, C., Goldberg, J.B. et al. (2004). Characterization of the *Azotobacter vinelandii algC* gene involved in alginate and lipopolysaccharide production. *FEMS Microbiology Letters* 238: 199–206.

114 Mejía-Ruíz, H., Guzmán, J., Moreno, S. et al. (1997). The *Azotobacter vinelandii alg8* and *alg44* genes are essential for alginate synthesis and can be transcribed from an *algD*-independent promoter. *Gene* 199: 271–277.

115 Vazquez, A., Moreno, S., Guzmán, J. et al. (1999). Transcriptional organization of the *Azotobacter vinelandii algGXLVIFA* genes: characterization of *algF* mutants. *Gene* 232: 217–222.

116 Qiu, D., Eisinger, V.M., Rowen, D.W., and Yu, H.D. (2007). Regulated proteolysis controls mucoid conversion in *Pseudomonas aeruginosa. Proceedings of the National Academy of Sciences of the United States of America* 104: 8107–8112.

117 Yin, Y., Damron, F.H., Withers, T.R. et al. (2013). Expression of mucoid induction factor MucE is dependent upon the alternate sigma factor AlgU in *Pseudomonas aeruginosa. BMC Microbiology* 13: 232.

118 Gottesman, S. and Storz, G. (2011). Bacterial small RNA regulators: versatile roles and rapidly evolving variations. *Cold Spring Harbor Perspectives in Biology* 3: a003798.

119 Chambers, J.R. and Sauer, K. (2013). Small RNAs and their role in biofilm formation. *Trends in Microbiology* 21: 39–49.

120 Jørgensen, M.G., Nielsen, J.S., Boysen, A. et al. (2012). Small regulatory RNAs control the multi-cellular adhesive lifestyle of *Escherichia coli. Molecular Microbiology* 84: 36–50.

121 Ogasawara, H., Yamamoto, K., and Ishihama, A. (2011). Role of the biofilm master regulator CsgD in cross-regulation between biofilm formation and flagellar synthesis. *Journal of Bacteriology* 193: 2587–2597.

122 López-Pliego, L., García-Ramírez, L., Cruz-Gómez, E.A. et al. (2018). Transcriptional study of the RsmZ-sRNAs and their relationship to the biosynthesis of alginate and alkylresorcinols in *Azotobacter vinelandii. Molecular Biotechnology* 60: 670–680.

123 Wolanin, P.M., Thomason, P.A., and Stock, J.B. (2002). Histidine protein kinases: key signal transducers outside the animal kingdom. *Genome Biology* 3: REVIEWS3013.

124 Hoch, J.A. (2000). Two-component and phosphorelay signal transduction. *Current Opinion in Microbiology* 3: 165–170.

125 Stock, A.M., Robinson, V.L., and Goudreau, P.N. (2000). Two-component signal transduction. *Annual Review of Biochemistry* 69: 183–215.

126 Manzo, J., Cocotl-Yañez, M., Tzontecomani, T. et al. (2011). Post-transcriptional regulation of the alginate biosynthetic gene algD by the Gac/Rsm system in *Azotobacter vinelandii. Journal of Molecular Microbiology and Biotechnology* 21: 147–159.

127 Quiroz-Rocha, E., Bonilla-Badía, F., García-Aguilar, V. et al. (2017). Two-component system CbrA/CbrB controls alginate production in *Azotobacter vinelandii. Microbiology* 163: 1105–1115.

128 Wang, X., Dubey, A.K., Suzuki, K. et al. (2005). CsrA post-transcriptionally represses pgaABCD, responsible for synthesis of a biofilm polysaccharide adhesin of *Escherichia coli. Molecular Microbiology* 56: 1648–1663.

129 Bellows, L.E., Koestler, B.J., Karaba, S.M. et al. (2012). Hfq-dependent, co-ordinate control of cyclic diguanylate synthesis and catabolism in the plague pathogen *Yersinia pestis. Molecular Microbiology* 86: 661–674.

130 Morgan, J.L., Mcnamara, J.T., and Zimmer, J. (2014). Mechanism of activation of bacterial cellulose synthase by cyclic di-GMP. *Nature Structural & Molecular Biology* 21: 489–496.

131 Morgan, J.L., Strumillo, J., and Zimmer, J. (2013). Crystallographic snapshot of cellulose synthesis and membrane translocation. *Nature* 493: 181–186.

132 Steiner, S., Lori, C., Boehm, A., and Jenal, U. (2013). Allosteric activation of exopolysaccharide synthesis through cyclic di-GMP-stimulated protein–protein interaction. *EMBO Journal* 32: 354–368.

133 Moradali, M.F., Ghods, S., and Rehm, B.H.A. (2017). Activation mechanism and cellular localization of membrane-anchored alginate polymerase in *Pseudomonas aeruginosa. Applied and Environmental Microbiology* 83: e03499–e03416.

134 Hay, I.D., Remminghorst, U., and Rehm, B.H. (2009). MucR, a novel membrane-associated regulator of alginate biosynthesis in *Pseudomonas aeruginosa. Applied and Environmental Microbiology* 75: 1110–1120.

135 Römling, U., Rohde, M., Olsén, A. et al. (2000). *AgfD*, the checkpoint of multicellular and aggregative behaviour in *Salmonella typhimurium* regulates at least two independent pathways. *Molecular Microbiology* 36: 10–23.

136 Zorraquino, V., García, B., Latasa, C. et al. (2013). Coordinated cyclic-di-GMP repression of *Salmonella* motility through YcgR and cellulose. *Journal of Bacteriology* 195: 417.

137 Ahmad, I., Cimdins, A., Beske, T., and Römling, U. (2017). Detailed analysis of c-di-GMP mediated regulation of *csgD* expression in *Salmonella typhimurium. BMC Microbiology* 17: 27.

138 Gerstel, U. and Römling, U. (2003). The csgD promoter, a control unit for biofilm formation in *Salmonella typhimurium. Research in Microbiology* 154: 659–667.

139 Gualdi, L., Tagliabue, L., and Landini, P. (2007). Biofilm formation-gene expression relay system in *Escherichia coli*: modulation of σ^S-dependent gene expression by the CsgD regulatory protein via σ^S protein stabilization. *Journal of Bacteriology* 189: 8034–8043.

140 Li, S., Bashline, L., Lei, L., and Gu, Y. (2014). Cellulose synthesis and its regulation. *The Arabidopsis Book* 12: e0169.

141 Ambri, F., Snoek, T., Skjoedt, M.L. et al. (2018). Design, engineering, and characterization of prokaryotic ligand-binding transcriptional activators as biosensors in yeast. *Methods in Molecular Biology* 1671: 269–290.

142 D'Ambrosio, V. and Jensen, M.K. (2017). Lighting up yeast cell factories by transcription factor-based biosensors. *FEMS Yeast Research* 17 (7): fox076, https://doi-org.lp.hscl.ufl.edu/10.1093/femsyr/fox076.

143 De Paepe, B., Peters, G., Coussement, P. et al. (2017). Tailor-made transcriptional biosensors for optimizing microbial cell factories. *Journal of Industrial Microbiology and Biotechnology* 44: 623–645.

144 Skjoedt, M.L., Snoek, T., Kildegaard, K.R. et al. (2016). Engineering prokaryotic transcriptional activators as metabolite biosensors in yeast. *Nature Chemical Biology* 12: 951–958.

145 Sutherland, I.W. (1990). *Biotechnology of Microbial Exopolysaccharides.* Cambridge: Cambridge University Press.

146 Sewalt, V., Shanahan, D., Gregg, L. et al. (2016). The generally recognized as safe (GRAS) process for industrial microbial enzymes. *Industrial Biotechnology* 12: 295–302.

147 Linton, J.D. (1990). The relationship between metabolite production and the growth efficiency of the producing organism. *FEMS Microbiology Reviews* 6: 1–18.

148 Linton, J.D. (1991). Metabolite production and growth efficiency. *Antonie van Leeuwenhoek* 60: 293–311.

149 Kell, D.B., Van Dam, K., and Westerhoff, H.V. (1989). Control analysis of microbial growth and productivity. In: *Microbial Products: New Approaches. Society for General Microbiology Symposium 44* (eds. S. Banmberg, I. Hunter and M. Rhodes), 61–93. Cambridge University Press.

150 García, A., Ferrer, P., Albiol, J. et al. (2018). Metabolic flux analysis and the $NAD(P)H/NAD(P)^+$ ratios in chemostat cultures of *Azotobacter vinelandii. Microbial Cell Factories* 17: 10.

151 de Oliveira, J.D., Carvalho, L.S., Gomes, A.M.V. et al. (2016). Genetic basis for hyper production of hyaluronic acid in natural and engineered microorganisms. *Microbial Cell Factories* 15: 119–119.

152 Vigetti, D. and Passi, A. (2014). Chapter 4: Hyaluronan synthases posttranslational regulation in cancer. In: *Advances in Cancer Research*, vol. 123 (eds. M.A. Simpson and P. Heldin), 95–119. Academic Press.

153 Marshall, S., Nadeau, O., and Yamasaki, K. (2004). Dynamic actions of glucose and glucosamine on hexosamine biosynthesis in isolated adipocytes – differential effects on glucosamine 6-phosphate, UDP-*N*-acetylglucosamine, and ATP levels. *Journal of Biological Chemistry* 279 (34): 35313–35319.

154 Chong, B.F., Blank, L.M., Mclaughlin, R., and Nielsen, L.K. (2005). Microbial hyaluronic acid production. *Applied Microbiology and Biotechnology* 66: 341–351.

155 Liu, L., Liu, Y., Li, J. et al. (2011). Microbial production of hyaluronic acid: current state, challenges, and perspectives. *Microbial Cell Factories* 10: 99.

156 Ringenberg, M., Lichtensteiger, C., and Vimr, E. (2001). Redirection of sialic acid metabolism in genetically engineered *Escherichia coli. Glycobiology* 11: 533–539.

157 Yu, L., Wu, J., Liu, J. et al. (2011). Enhanced curdlan production in *Agrobacterium* sp. ATCC 31749 by addition of low-polyphosphates. *Biotechnology and Bioprocess Engineering* 16: 34–41.

158 Ruffing, A. and Chen, R.R. (2006). Metabolic engineering of microbes for oligosaccharide and polysaccharide synthesis. *Microbial Cell Factories* 5: 25.

159 Liu, Z., Lu, Y., Zhang, J. et al. (2003). P1 Trisaccharide (Galα1,4Galβ1,4GlcNAc) synthesis by enzyme glycosylation reactions using recombinant *Escherichia coli. Applied and Environmental Microbiology* 69: 2110.

160 Priem, B., Gilbert, M., Wakarchuk, W.W. et al. (2002). A new fermentation process allows large-scale production of human milk oligosaccharides by metabolically engineered bacteria. *Glycobiology* 12: 235–240.

161 Dumon, C., Priem, B., Martin, S.L. et al. (2001). *In vivo* fucosylation of lacto-*N*-neotetraose and lacto-*N*-neohexaose by heterologous expression of *Helicobacter pylori* α-1,3 fucosyltransferase in engineered *Escherichia coli. Glycoconjugate Journal* 18: 465–474.

162 Antoine, T., Priem, B., Heyraud, A. et al. (2003). Large-scale *in vivo* synthesis of the carbohydrate moieties of gangliosides GM1 and GM2 by metabolically engineered *Escherichia coli. ChemBioChem* 4: 406–412.

163 Deng, M.D., Severson, D.K., Grund, A.D. et al. (2005). Metabolic engineering of *Escherichia coli* for industrial production of glucosamine and *N*-acetylglucosamine. *Metabolic Engineering* 7: 201–214.

164 Endo, T., Koizumi, S., Tabata, K., and Ozaki, A. (2000). Large-scale production of CMP-NeuAc and sialylated oligosaccharides through bacterial coupling. *Applied Microbiology and Biotechnology* 53: 257–261.

165 Kim, H., Jeong, H., Han, S. et al. (2017). Hyaluronate and its derivatives for customized biomedical applications. *Biomaterials* 123: 155–171.

166 Makwana, S.B., Patel, V.A., and Parmar, S.J. (2015). Development and characterization of in-situ gel for ophthalmic formulation containing ciprofloxacin hydrochloride. *Results in Pharma Sciences* 6: 1–6.

167 Sarker, B. and Boccaccini, A.R. (2018). Alginate utilization in tissue engineering and cell therapy. In: *Alginates and Their Biomedical Applications* (eds. B.H.A. Rehm and M.F. Moradali), 121–155. Springer Singapore: Singapore.

168 Xu, Z. and Lam, M.T. (2018). Alginate application for heart and cardiovascular diseases. In: *Alginates and Their Biomedical Applications* (eds. B.H.A. Rehm and M.F. Moradali), 185–212. Springer Singapore: Singapore.

169 Crupi, R. and Cuzzocrea, S. (2018). Mannuronic acid as an anti-inflammatory drug. In: *Alginates and Their Biomedical Applications* (eds. B.H.A. Rehm and M.F. Moradali), 257–268. Singapore: Springer Singapore.

170 Park, D., Kim, Y., Kim, H. et al. (2012). Hyaluronic acid promotes angiogenesis by inducing RHAMM-TGFβ receptor interaction *via* CD_{44}-PKCδ. *Molecules and Cells* 33: 563–574.

171 Rye, P.D., Tøndervik, A., Sletta, H. et al. (2018). Alginate oligomers and their use as active pharmaceutical drugs. In: *Alginates and Their Biomedical Applications* (eds. B.H.A. Rehm and M.F. Moradali), 237–256. Springer Singapore: Singapore.

172 Safrankova, B., Gajdova, S., and Kubala, L. (2010). The potency of hyaluronan of different molecular weights in the stimulation of blood phagocytes. *Mediators of Inflammation* 2010: 380948.

173 Stern, R., Asari, A.A., and Sugahara, K.N. (2006). Hyaluronan fragments: an information-rich system. *European Journal of Cell Biology* 85: 699–715.

174 Zeng, C., Toole, B.P., Kinney, S.D. et al. (1998). Inhibition of tumor growth in vivo by hyaluronan oligomers. *International Journal of Cancer* 77: 396–401.

175 Ariffin, S.H., Yeen, W.W., Abidin, I.Z. et al. (2014). Cytotoxicity effect of degraded and undegraded *kappa* and *iota* carrageenan in human intestine and liver cell lines. *BMC Complementary and Alternative Medicine* 14: 508.

176 Jiang, H.Y., Wang, F., Chen, H.M., and Yan, X.J. (2013). κ-carrageenan induces the disruption of intestinal epithelial Caco-2 monolayers by promoting the interaction between intestinal epithelial cells and immune cells. *Molecular Medicine Reports* 8: 1635–1642.

177 Breyer, W.A. and Matthews, B.W. (2001). A structural basis for processivity. *Protein Science* 10: 1699–1711.

178 Levengood, M.R., Splain, R.A., and Kiessling, L.L. (2011). Monitoring processivity and length control of a carbohydrate polymerase. *Journal of the American Chemical Society* 133: 12758–12766.

179 Craig, N.L., Green, R., Greider, C. et al. (2014). *Molecular Biology: Principles of Genome Function*. Oxford: Oxford University Press.

180 McClure, W.R. and Chow, Y. (1980). The kinetics and processivity of nucleic acid polymerases. In: *Methods in Enzymology* (ed. D.L. Purich), 277–297. Academic Press.

181 May, J.F., Splain, R.A., Brotschi, C., and Kiessling, L.L. (2009). A tethering mechanism for length control in a processive carbohydrate polymerization. *Proceedings of the National Academy of Sciences of the United States of America* 106: 11851–11856.

182 Douthit, S.A., Dlakic, M., Ohman, D.E., and Franklin, M.J. (2005). Epimerase active domain of *Pseudomonas aeruginosa* AlgG, a protein that contains a right-handed β-helix. *Journal of Bacteriology* 187: 4573–4583.

183 Gimmestad, M., Sletta, H., Ertesvåg, H. et al. (2003). The *Pseudomonas fluorescens* AlgG protein, but not its mannuronan C-5-epimerase activity, is needed for alginate polymer formation. *Journal of Bacteriology* 185: 3515–3523.

184 Bastin, D.A., Stevenson, G., Brown, P.K. et al. (1993). Repeat unit polysaccharides of bacteria: a model for polymerization resembling that of ribosomes and fatty acid synthetase, with a novel mechanism for determining chain length. *Molecular Microbiology* 7: 725–734.

185 Murray, G.L., Attridge, S.R., and Morona, R. (2003). Regulation of *Salmonella typhimurium* lipopolysaccharide O antigen chain length is required for virulence; identification of FepE as a second Wzz. *Molecular Microbiology* 47: 1395–1406.

186 Chen, W.Y., Marcellin, E., Hung, J., and Nielsen, L.K. (2009). Hyaluronan molecular weight is controlled by UDP-*N*-acetylglucosamine concentration in *Streptococcus zooepidemicus*. *Journal of Biological Chemistry* 284: 18007–18014.

187 Van den Koornhuyse, N., Libessart, N., Delrue, B. et al. (1996). Control of starch composition and structure through substrate supply in the monocellular alga *Chlamydomonas reinhardtii*. *Journal of Biological Chemistry* 271: 16281–16287.

188 Ventura, C.L., Cartee, R.T., Forsee, W.T., and Yother, J. (2006). Control of capsular polysaccharide chain length by UDP-sugar substrate concentrations in *Streptococcus pneumoniae*. *Molecular Microbiology* 61: 723–733.

189 Duan, X.J., Yang, L., Zhang, X., and Tan, W.S. (2008). Effect of oxygen and shear stress on molecular weight of hyaluronic acid. *Journal of Microbiology and Biotechnology* 18: 718–724.

190 Peña, C., Trujillo-Roldán, M.A., and Galindo, E. (2000). Influence of dissolved oxygen tension and agitation speed on alginate production and its molecular weight in cultures of *Azotobacter vinelandii*. *Enzyme and Microbial Technology* 27: 390–398.

191 Lee, I.Y., Kim, M.K., Lee, J.H. et al. (1999). Influence of agitation speed on production of curdlan by *Agrobacterium* species. *Bioprocess Engineering* 20: 283–287.

192 Lee, A., Ji-Hyun, A., Kim, J.H. et al. (2002). Effect of dissolved oxygen concentration and pH on the mass production of high molecular weight pullulan by *Aureobasidium pullulans*. *Journal of Microbiology and Biotechnology* 12 (1): 1–7.

193 Valentine, M.E., Kirby, B.D., Withers, T.R. et al. (2020). Generation of a highly attenuated strain of *Pseudomonas aeruginosa* for commercial production of alginate. *Microbial Biotechnology* 13 (1): 162–175.

194 Mohr, T.J., Liu, H., Yan, S. et al. (2008). Naturally occurring nonpathogenic isolates of the plant pathogen *Pseudomonas syringae* lack a type III secretion system and effector gene orthologues. *Journal of Bacteriology* 190: 2858.

195 Mamat, U., Woodard, R., Wilke, K. et al. (2013). Endotoxin-free protein production-ClearColi™ technology. *Nature Methods* 10: 916, https://doi.org/10.1038/nmeth.f.367.

196 Elshaghabee, F.M.F., Rokana, N., Gulhane, R.D. et al. (2017). *Bacillus* as potential probiotics: status, concerns, and future perspectives. *Frontiers in Microbiology* 8: 1490–1490.

197 Schallmey, M., Singh, A., and Ward, O.P. (2004). Developments in the use of *Bacillus* species for industrial production. *Canadian Journal of Microbiology* 50: 1–17.

198 Eppinger, M., Bunk, B., Johns, M.A. et al. (2011). Genome sequences of the biotechnologically important *Bacillus megaterium* strains QM B1551 and DSM319. *Journal of Bacteriology* 193: 4199–4213.

199 Stammen, S., Müller, B.K., Korneli, C. et al. (2010). High-yield intra- and extracellular protein production using *Bacillus megaterium*. *Applied and Environmental Microbiology* 76: 4037–4046.

200 Vary, P. (1992). Development of genetic engineering in *Bacillus megaterium*. *Biotechnology* 22: 251–310.

201 Vary, P.S. (1994). Prime time for *Bacillus megaterium*. *Microbiology* 140 (Pt 5): 1001–1013.

202 Grage, K., Mcdermott, P., and Rehm, B.H.A. (2017). Engineering *Bacillus megaterium* for production of functional intracellular materials. *Microbial Cell Factories* 16: 211.

203 van Kranenburg, R., Boels, I.C., Kleerebezem, M., and de Vos, W.M. (1999). Genetics and engineering of microbial exopolysaccharides for food: approaches for the production of existing and novel polysaccharides. *Current Opinion in Biotechnology* 10: 498–504.

204 Buldum, G., Bismarck, A., and Mantalaris, A. (2018). Recombinant biosynthesis of bacterial cellulose in genetically modified *Escherichia coli*. *Bioprocess and Biosystems Engineering* 41: 265–279.

205 Standal, R., Iversen, T.G., Coucheron, D.H. et al. (1994). A new gene required for cellulose production and a gene encoding cellulolytic activity in *Acetobacter xylinum* are colocalized with the *bcs* operon. *Journal of Bacteriology* 176: 665–672.
206 Widner, B., Behr, R., Von Dollen, S. et al. (2005). Hyaluronic acid production in *Bacillus subtilis*. *Applied and Environmental Microbiology* 71: 3747–3752.
207 Mao, Z., Shin, H.D., and Chen, R. (2009). A recombinant *E. coli* bioprocess for hyaluronan synthesis. *Applied Microbiology and Biotechnology* 84: 63–69.
208 Chien, L.J. and Lee, C.K. (2007). Hyaluronic acid production by recombinant *Lactococcus lactis*. *Applied Microbiology and Biotechnology* 77: 339–346.
209 Mao, Z. and Chen, R.R. (2007). Recombinant synthesis of hyaluronan by *Agrobacterium* sp. *Biotechnology Progress* 23: 1038–1042.
210 Smith, C.M., Fry, S.C., Gough, K.C. et al. (2014). Recombinant plants provide a new approach to the production of bacterial polysaccharide for vaccines. *PLoS One* 9: e88144.
211 Ghods, S. (2018). Investigation of biosynthesis of exopolysaccharides within the biofilm matrix of *Pseudomonas aeruginosa* and *Pseudomonas syringae* pv. actinidiae. PhD thesis. Doctor of Philosophy. Massey University.

4

Exploitation of Exopolysaccharides from Lactic Acid Bacteria

Tsuda Harutoshi

Hirosaki University, Faculty of Agriculture and Lifescience, 3 Bunkyo-cho, Hirosaki, Aomori 036-8561, Japan

4.1 Introduction

4.1.1 Lactic Acid Bacteria

Lactic acid bacteria (**LAB**) are used in many fermented foods, particularly fermented dairy products such as cheese, buttermilk, and fermented milk. LAB are anaerobic bacteria and are defined as organisms (i) producing lactic acid as a sole or main end product from glucose, (ii) gram-positive, (iii) catalase non-productive, (iv) non-spore forming, and (v) non-motile. A molecule of glucose is converted to two molecules of lactic acid (homofermentative) or to one molecule of lactic acid and other products such as carbon dioxide, ethanol, and acetate (heterofermentative). Three types of the bacterial morphology were observed: rods, cocci, and tetrad. LAB does not produce catalase, but a few of them can produce pseudo-catalase in a glucose-limited medium. Moreover, LAB does not formed endospores. However, there is a confusing genus, *Sporolactobacillus*, which produces lactic acid from glucose by homofermentation. They can also form endospores that are resistant to heating at 80 °C for 10 minutes while they move by peritrichous flagella [1]. Generally, *Sporolactobacillus* are considered to be spore-forming LAB.

4.1.2 Exopolysaccharides

LAB produce lactic acid, carbon dioxide, and diacetyl/acetoin. These substances contribute to the flavor, texture, and shelf life of fermented foods. Some LAB produce polysaccharides (PSs). Generally, PSs are present outside the cell wall, and they exhibit great diversity, not only in their sugar composition but also in their linkage, branching, and substitution [2]. PSs can be bound or unbound to the cell wall, and cell-bound PSs are distinguished into capsular polysaccharides (CPSs) [3]. PS production was confirmed by the Indian ink method [4]. LAB culture and Indian ink were mixed on a slide glass, and then a cover slip was placed over the mixture. The prepared slide was observed microscopically. A clear zone was observed around cells, if the LAB strain could produce PS (Figure 4.1).

Biopolymers for Biomedical and Biotechnological Applications, First Edition.
Edited by Bernd H. A. Rehm and M. Fata Moradali.

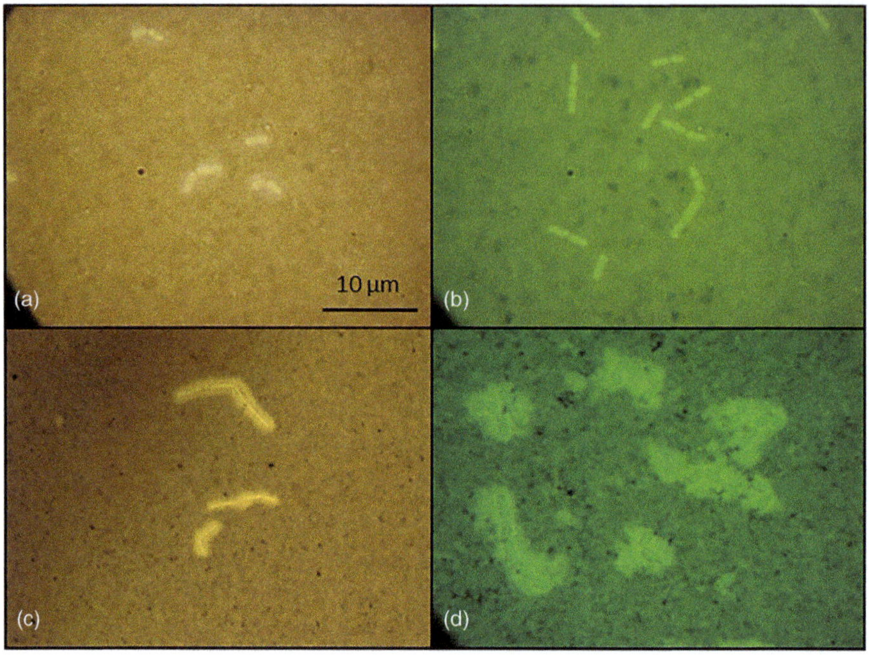

Figure 4.1 The Indian ink stain with non-PS-producible LAB (a, b) and with PS-producible LAB (c, d). Indian ink is a suspension of finely ground carbon particles. A clear zone observed around cells (c, d) is PS that India ink does not penetrate.

Generally, PS has been thought to protect the bacterial cell from various stresses such as desiccation, phage attack, bile salts, and low pH [5–8]. Dertli et al. [9] reported that PS-producible *Lactobacillus johnsonii* FI9785 showed a resistance to tetracycline and bile salt. Ruas-Madiedo et al. reported that *Bifidobacterium animalis* was induced to produce PS when 0.6% bovine bile was supplemented to DeMan-Rogosa-Sharpe (MRS) (peptone 10.0 g/l, Lab-Lemco powder 8.0 g/l, yeast extract 4.0 g/l, glucose 20.0 g/l, tween80 1.0 ml/l, dipotassium hydrogen phosphate 2.0 g/l, sodium acetate $3H_2O$ 5.0 g/l, triammonium citrate 2.0 g/l, magnesium sulfate $7H_2O$ 0.2 g/l, manganese sulfate $4H_2O$ 0.05 g/l, pH 6.2 ± 0.2) broth [5]. Alp and Aslim reported that there were positive correlations between PS production and the resistance to bile salts and low pH after investigating 31 *Bifidobacterium* strains [6].

PSs produced by LAB are divided into two classes: homo- and hetero-PS. A homo-PS is composed of single type of monosaccharide, whereas a hetero-PS consists of regular repeating units of three to eight different carbohydrate moieties synthesized from intracellular sugar nucleotide precursors [10]. The biosynthesis of homo-PS is different from that of hetero-PS. Homo-PS is made from sucrose using glucansucrase or levansucrase [11, 12], and hetero-PS is made as follows: sugar transportation, followed by sugar nucleotide synthesis, repeating unit synthesis, and polymerization of the repeating units [13]. Homo-PS consists of a single type of monosaccharide (α-D-glucans, β-D-glucans, fructans, and others represented by polygalactan), whereas hetero-PS is composed of different types of monosaccharides, mainly D-glucose, D-galactose, L-rhamnose, and their derivatives [14].

4.1.3 Importance of PS Produced by LAB

PS is exploited mainly by food industries. PS has the ability to increase the moisture content in the food product. PS plays a major role as natural texturizer in the industrial production of yogurt, cheese, and milk-based desserts. Rheological effects of PS produced by LAB were reviewed by Mende et al. [15]. The firmness, creaminess, and appearance, i.e. syneresis and viscosity, of the fermented milks are changed due to presence of PS. Most studies reported that PS reduced syneresis and enhanced viscosity of stirred yogurt. Similarly, the moisture content of low-fat mozzarella cheese made by using LAB-producing PS was higher than that made with non-PS-producing LAB [16].

Recently, PSs produced by LAB have received increasing attention, mainly because of their health benefits. Immune stimulation, anti-mutagenicity, and the antitumor activity of fermented dairy products prepared with PS-producing LAB or PSs in particular have been investigated [17–22]. Immunostimulatory effects of PS produced by LAB were reviewed by Laino et al. [23]. It has been found that PS induces production of interleukin-6 (IL-6), interleukin -1β (IL-1β), and tumor necrosis factor α (TNF-α) in murine RAW macrophages [24]. There are remarkable differences between acidic and neutral PS with respect to immunomodulatory activities [18, 25, 26]. The acidic PS, which contains phosphate, improved the macrophage function *in vivo*, whereas the neutral PS showed no effect. Acidic PS was found to be a potent stimulator of macrophages; however, this effect was reduced with the removal of phosphorus from acidic PS [23].

PS may be used as an inhibitor of some toxins. Tetrodotoxin is a poison accumulated in puffer fishes; the antidote for it has not yet been developed. Nguyen et al. [27] reported that PSs produced by *Leuconostoc mesenteroides* and *Lactobacillus rhamnosus* detoxified tetrodotoxin in mouse in the presence of cuprous oxide. Some bacterial pathogens produce a virulence CPS and build a biofilm [28]. Some PS produced by LAB showed the biofilm inhibition activity. PS produced by *Lactobacillus plantarum* exhibited an inhibitory activity on biofilm formation by *Escherichia coli* O157, *Shigella flexneri*, *Staphylococcus aureus*, and *Salmonella typhimurium* [29, 30]. PS produced by *Lactobacillus helveticus* MB2-1 exhibited inhibitory activity on biofilm formation by *E. coli* O157, *S. aureus*, and *Pseudomonas aeruginosa* [31]. PSs produced by *L. rhamnosus* and *Lactobacillus gasseri* inhibited biofilm formation by *S. typhimurium* and *Listeria monocytogenes*, respectively [32, 33]. On the other hand, lipopolysaccharides produced by gram-negative bacteria and CPSs produced by *Bacteroides fragilis* are important virulence factors in pathogenesis [28].

4.2 Homo-PS

4.2.1 Biosynthesis

Homo-PS is roughly divided into two groups including glucans and fructans. Homo-PS contains one sugar type, and glucans contain D-glucopyranose, and fructans contain D-fructofuranose. Homo-PS are synthesized by some LAB, and the LAB have glycosyltransferase or fructosyltransferase. A glycosyltransferase or fructosyltransferase is secreted outside by the cells, and then a homo-PS is synthesized from sucrose by this polymerase. Synthesis reaction is as follows: two sucrose molecules are converted to a trisaccharide (glucose–glucose–fructose) and monosaccharide (fructose). More glucose units coming from sucrose are built on the trisaccharide that is accompanied by fructose liberation [34].

4.2.2 Composition and Structure

The differences between the homo-PS arise mainly from the features of their primary structure such as the pattern of main chain bonds, molecular weight, and branch structure (Tables 4.1 and 4.2). Glucans include polymers with α-1,6 linkages (dextran), α-1,3 linkages (mutan), and both α-1,6 and α-1,3 linkages (alternan). The relative molecular weight of glucans from lactobacilli range from 1×10^6 to 5×10^7 Da [11]. Dextran is produced by *L. mesenteroides* subsp. *mesenteroides* and *L. mesenteroides* subsp. *dextranicum*, and mutan is produced by *Streptococcus mutans* and *Streptococcus sobrinus*. Fructans are mainly composed of β-2,6-linked fructose molecules and generally have a relative molecular weight exceeding 5×10^6 Da, such as levan produced by *Streptococcus salivarius* [35].

In the 1950s, the use of a cell-free enzyme solution permitted dextran synthesis under controlled conditions yielding a polymer of greater purity. A common feature of all dextrans is the preponderance of α-1,6-linkages with branch points at positions 2, 3, or 4 [35]. Some strains of *Leuconostoc amelibiosum* [36] and *Lactobacillus curvatus* [37] are reported to be dextran-producing strains. Dextran has a long history of clinical use as a blood extender [38].

Mutan is the glucan that is synthesized by various serotypes of *S. mutans* and differs from dextran as it contains a high percentage of α-1,3 linkages. Differences in solubility result from the proportions of different types of linkages. Water-soluble glucans are rich in α-1,6 linkages, whereas water-insoluble glucans are rich in α-1,3 linkages [35]. Ingestion of mutan has been linked with dental caries, as insoluble mutans can adhere to teeth, thus helping microorganisms adhere to the surface of teeth.

Alternan has alternate α-1,6 and α-1,3 linkages. This structure is thought to be responsible for its distinctive physical properties including high solubility and low viscosity. These characteristics provide this glucan with a potential for commercial application as a low viscosity texturizer in foods. *L. mesenteroides* NRRL B-1355 was the first to be reported as an alternan-producing strain [39].

Levan is a PS produced from sucrose. It is a fructan composed of β-2,6-linked fructose molecules with some β-2,1-linked branches. Incidentally, inulin is a fructan composed of β-2,1-linked fructose molecules with some β-2,6-linked branches. *S. salivarius*, *L. mesenteroides*, and *Lactobacillus reuteri* are known to be levan-producing strains [40, 41]. In addition, the PS produced by *Lactobacillus sanfranciscensis* TMW 1.392 has been reported to be fructan [42].

4.2.3 Instability of Homo-PS Production

L. mesenteroides subsp. *mesenteroides* produce dextran from sucrose. They often lose the ability to produce dextran when serial transfers are made in media with increasing salt concentrations. Non-dextran-producing strains of *Leuconostoc* sp. can revert to dextran production when they are inoculated into medium

Table 4.1 Homo-PS structure (glucans).

	Main linkage (branching linkage)	Structure	Organism
Dextran	α-1,6 (α-1,3)	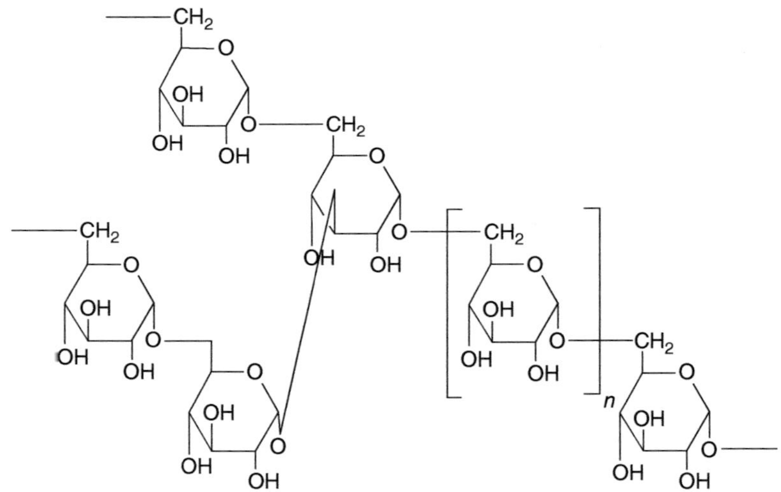	*L. mesenteroides* ssp. *mesenteroides* *L. mesenteroides* ssp. *dextranicum* *L. amelibiosum* *L. curvatus*
Mutan	α-1,3 (α-1,6)	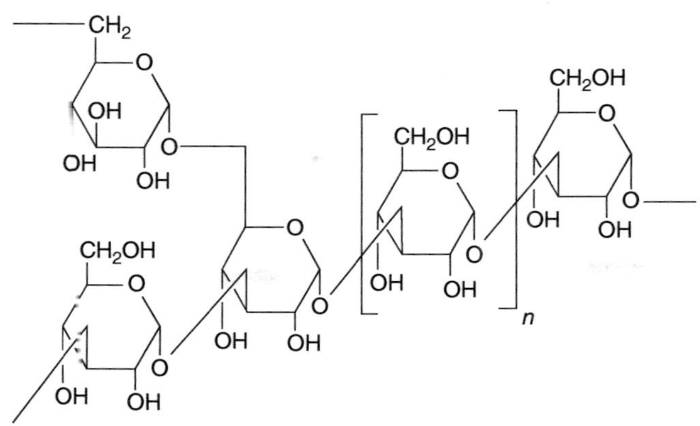	*S. mutans*

(*continued*)

Table 4.1 (Continued)

	Main linkage (branching linkage)	Structure	Organism
Alternan	α-1,3 and α-1,6	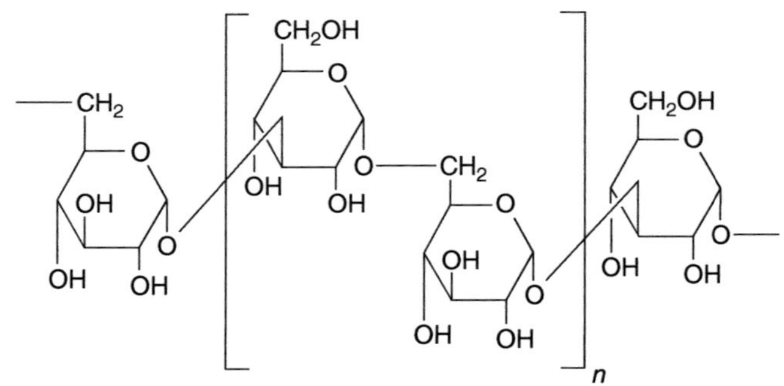	*L. mesenteroides*

Table 4.2 Homo-PS structure (fructans).

	Main linkage (branching linkage)	Structure	Organism
Levan	β-2,6 (β-2,1)	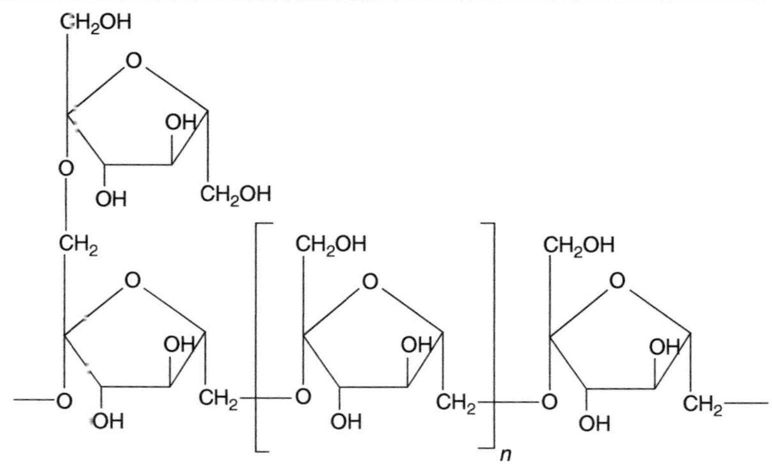	*L. mesenteroides* *L. reuteri*

(*continued*)

Table 4.2 (Continued)

	Main linkage (branching linkage)	Structure	Organism
Inulin	β-2,1 (β-2,6)	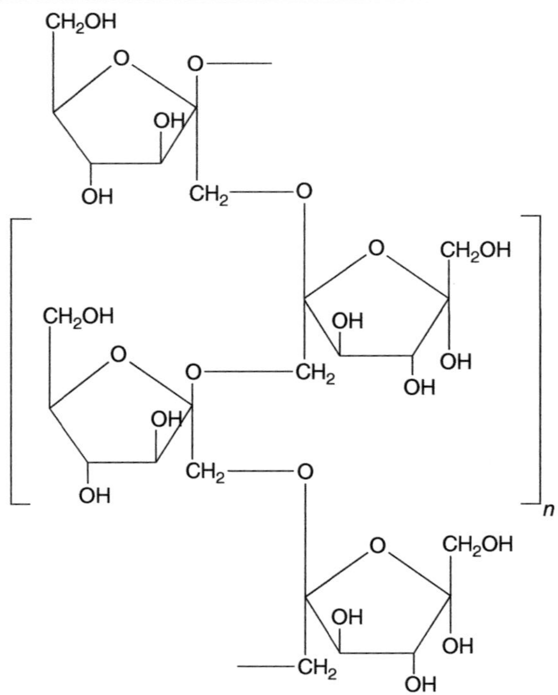	*S. mutans*

containing tomato or orange juice [43]. Although there is insufficient genetic evidence to explain this fact, a mutation possibly occurs in dextransucrase gene. Dextransucrases are used in the synthesis of dextran, and they transfer the glucosyl moiety from sucrose to acceptor molecules [44]. Dextransucrase genes usually are located on the chromosome of *Leuconostoc* species, and Jin et al. [45] successfully constructed a dextran-negative *Leuconostoc* mutant by disrupting the dextransucrase gene on the chromosome.

4.3 Hetero-PS

4.3.1 Biosynthesis

Hetero-PS are made by a complex sequence of interactions involving intracellular enzyme. First, sugar such as glucose and lactose are transported into the cell. In the cytoplasm, monosaccharides are converted to sugar nucleotides harboring uridine diphosphate (UDP), thymidine diphosphate (dTDP), or guanosine diphosphate (GDP) moieties by glycosylpyrophosphorylases, as active sugar donors in polymerization reactions. The sugar nucleotide is transferred to C55-polyprenyl phosphate, which is a carrier lipid and component of the membrane, by priming glycosyltransferases. Then the sugar nucleotides are combined and formed a specific unit. This reaction is catalyzed by glycosyltransferases. Therefore, the type and number of glycosyltransferases determine the composition and range of the unit. The unit, which is called repeating unit, is exported through the bacterial membrane and is extracellularly polymerized to form the final hetero-PS. Hence, several enzymes are involved in the biosynthesis and secretion of hetero-PS (Figure 4.2).

The genetic aspects of the hetero-PS biosynthesis are poorly understood, although several mechanisms have been suggested. In regard to the aspects, the review by Jiang and Yang [46] would be helpful for readers.

4.3.2 Monosaccharides Composition of Hetero-PS

The chemical composition of hetero-PS shows wide variability. Hetero-PS is polymerized by repeating the units mainly composed of D-glucose, D-galactose, and L-rhamnose. The composition of the monosaccharide subunits and the structure of the repeating units are not considered to be species specific, except in case of *Lactobacillus kefiranofaciens* subsp. *kefiranofaciens*. This species, isolated from kefir grain, a fermented dairy food from the North Caucasus region, produces large amounts of PSs [47]. Hetero-PS-producing strains of LAB have been identified, and a web-based database of PS is freely available online [48] (http://www.epsdatabase.com/). This database has 90 PSs of LAB, registered in 2018, and it will help to further the research of PS function and structure. Monosaccharide ratios of these PSs taken from the database and recent reports are shown in Table 4.3.

It is noteworthy to mention that crudely purified PSs may contain mannan from the yeast extract in media. This leads to a false high yield of PS by LAB and an

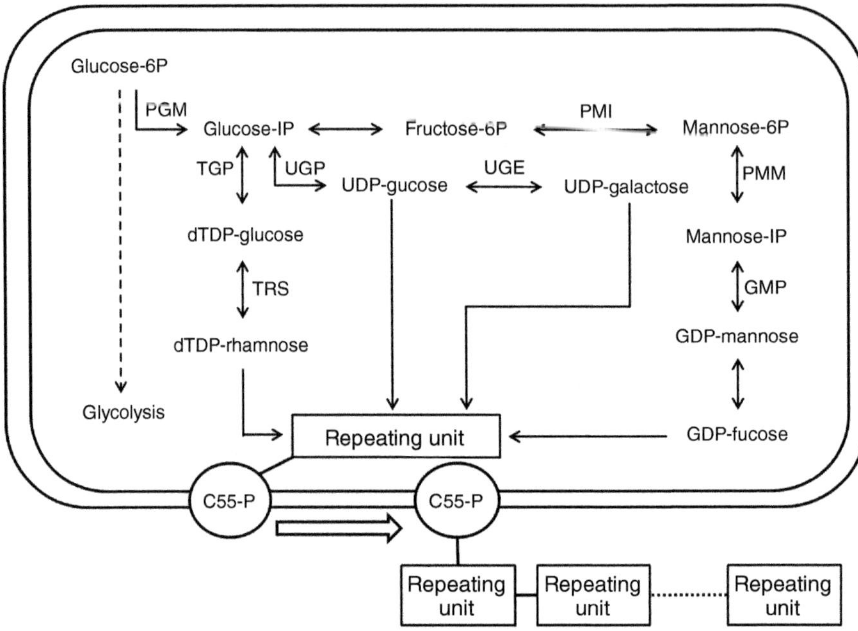

Figure 4.2 Outline of biosynthesis of hetero-PS in a bacterial cell. Glucose in the cell is converted to sugar nucleotides harboring UDP, dTDP, or GDP moieties. The sugar nucleotides on C55-polyprenyl phosphate are combined and formed a repeating unit. A repeating unit is extracellularly polymerized to form the final hetero-PS. PGM, α-phosphoglucomutase; UGP, UDP-glucose pyrophosphorylase; UGE, UDP-galactose 4-epimerase; TGP, dTDP-glucose pyrophosphorylase; TRS, dTDP-rhamnose synthetic enzyme system; PMI, phosphomannoisomerase; PMM, phosphomannomutase; GMP, GDP-mannose pyrophosphorylase.

addition of mannose to the monosaccharide composition of PS. The mannan was recovered at 50.8 mg/l from fresh media containing 0.5% yeast extract [54]. The mannan may be utilized by some bacteria such as *E. coli*, and this leads to difficulty with a correct PS utilization analysis.

4.3.3 Yield of Hetero-PS

The yields of hetero-PS produced by LAB vary greatly. PS production is 50–350 mg/l for *Streptococcus thermophilus*, 80–600 mg/l for *Lactococcus lactis* subsp. *cremoris*, 60–150 mg/l for *Lactobacillus delbrueckii* subsp. *bulgaricus*, 50–60 mg/l for *Lactobacillus casei* [54–58], 140 mg/l for *L. plantarum* [59, 60], over 200 mg/l for *Bifidobacterium bifidum* WBIN03 and some *Lactobacillus* spp. [61], 196 mg/l for *Lactobacillus mucosae* DPC 6426 [62], 304 mg/l for *L. plantarum* BC-25 [51], 68.69 mg/l for *Lactobacillus sicerae*, 37.98 mg/l for *Lactobacillus collinoides* [63], and 737 mg/l for *L. rhamnosus* [32]. The highest recorded yield of hetero-PS is 2800 mg/l for *L. plantarum* BR2 [64], 2775 mg/l for *L. rhamnosus* RW-9595M [65], and 2200–2500 mg/l for *L. kefiranofaciens* [49, 66]. These yields are much lower than those from other industrially important microorganisms such as *Xanthomonas campestris*, which produces

Table 4.3 Monosaccharide ratio in hetero-PS from PS database and recent studies.

	Glc	GlcA	Gal	Rha	Fru	Ara	Fuc	Man	Rib	Xyl	NAc Glu	NAc Gal
B. animalis ssp. *lactis*	1		1	2								
B. longum	1		3	1								
L. acidophilus	3		1									
L. acidophilus	3		1									
L. acidophilus [22]	1	3					1					
L. delbrueckii ssp. *bulgaricus*	2		1									
L. delbrueckii ssp. *bulgaricus*	1		3									
L. delbrueckii ssp. *bulgaricus*	1		1									
L. delbrueckii ssp. *bulgaricus*	1		3									
L. delbrueckii ssp. *bulgaricus*	1		3									
L. delbrueckii ssp. *bulgaricus*	1		2	2								
L. delbrueckii ssp. *bulgaricus*	1		1									
L. delbrueckii ssp. *bulgaricus*	3			2								
L. delbrueckii ssp. *bulgaricus*	1		3									
L. delbrueckii ssp. *bulgaricus*	1		3									
L. delbrueckii ssp. *bulgaricus*	1		3									
L. delbrueckii ssp. *bulgaricus*	1		3									
L. helveticus	2		3									
L. helveticus	4		1									
L. helveticus	1		1									
L. helveticus	3		1									
L. helveticus			1					2				
L. helveticus	1		1									
L. helveticus	1		1									
L. helveticus	2		1									
L. helveticus	3		2									
L. helveticus	1		1									
L. helveticus [31]	1.7		1					3.5				
L. johnsonii	2		1									
L. johnsonii	2		3									
L. johnsonii	1		4									
L. kefiranofaciens	2		3									
L. kefiranofaciens [49]	20		18	1		25		92				
L. paracasei			3									1
L. paracasei [50]			1	4								1
L. pentosus	3			2								
L. plantarum	2											1
L. plantarum [29]	4.2		4.1		1			8.2				
L. plantarum [30]	1.8		1							3.4		

(*continued*)

Table 4.3 (Continued)

	Glc	GlcA	Gal	Rha	Fru	Ara	Fuc	Man	Rib	Xyl	NAc Glu	NAc Gal
L. plantarum [51]	6		1.8					92				
L. plantarum [52]	1		3									
L. plantarum [53]	2.7		1									
L. rhamnosus	2			4								
L. rhamnosus	2			4								
L. rhamnosus	2		3									
L. rhamnosus			3	1							1	
L. rhamnosus	1			2								
L. rhamnosus	1		1				3					
L. rhamnosus	1		6									
L. rhamnosus	2		3									
L. rhamnosus	2		3									
L. rhamnosus	2		3									
L. sake	2			1								
L. lactis (genetically modified)	1		2									
L. lactis (genetically modified)	2		1									
L. lactis ssp. *cremoris*	2		1									
L. lactis ssp. *cremoris*	1		2	2								
L. lactis ssp. *cremoris*	2		1									
L. lactis ssp. *cremoris*	2		1									
L. lactis ssp. *cremoris*	2		1									
L. lactis ssp. *cremoris*	2		1									
L. lactis ssp. *cremoris*	2		1									
L. lactis ssp. *cremoris*	2		1									
L. lactis ssp. *cremoris*	2		1									
S. agalactiae type III	2		1									
S. macedonicus	2		1									
S. suis type 14	1		2									
S. suis type 2	1		1	1								
S. thermophilus	1		1									1
S. thermophilus			3	2								
S. thermophilus	2		1									
S. thermophilus	2		3	1								
S. thermophilus	1		1									1
S. thermophilus	1		2	2								
S. thermophilus	1		1									1
S. thermophilus	2		1									

(*continued*)

Table 4.3 (Continued)

	Glc	GlcA	Gal	Rha	Fru	Ara	Fuc	Man	Rib	Xyl	NAc Glu	NAc Gal
S. thermophilus			3	2								
S. thermophilus	2		1									
S. thermophilus	3		1									
S. thermophilus	1		1									
S. thermophilus	2		3						1			
S. thermophilus	1		1									1
S. thermophilus	1		1									1
S. thermophilus	1		1									1
S. thermophilus	1		1									1
S. thermophilus	1		1									1
S. thermophilus	2		1									
S. thermophilus	2		1									
S. thermophilus	1		1									1
S. thermophilus			3	2								
S. thermophilus			3	2								
S. thermophilus			3	2								

Glc, glucose; GlcA, glucuronic acid; Gal, galactose; Rha, rhamnose; Fru, fructose; Ara, arabinose; Fuc, fucose; Man, mannose; Rib, ribose; Xyl, xylose; NAc Glu, *N*-acetyl glucosamine; NAc Gal, *N*-acetyl galactosamine.

30 000–50 000 mg/l xanthan gum [67]. In spite of this, amounts of PS produced by LAB will be sufficient to exploit for in situ applications. LABs are generally recognized as safe (GRAS) microorganisms, and if the LAB could produce PS in edible and safe culture media such as whey, there would be no need to purify PS.

PS yield produced by LAB depends on the composition of the medium (carbon and nitrogen sources) and the growth conditions, i.e. temperature, pH, and incubation time. Under conditions of higher temperatures and slower growth, the production of the polymer per cell in *L. delbrueckii* subsp. *bulgaricus* NCFB 2772 was greater in milk [68]. The optimum culture condition for PS production was investigated with *L. delbrueckii* subsp. *bulgaricus* RR in semi-defined medium [69]; the optimum temperature and pH conditions for PS production were determined to be 36–39 °C and pH 4.5–5.5. The optimal temperature for PS production was approximately 40 °C for thermophilic LAB strains and around 25 °C for mesophilic LAB strains. Gamar et al. [70] reported increased slime production at lower incubation temperatures and an increase in the final PS concentration in *L. rhamnosus* following incubation at 25 °C instead of 30 °C. The effects of temperature on PS production in whey were investigated in *L. plantarum* [60], and the yield was found to be higher at 25 °C than at either 30 or 37 °C. Moreover, the lower the temperature, the higher the PS production per cell by *Lactobacillus sake* [71]. However, the growth rate in the exponential phase decreased at low temperatures. Therefore, temperature for the maximal production of PS is based on a balance of cell density and PS production per cell.

Maximal PS production by *L. sake* was obtained under anaerobic conditions at 20 °C, although PS production per cell was higher at 10 °C. Generally, it is well accepted that stress conditions trigger PS production as a protective mechanism.

The effects of nitrogen and carbon sources on PS production have also been investigated. According to early reports, neither LAB growth nor PS production was specifically linked to the presence of casein or whey proteins in the growth medium. Garcia-Garibay and Marshall [68] reported that PS production by *L. delbrueckii* subsp. *bulgaricus* NCFB 2772 increased during the early growth phase in the presence of hydrolyzed casein in milk, whereas the addition of hydrolyzed casein to MRS medium did not increase PS production. This strain produced 25 mg/l PS when cultured on fructose in a defined medium and 80 mg/l PS when cultured on glucose [72]. There was a distinct relationship between the Bacto-casitone concentration and PS production by *L. delbrueckii* subsp. *bulgaricus* RR: the higher the casitone concentration, the higher the PS yield was obtained [69]. For *L. plantarum* grown in whey, yeast extract was a more effective nitrogen source for PS production than soy peptide, tryptone, peptone, and Lab-Lemco powder; glucose was a more effective carbon source than galactose, sucrose, maltose, fructose, and raffinose [60]. PS production by *L. casei* CG11 was investigated in basal minimum medium containing galactose, glucose, lactose, sucrose, maltose, and melibiose. Glucose was found to be the most efficient carbon source, whereas lactose and galactose were the least efficient ones [73]. PS production by *L. rhamnosus* C83 was investigated in a chemically defined medium containing different carbon sources (glucose, fructose, mannose, and maltose) at different concentrations. Mannose at 40 g/l was by far the most efficient carbon source. The PS yield by *Lactobacillus buchneri* GM3701 was higher at 25 °C and 30 °C than at 37 °C, and no differences were observed among the four tested sugars including glucose, fructose, sucrose, and lactose at 25 °C [54]. Furthermore, increase in Mg^{2+}, Mn^{2+}, and Fe^{2+} concentrations stimulated PS production in synthetic media [70]. In addition, Macedo et al. [65] reported the importance of salts in culture media and the strong positive effect of salts and amino acids on *L. rhamnosus* RW-9595M growth and PS production. The addition of salts and amino acids largely increased PS production (to 2775 mg/l) in whey permeate supplemented with yeast extract, although the addition of amino acids alone had no effect on PS production.

4.3.4 Instability of Hetero-PS Production

PS-producible LAB possess PS biosynthesis gene *organized in operons. L. plantarum* was reported to possess most of the gene clusters in chromosome and some of the gene in plasmids for synthesis of PS [46]. The instability of hetero-PS production has been reviewed by de Vuyst et al. [13]. Briefly, a loss in the ability to produce PS may be caused by repeated subculture of bacterial strains or incubation at high temperatures. Generally, the loss of plasmids is the reason for loss of PS production. On the other hand, the gene clusters mediating PS synthesis in *S. thermophilus* are located on the chromosome [74]. Some of the gene clusters may contain mobile genetic elements such as insertion sequences and transposase genes. Although these transposable elements contribute to the diversity of PS gene clusters, they may induce the disappearance of PS productivity from PS-producing *S. thermophilus* [74]. Occasionally, this loss of PS productivity is recovered by repeated subculture in milk [13].

4.4 Prebiotic Activity

PS produced by LAB have been used as food additives as texturizer. Furthermore, PS has been shown to have various health benefits for humans and animals. These are mainly immunomodulatory properties, antiviral activity, antioxidant activity, and antihypertensive activity [17, 66, 75, 76]. These properties have been extensively reviewed [8, 13, 67, 77, 78]. Besides these properties, there are other reasons to expect that PS can be used in prebiotics that are discussed below.

The term "probiotic" was first proposed by Fuller [79], and its definition was further refined to "Live microorganisms which when consumed in adequate amounts as part of food confers a health benefit on the host" [80]. Probiotic LAB thus represent a class of live food ingredients that exert a beneficial effect on the health of the host. Beneficial microorganisms in the intestine are enhanced by "prebiotics," which are defined as "non-digestible food ingredients that beneficially affect the host by selectively stimulating the growth and activity of one or a limited number of bacterial species already resident in the colon, and thus improving host health" [81]. Most of the current prebiotics are low molecular weight except for inulin. As long as carbohydrate chains are metabolized more slowly than the short ones, PSs thus exert prebiotic effects in more distal colonic regions compared with oligosaccharides, which are more rapidly digested in the proximal colon [82]. Therefore, PS produced by LAB can be used as prebiotics.

Prebiotics are usually nondigestible oligosaccharides that selectively stimulate the growth and activity of a limited number of bacterial species in the colon, such as Bifidobacteria and lactobacilli, and therefore improve host health. Detrimental bacteria are defined as bacteria that form substances that are detrimental to humans, such as ammonia, hydrogen sulfide, indoles, and amines. On the other hand, beneficial bacteria such as Bifidobacteria and lactobacilli inhibit the proliferation of detrimental bacteria, and their cell components stimulate the host immune system [83]. Gastrointestinal microflora consists of approximately 10^{14} cfu/g of various types of detrimental and beneficial bacteria; the number and composition varies greatly along the gastrointestinal tract. The balance of the gastrointestinal microflora influences different aspects of host health such as bowel movement, flatulence, and the absorption of nutrients. Many factors may upset this balance including stress, consumption of antibiotics, infection, food poisoning, and natural aging process. To maintain this balance, growth and activities of the beneficial bacteria may be enhanced by specific ingredients in foods.

4.4.1 Commercial Prebiotic Oligosaccharides

Various oligosaccharides have been identified as prebiotics that can increase the density of *Bifidobacterium* in the host colon. Galacto-oligosaccharides (GOS) and fructo-oligosaccharides (FOS) are considered important prebiotics. GOS and FOS are well-known type of prebiotic oligosaccharides found in human milk [84]. Many studies reported that GOS or FOS supplementation increased bifidobacterial counts in feces of infants [85]. Other carbohydrates including gentio-oligosaccharides, isomalto-oligosaccharides, lactulose, nigero-oligosaccharides, and xylo-oligosaccharides are also considered prebiotics. Detailed research of the utilization of these oligosaccharides was done with intestinal bacteria including *Bifidobacterium, Clostridium, Ruminococcus, Eubacterium, Faecalibacterium, Bacteroides,* and *Collinsella* [86]. This research showed that intestinal bacteria can easily degrade the linkages of Fru–Glc, Fru–Fru, and Glc–Glc, but they have difficulty in degrading Gal–Gal linkage and xylo-oligosaccharides, though *Bifidobacterium* can utilize all oligosaccharides. These results suggest that GOS, which has Gal–Gal linkage, and xylo-oligosaccharides are selectively utilized by *Bifidobacterium* in the intestine and they fall under the category of prebiotics.

The number of fecal Bifidobacteria increased significantly after subjects consumed 2.5 g of GOS per day for three weeks [87]. GOS utilization by enterobacteria was further investigated *in vitro*. The trisaccharide forms of GOS were utilized by *Bifidobacterium, Lactobacillus acidophilus, L. reuteri, Bacteroides, Clostridium perfringens, Klebsiella pneumoniae,* and *Enterococcus faecium*; the tetrasaccharide forms were utilized by *Bifidobacterium adolescentis, Bifidobacterium breve, Bifidobacterium infantis,* and *E. faecium*. These results suggest that a higher degree of polymerization (DP) of GOS enhanced selectivity. Similarly, *Bifidobacterium lactis* DR10 utilizes trisaccharide and tetrasaccharide forms of GOS, whereas *L. rhamnosus* DR20 prefers disaccharides and monosaccharides [88]. Barboza et al. [89] reported that *B. breve* and *Bifidobacterium longum* subsp. *infantis* can utilize GOS with a DP ranging from 3 to 8. Furthermore, *B. longum* subsp. *infantis* preferentially utilizes GOS with a DP of 4, and *B. adolescentis* utilizes GOS with DP of 3. In addition, the structure of GOS influences its utilization by lactobacilli and Bifidobacteria [90]. Trisaccharides of 4′-GOS (β-1,4 linkage) and 6′-GOS (β-1,6 linkage) can be used as the sole carbon source. Almost all lactobacilli and Bifidobacteria tested preferred to utilize 4′-GOS, whereas *L. acidophilus, L. reuteri,* and *L. casei* could utilize both 4′- and 6′-GOS. GOS are not only used to stimulate beneficial bacteria but can also be utilized by *Bacteroides* and Clostridia [91]. GOS selectivity may be enhanced by altering the structure and increasing the DP.

Increased numbers of Bifidobacteria and/or lactobacilli in the colon have been shown to have beneficial effects, although the specific mixture of populations of these genera necessary to provide health-promoting effects has not yet been determined. This is because the beneficial effects are likely to be owed to balancing a colonic microflora, and the microflora in colon is different in each person. To function most effectively, prebiotics must be resistant to digestive processes in the stomach and small bowel so that they can come into contact with the bacteria growing in the large intestine.

4.4.2 Prebiotic Polysaccharides

Dietary fibers are nondigestible parts of plants, fruits, and sea grasses. The well-known health benefit of dietary fibers is to prevent or relieve constipation.

Besides, it could be an energy source for intestinal microflora, and dietary fibers are thought to be essential for maintaining a balance of colonic microflora. A dietary fiber comprises mainly PSs, and much has been published on the utilization of plant PSs by *Bifidobacterium*. Gum guar and gum locust bean, which are made of galactomannan, were utilized by *Bifidobacterium dentium*, and arabinogalactan and gum Arabic, ghatti, and tragacanth were utilized by *B. longum* [92]. *B. adolescentis* fermented arabinoxylan-oligosaccharides, isomalto-oligosaccharides, galacto-oligosaccharides, and arabinogalactan, and glycosyl hydrolase systems to utilize each oligosaccharide were found [93]. Yogurt containing PS produced by *Lactobacillus fermentum* Lf2 and *B. animalis* subsp. *lactis* showed bifidogenic effect as increasing viable cell counts of Bifidobacteria in faces of mice [94].

4.4.3 Prebiotics in Japanese FOSHU

The food for specified health use (FOSHU) system was introduced in Japan in 1991. FOSHU refers to foods containing ingredients that provide health benefits and have officially approved physiological effects on the human body. To be defined as FOSHU, it is important to assess the safety of the food as well as the effectiveness of health promotion, and this assessment must be approved by the Ministry of Health, Labor, and Welfare in Japan. In 2018, 1062 foods have been recognized as FOSHU, and of these, there are 99 available foods that provide regulation of bowel movements and gastrointestinal health benefits due to oligosaccharides and PSs. The functional ingredients are indigestible dextrin in 49 foods, fiber of psyllium seed gum in 14, GOS in 12, lactulose in 8, FOS in 4, low molecular sodium alginate in 3, isomalto-oligosaccharides in 3, fiber of wheat bran in 2, fiber of agar in 2, and polydextrose in 2 foods.

4.4.4 Prebiotics Produced by LAB

Oligosaccharides produced by beneficial bacteria or their enzymes may enhance the growth of beneficial bacteria. Oligosaccharides synthesized by the enzymes from *Bifidobacterium* strains were favored by the producer strains [95]. A novel GOS mixture produced using *B. longum* NCIBM 41171 galactosidases increased the proportion of Bifidobacteria in feces relative to commercial GOS [96]. These studies suggest that the oligosaccharides produced by beneficial bacteria are selectively utilized by the producer strain and are closely related species. In addition, glycosyltransferases may possess both hydrolytic and transglycosylation activities [97], and both glycosidases and glycosyltransferases may coexist in the same strain.

PSs are not usually utilized by microorganisms. However, for PS production by LAB, reduced yields were frequently observed after the maximal level had been reached, which might be caused by the enzymes produced by the bacteria [98]. In fact, *B. bifidum* DSM20456 can utilize the PS produced by *Pediococcus pentosaceus*, *L. plantarum*, *Weissella cibaria*, and *Weissella confusa*, and some growth is observed in case of *B. longum*, *B. adolescentis*, and *L. acidophilus* [99]. Tsuda and Miyamoto [60] investigated the prebiotic activity of PS produced by *L. plantarum* 301102S, a PS-producible mutant strain derived from non-PS-producible *L. plantarum* 301102. The *in vitro* prebiotic activities of GOS, inulin, and the PS were measured in 37 LAB strains, and the activity scores of the PS in strains 301102 and 301102S were highest. The PS produced by the mutant strain enhanced the growth of strain 301102S and the parental 301102 strain. This suggests that the parental strain has enzymes that can degrade the PS. Furthermore, the prebiotic activities of PSs produced by *L. buchneri* GM3701 and *L. plantarum* RB-3 were measured in *Lactobacillus*, *Staphylococcus*, and *Escherichia* strains, respectively [54]. PS from both strains were found to enhance the growth of *Lactobacillus farciminis* HM2001. This result suggests that the growth of some LAB can be enhanced by the supplementation with a PS produced by *Lactobacillus* strains.

PS produced by LAB may have potential as prebiotics. Studies on the production of PS by enzymes in beneficial microorganisms may lead to the production of highly selective prebiotics.

4.5 Conclusion

Some LAB can produce PS. The PS is divided into two classes: homo- and hetero-PS. Homo-PS contains one sugar type. Hetero-PS contains various sugars, and it is mainly composed of D-glucose, D-galactose, and L-rhamnose. A main function of PS is a protection of the cell. PS protects the bacterial cell from various stresses such as desiccation, phage attack, bile salts, and low pH, and otherwise it is exploited by food industries. PS plays a major role as natural texturizer in dairy products. Recently a health-promoting effect of PS has attracted a lot of attention. These are mainly immunomodulatory properties, antioxidant activity, and prebiotic activity. Although there are many studies about health-promoting effects, biosynthesis of PS is not fully understood on genetic aspects. In future, studies about the functionality of the gene related with PS production may lead us to regulate the yield and molecular mass and beneficial properties of PS.

References

1 de Vos, P., Garrity, G.M., Jones, D. et al. (2009). Sporolactobacillus. In: *Bergey's Manual of Systematic Bacteriology*, 2e, vol. 3 (eds. F. Yanagida and K. Suzuki), 386. New York: Springer.

2 Chapot-Chartier, M.P. and Kulakauskas, S. (2014). Cell wall structure and function in lactic acid bacteria. *Microbial Cell Factories* 13: S9.

3 Caggianiello, G., Kleerebezem, M., and Spano, G. (2016). Exopolysaccharides produced by lactic acid bacteria: from health-promoting benefits to stress tolerance mechanisms. *Applied Microbiology and Biotechnology* 100: 3877–3886.

4 Dubois, M., Gilles, K.A., Hamilton, J.K. et al. (1956). Colorimetric method for determination of sugars and related substances. *Analytical Chemistry* 28: 350–356.

5 Ruas-Madiedo, P., Gueimonde, M., Arigoni, F. et al. (2009). Bile affects the synthesis of exopolysaccharides by *Bifidobacterium animalis*. *Applied and Environmental Microbiology* 75: 1204–1207.

6 Alp, G. and Aslim, B. (2010). Relationship between the resistance to bile salts and low pH with exopolysaccharide (EPS) production of *Bifidobacterium* spp. isolated from infants feces and breast milk. *Anaerobe* 16: 101–105.

7 Ryan, P.M., Ross, R.P., Fitzgerald, G.F. et al. (2015). Sugar-coated: exopolysaccharide producing lactic acid bacteria for food and human health applications. *Food & Function* 6: 679–693.

8 Ruas-Madiedo, P., Hugenholtz, J., and Zoon, P. (2002). An overview of the functionality of exopolysaccharides produced by lactic acid bacteria. *International Dairy Journal* 12: 163–171.

9 Dertli, E., Mayer, M.J., and Narbad, A. (2015). Impact of the exopolysaccharide layer on biofilms, adhesion and resistance to stress in *Lactobacillus johnsonii* FI9785. *BMC Microbiology* 15: 8–16.

10 Ganzle, M., Michael, G., and Schwab, C. (2005). Exopolysaccharide production by intestinal lactobacilli. In: *Probiotics & Prebiotics: Scientific Aspects* (ed. G.W. Tannock), 83–96. Norfolk: Caister Academic Press.

11 Kralj, S., van Geel-Schutten, G.H., Dondorff, M.M.G. et al. (2004). Glucan synthesis in the genus *Lactobacillus*: isolation and characterization of glucansucrase genes, enzymes and glucan products from six different strains. *Microbiology* 150: 3681–3690.

12 van Hijum, S.A.F.T., Szalowska, E., van der Maarel, M.J.E.C., and Dijkhuizen, L. (2004). Biochemical and molecular characterization of a levansucrase from *Lactobacillus reuteri*. *Microbiology* 150: 621–630.

13 de Vuyst, L., de Vin, F., Vaningelgem, F., and Degeest, B. (2001). Recent developments in the biosynthesis and applications of heteropolysaccharides from lactic acid bacteria. *International Dairy Journal* 11: 687–707.

14 Mayo, B., Aleksandrzak-Piekarczyk, T., Fernandez, M. et al. (2010). Updates in the metabolism of lactic acid bacteria. In: *Biotechnology of Lactic Acid Bacteria* (eds. F. Mozzi, R.R. Raya and G.M. Vignolo), 3–33. Iowa: Blackwell Publishing.

15 Mende, S., Rohn, H., and Jaros, D. (2016). Influence of exopolysaccharides on the structure, texture, stability and sensory properties of yoghurt and related products. *International Dairy Journal* 52: 57–71.

16 Low, D., Ahlgren, J.A., Horne, D. et al. (1998). Role of *Streptococcus thermophilus* MR-1C capsular exopolysaccharide in cheese moisture retention. *Applied and Environmental Microbiology* 64: 2147–2151.

17 Chabot, S., Yu, H.L., de Leseleuc, L. et al. (2001). Exopolysaccharides from *Lactobacillus rhamnosus* RW-9595M stimulate TNF, IL-6 and IL-12 in human and mouse cultured immunocompetent cells, and IFN-γ in mouse splenocytes. *Le Lait* 81: 683–698.

18 Kitazawa, H., Harata, T., Uemura, J. et al. (1998). Phosphate group requirement for mitogenic activation of lymphocytes by an extracellular phosphopolysaccharide from *Lactobacillus delbrueckii* spp. *bulgaricus*. *International Journal of Food Microbiology* 40: 169–175.

19 Sreekumar, O. and Hosono, A. (1998). The antimutagenic properties of a polysaccharide produced by *Bifidobacterium longum* and its cultured milk against some heterocyclic amines. *Canadian Journal of Microbiology* 44: 1029–1036.
20 Tsuda, H., Hara, K., and Miyamoto, T. (2008). Binding of mutagens to exopolysaccharide produced by *Lactobacillus plantarum* mutant strain 301102S. *Journal of Dairy Science* 91: 2960–2966.
21 Kanmani, P., Albarracin, L., Kobayashi, H. et al. (2018). Exopolysaccharides from *Lactobacillus delbrueckii* OLL1073R-1 modulate innate antiviral immune response in porcine intestinal epithelial cells. *Molecular Immunology* 93: 253–265.
22 El-Deeb, N.M., Yassin, A.M., Al-Madboly, L.A., and El-hawiet, A. (2018). A novel purified *Lactobacillus acidophilus* 20079 exopolysaccharide, LA-EPS-20079, molecularly regulates both apoptotic and NF-κB inflammatory pathways in human colon cancer. *Microbial Cell Factories* 17: 29–43.
23 Laino, J., Villena, J., Kanmani, P., and Kitazawa, H. (2016). Immunoregulatory effects triggered by lactic acid bacteria exopolysaccharides: new insights into molecular interactions with host cells. *Microorganisms* 4: 27–42.
24 Liu, C.F., Tseng, K.C., Chiang, S.S. et al. (2011). Immunomodulatory and antioxidant potential of *Lactobacillus* exopolysaccharides. *Journal of the Science of Food and Agriculture* 91: 2284–2291.
25 Uemura, J., Itoh, T., Kasneko, T., and Noda, K. (1998). Chemical characterization of extracellular polysaccharide from *Lactobacillus delbrueckii* subsp. *bulgaricus* OLL1073R-1. *Milchwissenschaft* 53: 443–446.
26 Nishimura-Uemura, J., Kitazawa, H., Kawai, Y. et al. (2003). Functional alteration of murine macrophages stimulated with extracellular polysaccharides from *Lactobacillus delbrueckii* ssp. *bulgaricus* OLL1073R-1. *Food Microbiology* 20: 267–273.
27 Nguyen, H.K.T., Nghe, V.D., Le, V.C., and Doan, T.T.V. (2018). Detection of the potential inactivation of tetrodotoxin by lactic acid bacterial exopolysaccharide. *Toxins* 10: 288–298.
28 Tzianabos, A., Kasper, D.L., Cisneros, R.L. et al. (1995). Polysaccharide-mediated protection against abscess formation in experimental intra-abdominal sepsis. *Journal of Clinical Investigation* 96: 2727–2731.
29 Wang, J., Zhao, X., Yang, Y. et al. (2015). Characterization and bioactivities of an exopolysaccharide produced by *Lactobacillus plantarum* YW32. *International Journal of Biological Macromolecules* 74: 119–126.
30 Liu, Z., Zhang, Z., Qiu, L. et al. (2017). Characterization and bioactivities of the exopolysaccharide from a probiotic strain of *Lactobacillus plantarum* WLPL04. *Journal of Dairy Science* 100: 6895–6905.
31 Li, W., Ji, J., Rui, X. et al. (2014). Production of exopolysaccharides by *Lactobacillus helveticus* MB2-1 and its functional characteristics in vitro. *LWT Food Science and Technology* 59: 732–739.
32 Rajoka, M.S.R., Jin, M., Haobin, Z. et al. (2018). Functional characterization and biotechnological potential of exopolysaccharide produced by *Lactobacillus rhamnosus* strains isolated from human breast milk. *LWT Food Science and Technology* 89: 638–647.

33 Rani, R.P., Anandharaj, M., and Ravindran, A.D. (2018). Characterization of a novel exopolysaccharide produced by *Lactobacillus gasseri* FR4 and demonstration of its in vitro biological properties. *International Journal of Biological Macromolecules* 109: 772–783.

34 Hollo, J. and Laszlo, E. (1970). The synthesis of dextran, I. *Periodica Polytechnica, Chemical Engineering* 15: 35–42.

35 Cerning, J. (1990). Exocellular polysaccharides produced by lactic acid bacteria. *FEMS Microbiology Letters* 87: 113–130.

36 Dellaglio, F., Dicks, L.M.T., and Torriani, S. (1995). The genus *Leuconostoc*. In: *The Genera of Lactic Acid Bacteria* (eds. B.J.B. Wood and W.H. Holzapfel), 235–278. Glasgow: Blackie Academic and Professional.

37 Minervini, F., Angelis, M.D., Surico, R.F. et al. (2010). Highly efficient synthesis of exopolysaccharides by *Lactobacillus curvatus* DPPMA10 during growth in hydrolyzed wheat flour agar. *International Journal of Food Microbiology* 141: 130–135.

38 European Commission (2000). Opinion of the scientific committee on food on a dextran preparation, production, produced using *Leuconostoc mesenteroides, Saccharomyces cerevisiae* and *Lactobacillus* spp., as a novel food ingredient in bakery products. European Commission Health & Consumer Protection Directorate-General.

39 Cote, G.L. and Robyt, F.J. (1982). Isolation and partial characterization of an extracellular glucansucrase from *Leuconostoc mesenteroides* NRRl B-1355 that synthesizes an alternating (1→6), (1→3)-α-D-glucan. *Carbohydrate Research* 101: 57–74.

40 Uchida, K. (1996). Nyuusannkinn no kouzou to kinntaiseibunn. In: *Nyuusannkinn kennkyuu syuudannkai* (ed. Nyuusannkinn no kagaku to gijutsu), 59–88. Tokyo: Gakkai Syuppann Center. (In Japanese).

41 van Geel-Schutten, G.H., Faber, E.J., Smit, E. et al. (1999). Biochemical and structural characterization of the glucan and fructan exopolysaccharides synthesized by the *Lactobacillus reuteri* wild-type strain and by mutant strains. *Applied and Environmental Microbiology* 65: 3008–3014.

42 Korakli, M., Ganzle, M.G., and Vogel, R.F. (2002). Metabolism by bifidobacteria and lactic acid bacteria of polysaccharides from wheat and rye, and exopolysaccharides produced by *Lactobacillus sanfranciscensis*. *Journal of Applied Microbiology* 92: 958–965.

43 Pederson, C.S. and Albury, M. (1955). Variation among the heterofermentative lactic acid bacteria. *Journal of Bacteriology* 70: 702–708.

44 Quirasco, M., Lopez-Munguia, A., Remaud-Simeon, M. et al. (1999). Induction and transcription studies of the dextransucrase gene in *Leuconostoc mesenteroides* NRRL B-512F. *Applied and Environmental Microbiology* 65: 5504–5509.

45 Jin, Q., Li, L., Kim, Y.J., and Han, N.S. (2014). Construction of a dextran-free *Leuconostoc citreum* mutant by targeted disruption of the dextransucrase gene. *Journal of Applied Microbiology* 117: 1104–1112.

46 Jiang, Y. and Yang, Z. (2018). A functional and genetic overview of exopolysaccharides produced by *Lactobacillus plantarum*. *Journal of Functional Foods* 47: 229–240.

47 de Vos, P., Garrity, G.M., Jones, D. et al. (2009). Lactobacillus. In: *Bergey's Manual of Systematic Bacteriology*, 2e, vol. 3 (eds. W.P. Hammes and C. Hertel), 501. New York: Springer.

48 Birch, J., van Calsteren, M.R., Perez, S., and Svensson, B. (2019). The exopolysaccharide properties and structures database: EPS-DB. Application to bacterial exopolysaccharides. *Carbohydrate Polymers* 205: 565–570.

49 Jeong, D., Kim, D.H., Kang, I.-B. et al. (2017). Characterization and antibacterial activity of a novel exopolysaccharide produced by *Lactobacillus kefiranofaciens* DN1 isolated from kefir. *Food Control* 78: 436–442.

50 Balzaretti, S., Taverniti, V., Guglielmetti, S. et al. (2017). A novel rhamnose-rich hetero-exopolysaccharide isolated from *Lactobacillus paracasei* DG activates THP-1 human monocytic cells. *Applied and Environmental Microbiology* 83: 2702–2716.

51 Zhou, K., Zeng, Y., Yang, M. et al. (2016). Production, purification and structural study of an exopolysaccharide from *Lactobacillus plantarum* BC-25. *Carbohydrate Polymers* 144: 205–214.

52 Nambiar, R.B., Sellamuthu, P.S., Perumal, A.B. et al. (2018). Characterization of an exopolysaccharide produced by *Lactobacillus plantarum* HM47 isolated from human breast milk. *Process Biochemistry* 73: 15–22.

53 Wang, J., Zhao, X., Tian, Z. et al. (2015). Characterization of an exopolysaccharide produced by *Lactobacillus plantarum* YW11 isolated from Tibet kefir. *Carbohydrate Polymers* 125: 16–25.

54 Tsuda, H., Okuda, S., Haraguchi, T., and Kodama, K. (2019). Influence of exopolysaccharide on the growth of lactic acid bacteria. *Italian Journal of Food Science* 31: 233–242.

55 Cerning, J. (1995). Production of exopolysaccharides by lactic acid bacteria and dairy propionibacteria. *Le Lait* 75: 463–472.

56 Khanal, S.N. and Lucey, J.A. (2017). Evaluation of the yield, molar mass of exopolysaccharides, and rheological properties of gels formed during fermentation of milk by *Streptococcus thermophilus* strains St-143 and ST-10255y. *Journal of Dairy Science* 100: 6906–6917.

57 Makino, S., Sato, A., Goto, A. et al. (2016). Enhanced natural killer cell activation by exopolysaccharides derived from yogurt fermented with *Lactobacillus delbrueckii* ssp. *bulgaricus* OLL1073R-1. *Journal of Dairy Science* 99: 915–923.

58 Ali, K., Mehmood, M.H., Iqbal, M.A. et al. (2019). Isolation and characterization of exopolysaccharide-producing strains of *Lactobacillus bulgaricus* from curd. *Food Science & Nutrition* 7: 1207–1213.

59 Tallon, R., Bressollier, P., and Urdaci, M.C. (2003). Isolation and characterization of two exopolysaccharides produced by *Lactobacillus plantarum* EP56. *Research in Microbiology* 154: 705–712.

60 Tsuda, H. and Miyamoto, T. (2010). Production of exopolysaccharide by *Lactobacillus plantarum* and the prebiotic activity of the exopolysaccharide. *Food Science and Technology Research* 16: 87–92.

61 London, L.E.E., Price, N.P.J., Ryan, P. et al. (2014). Characterization of a bovine isolate *Lactobacillus mucosae* DPC 6426 which produces an exopolysaccharide composed predominantly of mannose residues. *Journal of Applied Microbiology* 117: 509–517.

62 Tidona, F., Zago, M., Corredig, M. et al. (2016). Selection of *Streptococcus thermophilus* strains able to produce exopolysaccharides in milk. *International Journal of Dairy Technology* 69: 569–575.

63 Puertas, A.I., Ibarburu, I., Elizaquivel, P. et al. (2018). Disclosing diversity of exopolysaccharide-producing lactobacilli from Spanish natural ciders. *LWT Food Science and Technology* 90: 469–474.

64 Sasikumar, K., Vaikkath, D.K., Devendra, L., and Nampoothiri, K.M. (2017). An exopolysaccharide (EPS) from a *Lactobacillus plantarum* BR2 with potential benefits for making functional foods. *Bioresource Technology* 241: 1152–1156.

65 Macedo, M.G., Lacroix, C., Gardner, N.J., and Champagne, C.P. (2002). Effect of medium supplementation on exopolysaccharide production by *Lactobacillus rhamnosus* RW-9595M in whey permeate. *International Dairy Journal* 12: 419–426.

66 Maeda, H., Zhu, X., Suzuki, S. et al. (2004). Structural characterization and biological activities of an exopolysaccharide kefiran produced by *Lactobacillus kefiranofaciens* WT-2B^{T}. *Journal of Agricultural and Food Chemistry* 52: 5533–5538.

67 de Vuyst, L. and Degeest, B. (1999). Heteropolysaccharides from lactic acid bacteria. *FEMS Microbiology Reviews* 23: 153–177.

68 Garcia-Garibay, M. and Marshall, V.M.E. (1991). Polymer production by *Lactobacillus delbrueckii* ssp. *bulgaricus*. *Journal of Applied Bacteriology* 70: 325–328.

69 Kimmel, S.A., Roberts, R.F., and Ziegler, G.R. (1998). Optimization of exopolysaccharide production by *Lactobacillus delbrueckii* subsp. *bulgaricus* RR grown in a semidefined medium. *Applied and Environmental Microbiology* 64: 659–664.

70 Gamar, L., Blondeau, K., and Simonet, J.M. (1997). Physiological approach to extracellular polysaccharide production by *Lactobacillus rhamnosus* strain C83. *Journal of Applied Microbiology* 83: 281–287.

71 van den Berg, D.J.C., Robijn, G.W., Janssen, A.C. et al. (1995). Production of a novel extracellular polysaccharide by *Lactobacillus sake* 0-1 and characterization of the polysaccharide. *Applied and Environmental Microbiology* 61: 2840–2844.

72 Grobben, G.J., Chin-Joe, I., Kitzen, V.A. et al. (1998). Enhancement of exopolysaccharide production by *Lactobacillus delbrueckii* subsp. *bulgaricus* NCFB 2772 with a simplified defined medium. *Applied and Environmental Microbiology* 64: 1333–1337.

73 Cerning, J., Renard, C.M.G.C., Thibault, J.F. et al. (1994). Carbon source requirements for exopolysaccharide production by *Lactobacillus casei* CG11 and partial structure analysis of the polymer. *Applied and Environmental Microbiology* 60: 3914–3919.

74 Cui, Y., Jiang, X., Hao, M. et al. (2017). New advances in exopolysaccharides production of *Streptococcus thermophilus*. *Archives of Microbiology* 199: 799–809.

75 Nagai, T., Makino, S., Ikegami, S. et al. (2012). Effects of oral administration of yogurt fermented with *Lactobacillus delbrueckii* ssp. *bulgaricus*

OLL1073R-1 and its exopolysaccharides against influenza virus infection in mice. *International Immunopharmacology* 11: 2246–2250.
76 Kodali, V.P. and Sen, R. (2008). Antioxidant and free radical scavenging activities of an exopolysaccharide from a probiotic bacterium. *Biotechnology Journal* 3: 245–251.
77 Kleerebezem, M., van Kranenburg, R., Tuinier, R. et al. (1999). Exopolysaccharides produced by *Lactococcus lactis*: from genetic engineering to improved rheological properties? *Antonie van Leeuwenhoek* 76: 357–365.
78 O'Connor, E.B., Barrett, E., Fitzgerald, G. et al. (2005). Production of vitamins, exopolysaccharides and bacteriocins by probiotic bacteria. In: *Probiotic Dairy Products* (ed. A.Y. Tamime), 167–194. Oxford: Blackwell Publishing Ltd.
79 Fuller, R. (1989). Probiotics in man and animals. *Journal of Applied Bacteriology* 66: 365–378.
80 FAO/WHO (2002). Guidelines for the Evaluation of Probiotics in Food. Report of a joint FAO/WHO working group on drafting guidelines for the evaluation of probiotics in food. London, ON, Canada.
81 Gibson, G.R. and Roberfroid, M.B. (1995). Dietary modulation of the human colonic microbiota: introducing the concept of prebiotics. *Journal of Nutrition* 125: 1401–1412.
82 Rastall, R. (2003). Enhancing the functionality of prebiotics and probiotics. In: *Functional Dairy Products* (eds. T. Mattila-Sandholm and M. Saarela), 301–315. Boca Raton, FL: CRC Press LLC.
83 Mitsuoka, T. (1992). Intestinal flora and aging. *Nutrition Reviews* 50: 438–446.
84 Kunz, C., Rudloff, S., Baier, W. et al. (2000). Oligosaccharides in human milk: structure, functional, and metabolic aspects. *Annual Review of Nutrition* 20: 699–722.
85 Roger, L.C. and McCartney, A.L. (2005). Prebiotics and the infant microbiota. In: *Probiotics & Prebiotics: Scientific Aspects* (ed. G.W. Tannock), 195–216. Norfolk: Caister Academic Press.
86 Yamada, T., Akiyama, T., Hatano, H., and Kimura, K. (2015). In vitro assessment of oligosaccharides assimilation by intestinal anaerobic bacteria. *Milk Science* 64: 87–98.
87 Ishikawa, F., Takayama, H., Matsumoto, K. et al. (1995). Effects of β 1-4 linked galactooligosaccharides on human fecal microflora. *Bifizusu* 9: 5–18. (in Japanese).
88 Gopal, P.K., Sullivan, P.A., and Smart, J.B. (2001). Utilisation of galacto-oligosaccharides as selective substrates for growth by lactic acid bacteria including *Bifidobacterium lactis* DR10 and *Lactobacillus rhamnosus* DR20. *International Dairy Journal* 11: 19–25.
89 Barboza, M., Sela, D.A., Pirim, C. et al. (2009). Glycoprofiling bifidobacterial consumption of galacto-oligosaccharides by mass spectrometry reveals strain-specific, preferential consumption of glycans. *Applied and Environmental Microbiology* 75: 7319–7325.
90 Cardelle-Cobas, A., Corzo, N., Olano, A. et al. (2011). Galactooligosaccharides derived from lactose and lactulose: influence of structure on *Lactobacillus*, *Streptococcus* and *Bifidobacterium* growth. *International Journal of Food Microbiology* 149: 81–87.

91 Macfarlane, G.T., Steed, H., and Macfarlane, S. (2008). Bacterial metabolism and health-related effects of galacto-oligosaccharides and other prebiotics. *Journal of Applied Microbiology* 104: 305–344.

92 Crociani, F., Alessandrini, A., Mucci, M.M., and Biavati, B. (1994). Degradation of complex carbohydrates by *Bifidobacterium* spp. *International Journal of Food Microbiology* 24: 199–210.

93 van den Broek, L.A.M., Hinz, S.W.A., Beldman, G. et al. (2005). Glycosyl hydrolases from *Bifidobacterium adolescentis* DSM20083. An overview. *Le Lait* 85: 125–133.

94 Ale, E.C., Bourin, M.J.B., Peralta, G.H. et al. (2019). Functional properties of exopolysaccharide (EPS) extract from *Lactobacillus fermentum* Lf2 and its impact when combined with *Bifidobacterium animalis* INL1 in yoghurt. *International Dairy Journal* 96: 114–125.

95 Rabiu, B.A., Jay, A.J., Gibson, G.R., and Rastall, R.A. (2001). Synthesis and fermentation properties of novel galacto-oligosaccharides by β-galactosidases from *Bifidobacterium* species. *Applied and Environmental Microbiology* 67: 2526–2530.

96 Depeint, F., Tzortzis, G., Vulevic, J. et al. (2008). Prebiotic evaluation of a novel galactooligosaccharide mixture produced by the enzymatic activity of *Bifidobacterium bifidum* NCIMB 41171, in healthy humans: a randomized, double-blind, crossover, placebo-controlled intervention study. *American Journal of Clinical Nutrition* 87: 785–791.

97 van den Broek, L.A.M., Hinz, S.W.A., Beldman, G. et al. (2008). *Bifidobacterium* carbohydrases – their role in breakdown and synthesis of (potential) prebiotics. *Molecular Nutrition & Food Research* 52: 146–163.

98 Pham, P.L., Dupont, I., Roy, D. et al. (2000). Production of exopolysaccharide by *Lactobacillus rhamnosus* R and analysis of its enzymatic degradation during prolonged fermentation. *Applied and Environmental Microbiology* 66: 2302–2310.

99 Hongpattarakere, T., Cherntong, N., Wichienchot, S. et al. (2012). In vitro prebiotic evaluation of exopolysaccharides produced by marine isolated lactic acid bacteria. *Carbohydrate Polymers* 87: 846–852.

5

Nanocellulose: A New Biopolymer for Biomedical Application

Hippolyte Durand, Megan Smyth, and Julien Bras

University of Grenoble Alpes, LGP2, 461 rue de la Papeterie, 38402 Saint Martin d'Hères, France

5.1 Trends of Biobased Polymers in Biomedical Application

Biobased polymers are produced from renewable resources in nature. Some of them can be biodegradable, but they can also be synthesized from fossil resources. Likewise, biobased polymers are not all biodegradable as in the case of bio-polyethylene (PE) [1]. The global production of biobased polymers is constantly rising and reached more than 2 million tons in 2017, which is still less than 1% of the 320 million tons of plastics produced annually [2].

However, the growing concern about the depletion of fossil resources drives the population and the research community toward more sustainable material resources and processes. In 2011, the European innovation policy called Lead Market Initiative (LMI) has identified biobased products as one of six important sectors that are supported by actions to bring new products or services onto the market [3]. Similarly, in North America, the BioPreferred® program is implemented by the US Department of Agriculture (USDA) and aims at increasing the purchase and use of biobased products. The program was extended recently by the 2014 Farm Bill [4].

The first generation of biobased polymers was mainly extracted from agricultural feedstocks such as corn and potatoes, but an important shift was made recently in order to move away from food resources. Nowadays, three principal ways are identified for the production of biobased polymers: (i) extracting natural polymers from plant and animal resources (cellulose, collagen, or chitin for instance) and applying partial modification; (ii) using biobased monomers produced by fermentation or conventional processes and carrying out classic polymerization, like for polylactic acid (PLA) or poly(butylene succinate) (PBS); or (iii) producing biobased polymers directly from bacteria as is the case for polyhydroxyalkanoates (PHAs) [5].

Tendencies of the past few decades regarding population aging and increased incidence of chronic diseases or disabilities call for the development of novel medical-grade materials [6]. The need for highly functional systems that are able to reproduce the biological conditions of living organisms is one of the

Biopolymers for Biomedical and Biotechnological Applications, First Edition.
Edited by Bernd H. A. Rehm and M. Fata Moradali.

key points in order to address the twenty-first century diseases. Nowadays, nature-derived therapeutics are of great relevance in the current biomedical sector. Most biobased materials present several complementary properties such as unique chemical structure, bioactivity, nontoxicity, and biocompatibility that make them good candidates for the use in medical applications. Some polymers from natural sources are known to mimic the extracellular matrix (ECM) and promote cell adhesion, interaction, and differentiation. Such functionalities match with the requirements of tissue engineering materials [7].

This first section aims at describing the landscape of existing biobased materials that are used in the biomedical field.

5.1.1 Introduction to Biomedical Engineering

Biomedical engineering has emerged over the last 50 years and encompasses interdisciplinary scientific fields. It aims at applying engineering principles and methods to the understanding of living tissues and the design and development of products that maintain, restore, or improve tissue functions. It includes a wide range of disciplines such as diagnostics and physiologic instrumentation, medical imaging, biological system analysis (modeling, simulation, control), materials for medical devices, artificial organs and rehabilitation, controlled drug delivery, and tissue engineering [8].

Tissue engineering and drug delivery are among the most investigated fields regarding emerging uses of biobased polymers.

Tissue engineering has a very wide scope. The International Union of Pure and Applied Chemistry (IUPAC) describes tissue engineering as "Use of a combination of cells, engineering and materials methods, and suitable biochemical and physicochemical factors to improve or replace biological functions" [9]. In other words, it aims at building up new functional living tissues by using cells that are grown in combination with a matrix or scaffolds that guide their development. The new tissue is grown *in vivo* or *in vitro* and then implanted in the human body. Tissue engineering thus also encompasses wound healing field in which implants are used. The main challenges associated with tissue engineering are (i) cell handling for *in vitro* or *in vivo* growth of the new tissue and its blood supply; (ii) materials used for scaffolding that can be permanent or biodegradable, natural, synthetic, or hybrid and tuned according to the targeted application; (iii) cell-type combination for complex structure; (iv) stem cell use; (v) integration of digital information and computer-assisted design; (vi) time dimension of the implantation (acceptation, degradation of scaffold); and (vii) social and regulatory challenges [10]. To sum up the challenges with tissue engineering, three pillars can be identified and are depicted in Figure 5.1. A relevant choice of biomaterials will lead to regulatory signals that induce cell differentiation and tissue formation (for instance growth factors) onto the scaffold.

The requirement for a material to be a good scaffold candidate for tissue engineering applications is a highly porous 3D structure that provides good cell attachment, guides the production of ECM by the cells, and ensures the mechanical strength before the new tissue has grown enough to support itself and sometimes to deliver bioactive molecules [11]. These induce proliferation

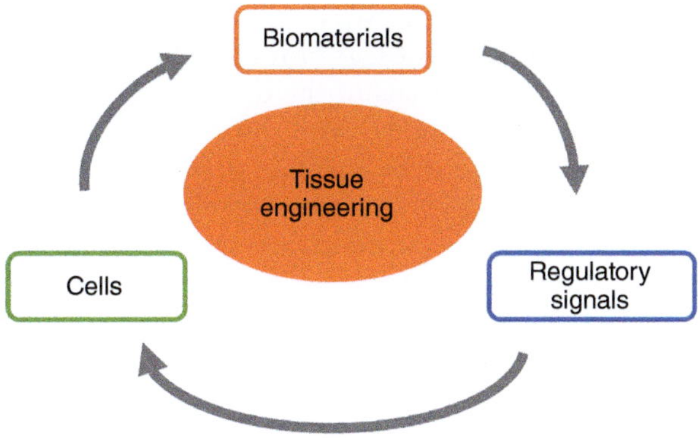

Figure 5.1 Illustration of tissue engineering principle triad composed of the biomaterial used for scaffolding, the cells, and the biological signals.

and differentiation of cells and finally neo-tissue-genesis. Key physical properties for the scaffold candidates are pore size, water content, mechanical strength, cell adhesiveness, biocompatibility, and degradability. Also, it should be noted that tissue engineering application does not require the same cell proliferation process as for natural development. An accelerated regeneration process is targeted, and naturally occurring ECM or derivatives may actually be detrimental for the success of the tissue engineering procedure. For instance, natural tissue matrices often show a lack of macro/micropore structure, which is necessary for uniform cell proliferation [12].

Drug delivery is defined as *a process of administration of a bioactive substance of pharmacological interest* by the IUPAC. The report also indicates that a drug delivery system can be a stationary implant or an active or passive moving system, with or without targeting abilities. A slight difference is also identified between controlled and sustained drug delivery. A drug delivery system is described as *sustained* if it shows slow release abilities for therapeutic effects. If the system also achieves the pharmacokinetic requirements, it can be described as *controlled* drug delivery system [9]. Typical drug concentration profiles are illustrated in Figure 5.2, which compares conventional release profiles with sustained and controlled release. Minimum and maximum therapeutic concentrations form the

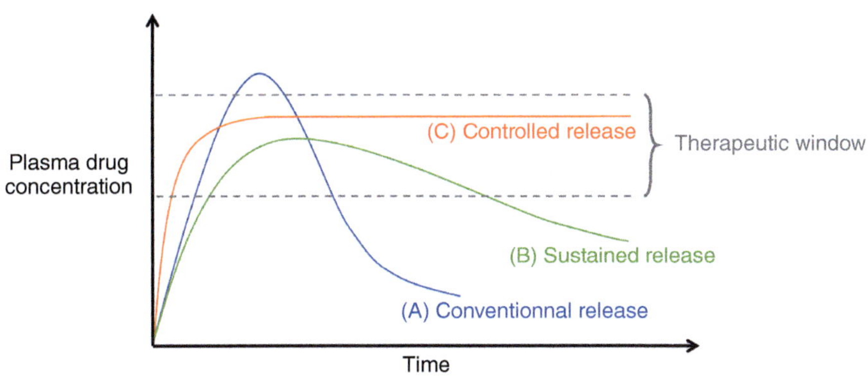

Figure 5.2 Typical plasma drug concentration profiles: conventional dosage forms (A) and sustained (B) and controlled release dosage forms (C).

therapeutic window in which the drug is supposed to have the most suitable effect. Conventional dosage forms usually show peaks out of the therapeutic window. These fluctuations of the drug level can lead to undesirable side effects. Sustained release dosage forms offer a slower drug release with a drug concentration that remains for a certain period of time inside the therapeutic window before losing its potency. Only controlled drug release profiles provide both slow and prolonged levels inside the therapeutic window. A zero-order kinetic release rate is the first key to achieve controlled drug release. The second objective is to maintain the drug concentration stable by the use of intelligent carriers that tackle the unpredictable depletion rate of the drug *in vivo* [13].

The biomedical field and associated topics are described more in detail in aforementioned books and reviews. Many materials are available for building scaffolds for tissue engineering or designing controlled drug release systems.

5.1.2 Overview of Biobased Materials for Biomedical Applications

5.1.2.1 Biomaterials: A Definition

In the context of scientific research, consistent terminology is crucial, especially for fast-growing fields of investigation. For instance, patenting process can be affected by the different uses of inaccurate words, often leading to litigation [14]. The use of words with the prefix bio- can imply different meaning. For instance, *biomaterial* is a word that can have two meanings. It can refer to a material that promotes life processes as in the field of biomedical engineering where it is in contact with living tissues, or it can refer to a material that is extracted from the living organisms or biomass (plants and animals). One could misuse *biomaterial* to describe these two topics. When it comes to building synergies in between the biomedical world and nature-extracted material experts, a clear definition is required. According to the IUPAC definition of 2012, a biomaterial is a material that is exploited in contact with living tissues, organisms, or microorganisms. The notion of exploitation includes utility for applications and fundamental research. The use of *polymeric biomaterial* is recommended when one wants to deal with polymer for medical purpose [9]. The definition of the IUPAC is chosen for this entire manuscript.

Biomaterials encompass various types of materials ranging from metals to ceramics and synthetic and biobased polymers as depicted in Figure 5.3.

Metals and metal alloy biomaterials have been used for temporary or permanent medical implant, prosthetics, and instrumentation for many decades due to their biocompatibility. Metallic biomaterials have a good *in vivo* mechanical, ionic, and electrical response, and they allow for the design of net-shape parts. Alloys such as stainless steel, cobalt–chromium, titanium alloys, and other precious metal alloy are now developed for such applications. Nowadays, the main areas where metal biomaterials are used are hard tissue applications such as bone and dental repair and replacement [16].

Ceramics are produced by the combination of groups of ions and cations to form crystals with regular repeating structure and unit cell shape and dimensions. Calcium hydroxyapatite (CaHAP), aluminum oxide Al_2O_3, zirconia ZrO_2, and porcelains (combination of SiO_2 and Al_2O_3) are examples of widely used ceramic

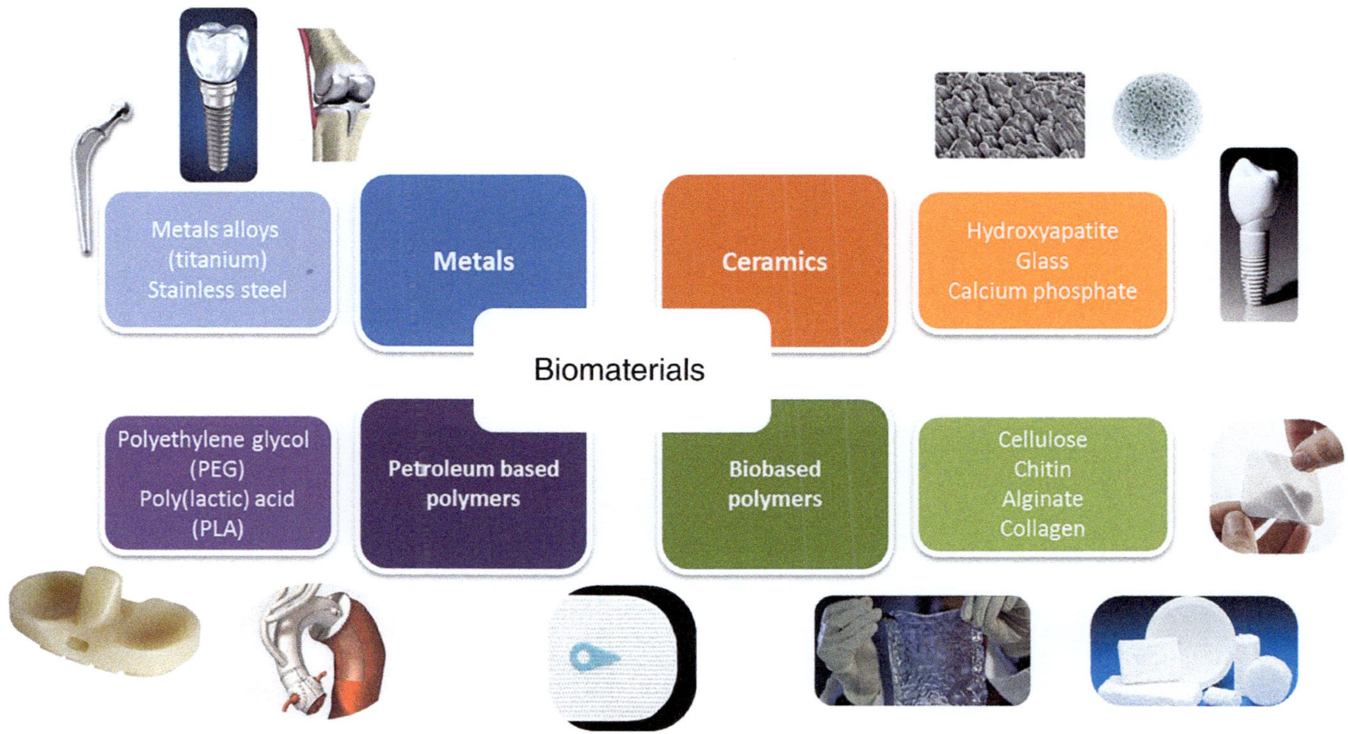

Figure 5.3 Commonly used biomaterials. *Source:* Adapted from Smyth 2017 [15].

biomaterials. They can be used as consolidated monoliths, coatings, or fillers in matrices. In the form of monoliths, they are mechanically comparable to metal that are used in hard tissue application. Coating of reactive metal structures is another field of application since they are highly biocompatible [16].

Petroleum-derived polymers can also be used as biomaterials. This category encompasses polyolefins (mainly PE and polypropylene [PP]), acrylates (polymethyl methacrylates [PMMAs], bis-glycidyl methacrylate [BisGMA]), polyamides (nylon 6,6, polyamide 6.10), polyesters like polyethylene terephthalate (PET), polycarbonates (PLA), polyglycolic acid (PGA), polycaprolactone (PCL), polyethers (poly(ethylene oxide) [PEO] and polypropylene oxide [PPO]), and silicones. Compared with metals and ceramics, the principal advantage of synthetic polymers is their high flexibility and low melt temperature that allow for easy melt processing and complex structure design. Therefore, numerous biomedical applications can be addressed with synthetic polymers that are summarized in Table 5.1. Vascular graft prosthesis design is one example, among many others, where melt processing is a key feature.

Table 5.1 Medical uses of synthetic petroleum-derived polymers.

Polymer	Applications
Polyvinyl chloride	Extracorporeal devices; hemodialysis or hemoperfusion, blood tubing, cardiac catheters, blood bag and IV infusion set, endotracheal tubes surgical tapes, sheet oxygenator, artificial heart, blood pump, artificial limb
Ultrahigh molecular weight (MW) polyethylene	Acetabulum in total hip prostheses, artificial knee prostheses
Polypropylene	Membrane oxygenator, finger joint prostheses, IV cannulae, unabsorbable sutures
Silicone rubber	Hydrocephalus shunts, catheters, membrane for oxygenator, artificial heart, heart-assisted pump, drug release system, atrioventricular shunts, ear prostheses, facial prostheses, artificial heart valve, tendon, finger joint repair, tracheal prostheses, bladder prostheses, retinal detachment, impressing materials, heart pacemaker leads
Polycarbonates	Membrane for oxygenator, hemodialyzer, plasmapheresis membrane
Polyester	Vascular graft prostheses, fixation device for tissue, hernia repair, patches for heart, bladder, arteries, suture
Polytetrafluoroethylene	Vascular graft prostheses, heart patch, retinal detachment, femoral stems
Polyurethane	Artificial heart pump material, balloon, heart valve prostheses, vascular graft prostheses, coating for blood compatibility
Polymethyl methacrylates	Bone cement, artificial teeth, denture material, bone prostheses, cranial bone replacement, intraocular lenses, membrane for dialysis

Source: Srichana and Domb 2009 [17]. Reproduced with permission of Springer Nature.

However, the drawback of synthetic polymer use is the release of by-products or plasticizers, upon degradation in physiological conditions, that can trigger an inflammatory response [16].

Nature-derived polymers (biobased polymers) have attracted increased attention since petroleum resources are finite. Moreover, many biobased polymers present inherent biocompatibility properties. Biomimetic approaches are used to harvest the technological advances made by nature over millions of years of development in order to design new innovative materials. Recent biomimetic approaches are strongly related to biobased polymers, such as cellulose [18, 19]. Consequently, biobased polymers are of particular interest in these approaches. The next part will deal with the novel use of biobased polymers in the field of biomedical engineering. The objective is to show how they can replace existing biomaterials and especially how they can allow for the development of new combinations of properties for innovative applications [1].

5.1.2.2 Biobased Polymers

Biobased polymers can be of two sorts: synthetic or naturally occurring. Synthetic biobased polymers are chemically equivalent to synthetic petroleum-derived polymers such as bioPE. Biomass is exploited, sometimes de-structured (mainly by bacterial fermentation) and reused through traditional processing routes to end up with biobased polymers. The second type of biobased polymers is directly harvested from biomass such as starch or cellulose, which are renewable and biodegradable. These polymers are naturally occurring polymers. Both synthetic biobased and naturally occurring polymers have already been used as biomaterials.

The thermoplastic polyester PLA was first synthetized in 1845 by Theophile-Jules Pelouze through the polycondensation of lactic acid (LA) into low molecular weight (800–5000 g/mol). The production process was improved over the twentieth century, especially by DuPont chemists in the 1950s, to reach 100 000 g/mol; PLA mechanical properties were suitable for competition with other commercial polymers even if its high cost narrowed its use to biomedical applications. A less expensive commercial process was invented in the late 1980s by Dr. P.R. Gruber, expanding the use of the polymer to low added-value products such as disposable bags. PLA is now the first most traded biobased polymer worldwide.

Interesting properties of PLA for biomedical applications are related to its good biocompatibility and its biodegradability into well-tolerated and safe degradation products [11]. It is also moldable, which allows for a numerous shapes design like scaffolds, sutures, rods, films, nanoparticles, and micelles. According to a recent review [20], PLA is used in many medical fields:

- *Plastic surgery*: Suture, reconstructive surgery, dermal fillers, skin grafts
- *General surgery*: Hernia mesh
- *Gynecology*: Stress incontinence mesh
- *Radiology*: Theranostic imaging
- *Oncology*: Drug delivery (intracranial, intranasal, or transdermal delivery), nanoparticles (micelles, thermoresponsive hydrogels, vaccines)

- *Orthopedic*: Peripheral nerve and spinal cord injury regeneration, bioabsorbable screws, meniscus repair, guided bone regeneration
- *Cardiac*: Chest wall reconstruction, stents
- *Dentistry*: Guided tissue regeneration, biocompatible space fillers.

PHAs are composed of biodegradable polyesters that are synthesized by a large amount of microorganisms as an intracellular storage of carbon and energy [21]. An unbalanced nutritional supply leads the microorganisms to produce extensive amounts of PHAs in the form of granules. PHAs can then be recovered by fermentation, isolation, and purification of the fermentation broth [1].

The first and simple PHA was discovered in 1926 by Maurice Lemoigne. Polyhydroxybutyrate (PHB) was found in *Bacillus megaterium* [1]. More than 150 monomers are identified for the PHA family, which gives a broad range of properties to develop new structures and products. Some drawbacks remain in these polymers such as brittleness, tendency to acquire a high degree of crystallinity, poor stiffness, slow degradation rate, and hydrophobic character. However, PHAs are also biocompatible, are biodegradable, and have piezoelectric properties that make them suitable for tissue engineering applications. They support cell adhesion, cell growth and communication, and cell organization to design the intended tissue scaffolds [7].

PBS is a family of biodegradable thermoplastic polymers that are synthetized via polycondensation of succinic acid or dimethyl succinate and 1,4-butanediol (BDO). Pioneering work on PBS synthesis is attributed to Carothers in 1931. Succinic acid can be derived from fossil sources but can also be isolated from the fermentation of renewable feedstocks or by harvesting bacterial productions. BDO is usually derived from petrochemical sources, but direct fermentation of sugar and even catalytic reduction of succinic acid are possible alternative routes [22]. PBS exists in the form of various copolymers that allow for the tuning of its physical properties [23].

PBS was originally studied for biodegradable packaging development, but recently its use for medical industry was reconsidered by researchers. Its excellent biodegradability is well known, and its biocompatibility with animal and human cells has been demonstrated. PBS is used in various applications such as bone repair, scaffolding, composites, tissue engineering, and drug delivery [24].

Starch can also be used in biomedical products. The key feature of these polymers is their biodegradability; they were used as temporary composite scaffolds for several medical applications such as orthopedic implants [25], bone cements or drug delivery carriers [26], and tissue engineering [27]. This research shows that starch is a functional biomaterial.

Alginate is a naturally occurring anionic polysaccharide obtained from brown seaweed (*Phaeophyceae*) by chemical treatment with alkali solutions, mostly NaOH, before filtration and precipitation with sodium or calcium chloride addition. The powder obtained is water soluble. Chemically speaking, alginate is a linear copolymer containing blocks of (1,4)-linked β-D-mannuronate (M) and α-L-guluronate (G) residues in various ratios.

Alginate is biocompatible, has low toxicity, is relatively low cost, and forms gels with the aid of divalent cations such as Ca^{2+}. This ability of alginate to form solid hydrogels is the reason for its interest for medical application because such hydrogels closely resemble ECMs. Research area such as wound healing, drug delivery, *in vitro* cell culture, and tissue engineering benefits from alginate hydrogels and their tunable crosslinking strategies as indicated in the extensive review by Lee and Mooney in 2012 [28]. Alginates have been used for more than 40 years for raft-forming formulations in order to treat heartburn and esophagitis: gastric acid is converted to carbon dioxide by the bicarbonate-containing alginate formulation, which turns into a foam that floats on the gastric content, like a raft on water, and acts like a neutral-pH barrier. The commercial brand Gaviscon has been on the market for several decades for the treatment of reflux symptoms in infants and children and the management of heartburn and reflux during pregnancy [29].

Epidermal and dermal wounds can also benefit from calcium-alginate dressings in the context of diabetic foot lesions or postsurgical wounds where the immunological properties of alginate itself, in combination with Ca^{2+} healing promotion, contributed to better treatment results in comparison with conventional products [30].

Calcium-alginate gels have also great potential in the immobilization of living cells. Calcium-alginate gel spheres can be designed as a vessel for cells in cell transplantation, for instance. Alginate solution is mixed with cells, and the mixture is then added dropwise in a Ca^{2+} solution in order to crosslink the surface of the drops (through egg-box system) and immobilize the sphere structure. Cells can be released from the spheres by exposure to a calcium ion scavenger like phosphate or citrate. Low G content alginate quickly releases the cells, while high G content alginate requires hours [31].

Chitin is the second most occurring polymer on Earth after cellulose. It is found in shell of crustaceans, mainly crabs, shrimps, and lobsters. The shells can be ground and treated with HCl for demineralization, and chitin can be transformed into chitosan through deacetylation with concentrated NaOH [17]. Chitosan is a linear polymer composed of $\alpha(1\rightarrow4)$ linked *N*-acetyl-D-glucosamine residues. A common molecular weight of chitosan is in the range of 50 000–1 000 000 g/mol. Chitosan is normally insoluble in aqueous solution above pH 7 but is readily soluble in dilute acid (pH < 5) where free amino groups are protonated. At pH above five, amino groups start to be deprotonated and become available for hydrogen bonding, which will establish the gel structure of chitosan at some critical pH, which depends on the degree of deacetylation and average molecular weight. The adhesive nature of chitin and chitosan and their permeability to oxygen and their antifungal and bactericidal character are very relevant properties for the treatment of wounds and burns. Hydrogels, fibers, membranes, scaffolds, and sponges based on chitin and chitosan have been successfully prepared for such biomedical applications [32].

Tissue engineering and regenerative medicine is also a wide field of application for chitin and chitosan. For instance, electrospun nanofibers based on chitosan, collagen, and hydroxyapatite can mimic the ECM and promote bone regeneration [33]. Liver tissue engineering was investigated with chitosan–gelatin 3D scaffolds crosslinked with genipin, which demonstrated suitable porosity and biocompatibility [34].

5.1.2.3 Cellulose as a Biomaterial

Nature produces an estimated quantity of 300×10^9 tons of biomass each year through photosynthesis, in which about 90% of it is wood. Cellulose roughly corresponds to 50% of wood components, i.e. an annual natural production of cellulose of 135×10^9 tons each year [35]. In comparison, 4 to 5×10^9 tons of oil was extracted in 2017 (includes crude oil, shale oil, oil sands, and natural gas liquids [NGLs] and excludes liquid fuels from other sources such as biomass and derivatives of coal and natural gas) according to British Petroleum statistical review on world energy released in June 2018, almost 30 times less than the amount of cellulose. With the current global concerns about environment and raw materials extraction, cellulose appears to be a potential alternative to oil-based materials. It is the most abundant natural polymer on Earth. It is also renewable, biodegradable, and biocompatible.

Cellulose is mainly found in the cell wall of plants but also in a wide range of sources. Wood is the most common source of cellulose, but biomass is far more than wood-derived products: annual plants such as cotton or flax, agricultural wastes, and marine biomass like algae but also animals like tunicates or even bacteria and fungi are all used as cellulose sources [36].

The establishment of cellulose chemical and physical structure took almost a century [37]. Its chemical identification is attributed to the French chemist Anselme Payen in 1838. In the beginning of the twentieth century, many scientists still doubted that it was a polymer, but the molecular structure was revealed by Haworth et al. in 1930. This work confirmed that cellulose was a linear polymer composed of D-glucopyranose units linked by $\beta(1\rightarrow4)$ glycosidic bonds, resulting in anhydroglucose units (AGUs) ($C_6H_{10}O_5$). The repeating unit is the cellobiose that is composed of two AGUs. Nevertheless, the degree of polymerization (DP) of cellulose is commonly expressed as a number of AGUs. The degree of polymerization varies with the cellulose source, which ranges from 10 000 in wood-derived cellulose up to 20 000 in cotton. A 10-fold decrease is usually observed after industrial isolation. The pyranose rings have been found to be in the chair conformation with hydroxyl groups in equatorial position. All along the polymer chain, from one glucopyranose unit to the next one, a rotation of 180° of the ring around the β link is observed. Figure 5.4a describes the three types of AGUs: (i) the internal AGUs, (ii) a reducing end with a free hemiacetal or aldehyde group in equilibrium, and (iii) the nonreducing end with a free alcohol group.

The hydroxyl groups at the edge of the polymer chain provide cellulose with intra- and interchain hydrogen bonding, which contribute to many of its peculiar properties (see Figure 5.4b). Hydrogen bonding has quite weak energy in comparison with a covalent bond: about 20 kJ/mol for O–H–O systems are found in cellulose vs. 400 kJ/mol for common covalent bonds in cellulose. But the repetition of these weak energy hydrogen bonds plays a key role in cellulose structure and its physical organization. They allow for the arrangement of cellulose polymeric chains into parallel arrays to form fibrils on the nano- and microscale, which will be described later on. These nano- and microfibers further assemble to form macroscopic wood fibers, for instance, as described in Figure 5.5.

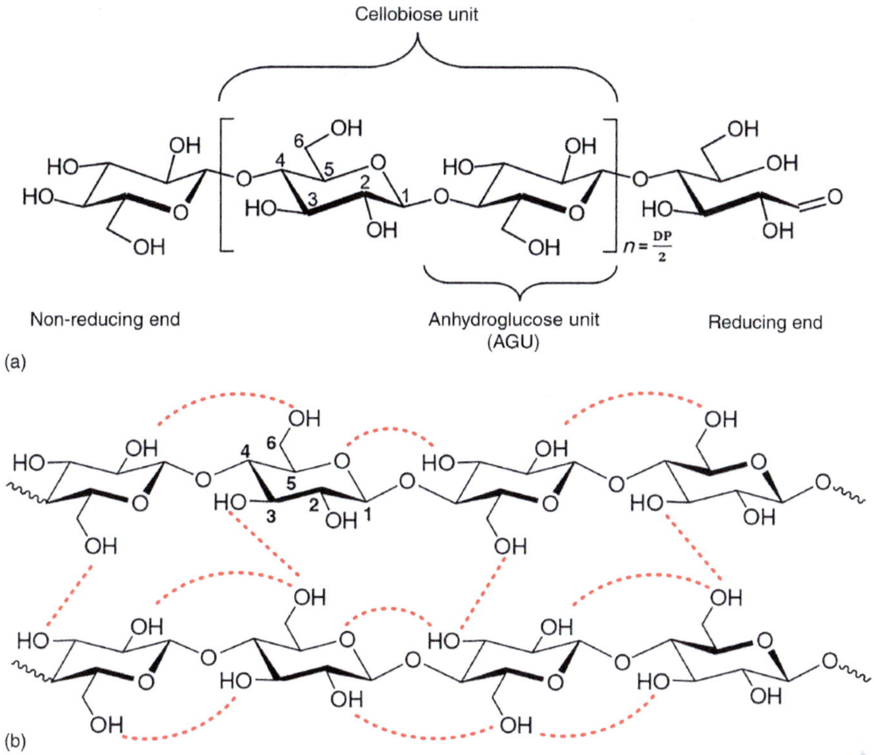

Figure 5.4 Cellulose polymer chemical structure: (a) molecular structure of cellulose polymer chain, carbon atoms of the anhydroglucose unit are numbered from 1 to 6, and (b) hydrogen bonding in cellulose.

Cellulose has been the subject of intensive research because of its sustainability, biodegradability, widespread availability, and biosafety and has been used extensively in recent years in the biomedical field [38]. One of the first industrial-scale uses of cellulose in the medical industry appeared with the development of cellulose membranes for the treatment of renal failure in the beginning of the twentieth century. Cellulose membranes were used for hemodialysis that purifies patient's blood and act as artificial kidney [39, 40]. Nowadays the trend for hemodialysis membranes is moving to synthetic polymers with a better controlled molecular cutoff.

Microcrystalline cellulose (MCC) is derived from high-purity cellulose by hydrochloric acid treatment until a very low degree of polymerization is reached (a few hundred compared with starting cellulose of at least 1000). Avicel© is one of the most known brands of MCC. It is white, fine, and odorless crystalline nonfibrous powder with particle size in the range of 20–200 μm depending on the production process [41]. In the pharmaceutical industry MCC is mainly used as an excipient for tablet production in drug delivery. It provides good interactions with the active pharmaceutical ingredients (APIs) and allows for the preparation of stable pharmaceutical ingredients. MCC also provides a better disintegration in the stomach [42].

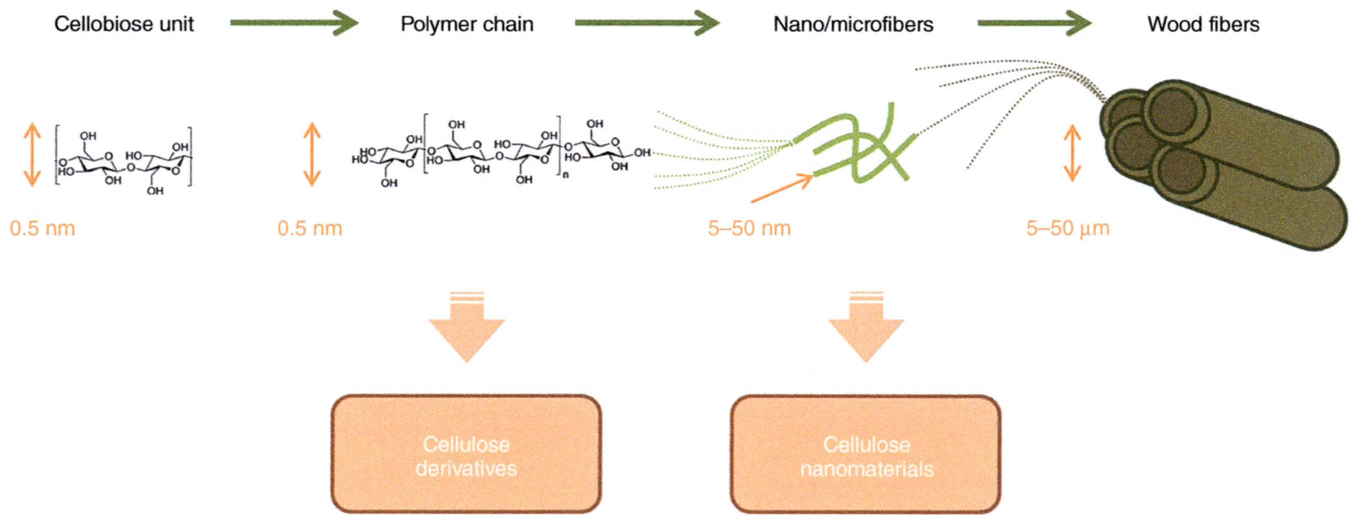

Figure 5.5 Hierarchical organization of cellulose polymer chain into microfibers and wood fibers.

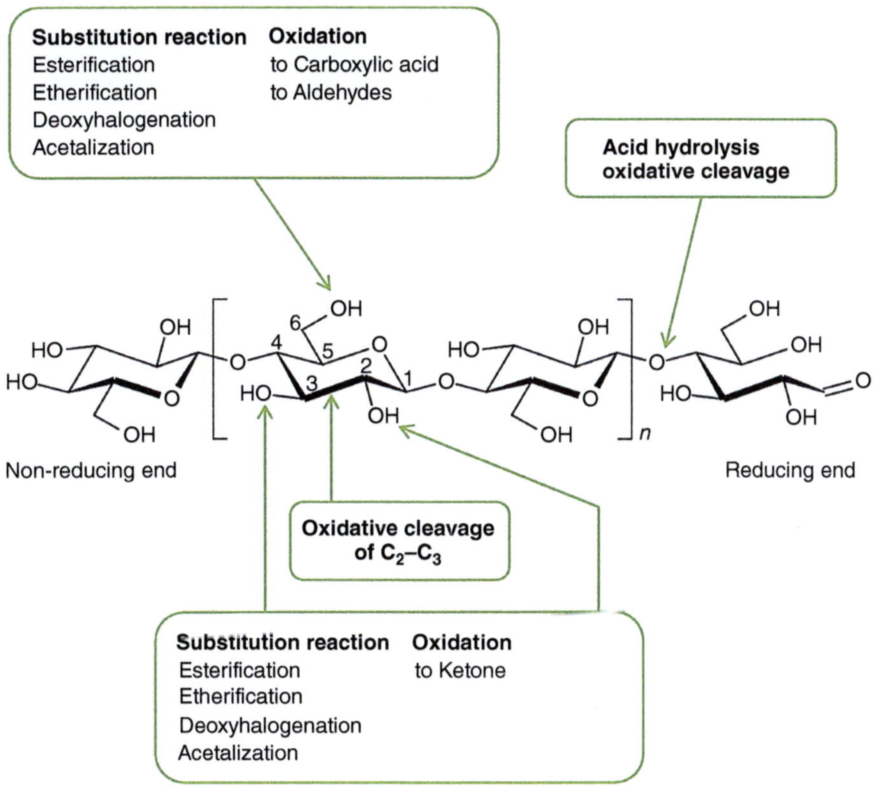

Figure 5.6 Suitable positions for cellulose chemical modifications.

Furthermore, the chemical modification of cellulose creates new polymers that are also used in the medical field, the so-called cellulosics or cellulose derivatives, which are mostly composed of oxidized cellulose, ether, and ester of cellulose. The hydroxyl groups of C_2 and C_3 (secondary) and C_6 (primary) are available for numerous chemical modifications as described in Figure 5.6. The hydroxyl groups of the amorphous regions of cellulose are more accessible for chemical modification, whereas those of crystalline regions are very hard to reach due to the close packing of cellulose chains and strong interchain hydrogen bonding [43]. Cellulosics are known to be strong, of low cost, reproducible, recyclable, and biocompatible, making some of them suitable for medical applications [44].

For instance, the cellulose ethers used in drug tableting industry, like hydroxymethylpropylcellulose (HPMC), are able to form matrices that swell in aqueous media and create a tunable diffusional barrier to further control drug delivery, with the advantage of eventually solubilizing in the medium. This diffusional barrier was observed by colorimetric techniques and fitted to mathematical models that successfully described the drug release [45–47]. Cellulose esters such as cellulose acetate phthalate (CAP) were used to prepare enteric-coated granules. The microencapsulation of the granules with CAP significantly prolonged the release of the embedded drug and enhanced the effectiveness of the enteric coating.

These formulations are able to go through the stomach and only dissolve and release the drug in a gastric environment [48]. The overall use of cellulosics in pharmaceutical preparations is widely described in the literature like the book chapter by Tohru Shibata [40].

Likewise, due to their water-soluble properties, hydrogels of cellulose derivatives can also be formed for medical applications, like hemostat and drug release [49]. Silated hydroxypropylmethylcellulose hydrogels were successfully prepared in neutral pH conditions. Such conditions are suitable for their subsequent use for cell culture, bone defect treatment, or model cartilage development [50]. Chemically crosslinked sodium-carboxymethylcellulose (Na-CMC) and hydroxyethylcellulose (HEC) hydrogels were also produced with nontoxic coupling agents such as citric acid, resulting in a super-absorbent material with possible hemostat applications [51]. Chemical crosslinking of cellulose derivative hydrogels can also be achieved by irradiation techniques. Both degradation and crosslinking of polymers occur upon such high-energy treatments, but concentrated polymer solutions (above 50 wt% for carboxymethylcellulose) seem to favor crosslinking [52]. Indeed, hydrogels were produced from CMC and acrylamide monomer (AM) through γ-radiation treatment, resulting in materials with greater swelling capacities than pure AM gels. Drug release from such a material was also investigated and revealed a pH-dependent release behavior, with slower release in acidic conditions [53].

The widespread medical uses of cellulosics are due to their availability, physical properties, and physiological inertness, which gave rise to numerous commercial products in the twentieth century. Endless possibilities of cellulose derivative polymers and cellulose-derived materials (tablets, coating, membranes, and hydrogels) provide cellulose with a consistent and promising future in the medical field.

Many different conventional and biobased biomaterials have been introduced, and their properties and applications have been described. Among cellulosic materials, a new class of high potential derivatives was discovered in the 1980s: nanocellulose. It is a new tremendous field of research and has applications in many different areas, not only from papermaking to composites but also in medical industry.

5.2 Nanocellulose: Production, Characterization, Application, and Commercial Aspects

Nanocellulose refers to cellulose particles with at least one dimension in the nanometric range of 1–100 nm. Two main families of nanocellulose have been identified, cellulose nanofibrils (CNF) and cellulose nanocrystals (CNC). Nanocellulose can be extracted from trees, plants, and cellulose-containing microorganisms and animal species. Although commercial applications are still limited, the exponentially growing interest of researchers and companies for these materials (see Figure 5.7) makes nanocellulose one of the most attractive natural and renewable polymers for advanced applications [54]. Even if we

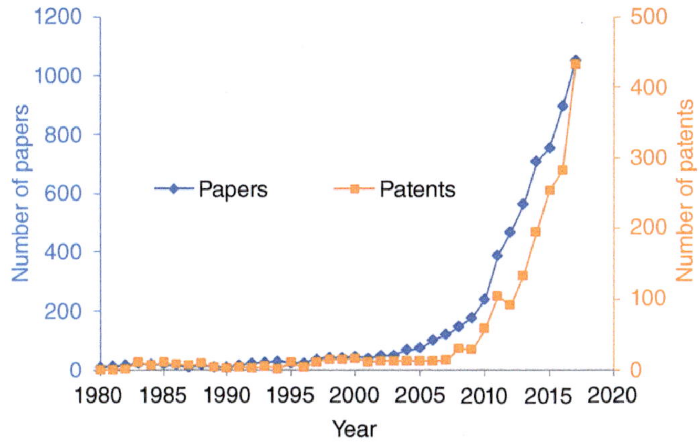

Figure 5.7 Number of papers and patents released each year dealing with CNFs and CNCs until 2017. Descriptors are cellulose nanofibrils/cellulose microfibril/microfibrillated cellulose/nanofibrillated cellulose/cellulose nanocrystals/cellulose nanowhiskers/nanocrystalline cellulose/cellulose whiskers. *Source:* Extracted from SciFinder July 2018.

correlate these figures to the growing number of scientific journals and the increasing number of researchers that work with nanocellulose, the tendency remains the same. Nanocellulose really differs from traditional cellulose-derived products (paper industry and cellulose-based polymers) and gives rise to new fields of application. The main characteristics of nanocellulose are related to its widespread availability, biodegradability, biocompatibility, low density, excellent mechanical properties, and tunable surface chemistry [55].

The isolation techniques of different cellulose nanomaterial will be explored, especially for CNF. Industrialization aspects will be discussed in terms of health and safety, and upcoming and first commercial products will be presented, especially those related to the biomedical field.

5.2.1 Isolation and Characterization of Nanocellulose Materials

Nanocellulosic materials are mainly extracted from wood even if other sources are available such as annual plant wastes, animals, and microorganisms. Compared to previously shown in Figure 5.5, a more detailed description of the hierarchical structure of wood is provided in Figure 5.8, from the tree trunk until the cellulose polymeric chain. It also indicates where cellulose nanomaterials are located. In order to isolate cellulose nanomaterials, purified cellulose fibers must be first obtained. Traditional cooking processes of the paper industry are able to produce high-purity cellulose fibers from softwood or hardwood by removing hemicelluloses and lignin, resulting in suitable material for cellulosic nanomaterial (CNM) isolation. Harsh chemical and mechanical treatments are used on purified cellulosic fibers, sometimes both combined, in order to decrease mechanical energy consumption, which was the main drawback for nanocellulose to reach the commercial production level during the last decades.

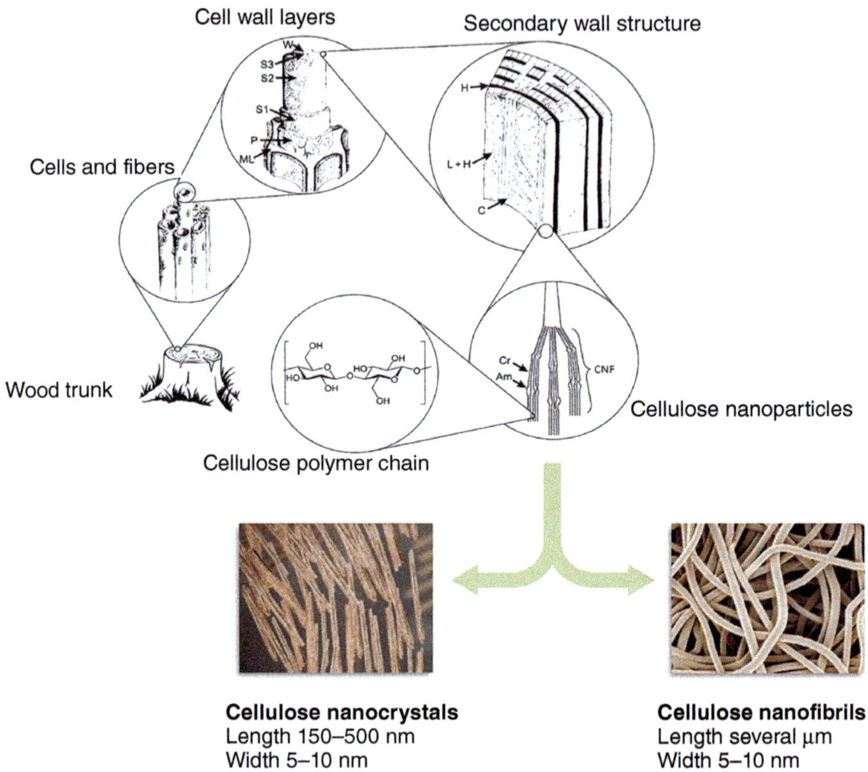

Figure 5.8 From wood trunk to cellulose nanomaterials, the cell wall layer is composed of middle lamella (ML); the primary wall (P), the outer (S1), middle (S2), and inner (S3) layers of secondary wall; and the warty layer (W). Cellulose (C), lignin (L), and hemicellulose (H) compose the secondary wall structure. On cellulose nanoparticles, crystalline (Cr) and amorphous (Am) domains are shown. *Source:* Nechyporchuk et al. 2016 [56]. Adapted with permission of Elsevier.

5.2.1.1 Cellulose Nanocrystals

CNC, sometimes referred as cellulose whiskers or nanowhiskers or even nanocrystalline cellulose, are composed of the crystalline part of cellulose nanoparticles. It is a rodlike nano-object with dimensions ranging from 150 to 500 nm in length and 5 to 10 nm in width, resulting in high aspect ratios (see representation in Figure 5.9). The discovery of CNC is attributed to Rånby et al. in 1950 [58, 59]. Researchers were inspired by the work of Nickerson and Habrle in the 1940s who observed the limitation of degradation of cellulose fibers by boiling in acidic conditions [60].

Isolation of CNC is thus currently done by strong acid hydrolysis of purified cellulosic material under controlled temperature and time conditions followed by sonication treatment. The amorphous regions of the cellulose fibers have a lower density than the crystalline regions and thus are much more sensitive to the acidic treatment. Acidic ions can penetrate more easily the amorphous regions, while crystalline regions remain untouched and eventually result in CNC [61]. Figure 5.9 describes the general procedure to obtain CNC from wood. We can

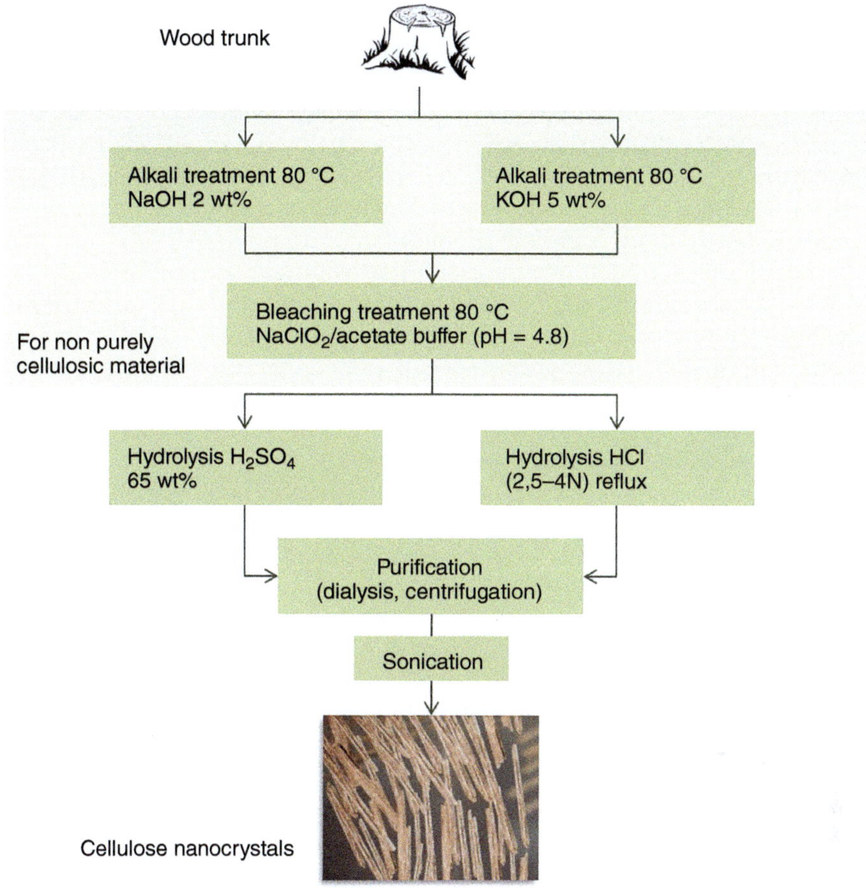

Figure 5.9 General procedure to obtain cellulose nanocrystals. *Source:* Dufresne 2017 [57]. Reproduced with permission of Walter de Gruyter.

note that very similar procedure can be used on cotton or also industrial crop wastes.

5.2.1.2 Cellulose Nanofibrils

Meanwhile CNF can also be extracted from cellulosic fiber, but they really differ from CNC. Composed of both crystalline and amorphous regions of cellulose, CNFs are long and flexible nanofibrils with dimensions of 5–50 nm in width and several micrometers in length, resulting in a higher aspect ratio than CNC. The terminology evolved over the time, and terms like cellulose microfibrils, microfibrillated cellulose (MFC), and nanofibrillated cellulose (NFC) all refer to CNF that has been selected by the scientific community as a standard terminology in 2012 [62].

In the 1980s, the pioneering works of Turbak et al. and Herrick et al. unveiled a new component of cellulose fibers with lateral dimensions in the nanometer range due to mechanical treatment. Softwood pulp was subjected to a high-pressure homogenizer (at least 3000 psi and several passes) so that the entangled networks of nanofibrils inside the cellulose fiber are isolated from each other under the high shearing forces. The treatment resulted in a gel-like suspension composed of interconnected microfibrils and nanofibrils of 10–100 nm diameter at only 2 wt% in water with shear thinning behavior [63]. CNF can be isolated from wood and annual plants like cotton by mechanical disintegration treatment. The high shear forces produce a longitudinal cleavage of the cellulose fibers and isolate to some extent the nanofibrils.

High energy consumption is often associated with first work on isolation of CNF [63, 64]. In order to obtain viable commercial grades of CNF and upscale production facilities, researchers investigated many ways to decrease the energy needs of CNF isolation by proposing specific pretreatments or development of new mechanical treatments. The most commonly used processes to induce high shear forces to cellulose fiber suspensions were originally grinding, microfluidization, and homogenization as depicted in Figure 5.10 [65].

These techniques are efficient to delaminate the cellulose fiber and are also suitable for upscaling. They are used nowadays for pioneering industrial production of CNF [56]. However, it is the use of a combination of mechanical and chemical or biochemical pretreatments that really helps to decrease energy consumption from 100 kWh/kg for un-pretreated cellulosic materials to as less as 1–2 kWh/kg [66]. Seminal works proposed successive refining, enzymatic hydrolysis, refining again, and finally homogenization [67], (2,2,6,6-tetramethylpiperidin-1-yl)oxidanyl-mediated (TEMPO) oxidation in combination with blending [68], or in combination with homogenization [69], carboxymethylation, or quaternization, followed by homogenization [70]. Thus, a wide number of procedures lead to a very wide range of CNF grades. The recent review of Nechyporchuk et al. extensively covered conventional and less conventional CNF production methods. Figure 5.11 summarizes the general procedures to obtain CNF suspension. As for CNC, the first step deals with the purification of the raw material in order to retrieve the maximum amount of cellulose and get rid of lignin and hemicellulose. The second and third steps are mechanical or (bio)chemical pretreatments that come before the fourth step of principal mechanical treatment. The final step aims at purifying the obtained CNF and further expands their properties. It clearly shows how a unique raw material such as cellulosic fibers extracted from biomass can lead to the isolation of more than 50 types of CNF. A very recent review extensively describes all the possibilities used these days or under development for cellulose functionalization as a pretreatment for CNF production [71].

Among all these procedures, enzymatic hydrolysis and TEMPO-mediated oxidation pretreatments deserve a particular consideration since they are the most studied procedures of the last decade. A comprehensive description of the state of the art associated with their production procedures is necessary to be able to understand the Section 5.2.5.

Cellulose Nanofibril Production Through Enzymatic Pretreatment The purpose of enzymatic pretreatment is to catalyze the hydrolysis of cellulose in order to favor the defibrillation process upon subsequent mechanical shearing. This process was patented in the paper industry in the 1960s to decrease the energy consumption of refining. Researchers also investigated enzymatic hydrolysis for

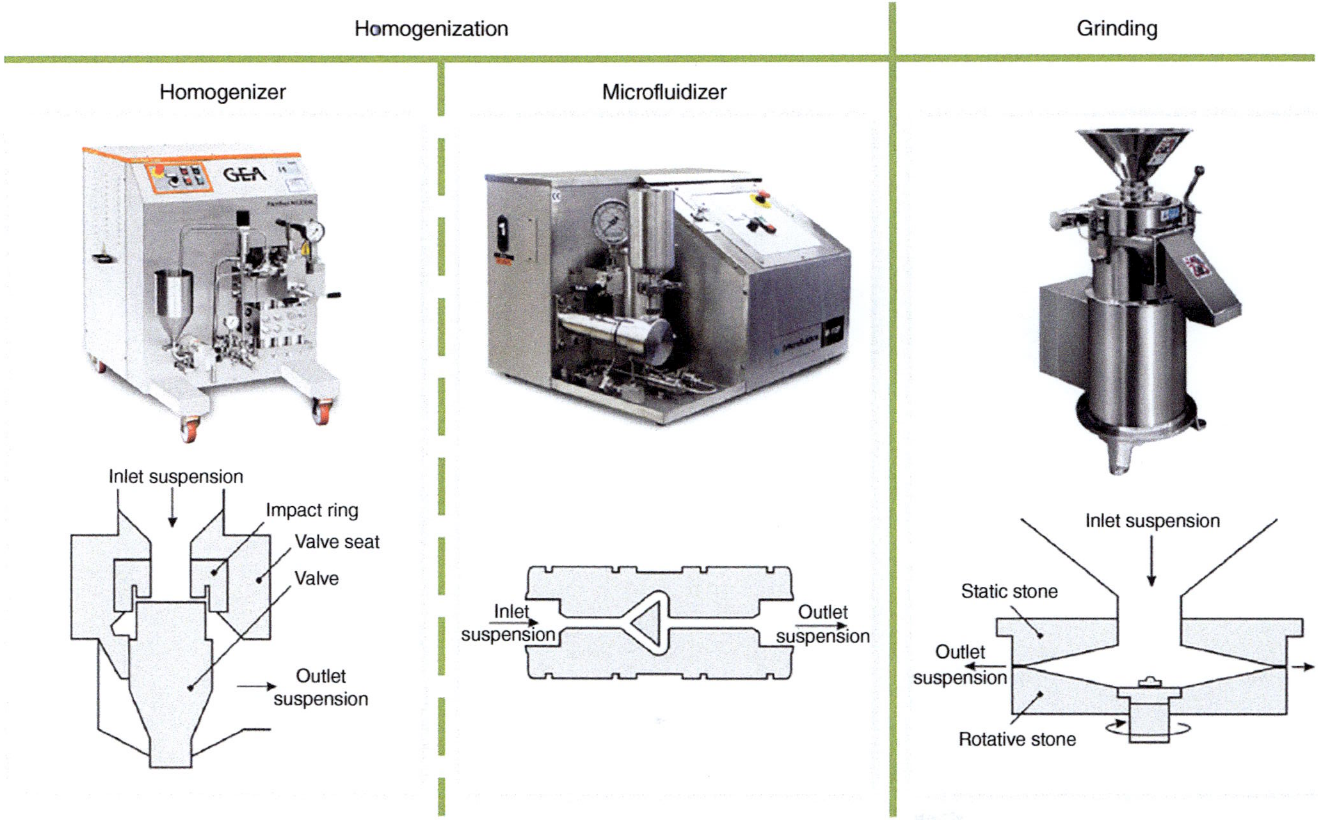

Figure 5.10 Main mechanical treatment device for production of CNF suspension. *Source:* Nechyporchuk et al. 2016 [56]. Reproduced with permission of Elsevier.

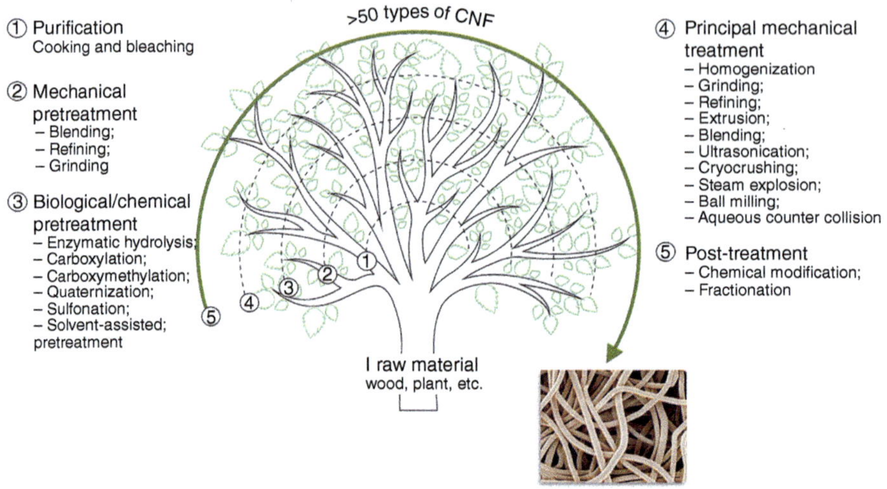

Figure 5.11 General procedures to obtain cellulose nanofibrils. *Source:* Nechyporchuk et al. 2016 [56]. Adapted with permission of Elsevier.

the production of bioethanol through the conversion of cellulose to glucose and ethanol. But the pioneering results of CNF production assisted by enzymatic pretreatments are attributed to Pääkkö et al. and Henriksson et al. in 2007 [67, 72]. Pääkkö et al. proposed a combination of refining, which increases cellulose swelling and accessibility of the enzyme, followed by the enzymatic treatment and washing plus a second step of refining before the actual homogenization process in a microfluidizer (Figure 5.10) for eight passes. Henriksson et al. used a very similar procedure and compared enzymatic concentrations. The fibrillation process was performed with a homogenizer for 20 passes.

The enzymatic process produces CNF with a decreasing DP and an increase of the crystallinity index [73]. Cellulose-degrading enzymes are referred as cellulase and can be divided into three subfamilies: (i) endoglucanases (or endocellulases) that hydrolyze amorphous regions of cellulose, (ii) cellobiohydrolases (exoglucanases) that degrade the ends of cellulose structures releasing cellobiose or cellobiose dimer, and (iii) β-glucosidases (cellobioses) that hydrolyze the later small molecules into glucose. Figure 5.12 shows the mode of action of each of the abovementioned cellulases on cellulose. In the end, these enzymatic isolated CNF have the same surface chemistry compared with cellulose fibers, with hydroxyl groups on their surface.

Cellulose Nanofibril Production Through TEMPO-Mediated Oxidation Pretreatment TEMPO-mediated oxidation of cellulose fibers allows the introduction of negatively charged groups (carboxylate groups) at the surface of cellulose chains, which promote ionic repulsion between cellulose chains and eventually help the defibrillation process under high shear mechanical treatment. It has become one of the most investigated pretreatment strategies for cellulose oxidation and cellulose nanomaterial production. The first work on the TEMPO-mediated oxidation of sugar molecules was published by Davis and Flitsch in 1993 [75].

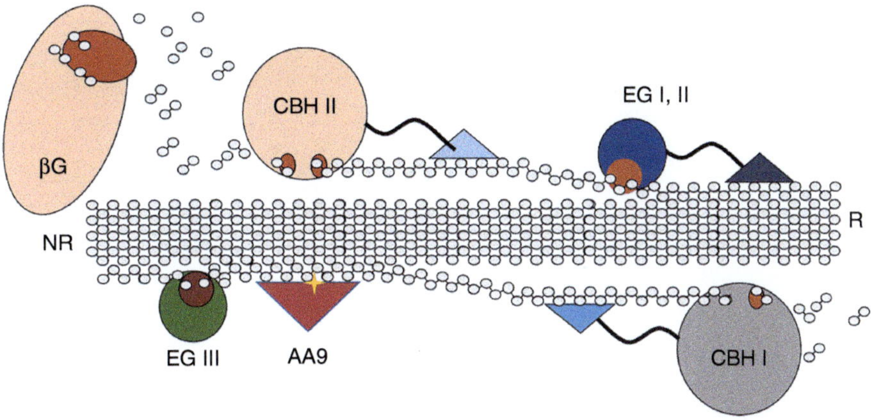

Figure 5.12 Illustration of cellulase mode of action. β-Glycosidase (βG) releases free glucose from cellobiose; cellobiohydrolase II (CBH II) releases cellobiose from the nonreducing (NR) end of the cellulose chain; cellobiohydrolase I (CBH I) releases cellobiose from the reducing (R) end of the cellulose chain; and the endoglucanases EG I, II, and III break the internal glycosidic bond, creating action points for CBH I and CBH II. The LPMO (AA9) is shown in red. Some cellulases contain cellulose-binding modules, depicted here as blue triangles. *Source:* Johansen 2016 [74]. Reproduced with permission of Elsevier.

They presented a method to selectively oxidize the primary hydroxyl group of monosaccharides using sodium hypochlorite in the presence of catalytic amount of TEMPO radical. The research group of Pr. Akira Isogai further studied this reaction and came up with CNF isolated with a blending apparatus from never-dried cellulose from sulfite wood pulp, cotton, tunicate, and bacterial cellulose. They achieved a carboxyl content of 1.5 mmol/g [68] and obtained highly individualized nanofibrils, while the blending apparatus was of much lower energy input than conventional mechanical treatment. This research group also investigated many different reaction conditions in order to determine the most energy-efficient procedure to obtain CNF through TEMPO-mediated oxidation pretreatments of cellulosic raw materials. Different amounts of NaClO and reaction times were tested and compared against each other by following carboxyl and aldehydes group content and the DP evolution [76]. Figure 5.13 exposes the proposed mechanisms for the TEMPO-mediated oxidation using TEMPO/NaBr/NaClO in basic conditions or TEMPO/NaClO/$NaClO_2$ in neutral or slightly acidic pH. The later system has been proved to limit the depolymerization of cellulose and the number of formed aldehyde groups [77].

Such TEMPO-oxidized CNF bring new opportunities of functionalization with the presence of carboxylic acid groups, together with aldehyde groups (not fully converted hydroxyls) and the remaining non-oxidized hydroxyl groups, on its surface.

5.2.1.3 Bacterial Nanocellulose (BNC)

Bacterial nanocellulose (BNC) is in competition with wood-derived nanocellulose for medical applications and thus deserves a short introduction and comparison. Populations of bacteria are generally divided into two types: planktonic,

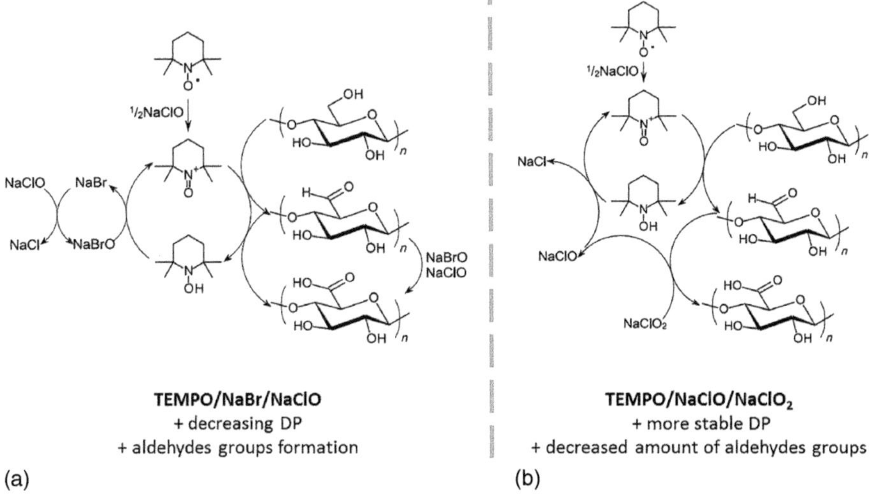

Figure 5.13 Schematic mechanisms of regioselective oxidation of primary hydroxyl groups of cellulose by TEMPO/NaBr/NaClO in water in basic pH conditions (a) and TEMPO/NaClO/$NaClO_2$ in water at neutral or slightly acidic pH (b).

which refers to freely dispersed bacteria in bulk solution, and sessile, which refers to a united colony of bacteria embedded in a *biofilm*. Many different microorganisms produce this so-called biofilm upon their development on every kind of surface [78]. It is acting as a matrix that contains the microorganisms and provides them with a safe environment, preventing them from drying out, which also gives protection from enemies, irradiation, and lack of food or oxygen [79]. The biofilm is usually composed of extracellular polymeric substances.

In parallel to plant cellulose production, it appears that some bacterial strains are capable of producing biofilms that contain high amounts of cellulose in the form of nanofibers: *Acetobacter, Agrobacterium, Alcaligenes, Pseudomonas, Rhizobium, Aerobacter, Achromobacter, Azotobacter, Salmonella*, and *Sarcina* [80]. The most efficient and the most studied strain appear to be *Acetobacter xylinum*, recently reclassified as *Gluconacetobacter xylinus*. It is a member of the acetic acid bacteria (AAB) group, which is known to perform specific oxidation reactions exploited for the production of numerous compounds including cellulose [81]. Figure 5.14 shows *G. xylinus* producing cellulose with nanostructured architecture (black arrow).

The efficiency of BNC production is highly dependent on the strain but also on the culture medium: temperature, pH, carbon source (such as -glucose), and most importantly oxygen supply during the process (since AAB are aerobic bacteria) [40, 41]. The bacteria synthesize the cellulose polymer at the air–liquid interface of the culture medium and always as a membrane. Two major systems are mentioned, static and agitated culture media, and both show specific advantages and disadvantages. The main difference is the oxygen supply and diffusion in the culture medium.

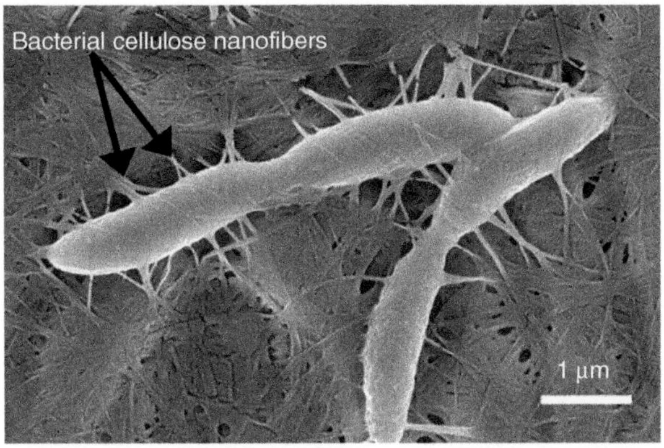

Figure 5.14 *Komagataeibacter xylinus* (also known as *Acetobacter xylinum*) producing bacterial cellulose nanofibers (black arrow). *Source:* From https://vtnews.vt.edu/content/vtnews_vt_edu/en/articles/2008/11/2008-693.html [82].

The most important feature of BNC is its inherent purity in comparison with wood-derived nanocellulose. The latter contains hemicelluloses and lignin that requires complex chemical treatment to be removed. The nanostructure organization of BNC makes it a very good candidate for various biomedical applications such as tissue engineering. In addition, its *in vivo* biocompatibility and potential use in biomedical field have been proven many times [83–85].

However, industrialization is still limited and expensive compared with the development of wood-derived nanocellulose pilot plants. Moreover, BNC seems quite difficult to obtain as a suspension or as a re-dispersible powder, which can be useful in formulation processes.

These three types of cellulose nanomaterials (CNC, CNF, and BNC) can be subdivided in different grades in terms of nanomaterial proportion, morphology, or even surface chemistry. Even if they are all cellulose on the molecular level, different forms of nanocellulose can present specific features, thus becoming suitable for different applications. Consequently, the next paragraph focuses on the characterization techniques of such materials.

5.2.2 Characterization of Cellulosic Nanomaterials (CNMs)

The development of reliable and accurate characterization methods for CNMs is driven by both the need for deeper scientific understanding of these materials by research communities and the growing industrial commercialization. The available grades of CNMs on the market vary. Industrial partners should be provided with clear guidelines for the choice of the most relevant CNMs according to the intended application.

In 2018, a wide conglomerate of expert researchers from all around the world gathered to establish the best practices, methods, and techniques for characterizing CNM particles. It covers morphology, surface chemistry, surface charge, purity, crystallinity, rheological properties, mechanical properties, and toxicity. CNC and CNF were chosen to be the CNMs for the basis of this work [86]. Numerous decision tree schemes were proposed to answer every potential needs a researcher from academic or industrial community would have. Figure 5.15 describes the pathway one should follow when working with CNMs and associated characterization tools.

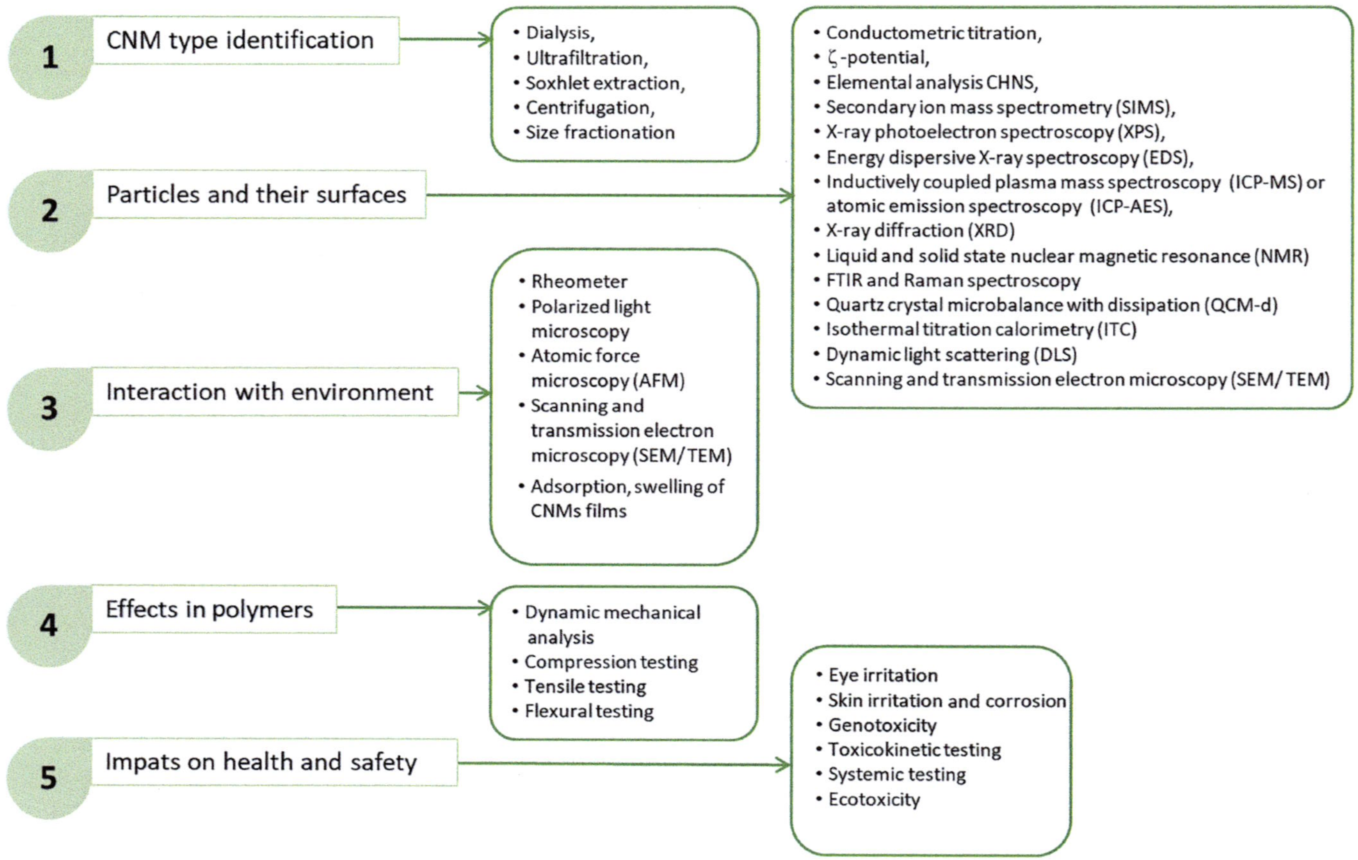

Figure 5.15 Five topics for the characterization of CNMs and associated characterization tools. *Source:* Adapted from Moon and Foster 2018 [86].

The first point emphasizes the critical importance of knowing the starting material. As already mentioned, hundreds of CNM grades are available and differ regarding if they are in wet or dry state (and concentration and ability to be redispersed), presence of impurities, and proportion of nano- and microsized material. The second topic deals with the shape and surface of CNMs. Surface charge, morphology, elemental analysis study, crystallinity, and surface chemical structure measurement methods are described. Topic three focuses on rheology of CNM suspensions and adsorption mechanisms of CNM particles. Topic four is dedicated to the development of CNM composites and characterization of CNM dispersions. Finally, health and safety aspects of CNMs are discussed in the fifth and last topic.

Other works especially focused on CNF quality characterization. For example, a review of different techniques used to characterize CNF was done by Kangas et al., while a quality index was developed based on eight different characterization techniques to assess the homogeneity of CNF dispersion by Desmaisons et al. [87, 88].

5.2.3 Industrialization of Nanocellulose: First and Upcoming Applications

Global enthusiasm for nanocellulose is described in Figure 5.7, which shows the number of publication and patents regarding nanocellulose. To deepen this overview, it would be interesting to consider the number of patents published by countries (Figure 5.16a). These data clearly indicate that the strong industrial interest for nanocellulose is worldwide. Furthermore, when these data are sorted according to the field of application, the multiple end uses of nanocellulose improved products are exposed, with composites being the field that holds the most patents.

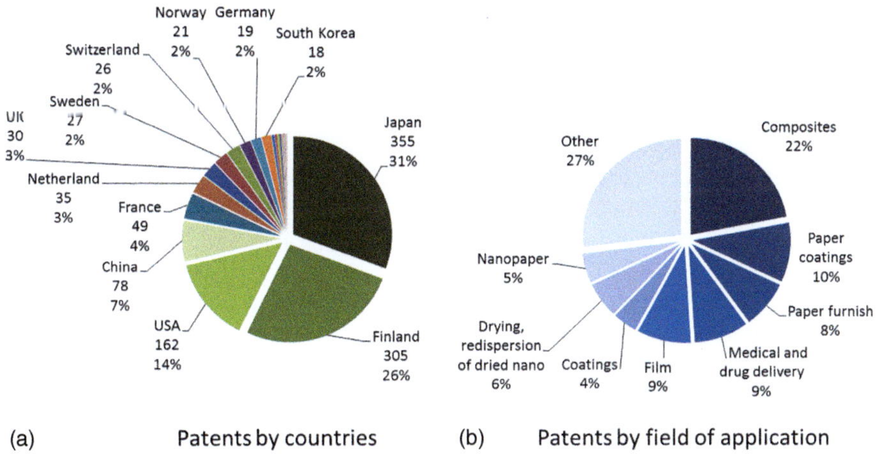

Figure 5.16 (a) Number of patents dealing with CNF materials. Descriptors: cellulose nanofibrils/cellulose microfibrils/microfibrillated cellulose/nanofibrillated cellulose, (b) nanocellulose (CNF, CNC, and BNC) patents repartition in field of application. *Source:* Extracted from SciFinder October 2018.

However, the first years of industrialization progress were limited by several factors such as (i) a high production energy cost, (ii) the lack of high added-value and/or cutting-edge applications, (iii) competition with the existing renewable and nonrenewable materials, and (iv) precaution about the "nano" aspect regarding health and safety. This explains why a relatively low number of patents were published before 2008.

In the last decade, these challenges were intensively addressed by the scientific community in order to unlock the industrialization of nanocellulose. The development and optimization of cellulosic source pretreatments allow for the decrease of the energy consumption of nanocellulose production. The current identified producers of CNF and CNC are shown in Table 5.2.

5.2.4 Health and Toxicology: A Concern for CNM Development in Biomedical Field

As with any emerging material, nanocellulose must undergo a health and toxicology assessment in order to facilitate the development of industrial processes and ensure the safety of nanocellulose-based products. When it comes to use nanocellulose as a biomaterial that is supposed to be in contact with living tissues and microorganisms, biocompatibility evaluation is of first importance. Also, the "nano" aspect of these materials raises concerns since it is sometimes associated with more risks than traditional materials in the mind of the general public. Moreover, it is well known that the human body does not possess cellulose-degrading enzymes. Consequently, a toxicological study has been carried out starting from the manufacturing processes that are used to produce CNMs, through the ways CNMs are handled and integrated or combined with other materials, until the fate of the CNMs in the product's end of life.

Celluloses in the form of cellulosic fibers used in papermaking or in the form of MCC are known to be safe [89–93]. But CNMs differ in terms of particle size and shape, surface charge and specific surface area, chemical properties, and degree of agglomeration, so health and toxicology risks must be assess once again.

Attempts to evaluate CNMs toxicity are fairly recent. A review from Lin and Dufresne published in 2014 listed the preliminary works for the toxicological evaluation of nanocellulose [94]. Initial work for CNC was conducted by Kovacs et al. in 2010 [95]. A group of Canadian researchers assessed the ecotoxicology of CNC on aquatic species. Low toxicity potential and environmental risk were confirmed. Among Lin and Dufresne reviewed evaluations, only potential respiratory toxicity was pointed out. Aerosolized CNCs were compared with carbon nanotubes (CNTs) and asbestos fibers on a model human lung cell structure. CNC revealed to be dose-dependent pro-inflammatory but of significantly lower impact than the two other materials. Stiffness, length, and aspect ratio clearly influence the nano-object/cell interactions [96]. Slight pulmonary inflammation of mice was also noted for CNF in the European SUNPAP project, which ended in 2012 [97]. However, for dozens of other works collected by Lin and Dufresne, no evidence of cytotoxicity is detected *in vitro* nor *in vivo*, especially on human cells (osteoblast, fibroblast, and endothelial cells).

One of the preliminary health and environmental safety studies about CNF was published in 2011 by Vatiainen and colleagues. They assessed the exposures of workers to particles in air during production of CNF through grinding

Table 5.2 Current industrial producers of CNF and CNC, including country of origin, yearly production capacity (tons per years), product type and starting year.

Company	Country	Capacity (yr^{-1})	Product type	Approximately since
Cellulose nanofibril production				
FiberLean	France/ Switzerland	10 000 t	CNF with fillers	2013
Performance BioFilament	Canada	2 000 t	Nanofilament	2014
Kruger/FPInnovations	Canada	1 000 t	Nanofilament	2015
Borregaard	Norway	1000 t	CNF	2016
Paperlogic	USA	730 t	CNF	—
Nippon Paper	Japan	300 t	CNF Tempo	2017
American Process Inc.	USA	180 t	CNF	2015
University of Maine	USA	110 t	CNF	2014
Chuetsu Pulp & Paper	Japan	50 t	CNF	2016
DKS	Japan	50 t	CNF Tempo	2016
Sugino Machine	Japan	50 t	CNF	2017
Oji Holdings	Japan	40 t	Phosphorylated CNF	2016
Suzano	Brazil	36 t	CNF	2018
Innventia	Sweden	35 t	CNF	2011
CTP	France	35 t	CNF	2011
PFI	Norway	35 t	CNF	2013
Seiko PMC	Japan	30 t	CNF	2018
Tokushu Tokai Paper	Japan	30 t	CNF	2018
VTT	Finland	15 t	CNF	2010
Inofib	France	0.5 t	CNF	2013
Icar-Circot	India	11 t	NA	2016
EMPA	Switzerland	5 t	CNF	2011
SAPPI	Netherlands	8 t	CNF	2016
Daio Paper Corporation	Japan	NA	CNF	2016
DIC Corporation	Japan	NA	CNF	2017
Daicel Corporation	Japan	NA	CNF	2000
Stora Enso	Finland	NA	CNF	2013
UPM	Finland	NA	CNF	2011
Bettulium	Finland	NA	CNF Tempo	2013
Norske Skog	Sweden	NA	CNF	—
Weidmann	Switzerland	NA	MFC	2016
CelluComp	UK	NA	CNF	2015
BioNC	Spain	NA	CNF	2018

(*continued*)

Table 5.2 (Continued)

Company	Country	Yearly capacity	Production type	Starting year
Cellulose nanocrystal plant				
CelluForce	Canada	365 t	Sulfated CNC	—
American Process Inc.	USA	200 t	NA	—
Holmen/Melodea	Sweden	35 t	Sulfated CNC	—
Icar-Circot	India	10 t	NA	—
Alberta Innovates	Canada	7 t	Sulfated CNC	—
Blue Goose Biorefineries	Canada	4 t	Sulfated CNC	—
FPInnovations	Canada	4 t	Sulfated CNC	
University of Maine	USA	4 t	Sulfated CNC	
Melodea	Israel	NA/pilot	Sulfated CNC	

and subsequent use by spray drying. Exposure to particles was very low or nonexistent during the fabrication process.

More recently a nano life cycle risk assessment (nanoLCRA) testing guide was proposed by J.A. Shatkin, K.J. Ong, and colleagues from American Process and can help the community to better assess the toxicity of CNMs [98]. The guide is divided into three endpoints: (i) physicochemical characterization, (ii) human health effects testing, and (iii) environmental effects testing. The physicochemical characterization is necessary because various grades of CNMs are now available and their impact on the two second endpoints may vary with the particle size, aspect ratio, surface chemistry, etc. The details of the guide are exposed in Table 5.3.

The study of Ong et al. evaluated the toxicity of lignin-coated CNF and CNC. Neither of the two materials caused acute oral, eye, or dermal inflammation or irritation. But environmental testing with high concentrations (>1000 times of maximum predicted emission values) demonstrated a very low potential toxicity [98].

However, for biomedical purposes, further characterizations and testing are necessary since risks associated with medical use of cellulose nanomaterials are beyond the eye or skin irritations. Indeed, nanocellulose-based medical devices could be used to treat wounds or even be implanted inside the human body, where physiological responses greatly differ from external contact. Even ingestion of materials and their way through the human digestive system can still be considered as external in comparison with implants. New aspects ranging from the simple cytotoxicity testing to the more complex genotoxicity, inflammatory response, reproductive, carcinogenic, and toxicokinetic effects must be addressed to better evaluate the biocompatibility of CNMs [38, 86].

From now on, the discussion will only focus on CNF. A few comparisons will be established again with BNC owing to its previously mentioned particular properties for biomedical use.

The genotoxicity of CNF was assessed by Hannukainen et al., where enzymatically pretreated CNF and TEMPO-oxidized CNF were used with human

Table 5.3 Testing plan to characterize safety of CNMs.

Endpoint	Protocol
Physicochemical characterization	
Particle size distribution, shape, and aspect ratio	Transmission electron microscopy
	Scanning electron microscopy
Composition	Infrared spectroscopy
	Inductively coupled plasma optical emission spectroscopy
Crystal structure	X-ray powder diffraction
Explosibility	Explosion severity test (ASTM E1226)
	Minimum ignition energy test (ASTM E2019)
Biodegradability	Aerobic biodegradation (OPPTS 835.3140)
	Anaerobic biodegradation (OECD 311)
Human health effects testing	
Acute oral toxicity	Up-and-down procedure in rats (OPPTS 870.1100)
Skin irritation	MatTek EpiDerm skin irritation test (OECD 439)
Eye irritation	Hen's egg test chorioallantoic membrane test
	Bovine corneal opacity and permeability test (OECD 437)
Environmental effects testing	
Microbial bioassay	Kinetic luminescent bacteria test (ISO 11 348-3)
Algal bioassay	Freshwater algal growth inhibition test (OECD 201)
Aquatic invertebrate bioassay	*Daphnia* sp. acute immobilization test (OECD 202)
Aquatic vertebrate bioassay	Acute toxicity to embryonic zebrafish

Source: Ong et al. 2017 [98]. Reproduced with permission of Elsevier.

bronchial epithelial cells (BEAS 2B). Both types of CNF caused possible DNA damage, but authors indicated that the testing procedure (enzyme comet assay) might have been incorrectly adapted to these new types of fibrous nanomaterials confirming the need for further analysis [99].

Another study was carried out by the group of G. Chinga-Carrasco and K. Syverud in 2013 on CNF. Dense films and 3D porous structures of oxidized CNF (high amount of anionic charge) and neat CNF were evaluated in terms of cytotoxicity against 3T3 human fibroblast cells. Cell membrane, cell mitochondrial activity, or viability, and DNA replication were not affected by direct and indirect contact with CNF materials [100].

Other works investigated the relation between structure, surface charge, and biological response of CNF. To this end, anionic and cationic CNF were obtained through chemical modifications including carboxymethylation or condensation of glycidyltrimethylammonium chloride (EPTMAC), respectively. Films were then produced by vacuum filtration, and indirect contact cytotoxicity test was performed with human fibroblasts. No cytotoxic effect was detected, and cell adhesion was promoted, due to CNF alignment [101].

An extensive overview of the recent papers on the toxicological assessment of CNFs is presented in Table 5.4. This table consolidates studies that were

Table 5.4 Extensive overview of toxicological assessment of cellulose nanofibrils in 2019.

CNF type and form	CNF source	Testing procedure	Biological impact				Comments	Authors	Year	Refs
			Cytotoxicity	Inflammatory response	Genotoxicity	Ecotoxicity				
Unmodified CNF	Bleached birch kraft pulp, fir/beech woods	Cytotoxicity on mouse hepatoma and human keratinocyte and cervix carcinoma	No	n/a	**No**	n/a	—	Pitkänen et al.	2010	[102]
CNF suspension and powder	Never-dried elementary chlorine free (ECF)-bleached birch kraft pulp	Air particle concentration, cytotoxicity on mouse and human macrophages, ecotoxicity test	No	**No**	n/a	Mild	Testing procedure not really adapted to CNF	Vartiainen et al.	2011	[103]
Commercial CNF (Booregard) suspension	n/a	Cytotoxicity assay on macrophage-like THP-1 cells	No	n/a	n/a	n/a	—	Kollar et al.	2011	[104]
Enzymatic and TEMPO CNF suspension	n/a	Cytotoxicity (propidium iodine staining and luminometric assay), *in vitro* genotoxicity in human bronchial epithelial BEAS-2B cells	No	n/a	**Yes**	n/a	Unclear success of genotoxicity assay	Hannukainen et al.	2012	[099]
Various grades of enzymatic and TEMPO CNF	n/a	Cytotoxicity on human cervical HeLa cells Genotoxicity on human bronchial BEAS 2B cells Immunotoxicity, tumor necrosis factor (TNF)-α, and interleukin (IL)-1β detection *In vivo* against Nematode model Pharyngeal aspiration exposure of mice	Low	No	**No**	Mild	Pulmonary inflammation detected	Norppa et al.	2012	[097]

Neat and TEMPO CNF films and aerogels	*Eucalyptus* and *Pinus radiata*	Cytotoxicity on 3T3 fibroblast cells	No	n/a	n/a	n/a	—	Alexandrescu et al.	2013	[100]
Unmodified, anionic and cationic CNF films	*Cladophora* green algae and softwood dissolving pulp	Cytotoxicity on human dermal fibroblasts Cell adhesion and viability	No	n/a	n/a	n/a	Surface properties influence to be noted	Hua et al.	2014	[101]
Neat CNF suspension	Bleached hardwood kraft pulp	Cytotoxicity with highest tolerated dose (HTD), total protein content (TPC), and RNA synthesis inhibition, genotoxicity with Ames test *in vitro*, ecotoxicity on nematodes	Mild	n/a	**No**	**No**	Effect of biocide to clarify	Pitkänen et al.	2014	[105]
Commercial CNF (USDA) suspension	n/a	Biodurability on *in vitro* cell-free and RAW 264.7 cell line models	n/a	n/a	n/a	Mild	CNF are biodurable in human lung	Stefaniak et al.	2014	[106]
Neat CNF suspension	Norway spruce *Picea abies*	Cytotoxicity on L929 cells, rat thymocytes, or human peripheral blood mononuclear cells Inflammatory by cytokines detection	No	**No**	n/a	n/a	—	Čolić et al.	2015	[107]
Unmodified, anionic and cationic CNF films	Never-dried bleached sulfite softwood dissolving pulp	Macrophage/monocyte behavior Immunotoxicity, TNF-α, IL-10, and IL-1ra detection	n/a	**No**	n/a	n/a	—	Hua et al.	2015	[108]
CNF suspension and films, CNF in paper	Never-dried bleached birch kraft pulp	Ecotoxicity, bioluminescence test with *Vibrio fischeri*	n/a	n/a	n/a	No	—	Vikman et al.	2015	[109]
Unmodified, anionic and cationic CNF films	*Cladophora* green algae	Inflammatory response of THP-1 cells (monocytes, macrophages) w/ or w/o lipopolysaccharide	n/a	**No**	n/a	n/a	—	Hua et al.	2015	[108]

(*continued*)

Table 5.4 (Continued)

CNF type and form	CNF source	Testing procedure	Biological impact				Comments	Authors	Year	Refs
			Cytotoxicity	Inflammatory response	Genotoxicity	Ecotoxicity				
TEMPO CNF suspension and aerogels	Never-dried bleached *P. radiata* pulp	Cytotoxicity on normal human dermal fibroblasts and human epidermal keratinocytes inflammatory by 27 cytokines detection	No	**No**	n/a	n/a	—	Nordli et al.	2016	[110]
Unmodified and TEMPO commercial CNF suspensions (USDA & Maine)	n/a	Ecotoxicology on embryonic zebrafish	No	n/a	n/a	**No**	No influence of aspect ratio	Harper et al.	2016	[111]
CNF suspension	Bleached dissolving pulp from the Norway spruce (*Picea abies*)	Cytotoxicity on human monocyte-derived dendritic cells	No	**No**	n/a	n/a	—	Tomić et al.	2016	[112]
TEMPO CNF suspension	Spruce sulfite dissolving pulp	Comet assay in bronchoalveolar lavage (BAL) and lung cells Bone marrow erythrocyte micronucleus assay	No	Yes	**Yes**	**Yes**	Toxicity on lung cells	Catalán et al.	2017	[113]
Commercial Neat CNF gels and powder (USDA)	n/a	Cytotoxicity, inflammatory, and oxidative stress on human lung epithelial cells (A549)	Mild	**No**	n/a	n/a	CNF are not taken up by cells, like CNC	Menas et al.	2017	[114]

Unmodified, anionic, and cationic CNF suspensions	Never-dried bleached sulfite softwood dissolving pulp	Cytotoxicity on human dermal fibroblasts, human medical research council (MRC)-5 lung fibroblast cell line and human THP-1 monocytic cell line, alamarBlue assay, reactive oxygen species (ROS) production, Inflammatory TNF-α and IL-1β detection	No	Mild	n/a	n/a	Surface charge influence	Lopes et al.	2017	[115]
Lignin-coated CNF (Bioplus)	n/a	Acute oral toxicity Skin and eye irritation Ecotoxicology with algal, microbial, and aquatic assay	No	**No**	n/a	No	—	Ong et al.	2017	[098]
TEMPO and carboxymethylated CNF hydrogels	n/a	Cytotoxicity on mouse fibroblasts L929 cells utilizing crystal violet, 3-(4,5-dimethylthiazol-2-yl)-2,5-diphenyltetrazolium bromide (MTT), and LDH assays Cell–hydrogel interaction	No	n/a	n/a	n/a	Good cell proliferation and migration	Rashad et al.	2017	[116]
Commercial neat and carboxymethylated CNF (Innventia)	Softwood sulfite dissolving pulp	(1,3)-β-Glucan/endotoxin-specific limulus amebocyte lysate (LAL) endpoint test	n/a	**No**	n/a	n/a	No immunogenic contaminants detected	Liu et al.	2018	[117]
TEMPO CNF	Bleached *Eucalyptus globulus* kraft pulp	Cytotoxic, immunotoxic, and genotoxic effects on a co-culture of lung epithelial alveolar (A549) cells and monocyte-derived macrophages (THP-1 cells)	Mild	**No**	Mild	n/a	Concentration-dependent effects	Ventura et al.	2018	[118]
Neat and fluorophore-grafted CNF suspension	Bleached never-dried softwood sulfite dissolving pulp	Ecotoxicity on feeding and life-history traits of *Daphnia magna*	Mild	n/a	n/a	**No**	Concentration-dependent effects	Ogonowski et al.	2018	[119]

(*continued*)

Table 5.4 (Continued)

CNF type and form	CNF source	Testing procedure	Biological impact				Comments	Authors	Year	Refs
			Cytotoxicity	Inflammatory response	Genotoxicity	Ecotoxicity				
Neat, car-boxymethylated and carboxylated CN	n/a	Cytotoxicity by lactate dehydrogenase (LDH) release assay, Inflammatory cytokines detection in THP-1 cells, Pulmonary effects and biopersistence of the materials in mice	Mild	Mild	n/a	n/a	CNF still present after one months in lung	Ilves et al.	2018	[120]
Neat CNF suspension	Softwood bleached kraft fiber	Sterility/biological assessment Endotoxin assessment	n/a	n/a	n/a	n/a	Free of endotoxins and bacteria	Pyrgiotakis et al.	2018	[121]
Neat CNF suspension	Curaua leaf fibers *Ananas erectifolius*	Cytotoxicity on fibroblast Vero cells Adhesion test	No	n/a	n/a	n/a	High affinity of cells to CNF	Souza et al.	2018	[122]

mentioned in relevant reviews such as Lin and Dufresne in 2014 [94], Endes et al. published in 2016 [123], and Kangas and Pitkänen published in 2016 [124]. The objective is to summarize the different types and forms of CNF that were already investigated in terms of biological response over five endpoints: cytotoxicity, inflammatory response (or immunotoxicity), oxidative stress, genotoxicity, and ecotoxicity. In the literature, many publications combine different types of cells with CNF without studying its potential toxicity as a principal objective. These works are not described in Table 5.4, but one should keep in mind that they also indirectly give information on CNF toxicity.

The overall toxicity of CNF toward cells from various lines appears to be very limited. However, the following endpoints should be kept in mind when trying to characterize the toxicity of CNF.

One should note the importance of careful sample characterization and exclusion of interfering factors like endotoxins or toxic chemical impurities that were used to prepare the samples (chemical agents or biocides). The objective is to avoid inaccurate conclusions from toxicity assessment [125]. Moreover, the biodurability of CNF in *in vivo* conditions allows extended period of time for such contaminants to leach out. Regarding this endpoint, the case of BNC is of particular interest. Synthetized by bacteria, no synthetic chemical contaminants are to be removed; due to this high purity, a very good biocompatibility was confirmed several times by *in vivo* testing since the end of the 1990s. Commercial products based on BNC such as Biofill® were implanted in animals [126], and artificial blood vessels made out of BNC were also successfully implanted in rats [127]. Reconstructive microsurgery of rat carotid artery was performed during one year with BNC model vascular system proving the high biocompatibility of BNC [128].

Also, even if CNFs mostly do not reveal any strong hazardous effect on human-derived cells and the environment or when it is used as a biomaterial, one should keep in mind that the surface modification of these materials could give different results. In the publication of G. Chinga-Carrasco and K. Syverud previously mentioned, the introduction of polyethylenimine (PEI) as a crosslinker or cetyltrimethylammonium bromide (CTAB) as debonding agent on the CNF revealed toxic behavior with detrimental effects on survival, viability, and proliferation of cells [100]. In comparison, a more recent work by Harper et al. revealed that various surface chemistries implemented on CNFs appeared to be of low toxicity to the development of zebrafish [111]. This indicates that careful biocompatibility studies must be undertaken when CNF surface is modified, either by covalent bonding or just adsorption.

To conclude on health and toxicology concerns, sporadic toxic response can be detected depending on the morphology, the presence of contaminants from the raw materials or the isolation process, and the surface chemistry of CNF. However, the inexistent or low biological response demonstrated in the above-mentioned studies are relevant proof that CNF are cytocompatible and mostly not hazardous to human health. Moreover, good cell proliferation and migration was also observed. Nonetheless, precautionary principle must be respected, and the best way to clarify the potential toxicity of CNF is to implement testing conditions that are as close as possible to the intended end application.

5.2.5 Cellulose Nanofibrils and Medical Applications

Since it is synthetized by bacteria, BNC is free of hazardous chemicals and thus led to the development of the first commercial clinical applications of nanocellulose. Namely, Bioprocess®, XCell®, and Biofill are used for topical applications such as burns treatment as depicted in Figure 5.17. The remarkable conformability and the ability to maintain a moist environment allow for the reduction of patient pain and contributed to the success of these products.

Although enthusiasm for the use of BNC for medical application is large, industrial-scale production facilities are still missing, and improvements have to be made regarding the time required for the production, which is commonly up to two weeks as pointed out in several reviews and papers [79, 129, 130].

Wood-derived nanocellulose is produced at a larger scale and faster, so its use in the biomedical field would be preferable. Many examples of the use of CNF for biomedical application in several applications are available in the literature [94, 131].

As an introduction to the wide overview of the strong academic research that involves CNF for medical application, it should be noted that an industrial grade of CNF designed for biomedical purposes appeared recently. It is produced by the Finnish company UPM under the commercial name of GrowDex® and is extracted from birch hardwood. It is said to be biocompatible and to mimic the ECM that support the growth and differentiation of cells. Moreover, the suspension is dispersed in ultrapure water and is then easily mixed with culture media. Specifically designed to be used in 2D and 3D cell culture, GrowDex CNF hydrogels can also be used in combination with microfluidics devices to mimic organs. At high concentrations, the thick gel acts as a barrier system preventing cell migration without hindering the migration of smaller compounds like nutrients and drugs. Drug delivery systems also benefit from this barrier effect. This commercial product is for research use only, but increasing investigation might soon end up with diagnostic and therapeutic products. For instance, GrowDex CNFs have been proven to better promote the maturation of hepatocyte-like cells [132]. Finally, GrowDex CNF-based medical prototypes were developed such as wound healing, antimicrobial film and hydrogel, scaffold, injectable hydrogels, and bone tissue engineering as described by P. Laurén in a recently released PhD thesis and from which Table 5.5 is extracted [133].

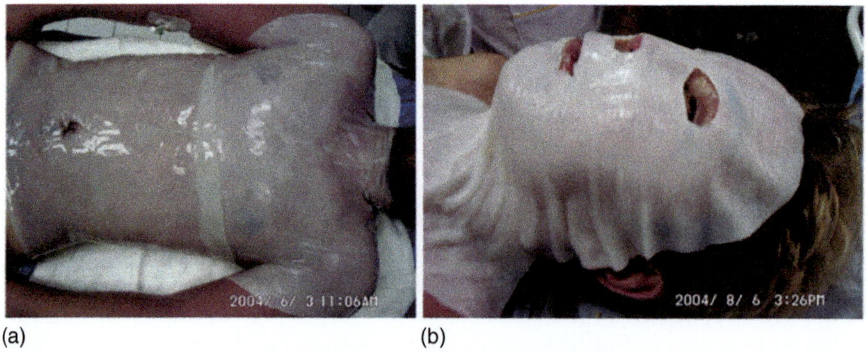

(a) (b)

Figure 5.17 Shape conformability and moist environment for reduction of pain on (a) upper body parts and (b) the facial area suffering from severe burn injuries. *Source:* Czaja et al. 2007 [129]. Reproduced with permission of American Chemical Society.

Table 5.5 Plant-derived CNF-based materials for biomedical applications.

Polymer composition	Formulation	Application
NFC	Aerogel and film	Wound healing
NFC-hemicellulose	Composite hydrogel	Wound healing
NFC[a)]	Surface-modified film	Antimicrobial film
NFC	Crosslinked hydrogel	Antimicrobial hydrogel
NFC[b)]	Hydrogel bioink	Tailor-made wound dressings
NFC	NFC wound dressing	Wound healing (clinical study)
NFC–alginate	Composite hydrogel bioink	Cell-laden ear cartilage scaffold
NFC–carbon nanotube	Conductive hydrogel bioink	Neural tissue engineering
NFC–polyvinyl acetate	Composite polymer film	Self-softening in situ implantation
Plant cellulose tissue	Decellularized scaffold	Subcutaneous implantation (*in vivo study*)
NFC	Hydrogel	Injectable in situ implantation (*in vivo study*)
Anionic nanofibrillated cellulose (ANFC)–chitosan	Hydrogel	Injectable in situ implantation (*in vivo study*)
NFC	Hydrogel	Injectable hydrogel for localized chemotherapy
NFC–alginate	Composite hydrogel	Suture coating for cell therapy (*ex vivo study*)
NFC[c)]	Crosslinked thread	Stem cell delivery (*ex vivo study*)
NFC	Hydrogel	3D organoid development
NFC	Hydrogel	3D cell culture scaffold
NFC	Hydrogel	3D culturing of pluripotent stem cells
NFC-RS/P	Composite film	Bioadhesive film
NFC-PEG	Composite hydrogel	Mucoadhesion
NFC	Aerogel	Gastroretentive drug delivery system
(A)NFC-polymer[d)]	Composite film	Bioadhesive film
NFC-chitin	Composite scaffold	Bone tissue engineering
NFC-gelatin	Composite scaffold	Bone tissue engineering
CNC-GIC[e)]	Composite dental cement	Restorative dentistry

a) Octadecyldimethyl(3-trimethoxysilylpropyl)ammonium chloride modified.
b) Carboxymethylated and periodate oxidated.
c) Glutaraldehyde crosslinked.
d) ANFC- and NFC-pectin, NFC-mucin and NFC-chitosan composites.
e) CNC acquired from NFC and reinforced with glass ionomer cement (GIC).
Source: From Laurén 2018 [133].

One of the pioneering works that involves CNF for drug delivery applications used 2D structures; a matrix carrier in the form of films was produced by filtration and showed drug loadings up to 20–40%. The mechanical properties of the films were suitable for easy handling and shape tailoring. The release was sustained for up to three months with a close to zero-order kinetic due to the tight nanofiber network in which the drug compound was entrapped. This result indicates that CNF can help to extend the release period [134]. Figure 5.18 shows scanning electron microscopy (SEM) pictures of the drug-loaded CNF film before and after the release experiment and the cumulative amount of drug released over time that demonstrates the sustained release. Further characterization of the interaction of different types of drug (size and charge) with CNF was conducted. Permeation through CNF films, binding abilities, and thermodynamics of the binding process to CNF were monitored with high-performance liquid chromatography (HPLC) and isothermal titration calorimetry (ITC). The results suggested a size dependence for the permeation study and that pH and electrostatic forces governed the binding abilities and thermodynamics [135].

CNF 3D structures such as drug-loaded aerogels were also designed for oral drug delivery applications. Four types of CNF were compared with MCC. Hydrophobic drugs were successfully included in these highly porous reservoirs due to amphiphilic hydrophobin proteins. Sustained drug release was obtained for TEMPO CNF and explained by the nanostructured organization and interactions with the protein-coated drug [136]. Similarly, pH-responsive 3D microporous hydrogel structures were produced from periodate-oxidized and carboxymethylated CNF. The ionizable functional groups available at the surface of the material explained the pH-dependent swelling degree (higher at neutral and alkaline pH than in acidic conditions). These hydrogels can achieve controlled and intelligent release of API when exposed to specific pH conditions [137].

When combined with currently used polymers, CNF can improve the properties of existing controlled drug release systems. For instance, CNF has been used to enhance conventional MCC tablets loaded with paracetamol. CNFs allow for a better flowability of powders and lower porosity upon packing even if high compacting forces were required since CNFs are more ductile than MCC. Release was

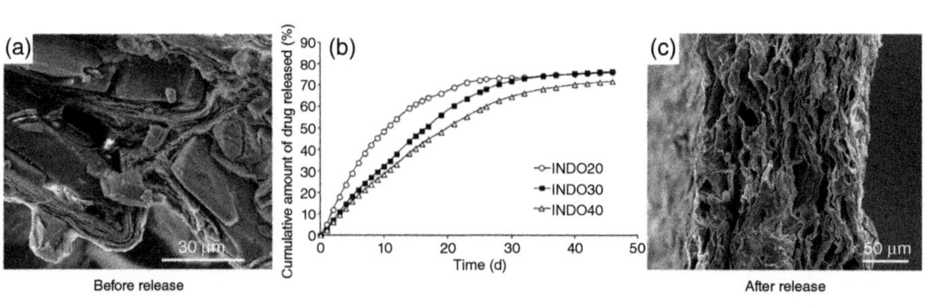

Figure 5.18 Release study of indomethacin from CNF films produced by filtration, (a) SEM pictures of the drug-loaded film before the release showing the drug crystals, (b) cumulative amount of drug released over time for three drug loadings, and (c) SEM pictures of the CNF film after the release. *Source:* Kolakovic et al. 2012 [134]. Reproduced with permission of Elsevier.

also quicker with CNF-enhanced CMC tablets. Depending on the target of the delivery, this might be an advantage [138]. Very recently, CNFs were also used with other biobased polymers like chitosan to prepare a CNF/chitosan transdermal film for the delivery of ketorolac tromethamine (KT) (antipain). The KT CNF/chitosan films exploited the CNF nanobarrier to help to sustain the release of the KT compound as proven by drug release profile and kinetics [139].

In order to combine CNF with other compounds, chemical surface modification can be explored. Polyethyleneimine polymeric chains were covalently bound to CNF surface before freeze drying and production of PEI/CNF aerogels. Those were successfully loaded with a model drug. The release study indicated a pH and temperature dependence of the release profiles. This work resulted in pH- and temperature-responsive release systems based on CNF offering a new innovative alternative to conventional release systems of the pharmaceutical industry [140]. Another interesting publication from P. Laurén used technetium-99m-labeled CNF (^{99m}Tc-CNF) injectable hydrogel for *in vivo* drug release application. The ^{99m}Tc-CNF were able to be traced after subcutaneous injection in the pelvic region of mice. The CNF did not migrate or disintegrate during the study, whereas the animal was awake and free to move as illustrated in Figure 5.19.

Such CNF hydrogels also allowed for a better controlled release of compounds compared with CNF-free injections [141].

Wound healing is another topic where CNF could bring innovative formulations. As a 100% CNF structures, films, and aerogels were used to develop chronic wound dressings and evaluate their response to the growth of *Pseudomonas aeruginosa*. A dose-dependent inhibition of the bacterial growth was observed, and the biofilm formation was hindered by the decreasing porosity and surface roughness of aerogels. These results demonstrate the potential of CNF in the design of novel wound dressings [142]. Similar results were obtained for ion-crosslinked TEMPO CNF aerogels exposed to *Staphylococcus epidermidis*

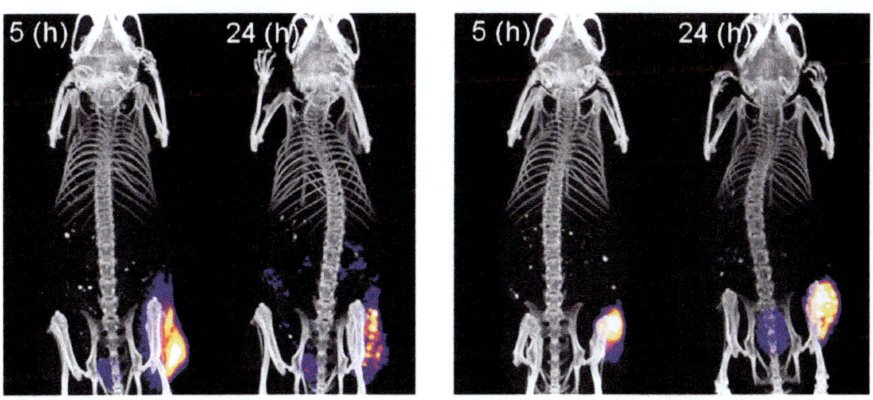

(a) (b)

Figure 5.19 SPECT/CT images of ^{99m}Tc-labeled human serum albumin injected in mice without (a) or with (b) CNF hydrogel. Over 24 hours, the human serum albumin was maintained around the injection site when CNF were used. *Source:* Laurén et al. 2014 [141]. Reproduced with permission of Elsevier.

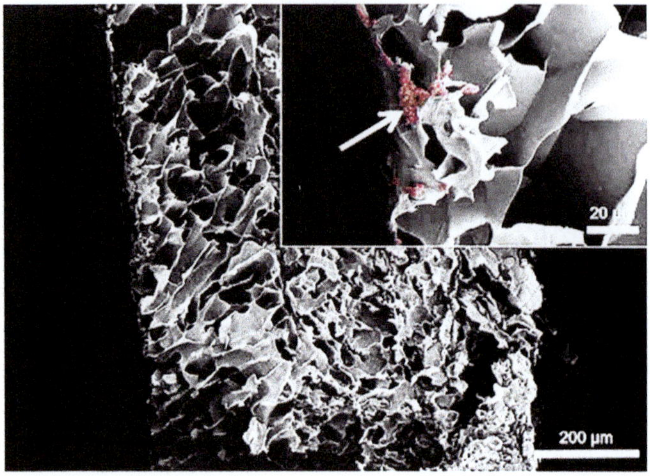

Figure 5.20 SEM pictures of calcium crosslinked TEMPO CNF aerogels exposed to *S. epidermidis* and visualization of bacterial barrier effect on the inset where bacteria grow on the surface of the aerogel but do not penetrate. *Source:* Basu et al. 2018 [143]. Reproduced with permission of Elsevier.

and *P. aeruginosa* bacterial strains [143]. A bacterial barrier effect was also detected with SEM pictures as shown in Figure 5.20.

Researchers also recently evaluated the elastic modulus of CNF hydrogel scaffolds for application in tissue engineering. TEMPO-oxidized CNFs, which are known to bear carboxyl groups but also a few amount of aldehydes groups, were crosslinked with primary diamine compounds of different chain length through the Schiff's base reaction (aldehydes and primary amine). Thus, reversible unstable gels were converted to stable irreversible gels. Moreover, CNF gel structure can be controlled by adjusting the amount of crosslinker or its chain length. This work further promotes the use of CNF-based systems for tissue engineering applications [144].

Similarly to drug release topic, combining CNF with other biobased materials is of importance to beneficiate from the peculiar properties of both materials. CNF hydrogels were reinforced with alginate and were assessed in terms of mechanical properties such as rupture strength, compressibility, and gel rigidity. The best combination used oxidized CNF with alginate hydrogels ending up with mechanically tunable systems and more stable physiological conditions. Such combination of two biobased polymers is of high interest in the scaffolding technology for tissue engineering and bioprinting [145].

Latest research proposes the use of CNF-based inks to produce 3D scaffolds by bioprinting for neural tissue engineering applications. CNF and CNTs as ink constituents allow for the printing of guidelines of less than 1 mm with conductive properties. The results showed that neural cells attached, proliferated, and differentiated on the 3D printed guidelines. An overview of this work is shown in Figure 5.21. Promising application in mimicking brain environment is expected [146].

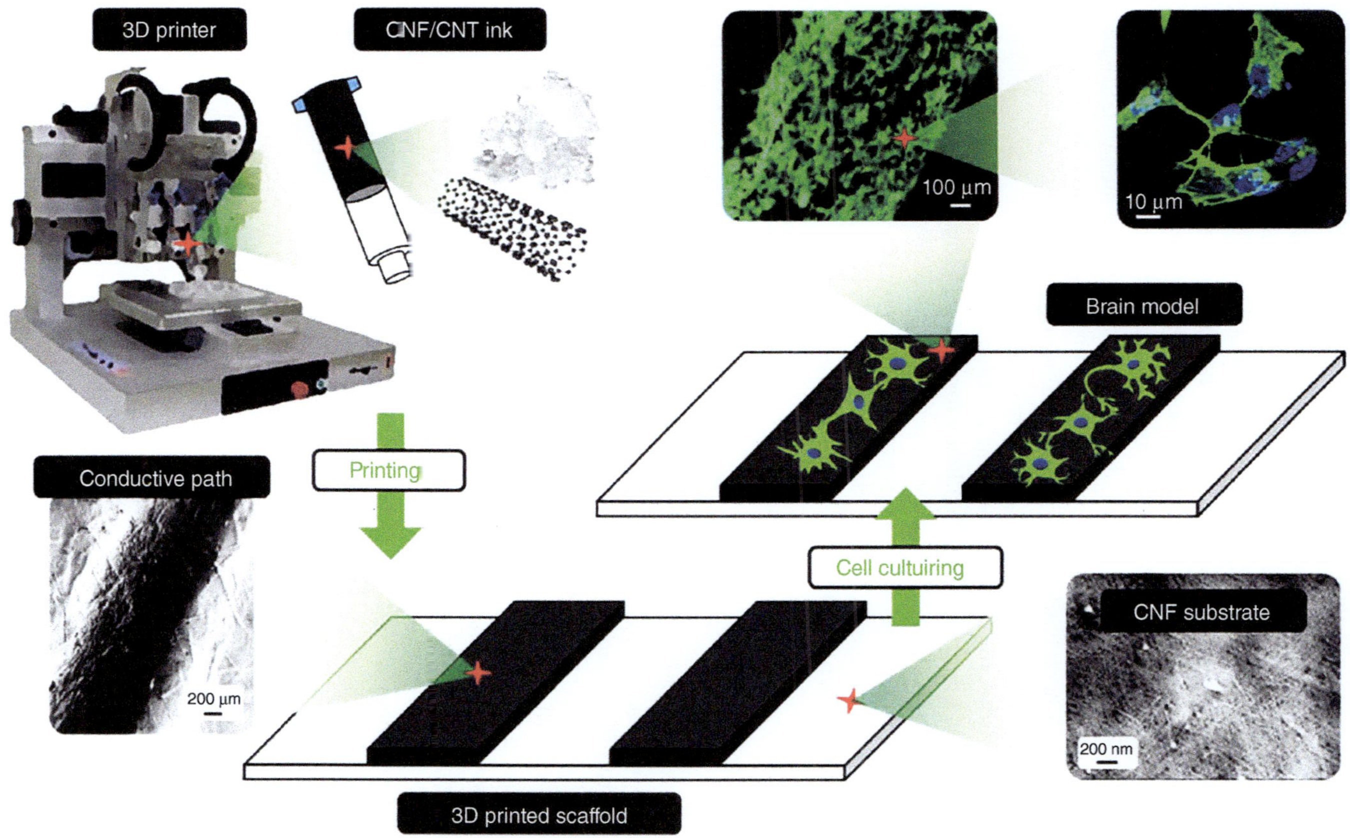

Figure 5.21 Overview of the fabrication of the cellulose-derived nanofibrous scaffolds for neural network development based on CNF-loaded inks. *Source:* Kuzmenko et al. 2018 [146]. Reproduced with permiss on of Elsevier.

5.3 Conclusions and Perspectives

As described in this chapter, a wide range of biobased polymers are suitable for biomedical applications. Among them cellulose deserves a particular attention. CNF forms 2D and 3D biocompatible structures that have high potential in drug release and wound healing or tissue engineering applications. The general context of nanomaterials in medical devices appeals for health and toxicology assessment. For CNF, a low or nonexistent toxicity has been proven and commonly regarded by the scientific community. However, complementary studies are still necessary for long-term impacts since cellulose does not degrade in the human body and could remain within the pulmonary system. In order to be widely used in the medical industry or any other field, nanocellulose often requires enhancement through surface modification for a better compatibility or new functionalities. The growing market of nanocellulose was detailed in terms of patents and field of application and confirmed the tendency for the transition to these biobased compounds. In conclusion, CNF are very promising materials for medical applications. The numerous and diverse research works about its use to improve existing medical formulation or to help designing new healing strategies show the wide potential of such a material.

References

1 Babu, R.P., O'connor, K., and Seeram, R. (2013). Current progress on biobased polymers and their future trends. *Progress in Biomaterials* 2 (1): 8.

2 European Bioplastic (2017). European Bioplastics, Bioplastics Market Data. https://docs.european-bioplastics.org/publications/market_data/2017/Report_Bioplastics_Market_Data_2017.pdf (accessed 10 April 2020).

3 European Commission 2011 Lead market initiative – speed up time-to-market of innovations and pilot new innovation policy in Europe. Growth - European Commission, *Growth*. hhttps://ec.europa.eu/growth/content/lead-market-initiative-–-speed-time-market-innovations-and-pilot-new-innovation-policy-0_en (accessed 07 August 2018).

4 US Department of Agriculture (2014) Agricultural Act of 2014. https://www.govinfo.gov/content/pkg/BILLS-113hr2642enr/pdf/BILLS-113hr2642enr.pdf (accessed 07 August 2018).

5 Avérous, L. and Pollet, E. (2012). Biodegradable polymers. In: *Environmental Silicate Nano-Biocomposites* (eds. L. Avérous and E. Pollet), 13–39. London: Springer.

6 Reis, R.L., Neves, N.M., Mano, J.F. et al. (2008). *Natural-Based Polymers for Biomedical Applications.* Elsevier.

7 Bedian, L., Villalba-Rodríguez, A.M., Hernández-Vargas, G. et al. (2017). Bio-based materials with novel characteristics for tissue engineering applications – a review. *International Journal of Biological Macromolecules* 98: 837–846.

8 Hıncal, A.A. and Kaş, H.S. (1998). *Biomedical Science and Technology: Recent Developments in the Pharmaceutical and Medical Sciences.* Boston, MA: Springer US.

9 Vert, M., Doi, Y., and Hellwich, K.H. (2012). Terminology for biorelated polymers and applications (IUPAC Recommendations 2012). *Pure and Applied Chemistry* 84 (2): 377–410.
10 Lanza, R., Langer, R., and Vacanti, J. (2013). *Principles of Tissue Engineering*, 4e. Elsevier Science.
11 Rice, M.A., Dodson, B.T., Arthur, J.A., and Anseth, K.S. (2005). Cell-based therapies and tissue engineering. *Otolaryngologic Clinics of North America* 38 (2): 199–214.
12 Ma, P.X. (2008). Biomimetic materials for tissue engineering. *Advanced Drug Delivery Reviews* 60 (2): 184–198.
13 Pişkin, E. (1998). Polymer based drug delivery systems. In: *Biomedical Science and Technology: Recent Developments in the Pharmaceutical and Medical Sciences* (eds. A.A. Hıncal and H.S. Kaş). Springer.
14 Williams, D.F. (2009). On the nature of biomaterials. *Biomaterials* 30 (30): 5897–5909.
15 Smyth, M. (2017). Nanocellulose based materials for cell culture. PhD thesis. Université Grenoble Alpes.
16 Love, B.J. (2017). *Biomaterials : A Systems Approach to Engineering Concepts*. Elsevier Science.
17 Srichana, T. and Domb, A.J. (2009). Polymeric biomaterials. In: *Biomedical Materials*, 83–119. Boston, MA: Springer.
18 Almeida, A.P.C., Canejo, J.P., Fernandes, S.N. et al. (2018). Cellulose-based biomimetics and their applications. *Advanced Materials* 30 (19): 1703655.
19 Teeri, T.T., Brumer, H., Daniel, G., and Gatenholm, P. (2007). Biomimetic engineering of cellulose-based materials. *Trends in Biotechnology* 25 (7): 299–306.
20 Tyler, B., Gullotti, D., Mangraviti, A. et al. (2016). Polylactic acid (PLA) controlled delivery carriers for biomedical applications. *Advanced Drug Delivery Reviews* 107: 163–175.
21 Li, Z., Yang, J., and Loh, X.J. (2016). Polyhydroxyalkanoates: opening doors for a sustainable future. *NPG Asia Materials* 8 (4): e265–e265.
22 Jambunathan, P. and Zhang, K. (2016). Engineered biosynthesis of biodegradable polymers. *Journal of Industrial Microbiology and Biotechnology* 43 (8): 1037–1058.
23 Xu, J. and Guo, B.-H. (2010). Poly(butylene succinate) and its copolymers: research, development and industrialization. *Biotechnology Journal* 5 (11): 1149–1163.
24 Gigli, M., Fabbri, M., Lotti, N. et al. (2016). Poly(butylene succinate)-based polyesters for biomedical applications: a review. *European Polymer Journal* 75: 431–460.
25 Salgado, A.J., Coutinho, O.P., Reis, R.L., and Davies, J.E. (2007). In vivo response to starch-based scaffolds designed for bone tissue engineering applications. *Journal of Biomedical Materials Research Part A* 80A (4): 983–989.
26 Pereira, C.S., Cunha, A.M., Reis, R.L. et al. (1998). New starch-based thermoplastic hydrogels for use as bone cements or drug-delivery carriers. *Journal of Materials Science: Materials in Medicine* 9 (12): 825–833.

27 Gomes, M.E., Godinho, J.S., Tchalamov, D. et al. (2002). Alternative tissue engineering scaffolds based on starch: processing methodologies, morphology, degradation and mechanical properties. *Materials Science and Engineering: C* 20 (1): 19–26.
28 Lee, K.Y. and Mooney, D.J. (2012). Alginate: properties and biomedical applications. *Progress in Polymer Science* 37 (1): 106–126.
29 Mandel, K.G., Daggy, B.P., Brodie, D.A., and Jacoby, H.I. (2000). Review article: alginate-raft formulations in the treatment of heartburn and acid reflux. *Alimentary Pharmacology & Therapeutics* 14 (6): 669–690.
30 Lalau, J.D., Bresson, R., Charpentier, P. et al. (2002). Efficacy and tolerance of calcium alginate versus vaseline gauze dressings in the treatment of diabetic foot lesions. *Diabetes & Metabolism* 28 (3): 223–229.
31 Smidsrød, O. and Skjåk-Bræk, G. (1990). Alginate as immobilization matrix for cells. *Trends in Biotechnology* 8: 71–78.
32 Jayakumar, R., Prabaharan, M., Sudheesh Kumar, P.T. et al. (2011). Biomaterials based on chitin and chitosan in wound dressing applications. *Biotechnology Advances* 29 (3): 322–337.
33 Xie, J., Peng, C., Zhao, Q. et al. (2016). Osteogenic differentiation and bone regeneration of iPSC-MSCs supported by a biomimetic nanofibrous scaffold. *Acta Biomaterialia* 29: 365–379.
34 Zhang, Y., Wang, Q.-S., Yan, K. et al. (2016). Preparation, characterization, and evaluation of genipin crosslinked chitosan/gelatin three-dimensional scaffolds for liver tissue engineering applications. *Journal of Biomedical Materials Research Part A* 104 (8): 1863–1870.
35 Gandini, A. and Belgacem, M.N. (2008). The state of the art. In: *Monomers, Polymers and Composites from Renewable Resources*, Chapter 1 (eds. M.N. Belgacem and A. Gandini), 1–16. Amsterdam: Elsevier.
36 O'sullivan, A.C. (1997). Cellulose: the structure slowly unravels. *Cellulose* 4 (3): 173–207.
37 Hon, D.N.-S. (1994). Cellulose: a random walk along its historical path. *Cellulose* 1 (1): 1–25.
38 Jorfi, M. and Foster, E.J. (2015). Recent advances in nanocellulose for biomedical applications. *Journal of Applied Polymer Science* 132 (14): 41719.
39 Hoenich, N.A. (2007). Cellulose for medical applications: past, present, and future. *BioResources* 1 (2): 270–280.
40 Shibata, T. (2011). Cellulose and its derivatives in medical use. In: *Renewable Resources for Functional Polymers and Biomaterials* (ed. P.A. Williams), 48–87. *RCS Royal Chemistry Society*, Chapter 3.
41 Ek, R., Alderborn, G., and Nyström, C. (1994). Particle analysis of microcrystalline cellulose: differentiation between individual particles and their agglomerates. *International Journal of Pharmaceutics* 111 (1): 43–50.
42 Dumitriu, S. (1996). *Polysaccharides in Medicinal Applications*. CRC Press.
43 Sjostrom, E. (2013). *Wood Chemistry: Fundamentals and Applications*. Elsevier.
44 Kamel, S., Ali, N., Jahangir, K. et al. (2008). Pharmaceutical significance of cellulose: a review. *Express Polymer Letters* 2 (11): 758–778.

45 Siepmann, J., Kranz, H., Bodmeier, R., and Peppas, N.A. (1999). HPMC-matrices for controlled drug delivery: a new model combining diffusion, swelling, and dissolution mechanisms and predicting the release kinetics. *Pharmaceutical Research* 16 (11): 1748–1756.

46 Colombo, P., Bettini, R., and Peppas, N.A. (1999). Observation of swelling process and diffusion front position during swelling in hydroxypropyl methyl cellulose (HPMC) matrices containing a soluble drug. *Journal of Controlled Release* 61 (1): 83–91.

47 Siepmann, J. and Peppas, N.A. (2001). Modeling of drug release from delivery systems based on hydroxypropyl methylcellulose (HPMC). *Advanced Drug Delivery Reviews* 48: 139–157.

48 Lin, S.-Y. and Kawashima, Y. (1987). Drug release from tablets containing cellulose acetate phthalate as an additive or enteric-coating material. *Pharmaceutical Research* 4 (1): 70–74.

49 Chang, C. and Zhang, L. (2011). Cellulose-based hydrogels: present status and application prospects. *Carbohydrate Polymers* 84 (1): 40–53.

50 Bourges, X., Weiss, P., Daculsi, G., and Legeay, G. (2002). Synthesis and general properties of silated-hydroxypropyl methylcellulose in prospect of biomedical use. *Advances in Colloid and Interface Science* 99 (3): 215–228.

51 Demitri, C., Sole, R.D., Scalera, F. et al. (2008). Novel superabsorbent cellulose-based hydrogels crosslinked with citric acid. *Journal of Applied Polymer Science* 110 (4): 2453–2460.

52 Wach, R.A., Mitomo, H., Yoshii, F., and Kume, T. (2001). Hydrogel of biodegradable cellulose derivatives. II. Effect of some factors on radiation-induced crosslinking of CMC. *Journal of Applied Polymer Science* 81 (12): 3030–3037.

53 Nizam El-Din, H.M., Abd Alla, S.G., and El-Naggar, A.W.M. (2010). Swelling and drug release properties of acrylamide/carboxymethyl cellulose networks formed by gamma irradiation. *Radiation Physics and Chemistry* 79 (6): 725–730.

54 Abitbol, T., Rivkin, A., Cao, Y. et al. (2016). Nanocellulose, a tiny fiber with huge applications. *Current Opinion in Biotechnology* 39: 76–88.

55 Kargarzadeh, H., Mariano, M., Gopakumar, D. et al. (2018). Advances in cellulose nanomaterials. *Cellulose* 25 (4): 2151–2189.

56 Nechyporchuk, O., Belgacem, M.N., and Bras, J. (2016). Production of cellulose nanofibrils: a review of recent advances. *Industrial Crops and Products* 93: 2–25.

57 Dufresne, A. (2017). *Nanocellulose, From Nature to High Performance Tailored Materials*. Berlin: De Gruyter.

58 Rånby, B.G. and Ribi, E. (1950). Über den Feinbau der Zellulose. *Experientia* 6 (1): 12–14.

59 Rånby, B.G. (1951). Fibrous macromolecular systems. Cellulose and muscle. The colloidal properties of cellulose micelles. *Discussions of the Faraday Society* 11: 158–164.

60 Nickerson, R.F. and Habrle, J.A. (1947). Cellulose intercrystalline structure. *Industrial and Engineering Chemistry* 39 (11): 1507–1512.

61 de Souza Lima, M.M. and Borsali, R. (2004). Rodlike cellulose microcrystals: structure, properties, and applications. *Macromolecular Rapid Communications* 25 (7): 771–787.
62 ISO/TS 20477:2017. *Nanotechnologies – standard terms and their definition for cellulose nanomaterial*. https://www.iso.org/standard/68153.html (accessed 09 October 2018).
63 F.W. Herrick, R.L. Casebier, J.K. Hamilton, and K.R. Sandberg (1982). Microfibrillated cellulose: morphology and accessibility. Conference: 9. cellulose conference, Syracuse, NY, USA, 24 May 1982, vol. 37.
64 A.F. Turbak, F.W. Snyder, and K.R. Sandberg (1982). Microfibrillated cellulose, a new cellulose product: properties, uses, and commercial potential. Conference: 9. cellulose conference, Syracuse, NY, USA, 24 May 1982, vol. 37.
65 Spence, K.L., Venditti, R.A., Rojas, O.J. et al. (2011). A comparative study of energy consumption and physical properties of microfibrillated cellulose produced by different processing methods. *Cellulose* 18 (4): 1097–1111.
66 Tejado, A., Alam, M.N., Antal, M. et al. (2012). Energy requirements for the disintegration of cellulose fibers into cellulose nanofibers. *Cellulose* 19 (3): 831–842.
67 Pääkkö, M., Ankerfors, M., Kosonen, H. et al. (2007). Enzymatic hydrolysis combined with mechanical shearing and high-pressure homogenization for nanoscale cellulose fibrils and strong gels. *Biomacromolecules* 8 (6): 1934–1941.
68 Saito, T., Nishiyama, Y., Putaux, J.-L. et al. (2006). Homogeneous suspensions of individualized microfibrils from TEMPO-catalyzed oxidation of native cellulose. *Biomacromolecules* 7 (6): 1687–1691.
69 Besbes, I., Alila, S., and Boufi, S. (2011). Nanofibrillated cellulose from TEMPO-oxidized eucalyptus fibres: effect of the carboxyl content. *Carbohydrate Polymers* 84 (3): 975–983.
70 Aulin, C., Johansson, E., Wågberg, L., and Lindström, T. (2010). Self-organized films from cellulose I nanofibrils using the layer-by-layer technique. *Biomacromolecules* 11 (4): 872–882.
71 Rol, F., Belgacem, M.N., Gandini, A., and Bras, J. (2018). Recent advances in surface-modified cellulose nanofibrils. *Progress in Polymer Science* 88: 241–264.
72 Henriksson, M., Henriksson, G., Berglund, L.A., and Lindström, T. (2007). An environmentally friendly method for enzyme-assisted preparation of microfibrillated cellulose (MFC) nanofibers. *European Polymer Journal* 43 (8): 3434–3441.
73 Nechyporchuk, O., Pignon, F., and Belgacem, M.N. (2015). Morphological properties of nanofibrillated cellulose produced using wet grinding as an ultimate fibrillation process. *Journal of Materials Science* 50 (2): 531–541.
74 Johansen, K.S. (2016). Lytic polysaccharide monooxygenases: the microbial power tool for lignocellulose degradation. *Trends in Plant Science* 21 (11): 926–936.
75 Davis, N.J. and Flitsch, S.L. (1993). Selective oxidation of monosaccharide derivatives to uronic acids. *Tetrahedron Letters* 34 (7): 1181–1184.

76 Isogai, A., Saito, T., and Fukuzumi, H. (2011). TEMPO-oxidized cellulose nanofibers. *Nanoscale* 3 (1): 71–85.

77 Saito, T., Hirota, M., Tamura, N. et al. (2009). Individualization of nano-sized plant cellulose fibrils by direct surface carboxylation using TEMPO catalyst under neutral conditions. *Biomacromolecules* 10 (7): 1992–1996.

78 Garrett, T.R., Bhakoo, M., and Zhang, Z. (2008). Bacterial adhesion and biofilms on surfaces. *Progress in Natural Science* 18 (9): 1049–1056.

79 Klemm, D., Kramer, F., Moritz, S. et al. (2011). Nanocelluloses: a new family of nature-based materials. *Angewandte Chemie International Edition* 50 (24): 5438–5466.

80 Mohanty, S., Nayak, S.K., Kaith, B.S., and Kalia, S. (2015). *Polymer Nanocomposites Based on Inorganic and Organic Nanomaterials*. Wiley.

81 Mamlouk, D. and Gullo, M. (2013). Acetic acid bacteria: physiology and carbon sources oxidation. *Indian Journal of Microbiology* 53 (4): 377–384.

82 Invention controls weavers of nanoscale biomaterials. https://vtnews.vt.edu/content/vtnews_vt_edu/en/articles/2008/11/2008-693.html (accessed 30 January 2018).

83 Helenius, G., Bäckdahl, H., Bodin, A. et al. (2006). In vivo biocompatibility of bacterial cellulose. *Journal of Biomedical Materials Research* 76A (2): 431–438.

84 Petersen, N. and Gatenholm, P. (2011). Bacterial cellulose-based materials and medical devices: current state and perspectives. *Applied Microbiology and Biotechnology* 91 (5): 1277–1286.

85 Rajwade, J.M., Paknikar, K.M., and Kumbhar, J.V. (2015). Applications of bacterial cellulose and its composites in biomedicine. *Applied Microbiology and Biotechnology* 99 (6): 2491–2511.

86 Moon, R.J. and Foster, E.J. (2018). Current characterization methods for cellulose nanomaterials. *Chemical Society Reviews* 47 (8): 2609–2679.

87 Kangas, H., Lahtinen, P., Sneck, A. et al. (2014). Characterization of fibrillated celluloses. A short review and evaluation of characteristics with a combination of methods. *Nordic Pulp and Paper Research Journal* 29 (1): 129–143.

88 Desmaisons, J., Boutonnet, E., Rueff, M. et al. (2017). A new quality index for benchmarking of different cellulose nanofibrils. *Carbohydrate Polymers* 174: 318–329.

89 Märtson, M., Viljanto, J., Hurme, T. et al. (1999). Is cellulose sponge degradable or stable as implantation material? An in vivo subcutaneous study in the rat. *Biomaterials* 20 (21): 1989–1995.

90 Märtson, M., Viljanto, J., Hurme, T., and Saukko, P. (1998). Biocompatibility of cellulose sponge with bone. *European Surgical Research* 30 (6): 426–432.

91 Märtson, M., Viljanto, J., Laippala, P., and Saukko, P. (1998). Connective tissue formation in subcutaneous cellulose sponge implants in the rat. *European Surgical Research* 30 (6): 419–425.

92 Miyamoto, T., Takahashi, S., Ito, H. et al. (1989). Tissue biocompatibility of cellulose and its derivatives. *Journal of Biomedical Materials Research* 23 (1): 125–133.

93 Bowry, S.K. and Rintelen, T.H. (1998). Synthetically modified cellulose (SMC): a cellulosic hemodialysis membrane with minimized complement activation. *ASAIO Journal* 44 (5): M579–M583.

94 Lin, N. and Dufresne, A. (2014). Nanocellulose in biomedicine: current status and future prospect. *European Polymer Journal* 59: 302–325.

95 Kovacs, T., Naish, V., O'Connor, B. et al. (2010). An ecotoxicological characterization of nanocrystalline cellulose (NCC). *Nanotoxicology* 4 (3): 255–270.

96 Clift, M.J.D., Foster E.J., Vanhecke D. et al. (2011). Investigating the interaction of cellulose nanofibers derived from cotton with a sophisticated 3D human lung cell coculture. http://pubs.acs.org/doi/abs/10.1021/bm200865j (accessed 02 August 2018).

97 Norppa, H. (2012). Nanofibrillated cellulose: results of in vitro and in vivo toxicological assays, Paper presented at Sunpap conference (19–20 June 2012).

98 Ong, K.J., Shatkin, J.A., Nelson, K. et al. (2017). Establishing the safety of novel bio-based cellulose nanomaterials for commercialization. *NanoImpact* 6: 19–29.

99 Hannukainen, K.-S., Suhonen, S., Savolainen, K., and Norppa, H. (2012). Genotoxicity of nanofibrillated cellulose in vitro as measured by enzyme comet assay. *Toxicology Letters* 211: S71.

100 Alexandrescu, L., Syverud, K., Gatti, A., and Chinga-Carrasco, G. (2013). Cytotoxicity tests of cellulose nanofibril-based structures. *Cellulose* 20 (4): 1765–1775.

101 Hua, K., Carlsson, D., Ålander, E. et al. (2014). Translational study between structure and biological response of nanocellulose from wood and green algae. *RSC Advances* 4 (6): 2892–2903.

102 Pitkänen, M., Sneck A., Hentze H.-P. et al. (2010). Nanofibrillar cellulose – assessment of cytotoxic and genotoxic properties in vitro. Presented at the Tappi International Conference on Nanotechnology for the Forest Products Industry, Espoo, Finland.

103 Vartiainen, J., Pöhler, T., Sirola, K. et al. (2011). Health and environmental safety aspects of friction grinding and spray drying of microfibrillated cellulose. *Cellulose* 18 (3): 775–786.

104 Kollar, P., Závalová, V., Hošek, J. et al. (2011). Cytotoxicity and effects on inflammatory response of modified types of cellulose in macrophage-like THP-1 cells. *International Immunopharmacology* 11 (8): 997–1001.

105 Pitkänen, M., Kangas, H., Laitinen, O. et al. (2014). Characteristics and safety of nano-sized cellulose fibrils. *Cellulose* 21 (6): 3871–3886.

106 Stefaniak, A.B., Seehra, M.S., Fix, N.R., and Leonard, S.S. (2014). Lung biodurability and free radical production of cellulose nanomaterials. *Inhalation Toxicology* 26 (12): 733–749.

107 Čolić, M., Mihajlović, D., Mathew, A. et al. (2015). Cytocompatibility and immunomodulatory properties of wood based nanofibrillated cellulose. *Cellulose* 22 (1): 763–778.

108 Hua, K., Ålander, E., Lindström, T. et al. (2015). Surface chemistry of nanocellulose fibers directs monocyte/macrophage response. *Biomacromolecules* 16 (9): 2787–2795.

109 Vikman, M., Vartiainen, J., Tsitko, I., and Korhonen, P. (2015). Biodegradability and compostability of nanofibrillar cellulose-based products. *Journal of Polymers and the Environment* 23 (2): 206–215.

110 Nordli, H.R., Chinga-Carrasco, G., Rokstad, A.M., and Pukstad, B. (2016). Producing ultrapure wood cellulose nanofibrils and evaluating the cytotoxicity using human skin cells. *Carbohydrate Polymers* 150: 65–73.

111 Harper, B.J., Clendaniel, A., Sinche, F. et al. (2016). Impacts of chemical modification on the toxicity of diverse nanocellulose materials to developing zebrafish. *Cellulose* 23 (3): 1763–1775.

112 Tomić, S., Kokol, V., Mihajlović, D. et al. (2016). Native cellulose nanofibrills induce immune tolerance *in vitro* by acting on dendritic cells. *Scientific Reports* 6: 31618.

113 Catalán, J., Rydman, E., Aimonen, K. et al. (2017). Genotoxic and inflammatory effects of nanofibrillated cellulose in murine lungs. *Mutagenesis* 32 (1): 23–31.

114 Menas, A.L., Yanamala, N., Farcas, M.T. et al. (2017). Fibrillar vs crystalline nanocellulose pulmonary epithelial cell responses: cytotoxicity or inflammation? *Chemosphere* 171: 671–680.

115 Lopes, V.R., Sanchez-Martinez, C., Strømme, M., and Ferraz, N. (2017). In vitro biological responses to nanofibrillated cellulose by human dermal, lung and immune cells: surface chemistry aspect. *Particle and Fibre Toxicology* 14 (1): 1.

116 Rashad, A., Mustafa, K., Heggset, E.B., and Syverud, K. (2017). Cytocompatibility of wood-derived cellulose nanofibril hydrogels with different surface chemistry. *Biomacromolecules* 18 (4): 1238–1248.

117 Liu, J., Bacher, M., Rosenau, T. et al. (2018). Potentially immunogenic contaminants in wood-based and bacterial nanocellulose: assessment of endotoxin and (1,3)-β-D-glucan levels. *Biomacromolecules* 19 (1): 150–157.

118 Ventura, C., Lourenço, A.F., Sousa-Uva, A. et al. (2018). Evaluating the genotoxicity of cellulose nanofibrils in a co-culture of human lung epithelial cells and monocyte-derived macrophages. *Toxicology Letters* 291: 173–183.

119 Ogonowski, M., Edlund, U., Gorokhova, E. et al. (2018). Multi-level toxicity assessment of engineered cellulose nanofibrils in Daphnia magna. *Nanotoxicology* 12 (6): 509–521.

120 Ilves, M., Vilske, S., Aimonen, K. et al. (2018). Nanofibrillated cellulose causes acute pulmonary inflammation that subsides within a month. *Nanotoxicology* 12 (7): 1–18.

121 Pyrgiotakis, G., Luu, W., Zhang, Z. et al. (2018). Development of high throughput, high precision synthesis platforms and characterization methodologies for toxicological studies of nanocellulose. *Cellulose* 25 (4): 2303–2319.

122 Souza, S.F., Mariano, M., Reis, D. et al. (2018). Cell interactions and cytotoxic studies of cellulose nanofibers from Curauá natural fibers. *Carbohydrate Polymers* 201: 87–95.

123 Endes, C., Camarero-Espinosa, S., Mueller, S. et al. (2016). A critical review of the current knowledge regarding the biological impact of nanocellulose. *Journal of Nanobiotechnology* 14 (1): 78.

124 Kangas, H. and Pitkänen, M. (2016). Environmental, Health & Safety (EHS) aspects of cellulose nanomaterials (CN) and CN-based products. *Nordic Pulp and Paper Research Journal* 31 (2): 179–190.
125 Roman, M. (2015). Toxicity of cellulose nanocrystals: a review. *Industrial Biotechnology* 11 (1): 25–33.
126 Wouk, A.F.P.D.F., Diniz, J.M., CíRio, S.M. et al. (1998). Membrana biológica (Biofill) – estudo comparativo com outros agentes promotores da cicatrização da pele em suínos: aspectos clínicos, histopatológicos e morfométricos. *Archives of Veterinary Science* 3 (1): 31–37.
127 Klemm, D., Schumann, D., Udhardt, U., and Marsch, S. (2001). Bacterial synthesized cellulose – artificial blood vessels for microsurgery. *Progress in Polymer Science* 26: 1561–1603.
128 Schumann, D.A., Wippermann, J., Klemm, D.O. et al. (2009). Artificial vascular implants from bacterial cellulose: preliminary results of small arterial substitutes. *Cellulose* 16 (5): 877–885.
129 Czaja, W.K., Young, D.J., Kawecki, M., and Brown, R.M. (2007). The future prospects of microbial cellulose in biomedical applications. *Biomacromolecules* 8 (1): 1–12.
130 Sheykhnazari, S., Tabarsa, T., Ashori, A. et al. (2011). Bacterial synthesized cellulose nanofibers; effects of growth times and culture mediums on the structural characteristics. *Carbohydrate Polymers* 86 (3): 1187–1191.
131 Halib, N., Perrone, F., Cemazar, M. et al. (2017). Potential applications of nanocellulose-containing materials in the biomedical field. *Materials* 10 (8): 977.
132 Toivonen, S., Malinen, M.M., Küblbeck, J. et al. (2016). Regulation of human pluripotent stem cell-derived hepatic cell phenotype by three-dimensional hydrogel models. *Tissue Engineering Part A* 22 (13–14): 971–984.
133 Laurén, P. (2018). Biomedical applications of nanofibrillar cellulose. PhD thesis.
134 Kolakovic, R., Peltonen, L., Laukkanen, A. et al. (2012). Nanofibrillar cellulose films for controlled drug delivery. *European Journal of Pharmaceutics and Biopharmaceutics* 82 (2): 308–315.
135 Kolakovic, R., Peltonen, L., Laukkanen, A. et al. (2013). Evaluation of drug interactions with nanofibrillar cellulose. *European Journal of Pharmaceutics and Biopharmaceutics* 85 (3): 1238–1244.
136 Valo, H., Arola, S., Laaksonen, P. et al. (2013). Drug release from nanoparticles embedded in four different nanofibrillar cellulose aerogels. *European Journal of Pharmaceutical Sciences* 50 (1): 69–77.
137 Chinga-Carrasco, G. and Syverud, K. (2014). Pretreatment-dependent surface chemistry of wood nanocellulose for pH-sensitive hydrogels. *Journal of Biomaterials Applications* 29 (3): 423–432.
138 Kolakovic, R., Peltonen, L., Laaksonen, T. et al. (2011). Spray-dried cellulose nanofibers as novel tablet excipient. *AAPS PharmSciTech* 12 (4): 1366–1373.
139 Sarkar, G., Orasugh, J.T., Saha, N.R. et al. (2017). Cellulose nanofibrils/chitosan based transdermal drug delivery vehicle for controlled release of ketorolac tromethamine. *New Journal of Chemistry* 41 (24): 15312–15319.

140 Zhao, J., Lu, C., He, X. et al. (2015). Polyethylenimine-grafted cellulose nanofibril aerogels as versatile vehicles for drug delivery. *ACS Applied Materials & Interfaces* 7 (4): 2607–2615.
141 Laurén, P., Lou, Y.-R., Raki, M. et al. (2014). Technetium-99m-labeled nanofibrillar cellulose hydrogel for in vivo drug release. *European Journal of Pharmaceutical Sciences* 65: 79–88.
142 Jack, A.A., Nordli, H.R., Powell, L.C. et al. (2017). The interaction of wood nanocellulose dressings and the wound pathogen *P. aeruginosa*. *Carbohydrate Polymers* 157: 1955–1962.
143 Basu, A., Heitz, K., Strømme, M. et al. (2018). Ion-crosslinked wood-derived nanocellulose hydrogels with tunable antibacterial properties: candidate materials for advanced wound care applications. *Carbohydrate Polymers* 181: 345–350.
144 Syverud, K., Pettersen, S.R., Draget, K., and Chinga-Carrasco, G. (2015). Controlling the elastic modulus of cellulose nanofibril hydrogels – scaffolds with potential in tissue engineering. *Cellulose* 22 (1): 473–481.
145 Aarstad, O., Heggset, E.B., Pedersen, I.S. et al. (2017). Mechanical properties of composite hydrogels of alginate and cellulose nanofibrils. *Polymers* 9 (8): 378.
146 Kuzmenko, V., Karabulut, E., Pernevik, E. et al. (2018). Tailor-made conductive inks from cellulose nanofibrils for 3D printing of neural guidelines. *Carbohydrate Polymers* 189: 22–30.

6

Advances in Mucin Biopolymer Research: Purification, Characterization, and Applications

Matthias Marczynski, Benjamin Winkeljann, and Oliver Lieleg

Technical University of Munich, Department of Mechanical Engineering and Munich School of Bioengineering, Biomaterials and Bio-Interfaces Lab, Boltzmannstraße 11, 85748 Garching, Germany

6.1 Introduction

In our daily life, slimy substances are typically considered to be yucky and disgusting – or at least annoying. For some of them such as drainpipe goo (i.e. biofilm, a viscoelastic material produced by bacteria), this is certainly a valid categorization as this slime variant can cause lots of problems both in private households and in industry. However, slime produced by the human body (=mucus, e.g. saliva and snot) has lots of beneficial properties that are pivotal for many physiological processes to function properly. Examples for such processes include blinking, swallowing of food, and copulation. To prevent mechanical damage on the tissues participating in those processes and to guarantee a long-lasting functionality of the related body parts, proper lubrication and wear protection of the involved epithelia are required. This is accomplished by several variants of mucus, a viscoelastic substance that covers all mucosal membranes of mammalians (Figure 6.1a). In addition to acting as lubricants and protecting wet epithelial surfaces from dehydration, these mucus systems also serve as a host environment for the gut microbiome, regulate the passage of molecules and food particles toward the tissue in the gastrointestinal tract, and act as a defense layer against pathogens (Figure 6.1b).

In addition to water, the main constituent of all mucus systems is mucins, a class of high molecular weight (~2–3 MDa) glycoproteins. Those mucins are not only crucial components of tears, saliva, lung, and gastrointestinal mucus but also constitute the major building blocks of cervical mucus. Interestingly, mucins can also be found in breast milk [1–3], which suggests that newborn babies are partially supplied with these glycoproteins by their mother until the infant has produced sufficient quantities of these important macromolecules on its own. During the last 20 years, the number of mucin-related publications has been continuously growing so that, in 2016, about 1800 articles were published on this topic [4]. On the one hand, increasing interest in studying mucins is driven by the fact that they play a key role in many physiological processes [5]; alterations in the composition or function of mucosal systems often lead to

Biopolymers for Biomedical and Biotechnological Applications, First Edition.
Edited by Bernd H. A. Rehm and M. Fata Moradali.

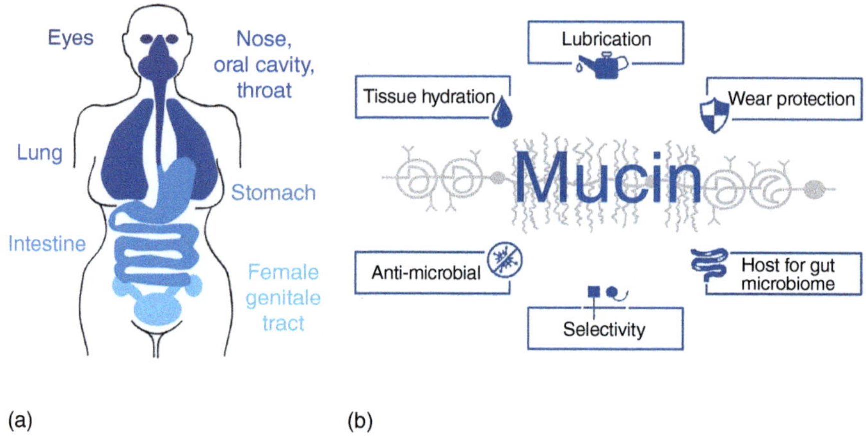

Figure 6.1 Mucins in the human body. (a) Mucus covers all wet epithelia and is associated with various organs. (b) Mucin glycoproteins are the macromolecular key components of mucus. Depending on the particular mucus system they are located in, mucins can fulfill a broad range of functions.

illnesses including dry eye/dry mouth, cystic fibrosis, or some forms of cancer. On the other hand, mucins have been shown to be very promising candidates for applications in the field of biomedical engineering [6], which will probably greatly increase the demand for purification or biotechnological production of mucins in the near future.

In this chapter, we briefly describe how mucins can be obtained from animal sources and summarize our current understanding of how the different chemical motifs of mucin glycoproteins contribute to the broad spectrum of mucin properties. Moreover, we list methods that are currently used in the field of mucin research to characterize mucins and mucin-based solutions and gels. Finally, we describe several case studies that indicate how their applications in the field of biomedical engineering could benefit from purified mucins as biomimetic lubricants, macromolecular coatings, or building blocks for developing hydrogel materials.

6.2 Mucin Sources and Purification Process

All vertebrates and some invertebrates produce some sort of mucus. Thus, in principle, there are many options for harvesting crude mucus from which mucins can be purified. Traditionally, as commonly adopted for commercial purposes, pigs and cows have served as the main sources for harvesting mucus, and both porcine gastric mucin (PGM) MUC5AC and bovine submaxillary mucin (BSM) MUC5B have been extensively used in many studies. One reason for this choice might be the abundant availability of bovine mucus samples in the United States (where beef is – in addition to chicken – the main item on the population's meat menu) and porcine mucus samples in Europe (especially in Germany, pork is very popular, so pigs are readily available from slaughterhouses). However, there are also other significant sources of mucus in nature such as snails, hagfish, and corals. Yet, to this point, mucins from those alternative sources have been characterized to a lesser extent; moreover, the cultivation of snails and corals for a large-scale biotechnological production of purified mucins is certainly more challenging compared with that of pigs and cows.

In recent years, several studies revealed that native and functional mucin glycoproteins exhibit promising properties, for example, in pathogen defense [7, 8], adsorption onto and lubrication of surfaces [9–12], and prevention of wear [13, 14], which make them attractive molecules for biomedical applications. Unfortunately, mucin research is still limited by the availability of such functional mucins. Although the poor quality of commercial mucins has been documented [15, 16], commercial mucin products have been largely applied in numerous studies by different research groups. It is likely that harsh conditions during industrial purification processes can damage the mucin glycoproteins, leaving corrupted molecules with altered physiochemical properties. Thus, results obtained with such damaged mucins are questionable, at best. Nevertheless, there are an increasing number of studies using in-lab purified mucins that have preserved their broad spectrum of biomedically important properties. Owing to the availability of those high-quality mucins – at least in small amounts – lots of progresses have been made in elucidating the structure–function relationship of mucin glycoproteins and evaluating the potential of using purified mucins in biomedical applications, and it seems likely that the demand for intact mucins will further increase.

In this section, we present a short overview of lab methods for the purification of different mucin variants as obtained from a variety of animal sources. Figure 6.2 depicts a schematic flow diagram of the principal steps required for a typical mucin purification process. The first step in every purification process chain is the harvesting of mucin-containing crude material from an animal tissue. Depending on the mucus source, different harvesting techniques are employed. For collecting mucus samples from stomachs and intestines (e.g. from pigs and mice), the mucus can be scraped, pulled, or squeezed from the internal surface of opened and rinsed organs [16–20]. Schömig et al. [21] used a solid/liquid/liquid extraction to purify mucins from homogenized pig stomachs while simultaneously removing lipids from the crude material. However, these methods are not applicable to obtain gastric or intestinal mucins from humans in large quantities, since the availability of these organs is limited to surgically resected specimens [22]. Instead, such human mucin variants are typically produced by cancer cell tumors that are grown in mice, and those mucus samples can be processed as outlined above [23]. Other human mucin sources such as saliva, tear fluid, and airways mucus are more easily accessible and can be collected directly from donors [24–28]. Nonmammalian mucin sources include fish and invertebrates. For instance, slime from hagfish can be obtained from live, anesthetized animals by electrical stimulation of superficial slime glands [29]. Jellyfish mucins ("qniumucins") can be harvested from sliced jellyfish mesoglea by means of precipitation with ethanol and subsequent recovery in water [30], whereas coral mucus is obtained by collecting the exudate of inverted corals exposed to air [31, 32].

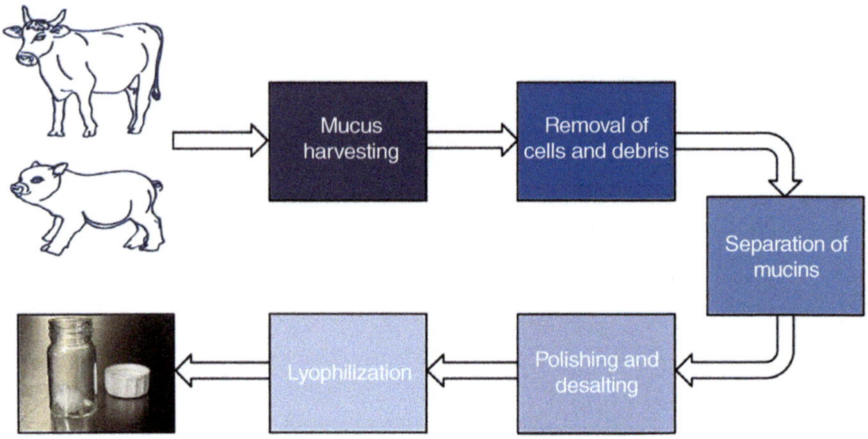

Figure 6.2 Purification of mucins from animal sources. A typical process for the purification of functional mucins comprises five steps. Mucus has to be harvested from animal tissues to obtain crude mucin-containing material that can be further processed. The removal of cells and cellular debris is followed by a technique that isolates the mucin glycoproteins from the majority of molecular impurities. Afterward, mucins are typically desalted and further purified to remove remaining molecular contaminants. Finally, lyophilization yields a protein powder that can be stored for extended time periods.

An inevitable consequence of harvesting mucus from animal tissues is the contamination of the crude material with cells, cellular debris, and tissue fragments. Therefore, after homogenization of the primary material in an appropriate buffer (which usually contains reducing agents or guanidinium chloride to ensure dissolution of mucin oligomers [27], as well sodium azide and protease inhibitors to prevent degradation of mucins by bacteria and proteases [2, 33, 34]), typically one or more (ultra)centrifugation steps are performed to guarantee the removal of macroscopic impurities [16, 24, 35, 36]. The purification of membrane-associated mucins requires an additional step before centrifugation can be conducted, i.e. the extraction of these proteins using detergents such as Triton X-100 [37, 38]. Some protocols even combine the removal of macroscopic impurities with a mucin isolation step, which can be accomplished by isopycnic cesium chloride density gradient centrifugation [19, 27, 31, 33, 34, 39]. An alternative for separating mucins from other constituents of the primary mucus material is using chromatography techniques [27]. Size exclusion chromatography (gel filtration) is frequently used, since the high molecular weight of mucins makes it relatively easy to separate them from other biomolecules [2, 16, 21, 36, 40]. Furthermore, attempts were made to purify mucins by means of ion exchange chromatography, and this strategy makes use of the polyanionic character of mucin glycoproteins [41–43].

Both CsCl gradient centrifugation and chromatography techniques are suitable for the isolation of mucins in good purity. After these process steps, however, mucin samples still contain considerable amounts of salts and minor impurities. These can be efficiently removed by dialyzing the mucin-containing solutions against ultrapure or deionized water [16, 19, 20, 31, 35]. Afterward, subsequent lyophilization is a convenient method to create a mucin powder, which can be easily stored and has an increased shelf life compared with frozen protein solutions.

6.3 Structure–Function Relation of Mucins

To date, around 20 mucin genes have been identified. The mucins discovered so far can be divided into three groups: secreted gel-forming mucins, secreted non-gel-forming mucins, and mucins that are attached to the cell membrane [8]. Secreted gel-forming mucins including MUC2, MUC5AC, MUC5B, MUC6, and MUC19 are the major constituents of all types of mucus. Depending on the specific mucus localization and function, the mucin concentration can vary in a range of 1% and 5%. Secreted gel-forming mucins are produced by specialized epithelial cells and by glands located in the submucosal connective tissues [44, 45]. Since those gel-forming mucins have been characterized in most detail, we mostly focus on those in this book chapter. However, also cell-bound mucins are critical components of the passive immune system and are very likely to harbor great potential for biomedical applications as well – once their purification is well established and their properties are understood in more detail.

Mucins comprise a linear and mostly unfolded polypeptide chain with a highly O-glycosylated central region. The high glycosylation density of mucins is probably a key factor that suppresses polypeptide folding and contributes to the bottle-brush-like structure of the glycoprotein (Figure 6.3a). This glycosylated area is flanked by cysteine-rich regions and von Willebrand factor-like domains at both termini of the macromolecule. The C-terminal cysteine knot domain is believed to contribute to the dimer formation of the glycoprotein via cystine (S–S) disulfide bonds, and these dimers can be further organized into larger oligomers. This assembly process increases the molecular weight of purified mucins from around 600 kDa (monomeric mucin [46]) to several MDa (oligomers). Although the detailed peptide sequence varies not only for different gel-forming mucins but also for a certain mucin variant when different mammalian sources are compared, the overall polarity map of the polypeptide chain seems to be quite conserved (Figure 6.3b). Whereas the central region is almost exclusively constituted by polar amino acids (and rich in serine and threonine as required for O-glycosylation), the unglycosylated termini exhibit a rather hydrophobic character.

This amphiphilic character of mucins appears to be a prerequisite for their ability to act as lubricants on a broad range of surfaces. The hydrophobic termini of the macromolecule allow mucins to bind to hydrophobic surfaces [12, 47], and efficient surface adsorption is necessary for the formation of a sacrificial layer (Figure 6.3c) – a key mechanism employed by macromolecular lubricants. The polar glycosylated region of mucins seems to serve two purposes at the same time: First, this area is thought to promote mucin adsorption to polar surfaces. Second, the high density of glycans in this area is responsible for the good hydration of mucins [11], a property that is pivotal for hydration

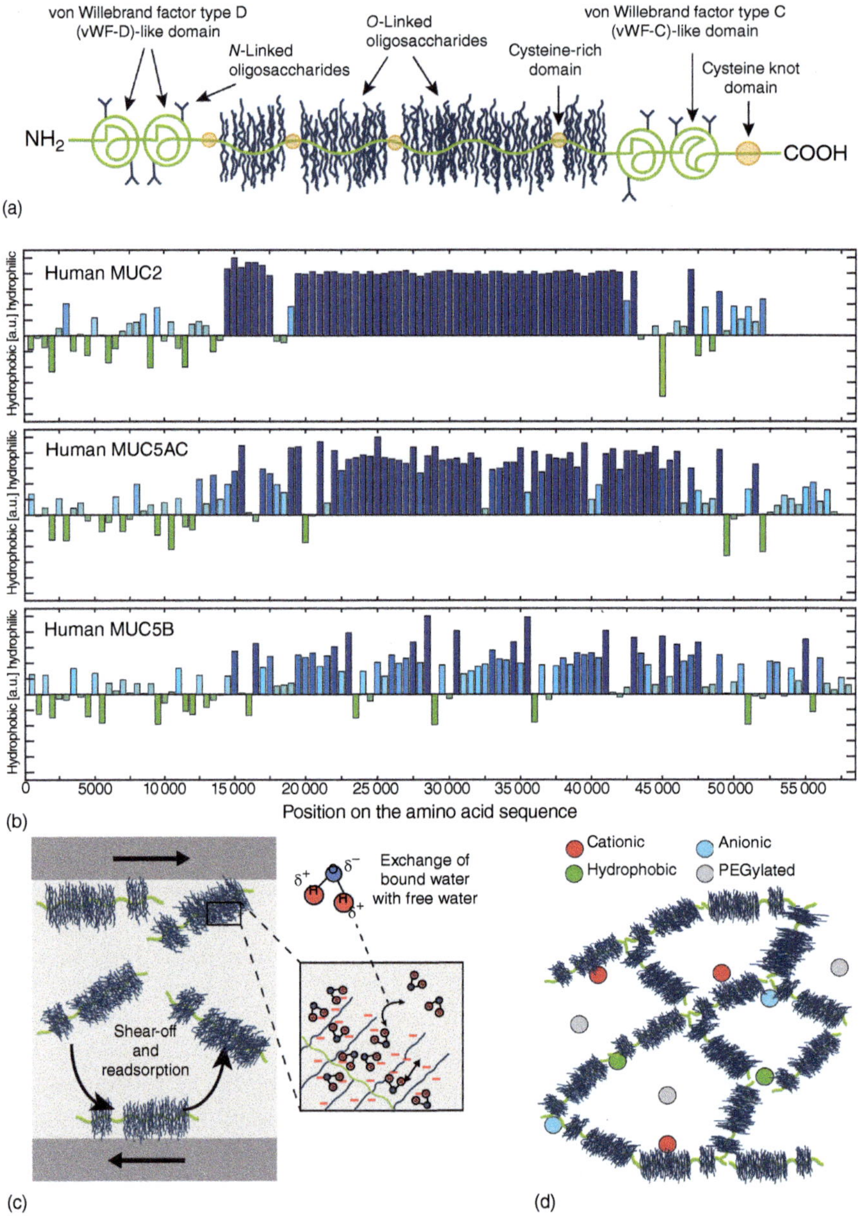

Figure 6.3 The complex molecular architecture of mucins is responsible for their broad range of functions. (a) Mucins are densely glycosylated and thus exhibit a bottle-brush-like structure. (b) The peptide sequences of intestinal (MUC2), gastric (MUC5AC), and salivary (MUC2) human mucin have a similar polarity profile giving rise to an amphiphilic character of mucins. The mucin sequences for the polarity analysis were obtained from UniProt protein database (MUC2: Q02817; MUC5AC: P98088; MUC5B: Q9HC84). The sequences were divided into sections of 50 amino acids each, and the number of amino acids with nonpolar side chains (i.e. Gly, Ala, Val, Met, Leu, Ile, Pro, Trp, Phe) and polar side chains (i.e. Arg, Lys, Asn, Asp, Gln, Glu, Tyr, His, Thr, Ser, Cys) was counted in each section and plotted as a bar. (c) Sacrificial layer formation and hydration lubrication are key mechanisms responsible for the lubricity of mucin solutions and make use of the amphiphilic structure of mucins. (d) Also gel formation and selective permeability of mucins gels are enabled by the amphiphilic character of mucin glycoproteins.

lubrication, i.e. the exchange of mucin-bound water with free water as a consequence of tribological shear stress.

Two additional properties of mucus that are critically linked to the amphiphilic character of mucins are gel formation and selective permeability. Through their hydrophobic termini, mucins can reversibly assemble into three-dimensional networks – a process whose details are still not fully understood, but it depends on many other factors including pH and buffer ionic strength [19, 48–50]. Once a mucin network is formed, it can selectively allow or deny the passage of particles and molecules by employing two distinct mechanisms: size filtering and interaction filtering [51]. The first mechanism can be understood as a microscopic sieving strategy where the spacing between mucin strands defines a maximal diameter for particles that can translocate through a mucin network [52, 53]. The second mechanism is based on attractive interactions between mucin glycoproteins and solutes from the fluid phase of mucus and is conceptually similar to that of the trapping strategy used by spider webs [54]. If a molecule or particle sticks to mucins, its diffusive spreading is prevented or at least retarded. Electrostatic forces, hydrophobic interactions, and specific binding events have been demonstrated to contribute to the selective trapping abilities of both mucin gels and native mucus samples. Due to diverse molecular interactions underlying mucoadhesion, mainly uncharged, polar molecules, and PEGylated nanoparticles [18, 55–57] seem to efficiently diffuse though mucins (Figure 6.3d). With the amphiphilic structure of mucins in mind, it seems logical that different areas of the mucin glycoprotein are responsible for establishing the distinct variants of mucoadhesive forces. Indeed, the trapping abilities of mucins toward influenza viruses can be traced back to sialic acid motifs in the glycosylated central region of the glycoprotein [58], and those glycan residues achieve virus binding by targeting its hemagglutinin receptor [59]. However, mucins possess much broader antiviral activity and also inhibit polyomaviruses [7], noroviruses [60], and – at least *in vitro* – the human immunodeficiency virus [61]. Since these other virus types lack hemagglutinin, it seems very unlikely that sialic acid residues are the primarily active mucin motifs here as well. It remains an open question which parts of the glycoprotein are crucial for the broad-spectrum antiviral properties of mucins and how the various glycan motifs contribute.

6.4 Characterizing Mucins and Mucin-Based Materials

Owing to the complex architecture and broad range of properties of mucins, characterizing these glycoproteins both on a molecular level and on a macroscopic scale (i.e. when studying mucin solutions or gels) requires a plethora of different techniques (Table 6.1). Since the molecular weight of mucins is very high, a classical electrophoretic analysis of mucus samples is extremely difficult; intact mucins do not travel well across polyacrylamide gels [27, 65]. However, gel electrophoresis is very useful for detecting mucin-associated molecules such as trefoil factor peptides [111, 112] or contaminants arising from the purification process, e.g. cellular debris. If binding affinities between mucins and other

Table 6.1 Overview of techniques employed for the characterization of mucins and mucin-based fluids or gels.

	Technique	Parameters studied	Example studies
	Light scattering	Size of mucin molecules and mucin aggregates	[62–64]
	high performance liquid chromatography (HPLC), gel electrophoresis	Molecular weight estimation of mucins and mucin fragments; purity analysis of mucin samples	[27, 37, 65–67]
	Microscale thermophoresis	Binding affinity between mucins and other molecules	[68–70]
	Fluorescence microscopy, electron microscopy (EM), atomic force microscopy (AFM)	Microarchitecture of mucin networks; mucin conformation	[71–76]
	QCM-D	Mucin adsorption to surfaces; antifouling properties of mucin coatings; hydration of mucin films (requires additional measurements, e.g. ellipsometry)	[77–80]
	Macrorheology	Gelation kinetics and viscoelastic properties of mucin gels; viscosity of mucin solutions and their shear thinning behavior	[44, 81–84]
	Microrheology	Nanoparticle mobility in mucin gels; microarchitecture and micromechanical properties of mucin solutions and gels	[18, 85–90]
	Microfluidics	Diffusive penetration of molecules into mucus and mucin gels	[91–94]
	Tribology	Friction measurements in stationary and migrating contact geometries: lubricity of mucin solutions and mucin coatings	[12, 25, 95–102],
	Profilometry	Optical quantification of surface structures: wear protection properties of mucin solutions and coatings	[13, 99, 103, 104]
	Microbiology	Antibacterial properties of mucin coatings	[105–107]
	Tissue culture	Biocompatibility of purified mucins; *in vitro* infection assays with eukaryotic cell lines	[2, 7, 108–110]

molecules are of interest, microscale thermophoresis is a suitable method [68, 69] and has recently been used to demonstrate the influence of MUC1 glycosylation on the binding efficiency of a therapeutic antibody to mucin [70]. Studying the conformation of mucins requires either light scattering techniques [62, 63] or atomic force microscopy (AFM) [71–73], and those methods also yield information on the degree of mucin aggregation [64, 113]. Fluorescently labeled mucins, on the other hand, allow for analyzing their distribution both *in vitro* and *in vivo* [74, 114, 115]. However, this approach requires either mucin-specific staining solutions or a conjugation of purified glycoproteins with an appropriate fluorophore.

As mentioned above, their amphiphilic character enables mucins to adsorb to a wide range of surfaces. This adsorption process can be quantified using a quartz crystal microbalance (QCM [116]). If this setup is also equipped with dissipation monitoring (QCM-D), additional information on the adsorbed mucin layer can be obtained, e.g. its viscoelastic properties [77, 78] or swelling state [79, 80]. Macroscopically, viscoelastic properties can be assessed with shear rheology. This mechanical characterization method has been employed intensively to quantify the gel formation abilities of mucins [44, 81] and the efficiency of mucolytic agents [82, 83] and to determine the material properties of cervical mucus [84], e.g. during ovulation [117] or as a risk assessment for preterm labor [118]. If local information on the viscoelastic properties of mucus is required or when the sample volume is insufficient for conducting macrorheological measurements, microrheological techniques [85, 86] can be used. Here, tracer particles are embedded into mucus samples or mucin gels, and the Brownian motion of those particles is recorded to study their micromechanical environment. Additionally, if the particle size is smaller than the mesh size of the sample [87], such tracking microscopy experiments can be used to investigate the permeability of native mucus [88, 89] or mucin gels [18, 90] toward nanoparticles. Such experiments give important insights for the design of mucus-penetrating drug carriers. For the penetration behavior of small molecules into or across mucus gels, however, this technique cannot be used. Instead, microfluidics experiments have been introduced and employed to quantify the diffusive entry and spreading of protons [91] or peptides [92] into and through mucin gels. Additionally, such microfluidics chips also allow for the analysis of mucus plug stability [93] and mucus–pathogen interactions [94].

Since a main physiological function of mucosal fluids is acting as macromolecular lubricants, there are also technical setups that can quantify the lubricity of saliva or other mucin-based fluids. There are different configurations for such tribological setups, i.e. linear and rotational tribometers as well as stationary and migrating contact geometries [12, 25, 95–99]. When conducting friction measurements that aim at studying the molecular principles that convey lubricity to mucin solutions, an artificial material pairing, e.g. steel-on-PDMS (polydimethylsiloxane), is well suited. With this material pairing, the relative difference between the (poor) lubricity of water/buffer and that of mucin solutions is very high, and thus intermediate levels of lubricity can be detected easily [11]. The same material pairing is also suitable for biotribological studies with saliva since steel mimics the hard properties of the palate and PDMS approximates the mechanical properties of the tongue. However, there are also setups where tissue samples [100, 101] or cellular monolayers [102] are chosen as one part of the material pairing. Of course, such a semi-biological material pairing offers are more realistic evaluation of biomimetic lubricants than when purely synthetic materials are used.

After friction experiments with biological or synthetic samples, the amount of surface damage induced by the tribological stress is of interest as well. Here, optical microscopy methods, especially white light profilometry (a variant of confocal light microscopy), have been proven very suitable. With this technique, it was demonstrated that mucin-based fluids not only reduce friction on a broad range of surfaces but also reduce or even fully prevent wear formation [13, 99, 104].

Of course, for purified mucins to be used in medical products, their biocompatibility needs to be assessed. Here, standard tissue culture experiments such as cell viability/cytotoxicity tests [108] are a good first choice. In combination with simple microbiological assays, tissue culture experiments also allow for quantifying the antibacterial [105–107, 109] and antiviral properties [2, 7, 110] of mucins.

6.5 Biomedical Applications of Purified Mucins

Having discussed our current understanding how the molecular architecture of mucins is linked to their broad range of properties and which techniques are used in the field of mucin research to characterize the properties of mucins and mucin-based systems, we now describe several areas of applications from the field of biomedical engineering where purified mucins could be part of novel and innovative products. Of course, the case studies we present are mostly still limited to lab tests. Two critical advances in the field of mucin research are necessary to turn these biomaterials used in those case studies into real products. First, the amount of purified, high-quality mucin needs to be increased so that mucin-based products become affordable. This entails revisiting the current mucin purification procedure and simplifying those steps in the purification chain that are time consuming and require lots of hands-on work. Second, as mucins are purified from animal tissue samples, we need to find a suitable sterilization method for the glycoproteins that does not damage the mucins – otherwise they will lose their highly attractive properties. However, although those two steps are still to be achieved, the examples we outline below already nicely illustrate the great potential mucins have as components of fluids, as gels, or as coatings (Figure 6.4).

6.5.1 Eye Drops or Contact Lens Coatings

When the natural eyesight becomes less efficient, wearing contact lenses can remedy the loss of visual accuracy. However, such contact lenses constitute foreign bodies that slowly move across the cornea, e.g. when the eyelid opens and closes during blinking [119]. Mucins in the ocular tear film serve as a molecular lubricant [27, 120] and ensure that this contact lens sliding does not damage the corneal epithelium. In the absence of a proper mucinous lubrication layer, however, increased friction and tissue damage induced by contact lens sliding lead to

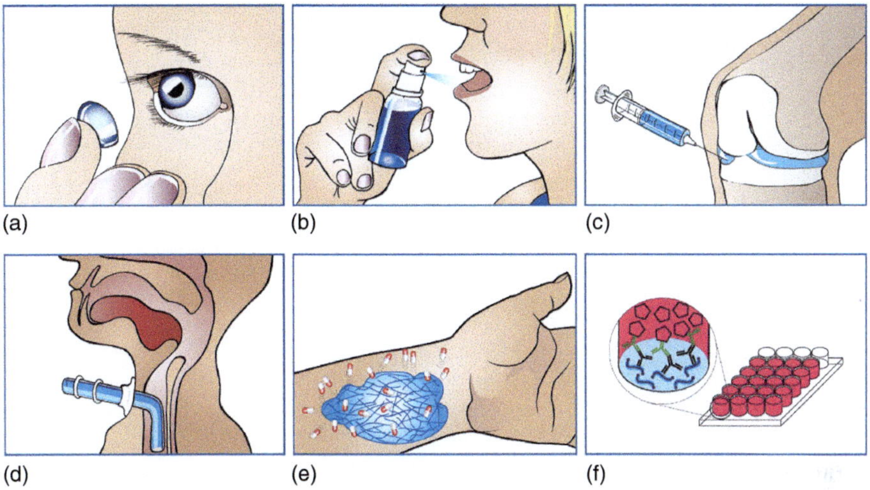

Figure 6.4 Biomedical applications of purified mucins. Mucin coatings or mucin solutions can improve hydration and reduce friction or tissue damage on the cornea (a), in the oral cavity (b), and in joints (c). Moreover, mucin-coated materials that reside in the human body for extended time periods better resist biofouling (d). Purified mucins can also be useful components of hydrogels for wound healing (e) and can serve as molecular standards for the analysis of clinical mucus samples in biochemical assays (f).

discomfort. Such insufficient corneal lubrication is a hallmark of *keratoconjunctivitis sicca* (commonly known as dry eye syndrome). With a prevalence of up to 30%, dry eye is one of the most frequent forms of ocular diseases [121, 122]; among contact lens wearers, this condition occurs even more often.

It was recently shown in *in vitro* experiments that purified gastric mucins can take over the role of native ocular mucins from the tear fluid and efficiently prevent damage on the cornea [13]. This protective effect can be achieved in two different ways: mucin solutions can be used as "artificial tears" (eye drops) for cornea lubrication or as a molecular coating on contact lenses. In either case, cornea samples that were subjected to friction measurements using commercial contact lenses as a counterpart did not show any topographical alterations compared with untreated tissue samples. The current gold standard for treating dry eye is applying a solution of hyaluronic acid (HA) [123], and indeed such HA solutions act as a good tear fluid proxy as well. However, this glycosaminoglycan does not form a spontaneous coating on contact lenses as mucins do and thus needs to be applied several times (in the form of eye drops) during the day. Thus, using mucin-coated contact lenses would increase the comfort of people wearing contact lenses. The mucin coating could be very easily achieved by the customer by using a contact lens storage solution enriched with purified mucins in which the protective mucin layer regenerates itself overnight.

6.5.2 Mouth Sprays

Especially among the elderly population, many people suffer from a dry mouthfeel (xerostomia) [124]. Furthermore, dry mouth can also be a side effect of different diseases or therapies, including chemotherapy treatment, the use of specific antidepressants, diabetes [125], AIDS, or some autoimmune diseases [126, 127]. Xerostomia not only leads to a permanent feeling of oral discomfort but also makes the ingestion of food difficult. Without a proper oral lubricant, chewing and swallowing becomes complicated and very unpleasant. The missing lubricity in the oral cavity of dry mouth patients is mostly due to a reduced or altered production of salivary mucins – an effect that healthy people can temporarily observe when consuming astringents, such as red wine or green tea.

Current xerostomia therapies either involve drugs that stimulate the production of saliva, e.g. pilocarpine [128], or make use of artificial saliva substitutes that are sprayed into the oral cavity. For the latter, using an aqueous mucin solution might seem obvious. However, although commercially available products comprising industrially purified mucins do exist, these mucins have lost many of their native properties and consequently do not improve the patients' situation compared with placebos [129]. Other salivary replacement products are solutions enriched with carboxymethylcellulose and aim at mimicking the viscosity of natural saliva – a strategy that is insufficient to create a good oral lubricant. Manually purified mucins, however, can reproduce the lubricity of native saliva very well – at least *in vitro* [25]. It is very likely that dry mouth patients would benefit from a mouth spray based on high-quality mucins. However, for such an improved product to become reality, the currently available number of high-quality mucins would need to be increased considerably.

6.5.3 Artificial Joint Fluids

Articular cartilage is a tissue that exhibits ultralow friction and little to no wear. Because of this sturdiness, the tissue can remain functional over several decades [130, 131]. However, in contrast to most other tissues, articular cartilage shows very poor self-renewing abilities, which is due to the low amount of cells in this particular tissue. Whereas this lack of regeneration properties does not matter much as long as the cartilage layer in joints is intact, it can become a serious issue once the cartilage is damaged, e.g. by trauma or disease. Such damaged cartilage then gives rise to persisting irritations and inflammations (osteoarthritis) that limit patient mobility and cause severe pain [132, 133]. Indeed, osteoarthritis and cartilage damage are a frequent problem, not only among the elderly population but also for young individuals engaging in intense sport activities.

Current noninvasive treatment of osteoarthritis includes the administration of analgesics and anti-inflammatory drugs or the injection of HA into the synovial cleft of affected joints [134]. The latter is motivated by the fact that, in contrast to the tear fluid that we discussed above, HA is a natural component of the synovial fluid and contributes to the outstanding tribological resistance of joints.

How this is achieved, however, is not fully clear yet. It has been suggested that HA increases the viscosity of the synovial fluid and thus improves the stability of the lubrication film [135]. Furthermore, HA is thought to interact with synovial macromolecules (e.g. lubricin – a glycoprotein that has strong structural similarity to mucins) to reduce friction [135, 136]. However, if the cartilage surface and also the surface-bound macromolecules are damaged, the lubricating potential of HA would become limited. Although mucins are physiologically not present in the synovial fluid, they can act as efficient lubricants on a wide range of surfaces. In fact, when used as the only macromolecular component of an artificial lubricant in *in vitro* tribology experiments on articular cartilage, mucins can outperform HA solutions in terms of wear protection [99]. Thus, injecting solutions of purified mucins into the synovial cleft of osteoarthritis patients could be a powerful alternative to current osteoarthritis treatment strategies. However, extensive research must still be conducted for assessing the biocompatibility of mucins regarding this specific application.

6.5.4 Coatings of Medical Devices

As part of surgery or post-surgery measures, catheters, ventilation tubes, or stents are inserted into the human body. Those tubes are typically made from synthetic polymeric materials such as polyethylene, polyvinyl chloride, or silicone elastomers. During their insertion, these devices cause friction on the epithelial tissue they rub across. This friction not only causes discomfort but may also damage the sensitive tissues [137, 138]. Moreover, if those artificial objects remain in the human body for extended time periods, i.e. for several days up to weeks, another issue starts to play a role: biofouling, i.e. the uncontrolled deposition of proteins or other biomolecules onto the surface of the tubes – a process that is typically followed by a colonization by bacteria or cells [139]. Biofouling negatively impacts the proper function of the tubes and can lead to inflammation and infections [140].

Under lab conditions, mucin coatings of synthetic materials such as polystyrene, polymethylmethacrylate, silicone, and glass could already be generated by passive adsorption. Indeed, such simple mucin coatings can reduce biofouling events as they successfully reduce the colonization of surfaces by bacteria [105, 141] or cells [142–144]. However, although mucins adsorb rather strongly onto most surfaces [145, 146], a passively adsorbed mucin layer might not be stable enough to resist the mechanical stresses that occur during the insertion of medical devices into the human body. Generating stable, covalent mucin coatings could not only increase the lubricity of the device surface and thus improve patient comfort but should also ensure long-term biocompatibility of the medical device. A covalent mucin coating has already been applied to vascular grafts, and this coating improved the hemocompatibility of the artificial material as well as the graft patency [147]. However, the specific coupling reaction chosen in this study cannot easily be transferred to a broader set of materials. This motivates that a more general chemical coupling method would be helpful for creating durable, covalent mucin coatings on different polymer materials to prevent their biofouling.

6.5.5 Components of Hydrogels for Drug Delivery

An intact skin provides an efficient barrier against viral and bacterial infections. However, when this natural barrier is damaged, e.g. by injuries, surgery, or bedsore, the connective tissue is exposed, and the infection risk suddenly becomes very high. Even today, such microbial infections represent a major cause of diseases that sometimes can have severe consequences including limb loss or even death [148]. On exposed body surfaces that are not covered by skin, i.e. on wet epithelia, the human body produces a protective mucus layer. This mucus layer either repels pathogens by blocking their passage or traps them within the mucin network, so they cannot reach the underlying epithelium [7, 8].

In addition to their antiviral and antibacterial properties, mucins can selectively bind a broad range of molecules and particles as described above. In gastrointestinal mucus, this makes it difficult for therapeutic molecules to reach the target epithelia. However, this molecule trapping ability of mucins can be harnessed to develop a mucin-based microenvironment that selectively stores and then slowly releases molecules. Indeed, mucins have been used as building blocks for drug-releasing materials both as multilayer coatings [149] and as hydrogels [114, 150]. Although pure mucin gels lack the mechanical stability required for hydrogels in wound healing applications, both covalent crosslinking of mucins and combining mucins with methylcellulose biopolymers can remedy this issue. Taking into account research on the interactions between pharmaceuticals/drug delivery particles and mucins [151–155], it should be possible to enrich mucin-based hydrogels with drugs or drug carrier particles such that a time-controlled release is achieved. Such mucin-based hydrogels with tailored drug release properties can be an important tool to improve the healing process of damaged tissues.

6.5.6 Molecular Standards for Lab Tests with Clinical Mucus Samples

For the biochemical analysis of biological mucus samples, e.g. as part of a clinical evaluation of mucus specimens from patients with cystic fibrosis, different lab tests are available. One of the most frequent lab tests used to identify certain biomolecules is enzyme-linked immunosorbent assay (ELISA) [156, 157], which makes use of a set of two antibodies to detect target molecules and to visualize them via a chromophore reaction. However, to give quantitative results, this well-established assay requires purified solutions of the target molecule as a standard. Currently, the compromised structure of industrially purified gastric mucins makes their detection by antibodies difficult – although the very same antibodies easily detect those gastric mucins if they were manually purified in the lab [13]. Thus, a better availability of highly functional purified mucins will simplify the analysis of clinical mucus samples and could contribute to a better detection of mucus-related diseases.

6.6 Outlook: Engineered Mucins and Mucin-Mimetic Polymers

Given the broad range of interesting biomedical applications of mucins and considering the difficulty of purifying mucins while maintaining their functionality, it is not surprising that there are increasing efforts for producing mucin-like molecules through biotechnological approaches. One major obstacle that needs to be overcome is to create the required mucin-like glycosylation pattern on proteins in *in vitro* cell cultures. Progress in this field has been made with both yeast [158] and plant cells [159], and small molecules with mucin-like motifs have been successfully synthesized [160–162]. Indeed, mucin-inspired macromolecules are able to reproduce some of the properties of native mucins. For instance, synthetic molecules [163] (and even nanoparticles [164, 165]) carrying multiple sialic acid groups bind to hemagglutinin and thus can inhibit influenza viruses. There are also synthetic, densely glycosylated macromolecules that mimic the amphiphilic properties of mucins and thus can be integrated into lipid membranes [166]. Engineered hydrogels, on the other hand, can exhibit similar mucoadhesive properties as native mucus [167] or prevent the diffusive spreading of viruses [168, 169]. Also, the surface adsorption and/or tribological properties of mucins can – to some extent – be recreated by synthetic macromolecules [12, 170] and mucin-type fusion proteins [171, 172]. However, a fully synthetic macromolecule that combines all (or at least most) of the different biomedically important properties of mucin glycoproteins does – to date – not exist. For such a fully synthetic mucin molecule to become reality in the future, our understanding of the structure–function relationship of mucins needs to advance far enough so that all of the properties of mucins can be traced back to the biochemical motifs responsible for them. Until then, purified mucins obtained from animal tissue will remain the key ingredient not only for mucin research but also for applications making use of the outstanding properties of those glycoproteins.

Acknowledgments

We thank Iris König-Decker and Carolin Rickert for providing schematics and drawings and Corinna Lieleg for proofreading the manuscript. Support from the Deutsche Forschungsgemeinschaft through grant LI 1902/6-1 is gratefully acknowledged.

References

1 Patton, S. (2001). MUC1 and MUC-X, epithelial mucins of breast and milk. In: *Bioactive Components of Human Milk* (ed. D.S. Newburg), 35–45. Boston, MA: Springer US.
2 Habte, H.H., Kotwal, G.J., Lotz, Z.E. et al. (2007). Antiviral activity of purified human breast milk mucin. *Neonatology* 92: 96–104. https://doi.org/10.1159/000100808.
3 Liu, B. and Newburg, D.S. (2013). Human milk glycoproteins protect infants against human pathogens. *Breastfeeding Medicine* 8: 354–362. https://doi.org/10.1089/bfm.2013.0016.
4 Crouzier, T. (2016). A call for better mucins. http://biopolymersforlife.org/a-call-for-better-mucins/ (accessed 7 August 2018).

5 Bansil, R. and Turner, B.S. (2006). Mucin structure, aggregation, physiological functions and biomedical applications. *Current Opinion in Colloid and Interface Science* 11: 164–170. https://doi.org/10.1016/j.cocis.2005.11.001.
6 Petrou, G. and Crouzier, T. (2018). Mucins as multifunctional building blocks of biomaterials. *Biomaterials Science* https://doi.org/10.1039/C8BM00471D.
7 Lieleg, O., Lieleg, C., Bloom, J. et al. (2012). Mucin biopolymers as broad-spectrum antiviral agents. *Biomacromolecules* 13: 1724–1732. https://doi.org/10.1021/bm3001292.
8 Linden, S.K., Sutton, P., Karlsson, N.G. et al. (2008). Mucins in the mucosal barrier to infection. *Mucosal Immunology* 1: 183. https://doi.org/10.1038/mi.2008.5.
9 Lee, S., Muller, M., Rezwan, K. et al. (2005). Porcine gastric mucin (PGM) at the water/poly(dimethylsiloxane) (PDMS) interface: influence of pH and ionic strength on its conformation, adsorption, and aqueous lubrication properties. *Langmuir* 21: 8344–8353. doi: https://doi.org/10.1021/la050779w.
10 Coles, J.M., Chang, D.P., and Zauscher, S. (2010). Molecular mechanisms of aqueous boundary lubrication by mucinous glycoproteins. *Current Opinion in Colloid and Interface Science* 15: 406–416. https://doi.org/10.1016/j.cocis.2010.07.002.
11 Crouzier, T., Boettcher, K., Geonnotti, A.R. et al. (2015). Modulating mucin hydration and lubrication by deglycosylation and polyethylene glycol binding. *Advanced Materials Interfaces* 2: 1500308. https://doi.org/10.1002/admi.201500308.
12 Käsdorf, B.T., Weber, F., Petrou, G. et al. (2017). Mucin-inspired lubrication on hydrophobic surfaces. *Biomacromolecules* 18: 2454–2462. https://doi.org/10.1021/acs.biomac.7b00605.
13 Winkeljann, B., Boettcher, K., Balzer, B.N. et al. (2017). Mucin coatings prevent tissue damage at the cornea–contact lens interface. *Advanced Materials Interfaces* 4: 1700186. https://doi.org/10.1002/admi.201700186.
14 Ramamoorthy, P. and Nichols, J.J. (2008). Mucins in contact lens wear and dry eye conditions. *Optometry and Vision Science* 85: E631–E642. https://doi.org/10.1097/OPX.0b013e3181819f25.
15 Kočevar-Nared, J., Kristl, J., and Šmid-Korbar, J. (1997). Comparative rheological investigation of crude gastric mucin and natural gastric mucus. *Biomaterials* 18: 677–681. https://doi.org/10.1016/S0142-9612(96)00180-9.
16 Schömig, V.J., Käsdorf, B.T., Scholz, C. et al. (2016). An optimized purification process for porcine gastric mucin with preservation of its native functional properties. *RSC Advances* 6: 44932–44943. https://doi.org/10.1039/C6RA07424C.
17 Celli, J.P., Turner, B.S., Afdhal, N.H. et al. (2007). Rheology of gastric mucin exhibits a pH-dependent sol–gel transition. *Biomacromolecules* 8: 1580–1586. https://doi.org/10.1021/bm0609691.
18 Lieleg, O., Vladescu, I., and Ribbeck, K. (2010). Characterization of particle translocation through mucin hydrogels. *Biophysical Journal* 98: 1782–1789. https://doi.org/10.1016/j.bpj.2010.01.012.

19 Meldrum, O.W., Yakubov, G.E., Bonilla, M.R. et al. (2018). Mucin gel assembly is controlled by a collective action of non-mucin proteins, disulfide bridges, Ca^{2+}-mediated links, and hydrogen bonding. *Scientific Reports* 8: 5802. https://doi.org/10.1038/s41598-018-24223-3.

20 Faure, M., Moënnoz, D., Montigon, F. et al. (2002). Development of a rapid and convenient method to purify mucins and determine their in vivo synthesis rate in rats. *Analytical Biochemistry* 307: 244–251. https://doi.org/10.1016/S0003-2697(02)00048-9.

21 Schömig, V., Isik, E., Martin, L. et al. (2017). Solid liquid liquid extraction of porcine gastric mucins from homogenized animal material. *RSC Advances* 7: 39708–39717. https://doi.org/10.1039/c7ra06594a.

22 Podolsky, D.K. and Isselbacher, K.J. (1983). Composition of human colonic mucin. Selective alteration in inflammatory bowel disease. *Journal of Clinical Investigation* 72: 142–153. https://doi.org/10.1172/JCI110952.

23 Gum, J.R., Byrd, J.C., Hicks, J.W. et al. (1989). Molecular cloning of human intestinal mucin cDNAs. Sequence analysis and evidence for genetic polymorphism. *Journal of Biological Chemistry* 264: 6480–6487.

24 Spurr-Michaud, S., Argüeso, P., and Gipson, I. (2007). Assay of mucins in human tear fluid. *Experimental Eye Research* 84: 939–950. https://doi.org/10.1016/j.exer.2007.01.018.

25 Biegler, M., Delius, J., Käsdorf, B.T. et al. (2016). Cationic astringents alter the tribological and rheological properties of human saliva and salivary mucin solutions. *Biotribology* 6: 12–20. https://doi.org/10.1016/j.biotri.2016.03.002.

26 Park, M.-S., Chung, J.-W., Kim, Y.-K. et al. (2007). Viscosity and wettability of animal mucin solutions and human saliva. *Oral Diseases* 13: 181–186. https://doi.org/10.1111/j.1601-0825.2006.01263.x.

27 Corfield, A.P., Carrington, S.D., Hicks, S.J. et al. (1997). Ocular mucins: purification, metabolism and functions. *Progress in Retinal and Eye Research* 16: 627–656.

28 Vukosavljevic, B., Murgia, X., Schwarzkopf, K. et al. (2017). Tracing molecular and structural changes upon mucolysis with *N*-acetyl cysteine in human airway mucus. *International Journal of Pharmaceutics* 533: 373–376. https://doi.org/10.1016/j.ijpharm.2017.07.022.

29 Fudge, D.S., Gardner, K.H., Forsyth, V.T. et al. (2003). The mechanical properties of hydrated intermediate filaments: insights from hagfish slime threads. *Biophysical Journal* 85: 2015–2027. https://doi.org/10.1016/S0006-3495(03)74629-3.

30 Ohta, N., Sato, M., Ushida, K. et al. (2009). Jellyfish mucin may have potential disease-modifying effects on osteoarthritis. *BMC Biotechnology* 9: 98. https://doi.org/10.1186/1472-6750-9-98.

31 Jatkar, A.A., Brown, B.E., Bythell, J.C. et al. (2010). Coral mucus: the properties of its constituent mucins. *Biomacromolecules* 11: 883–888. https://doi.org/10.1021/bm9012106.

32 Ducklow, H.W. and Mitchell, R. (1979). Composition of mucus released by coral reef coelenterates1. *Limnology and Oceanography* 24: 706–714. https://doi.org/10.4319/lo.1979.24.4.0706.

33 Mantle, M. and Allen, A. (1981). Isolation and characterization of the native glycoprotein from pig small-intestinal mucus. *Biochemical Journal* 195: 267–275. https://doi.org/10.1042/bj1950267.

34 Lafitte, G., Söderman, O., Thuresson, K. et al. (2007). PFG-NMR diffusometry: a tool for investigating the structure and dynamics of noncommercial purified pig gastric mucin in a wide range of concentrations. *Biopolymers* 86: 165–175. https://doi.org/10.1002/bip.20717.

35 Hill, H.D., Reynolds, J.A., and Hill, R.L. (1977). Purification, composition, molecular weight, and subunit structure of ovine submaxillary mucin. *Journal of Biological Chemistry* 252: 3791–3798.

36 Libao-Mercado, A.J. and de Lange, C.F.M. (2007). Refined methodology to purify mucins from pig colonic mucosa. *Livestock Science* 109: 141–144. https://doi.org/10.1016/j.livsci.2007.01.125.

37 Pallesen, L.T., Berglund, L., Rasmussen, L.K. et al. (2002). Isolation and characterization of MUC15, a novel cell membrane-associated mucin. *European Journal of Biochemistry* 269: 2755–2763. https://doi.org/10.1046/j.1432-1033.2002.02949.x.

38 Baeckström, D., Hansson, G.C., Nilsson, O. et al. (1991). Purification and characterization of a membrane-bound and a secreted mucin-type glycoprotein carrying the carcinoma-associated sialyl-Lea epitope on distinct core proteins. *Journal of Biological Chemistry* 266: 21537–21547.

39 Carlstedt, I., Lindgren, H., Sheehan, J.K. et al. (1983). Isolation and characterization of human cervical-mucus glycoproteins. *Biochemical Journal* 211: 13–22. https://doi.org/10.1042/bj2110013.

40 Parker, N., Finnie, I.A., Raouf, A.H. et al. (1993). High performance gel filtration using monodisperse highly cross-linked agarose as a one-step system for mucin purification. *Biomedical Chromatography* 7: 68–74. https://doi.org/10.1002/bmc.1130070204.

41 Raouf, A., Parker, N., Iddon, D. et al. (1991). Ion exchange chromatography of purified colonic mucus glycoproteins in inflammatory bowel disease: absence of a selective subclass defect. *Gut* 32: 1139–1145. https://doi.org/10.1136/gut.32.10.1139.

42 Shimamoto, C., Desmukh, G.D., Rigot, W.L. et al. (1989). Analysis of cancer-associated colonic mucsin by ion-exchange chromatography: evidence for a mucin species of lower molecular charge and weight in cancer. *Biochim Biophys Acta* 991: 284–295.

43 Thornton, D.J., Devine, P.L., Hanski, C. et al. (1994). Identification of two major populations of mucins in respiratory secretions. *American Journal of Respiratory and Critical Care Medicine* 150: 823–832. https://doi.org/10.1164/ajrccm.150.3.8087358.

44 Bansil, R., Celli, J., Hardcastle, J. et al. (2013). The influence of mucus microstructure and rheology in *Helicobacter pylori* infection. *Frontiers in Immunology* 4 https://doi.org/10.3389/fimmu.2013.00310.

45 McGuckin, M.A., Lindén, S.K., Sutton, P. et al. (2011). Mucin dynamics and enteric pathogens. *Nature Reviews Microbiology* 9: 265. https://doi.org/10.1038/nrmicro2538.

46 Mejías-Luque, R., Cobler, L., and de Bolós, C. (2010). MUC5AC (mucin 5AC, oligomeric mucus/gel-forming). *Atlas of Genetics and Cytogenetics in Oncology and Haematology* 14: 566–569.
47 Madsen, J.B., Svensson, B., Abou Hachem, M. et al. (2015). Proteolytic degradation of bovine submaxillary mucin (BSM) and its impact on adsorption and lubrication at a hydrophobic surface. *Langmuir* 31: 8303–8309. https://doi.org/10.1021/acs.langmuir.5b01281.
48 Cao, X.X., Bansil, R., Bhaskar, K.R. et al. (1999). pH-dependent conformational change of gastric mucin leads to sol-gel transition. *Biophysical Journal* 76: 1250–1258. https://doi.org/10.1016/S0006-3495(99)77288-7.
49 Taylor, C., Allen, A., Dettmar, P.W. et al. (2003). The gel matrix of gastric mucus is maintained by a complex interplay of transient and nontransient associations. *Biomacromolecules* 4: 922–927. https://doi.org/10.1021/bm025767t.
50 Wagner, C.E., Turner, B.S., Rubinstein, M. et al. (2017). A rheological study of the association and dynamics of MUC5AC gels. *Biomacromolecules* 18: 3654–3664. https://doi.org/10.1021/acs.biomac.7b00809.
51 Lieleg, O. and Ribbeck, K. (2011). Biological hydrogels as selective diffusion barriers. *Trends in Cell Biology* 21: 543–551. https://doi.org/10.1016/j.tcb.2011.06.002.
52 Kirch, J., Schneider, A., Abou, B. et al. (2012). Optical tweezers reveal relationship between microstructure and nanoparticle penetration of pulmonary mucus. *Proceedings of the National Academy of Sciences of the United States of America* 109: 18355–18360. https://doi.org/10.1073/pnas.1214066109.
53 Lai, S.K., Wang, Y.-Y., Hida, K. et al. (2010). Nanoparticles reveal that human cervicovaginal mucus is riddled with pores larger than viruses. *Proceedings of the National Academy of Sciences of the United States of America* 107: 598–603. https://doi.org/10.1073/pnas.0911748107.
54 Witten, J. and Ribbeck, K. (2017). The particle in the spider's web: transport through biological hydrogels. *Nanoscale* 9: 8080–8095. https://doi.org/10.1039/C6NR09736G.
55 Cu, Y. and Saltzman, W.M. (2009). Controlled surface modification with poly(ethylene)glycol enhances diffusion of PLGA nanoparticles in human cervical mucus. *Molecular Pharmaceutics* 6: 173–181. https://doi.org/10.1021/mp8001254.
56 Maisel, K., Reddy, M., Xu, Q. et al. (2016). Nanoparticles coated with high molecular weight PEG penetrate mucus and provide uniform vaginal and colorectal distribution in vivo. *Nanomedicine (London, England)* 11: 1337–1343. https://doi.org/10.2217/nnm-2016-0047.
57 Winkeljann, B., Käsdorf, B.T., Boekhoven, J. et al. (2018). Macromolecular coating enables tunable selectivity in a porous PDMS matrix. *Macromolecular Bioscience* 18: 1700311. https://doi.org/10.1002/mabi.201700311.
58 Matrosovich, M. and Klenk, H.-D. (2003). Natural and synthetic sialic acid-containing inhibitors of influenza virus receptor binding. *Reviews in Medical Virology* 13: 85–97. https://doi.org/10.1002/rmv.372.
59 Varghese, J.N., McKimm-Breschkin, J.L., Caldwell, J.B. et al. (1992). The structure of the complex between influenza virus neuraminidase and sialic

acid, the viral receptor. *Proteins: Structure, Function, and Bioinformatics* 14: 327–332. https://doi.org/10.1002/prot.340140302.

60 Tian, P., Brandl, M., and Mandrell, R. (2005). Porcine gastric mucin binds to recombinant norovirus particles and competitively inhibits their binding to histo-blood group antigens and Caco-2 cells. *Letters in Applied Microbiology* 41: 315–320. https://doi.org/10.1111/j.1472-765X.2005.01775.x.

61 Mall, A.S., Habte, H., Mthembu, Y. et al. (2017). Mucus and mucins: do they have a role in the inhibition of the human immunodeficiency virus? *Virology Journal* 14: 192. https://doi.org/10.1186/s12985-017-0855-9.

62 Varma, B.K., Demers, A., Jamieson, A.M. et al. (1990). Light scattering studies of the effect of Ca^{2+} on the structure of porcine submaxillary mucin. *Biopolymers* 29: 441–448. https://doi.org/10.1002/bip.360290214.

63 Watanabe, Y. and Inoko, Y. (2007). Small-angle light and X-ray scattering measurements of a protein–oligosaccharide complex mucin in solution. *Journal of Applied Crystallography* 40: s209–s212. https://doi.org/10.1107/S0021889807009247.

64 Bansil, R., Cao, X., Bhaskar, K.R. et al. (1996). Gelation and aggregation of mucin in relation to the stomach's protective barrier. *Macromolecular Symposia* 109: 105–113. https://doi.org/10.1002/masy.19961090110.

65 Paszkiewicz-Gadek, A., Gindzienski, A., and Halina, P. (1995). The use of preparative polycarylamide gel electrophoresis and electroelution for purification of mucus glycoproteins. *Analytical Biochemistry*: 263–267.

66 Hanisch, F.-G., Ragge, H., Kalinski, T. et al. (2013). Human gastric TFF2 peptide contains an *N*-linked fucosylated *N,N′*-diacetyllactosediamine (LacdiNAc) oligosaccharide. *Glycobiology* 23: 2–11. https://doi.org/10.1093/glycob/cws131.

67 Ishihara, K., Kameyama, J.-i., and Hotta, K. (1993). Development of an HPLC method to estimate hexosamines and its application to determine mucin content in rat and human gastric mucosa. *Comparative Biochemistry and Physiology Part B: Comparative Biochemistry* 104: 781–786. https://doi.org/10.1016/0305-0491(93)90213-O.

68 Swidergall, M., Ernst, A.M., and Ernst, J.F. (2013). *Candida albicans* mucin Msb2 is a broad-range protectant against antimicrobial peptides. *Antimicrobial Agents and Chemotherapy* 57: 3917–3922. https://doi.org/10.1128/aac.00862-13.

69 Recktenwald, C.V. and Hansson, G.C. (2016). The reduction-insensitive bonds of the MUC2 mucin are isopeptide bonds. *Journal of Biological Chemistry* 291: 13580–13590.

70 Movahedin, M., Brooks, T.M., Supekar, N.T. et al. (2017). Glycosylation of MUC1 influences the binding of a therapeutic antibody by altering the conformational equilibrium of the antigen. *Glycobiology* 27: 677–687. https://doi.org/10.1093/glycob/cww131.

71 Hong, Z., Chasan, B., Bansil, R. et al. (2005). Atomic force microscopy reveals aggregation of gastric mucin at low pH. *Biomacromolecules* 6: 3458–3466. https://doi.org/10.1021/bm0505843.

72 McMaster, T.J., Berry, M., Corfield, A.P. et al. (1999). Atomic force microscopy of the submolecular architecture of hydrated ocular mucins. *Biophysical Journal* 77: 533–541. https://doi.org/10.1016/S0006-3495(99)76910-9.

73 Round, A.N., Berry, M., McMaster, T.J. et al. (2002). Heterogeneity and persistence length in human ocular mucins. *Biophysical Journal* 83: 1661–1670. https://doi.org/10.1016/S0006-3495(02)73934-9.

74 Portal, C., Gouyer, V., Magnien, M. et al. (2017). In vivo imaging of the Muc5b gel-forming mucin. *Scientific Reports* 7: 44591. https://doi.org/10.1038/srep44591.

75 Fiebrig, I., Harding, S.E., Rowe, A.J. et al. (1995). Transmission electron microscopy studies on pig gastric mucin and its interactions with chitosan. *Carbohydrate Polymers* 28: 239–244. https://doi.org/10.1016/0144-8617(95)00105-0.

76 Wang, N., Zhang, M., Chen, X. et al. (2017). Mapping the interaction sites of Mucin 1 and DNA aptamer by atomic force microscopy. *Analyst* 142: 3800–3804. https://doi.org/10.1039/C7AN01119A.

77 Feiler, A.A., Sahlholm, A., Sandberg, T. et al. (2007). Adsorption and viscoelastic properties of fractionated mucin (BSM) and bovine serum albumin (BSA) studied with quartz crystal microbalance (QCM-D). *Journal of Colloid and Interface Science* 315: 475–481. https://doi.org/10.1016/j.jcis.2007.07.029.

78 Madsen, J.B., Sotres, J., Pakkanen, K.I. et al. (2016). Structural and mechanical properties of thin films of bovine submaxillary mucin versus porcine gastric mucin on a hydrophobic surface in aqueous solutions. *Langmuir* 32: 9687–9696. https://doi.org/10.1021/acs.langmuir.6b02057.

79 Feldötö, Z., Pettersson, T., and Dėdinaitė, A. (2008). Mucin–electrolyte interactions at the solid–liquid interface probed by QCM-D. *Langmuir* 24: 3348–3357. https://doi.org/10.1021/la703366k.

80 Znamenskaya, Y., Sotres, J., Gavryushov, S. et al. (2013). Water sorption and glass transition of pig gastric mucin studied by QCM-D. *Journal of Physical Chemistry B* 117: 2554–2563. https://doi.org/10.1021/jp311968b.

81 Pearson, J.P., Allen, A., and Hutton, D.A. (2000). Rheology of mucin. In: *Glycoprotein Methods and Protocols: The Mucins* (ed. A.P. Corfield), 99–109. Totowa, NJ: Humana Press.

82 Dippy, J.E. and Davis, S.S. (1969). Rheological assessment of mucolytic agents on sputum of chronic bronchitics. *Thorax* 24: 707–713. https://doi.org/10.1136/thx.24.6.707.

83 Khan, M.A., Wolf, D.P., and Litt, M. (1976). Effect of mucolytic agents on the rheological properties of tracheal mucos. *Biochimica et Biophysica Acta (BBA) - General Subjects* 444: 369–373. https://doi.org/10.1016/0304-4165(76)90380-9.

84 Clift, A.F. (1979). Early studies on the rheology of cervical mucus. *American Journal of Obstetrics and Gynecology* 134: 829–832. https://doi.org/10.1016/0002-9378(79)90956-6.

85 Lai, S.K., Wang, Y.-Y., Wirtz, D. et al. (2009). Micro- and macrorheology of mucus. *Advanced Drug Delivery Reviews* 61: 86–100. https://doi.org/10.1016/j.addr.2008.09.012.

86 Georgiades, P., Pudney, P.D.A., Thornton, D.J. et al. (2014). Particle tracking microrheology of purified gastrointestinal mucins. *Biopolymers* 101: 366–377. https://doi.org/10.1002/bip.22372.

87 Valentine, M.T., Perlman, Z.E., Gardel, M.L. et al. (2004). Colloid surface chemistry critically affects multiple particle tracking measurements of biomaterials. *Biophysical Journal* 86: 4004–4014. https://doi.org/10.1529/biophysj.103.037812.

88 Schuster, B.S., Suk, J.S., Woodworth, G.F. et al. (2013). Nanoparticle diffusion in respiratory mucus from humans without lung disease. *Biomaterials* 34: 3439–3446. https://doi.org/10.1016/j.biomaterials.2013.01.064.

89 Shan, W., Zhu, X., Liu, M. et al. (2015). Overcoming the diffusion barrier of mucus and absorption barrier of epithelium by self-assembled nanoparticles for oral delivery of insulin. *ACS Nano* 9: 2345–2356. https://doi.org/10.1021/acsnano.5b00028.

90 Runnsjö, A., Dabkowska, A.P., Sparr, E. et al. (2016). Diffusion through pig gastric mucin: effect of relative humidity. *PLoS One* 11: e0157596. https://doi.org/10.1371/journal.pone.0157596.

91 Li, L., Lieleg, O., Jang, S. et al. (2012). A microfluidic in vitro system for the quantitative study of the stomach mucus barrier function. *Lab on a Chip* 12: 4071–4079. https://doi.org/10.1039/C2LC40161D.

92 Li, L.D., Crouzier, T., Sarkar, A. et al. (2013). Spatial configuration and composition of charge modulates transport into a mucin hydrogel barrier. *Biophysical Journal* 105: 1357–1365. https://doi.org/10.1016/j.bpj.2013.07.050.

93 Abdula, D., Lerud, R., and Rananavare, S. (2017). Bubbling and foaming assisted clearing of mucin plugs in microfluidic Y-junctions. *Journal of Biomechanics* 64: 1–7. https://doi.org/10.1016/j.jbiomech.2016.04.028.

94 Shah, P., Fritz, J.V., Glaab, E. et al. (2016). A microfluidics-based in vitro model of the gastrointestinal human–microbe interface. *Nature Communications* 7: 11535. https://doi.org/10.1038/ncomms11535.

95 Bongaerts, J.H.H., Rossetti, D., and Stokes, J.R. (2007). The lubricating properties of human whole saliva. *Tribology Letters* 27: 277–287. https://doi.org/10.1007/s11249-007-9232-y.

96 Dunn, A.C., Urueña, J.M., Puig, E. et al. (2013). Friction coefficient measurement of an in vivo murine cornea. *Tribology Letters* 49: 145–149. https://doi.org/10.1007/s11249-012-0033-6.

97 Yakubov, G.E., McColl, J., Bongaerts, J.H.H. et al. (2009). Viscous boundary lubrication of hydrophobic surfaces by mucin. *Langmuir* 25: 2313–2321. https://doi.org/10.1021/la8018666.

98 Winkeljann, B., Bussmann, A.B., Bauer, M.G. et al. (2018). Oscillatory tribology performed with a commercial shear rheometer. *Biotribology* 14: 11–18. https://doi.org/10.1016/j.biotri.2018.04.002.

99 Boettcher, K., Winkeljann, B., Schmidt, T.A. et al. (2017). Quantification of cartilage wear morphologies in unidirectional sliding experiments: influence of different macromolecular lubricants. *Biotribology* 12: 43–51. https://doi.org/10.1016/j.biotri.2017.06.002.

100 Wilson, T., Aeschlimann, R., Tosatti, S. et al. (2015). Coefficient of friction of human corneal tissue. *Cornea* 34: 1179–1185. https://doi.org/10.1097/ico.0000000000000524.

101 Ranc, H., Elkhyat, A., Servais, C. et al. (2006). Friction coefficient and wettability of oral mucosal tissue: changes induced by a salivary layer. *Colloids and Surfaces A: Physicochemical and Engineering Aspects* 276: 155–161. https://doi.org/10.1016/j.colsurfa.2005.10.033.

102 Pitenis, A.A., Urueña, J.M., Hormel, T.T. et al. (2017). Corneal cell friction: survival, lubricity, tear films, and mucin production over extended duration in vitro studies. *Biotribology* 11: 77–83. https://doi.org/10.1016/j.biotri.2017.04.003.

103 Hara, A.T., González-Cabezas, C., Creeth, J. et al. (2008). The effect of human saliva substitutes in an erosion–abrasion cycling model. *European Journal of Oral Sciences* 116: 552–556. https://doi.org/10.1111/j.1600-0722.2008.00575.x.

104 Steiger-Ronay, V., Tektas, S., Attin, T. et al. (2019). Comparison of profilometric and microindentation analyses for determining the impact of saliva on the abrasion of initially eroded enamel. *Caries Research* 53: 33–40.

105 Co, J.Y., Crouzier, T., and Ribbeck, K. (2015). Probing the role of mucin-bound glycans in bacterial repulsion by mucin coatings. *Advanced Materials Interfaces* 2: 1500179. https://doi.org/10.1002/admi.201500179.

106 Iguchi, S.M.M., Aikawa, T., and Matsumoto, J.J. (1982). Antibacterial activity of snail mucus mucin. *Comparative Biochemistry and Physiology Part A: Physiology* 72: 571–574. https://doi.org/10.1016/0300-9629(82)90123-2.

107 Frenkel, E.S. and Ribbeck, K. (2015). Salivary mucins protect surfaces from colonization by cariogenic bacteria. *Applied and Environmental Microbiology* 81: 332–338. https://doi.org/10.1128/aem.02573-14.

108 Johnson, K.T., Fath, K.R., Henricus, M.M. et al. (2009). Self-assembly and growth of smart cell-adhesive mucin-bound microtubes. *Soft Materials* 7: 21–36. https://doi.org/10.1080/15394450802693969.

109 Parsons, C.L., Greenspan, C., Moore, S.W. et al. (1977). Role of surface mucin in primary antibacterial defense of bladder. *Urology* 9: 48–52. https://doi.org/10.1016/0090-4295(77)90284-9.

110 McAuley, J.L., Corcilius, L., Tan, H.X. et al. (2017). The cell surface mucin MUC1 limits the severity of influenza A virus infection. *Mucosal Immunology* 10: 1581. https://doi.org/10.1038/mi.2017.16.

111 Poulsom, R. and Wright, N.A. (1993). Trefoil peptides: a newly recognized family of epithelial mucin-associated molecules. *American Journal of Physiology-Gastrointestinal and Liver Physiology* 265: G205–G213. https://doi.org/10.1152/ajpgi.1993.265.2.G205.

112 Aihara, E., Engevik, K.A., and Montrose, M.H. (2017). Trefoil factor peptides and gastrointestinal function. *Annual Review of Physiology* 79: 357–380. https://doi.org/10.1146/annurev-physiol-021115-105447.

113 Bhaskar, K.R., Gong, D.H., Bansil, R. et al. (1991). Profound increase in viscosity and aggregation of pig gastric mucin at low pH. *American Journal of Physiology-Gastrointestinal and Liver Physiology* 261: G827–G832. https://doi.org/10.1152/ajpgi.1991.261.5.G827.

114 Nowald, C., Penk, A., Chiu, H.-Y. et al. (2016). A selective mucin/methylcellulose hybrid gel with tailored mechanical properties. *Macromolecular Bioscience* 16: 567–579. https://doi.org/10.1002/mabi.201500353.

115 Shumilov, D., Popov, A., Fudala, R. et al. (2014). Real-time imaging of exocytotic mucin release and swelling in Calu-3 cells using acridine orange. *Methods* 66: 312–324. https://doi.org/10.1016/j.ymeth.2013.09.004.

116 Dixon, M.C. (2008). Quartz crystal microbalance with dissipation moin-toring: enabling real-time characterization of biological materials and their interactions. *Journal of Biomolecular Techniques* 19: 151–158.

117 Tsiligianni, T., Amiridis, G.S., Dovolou, E. et al. (2011). Association between physical properties of cervical mucus and ovulation rate in superovulated cows. *Canadian Journal of Veterinary Research* 75: 248–253.

118 Critchfield, A.S., Yao, G., Jaishankar, A. et al. (2013). Cervical mucus properties stratify risk for preterm birth. *PLoS One* 8: e69528. https://doi.org/10.1371/journal.pone.0069528.

119 Dunn, A.C., Tichy, J.A., Urueña, J.M. et al. (2013). Lubrication regimes in contact lens wear during a blink. *Tribology International* 63: 45–50. https://doi.org/10.1016/j.triboint.2013.01.008.

120 Davidson, H.J. and Kuonen, V.J. (2004). The tear film and ocular mucins. *Veterinary Ophthalmology* 7: 71–77. https://doi.org/10.1111/j.1463-5224.2004.00325.x.

121 de Pinho Tavares, F., Fernandes, R.S., Bernardes, T.F. et al. (2010). Dry eye disease. *Seminars in Ophthalmology* 25: 84–93. https://doi.org/10.3109/08820538.2010.488568.

122 Moss, S.E., Klein, R., and Klein, B.K. (2000). Prevalence of and risk factors for dry eye syndrome. *Archives of Ophthalmology* 118: 1264–1268. https://doi.org/10.1001/archopht.118.9.1264.

123 Ali, M. and Byrne, M.E. (2009). Controlled release of high molecular weight hyaluronic acid from molecularly imprinted hydrogel contact lenses. *Pharmaceutical Research* 26: 714–726. https://doi.org/10.1007/s11095-008-9818-6.

124 Rhodus, N.L. and Brown, J. (1990). The association of xerostomia and inadequate intake in older adults. *Journal of the American Dietetic Association* 90: 1688–1692.

125 Moore, P.A., Guggenheimer, J., Etzel, K.R. et al. (2001). Type 1 diabetes mellitus, xerostomia, and salivary flow rates. *Oral Surgery, Oral Medicine, Oral Pathology, Oral Radiology, and Endodontology* 92: 281–291. https://doi.org/10.1067/moe.2001.117815.

126 Guggenheimer, J. and Moore, P.A. (2003). Xerostomia: etiology, recognition and treatment. *Journal of the American Dental Association* 134: 61–69. https://doi.org/10.14219/jada.archive.2003.0018.

127 Cassolato, S.F. and Turnbull, R.S. (2003). Xerostomia: clinical aspects and treatment. *Gerodontology* 20: 64–77. https://doi.org/10.1111/j.1741-2358.2003.00064.x.

128 Johnson, J.T., Ferretti, G.A., Nethery, W.J. et al. (1993). Oral Pilocarpine for Post-Irradiation Xerostomia in Patients with Head and Neck Cancer.

New England Journal of Medicine 329: 390–395. https://doi.org/10.1056/nejm199308053290603.

129 Sweeney, M., Bagg, J., Baxter, W. et al. (1997). Clinical trial of a mucin-containing oral spray for treatment of xerostomia in hospice patients. *Palliative Medicine* 11: 225–232. https://doi.org/10.1177/026921639701100307.

130 Lai, W.M., Hou, J.S., and Mow, V.C. (1990). A triphasic theory for the swelling properties of hydrated charged soft biological tissues. In: *Biomechanics of Diarthrodial Joints* (eds. A. Ratcliffe, S.L.Y. Woo and V.C. Mow). New York, NY: Springer.

131 Dowson, D. (1966). Paper 12: modes of lubrication in human joints. *Proceedings of the Institution of Mechanical Engineers, Conference Proceedings* 181: 45–54. https://doi.org/10.1243/pime_conf_1966_181_206_02.

132 Buckwalter, J.A., Mankin, H.J., and Grodzinsky, A.J. (2005). Articular cartilage and osteoarthritis. *Instructional Course Lectures-American Academy of Orthopaedic Surgeons* 54: 465.

133 Kurz, B., Lemke, A.K., Fay, J. et al. (2005). Pathomechanisms of cartilage destruction by mechanical injury. *Annals of Anatomy - Anatomischer Anzeiger* 187: 473–485. https://doi.org/10.1016/j.aanat.2005.07.003.

134 Altman, R.D., Hochberg, M.C., Moskowitz, R.W. et al. (2000). Recommendations for the medical management of osteoarthritis of the hip and knee: 2000 update. *Arthritis and Rheumatism* 43: 1905–1915. https://doi.org/10.1002/1529-0131(200009)43:9<1905::AID-ANR1>3.0.CO;2-P.

135 Bell, C.J., Ingham, E., and Fisher, J. (2006). Influence of hyaluronic acid on the time-dependent friction response of articular cartilage under different conditions. *Proceedings of the Institution of Mechanical Engineers, Part H: Journal of Engineering in Medicine* 220: 23–31. https://doi.org/10.1243/095441105x69060.

136 Bonnevie, E.D., Galesso, D., Secchieri, C. et al. (2015). Elastoviscous transitions of articular cartilage reveal a mechanism of synergy between lubricin and hyaluronic acid. *PLoS One* 10: e0143415. https://doi.org/10.1371/journal.pone.0143415.

137 Takashima, K., Shimomura, R., Kitou, T. et al. (2007). Contact and friction between catheter and blood vessel. *Tribology International* 40: 319–328. https://doi.org/10.1016/j.triboint.2005.10.010.

138 de Beer, B.A., Snik, A.F., Schilder, A.G.M. et al. (2004). Hearing loss in young adults who had ventilation tube insertion in childhood. *Annals of Otology, Rhinology and Laryngology* 113: 438–444. https://doi.org/10.1177/000348940411300604.

139 Bixler, G.D. and Bhushan, B. (2012). Biofouling: lessons from nature. *Philosophical Transactions of the Royal Society A: Mathematical Physical and Engineering Sciences* 370: 2381–2417. https://doi.org/10.1098/rsta.2011.0502.

140 Lawrence, E.L. and Turner, I.G. (2005). Materials for urinary catheters: a review of their history and development in the UK. *Medical Engineering & Physics* 27: 443–453. https://doi.org/10.1016/j.medengphy.2004.12.013.

141 Sandberg, T., Blom, H., and Caldwell, K.D. (2009). Potential use of mucins as biomaterial coatings. I. Fractionation, characterization, and model adsorption of bovine, porcine, and human mucins. *Journal of Biomedical Materials Research Part A* 91A: 762–772. https://doi.org/10.1002/jbm.a.32266.

142 Bavington, C.D., Lever, R., Mulloy, B. et al. (2004). Anti-adhesive glycoproteins in echinoderm mucus secretions. *Comparative Biochemistry and Physiology Part B: Biochemistry and Molecular Biology* 139: 607–617. https://doi.org/10.1016/j.cbpc.2004.07.008.

143 Sandberg, T., Carlsson, J., and Ott, M.K. (2007). Mucin coatings suppress neutrophil adhesion to a polymeric model biomaterial. *Microscopy Research and Technique* 70: 864–868. https://doi.org/10.1002/jemt.20489.

144 Crouzier, T., Jang, H., Ahn, J. et al. (2013). Cell patterning with mucin biopolymers. *Biomacromolecules* 14: 3010–3016. https://doi.org/10.1021/bm400447z.

145 Lundin, M., Sandberg, T., Caldwell, K.D. et al. (2009). Comparison of the adsorption kinetics and surface arrangement of "as received" and purified bovine submaxillary gland mucin (BSM) on hydrophilic surfaces. *Journal of Colloid and Interface Science* 336: 30–39. https://doi.org/10.1016/j.jcis.2009.03.061.

146 Shi, L. and Caldwell, K.D. (2000). Mucin adsorption to hydrophobic surfaces. *Journal of Colloid and Interface Science* 224: 372–381. https://doi.org/10.1006/jcis.2000.6724.

147 Janairo, R.R.R., Zhu, Y., Chen, T. et al. (2014). Mucin covalently bonded to microfibers improves the patency of vascular grafts. *Tissue Engineering Part A* 20: 285–293. https://doi.org/10.1089/ten.tea.2013.0060.

148 Edwards, R. and Harding, K.G. (2004). Bacteria and wound healing. *Current Opinion in Infectious Diseases* 17: 91–96.

149 Polak, R., Crouzier, T., Lim, R.M. et al. (2014). Sugar-mediated disassembly of mucin/lectin multilayers and their use as pH-tolerant, on-demand sacrificial layers. *Biomacromolecules* 15: 3093–3098. https://doi.org/10.1021/bm5006905.

150 Duffy, C.V., David, L., and Crouzier, T. (2015). Covalently-crosslinked mucin biopolymer hydrogels for sustained drug delivery. *Acta Biomaterialia* 20: 51–59. https://doi.org/10.1016/j.actbio.2015.03.024.

151 Svensson, O., Thuresson, K., and Arnebrant, T. (2008). Interactions between drug delivery particles and mucin in solution and at interfaces. *Langmuir* 24: 2573–2579. https://doi.org/10.1021/la702680x.

152 Griffiths, P.C., Cattoz, B., Ibrahim, M.S. et al. (2015). Probing the interaction of nanoparticles with mucin for drug delivery applications using dynamic light scattering. *European Journal of Pharmaceutics and Biopharmaceutics* 97: 218–222. https://doi.org/10.1016/j.ejpb.2015.05.004.

153 Aljayyoussi, G., Abdulkarim, M., Griffiths, P. et al. (2012). Pharmaceutical nanoparticles and the mucin biopolymer barrier. *BioImpacts: BI* 2: 173–174. https://doi.org/10.5681/bi.2012.029.

154 Pontremoli, C., Barbero, N., Viscardi, G. et al. (2015). Mucin–drugs interaction: the case of theophylline, prednisolone and cephalexin. *Bioorganic and*

Medicinal Chemistry 23: 6581–6586. https://doi.org/10.1016/j.bmc.2015.09.021.

155 Lafitte, G., Thuresson, K., and Söderman, O. (2007). Diffusion of nutrients molecules and model drug carriers through mucin layer investigated by magnetic resonance imaging with chemical shift resolution. *Journal of Pharmaceutical Sciences* 96: 258–263. https://doi.org/10.1002/jps.20749.

156 Engvall, E. and Perlmann, P. (1971). Enzyme-linked immunosorbent assay (ELISA) quantitative assay of immunoglobulin G. *Immunochemistry* 8: 871–874. https://doi.org/10.1016/0019-2791(71)90454-X.

157 Van Weemen, B.K. and Schuurs, A.H.W.M. (1971). Immunoassay using antigen—enzyme conjugates. *FEBS Letters* 15: 232–236. https://doi.org/10.1016/0014-5793(71)80319-8.

158 Amano, K., Chiba, Y., Kasahara, Y. et al. (2008). Engineering of mucin-type human glycoproteins in yeast cells. *Proceedings of the National Academy of Sciences of the United States of America* 105: 3232–3237. https://doi.org/10.1073/pnas.0710412105.

159 Yang, Z., Drew, D.P., Jørgensen, B. et al. (2012). Engineering mammalian mucin-type O-glycosylation in plants. *Journal of Biological Chemistry* 287: 11911–11923. https://doi.org/10.1074/jbc.M111.312918.

160 Marcaurelle, L.A. and Bertozzi, C.R. (2002). Recent advances in the chemical synthesis of mucin-like glycoproteins. *Glycobiology* 12: 69R–77R. https://doi.org/10.1093/glycob/12.6.69R.

161 Artigas, G., Hinou, H., Garcia-Martin, F. et al. (2017). Synthetic mucin-like glycopeptides as versatile tools to measure effects of glycan structure/density/position on the interaction with adhesion/growth-regulatory galectins in arrays. *Chemistry - An Asian Journal* 12: 159–167. https://doi.org/10.1002/asia.201601420.

162 Becker, T., Dziadek, S., Wittrock, S. et al. (2006). Synthetic glycopeptides from the mucin family as potential tools in cancer immunotherapy. *Current Cancer Drug Targets* 6: 491–517. https://doi.org/10.2174/156800906778194577.

163 Tang, S., Puryear, W.B., Seifried, B.M. et al. (2016). Antiviral agents from multivalent presentation of sialyl oligosaccharides on brush polymers. *ACS Macro Letters* 5: 413–418. https://doi.org/10.1021/acsmacrolett.5b00917.

164 Papp, I., Sieben, C., Ludwig, K. et al. (2010). Inhibition of influenza virus infection by multivalent sialic-acid-functionalized gold nanoparticles. *Small* 6: 2900–2906. https://doi.org/10.1002/smll.201001349.

165 Papp, I., Sieben, C., Sisson, A.L. et al. (2011). Inhibition of influenza virus activity by multivalent glycoarchitectures with matched sizes. *ChemBioChem* 12: 887–895. https://doi.org/10.1002/cbic.201000776.

166 Rabuka, D., Parthasarathy, R., Lee, G.S. et al. (2007). Hierarchical assembly of model cell surfaces: synthesis of mucin mimetic polymers and their display on supported bilayers. *Journal of the American Chemical Society* 129: 5462–5471. https://doi.org/10.1021/ja067819i.

167 Cook, M.T., Smith, S.L., and Khutoryanskiy, V.V. (2015). Novel glycopolymer hydrogels as mucosa-mimetic materials to reduce animal testing. *Chemical Communications* 51: 14447–14450. https://doi.org/10.1039/C5CC02428E.

168 Mahalingam, A., Jay, J.I., Langheinrich, K. et al. (2011). Inhibition of the transport of HIV in vitro using a pH-responsive synthetic mucin-like polymer system. *Biomaterials* 32: 8343–8355. https://doi.org/10.1016/j.biomaterials.2011.05.001.

169 Dey, P., Bergmann, T., Cuellar-Camacho, J.L. et al. (2018). Multivalent flexible nanogels exhibit broad-spectrum antiviral activity by blocking virus entry. *ACS Nano* 12: 6429–6442. https://doi.org/10.1021/acsnano.8b01616.

170 Authimoolam, S.P., Vasilakes, A.L., Shah, N.M. et al. (2014). Synthetic oral mucin mimic from polymer micelle networks. *Biomacromolecules* 15: 3099–3111. https://doi.org/10.1021/bm5006917.

171 An, J., Jin, C., Dėdinaitė, A. et al. (2017). Influence of glycosylation on interfacial properties of recombinant mucins: adsorption, surface forces, and friction. *Langmuir* 33: 4386–4395. https://doi.org/10.1021/acs.langmuir.7b00030.

172 An, J., Dėdinaitė, A., Nilsson, A. et al. (2014). Comparison of a brush-with-anchor and a train-of-brushes mucin on poly(methyl methacrylate) surfaces: adsorption, surface forces, and friction. *Biomacromolecules* 15: 1515–1525. https://doi.org/10.1021/bm500173s.

7

Advances in the Synthesis of Fibrous Proteins and Their Applications

Gang Wei[1,2], Xi Ma[1], Yaru Bai[1], Coucong Gong[2], and Yantu Zhang[1]

[1] Yan'an University, College of Chemistry & Chemical Engineering, Shengdi Road 580, Baota District, Yan'an, Shaanxi 716000, PR China
[2] University of Bremen, Faculty of Production Engineering, Am Fallturm 1, Bremen D-28359, Germany

7.1 Introduction

The formation of fibrous protein aggregates and superstructures is crucial for many of the functions of human body [1, 2]. In nature, the biocompatible fibrous proteins such as collagen, actin, and fibrinogen fibrils are related to bone growth, muscle formation, and tissue repair, respectively. However, amyloid protein nanofibrils caused by the aggregation of some proteins (such as lysozyme, α-synuclein, prion protein, and Tau protein) are related to more than 20 degenerative diseases including Parkinson's, Alzheimer's, type 2 diabetes, Huntington's, and others. Previously, great efforts have been devoted to investigate the conformation transition, self-assembly, aggregation, and formation of various protein fibrils [3, 4]. The main aim of scientific efforts has been to understand the mechanisms of both misfolded protein aggregation and organized protein self-assembly; however, the misfolded and organized amyloid protein nanofibrils are unique precursors for developing various novel nanomaterials [5, 6].

With molecular self-assembly, it is easy to produce fibrous protein nanostructures like nanofibrils, nanotubes, and nanoribbons by adjusting the pH, ionic strength, temperature, and other reaction parameters [3]. Besides the self-assembly method, other methods such as electrospinning [7], template-assisted synthesis [8, 9], solution blowing [10], and others [11] have also been utilized to create protein nanofibers, nanowires, nanorods, and microfibers.

Fibrous proteins exhibited several advantages compared with traditional inorganic or polymer fibrous materials. First, fibrous proteins such as collagen, elastin, and silk fibroin (SF) have shown higher biocompatibility, which are suitable for fabricating numerous fibrous protein-based biomaterials for biomedical and tissue engineering applications [12]. Second, a fibrous protein provides a large amount of functional groups on its surface, which promotes the easy modification and hybridization with other nanoscale building blocks. Third, such protein provides a high affinity to some biomolecules (e.g. enzymes and antigens), making it possible to fabricate novel biosensors with high selectivity and

Biopolymers for Biomedical and Biotechnological Applications, First Edition.
Edited by Bernd H. A. Rehm and M. Fata Moradali.

sensitivity. Fourth, some proteins show biomimetic and bioinspired functions to form uniform nanomaterials that are not easy to synthesize by traditional chemical methods. Because of these unique properties, fibrous protein materials have been widely used in the fields of biomedicine, tissue engineering, sensors, nanodevices, energy, and environment [13–17].

In this chapter, recent advances in the synthesis and applications of fibrous protein materials are presented. In Section 7.2, the synthesis methods, structures, and characterization techniques for fibrous protein materials are introduced, and then in Section 7.3 the potential applications of fibrous protein-based functional materials in different fields are demonstrated and discussed. It is expected that this work will be valuable for readers to understand the synthesis and applications of fibrous protein materials in one way and in another way to develop novel protein-based fibrous nanomaterials with advanced functions.

7.2 Synthesis, Structure, and Characterizations of Fibrous Protein Materials

Fibrous proteins exhibited various structures and functions based on their synthesis methods, which mediate specific applications of the fabricated fibrous protein materials in different fields. Herein, the synthesis methods, structures, and characterizations of fibrous protein materials are introduced in brief.

7.2.1 Synthesis Methods

Fibrous protein materials can be produced mainly by the synthesis methods including electrospinning, solution blowing, self-assembly, and template-assisted assembly (Figure 7.1).

Electrospinning is a facile electrostatic fabrication technique for producing polymer nanofibers, which has also been utilized to produce protein nanofibers for diverse applications. It has several advantages like low cost and high throughput to fabricate protein or protein/polymer nanofibers with controllable structure and properties. In an electrospinning process, a critical potential is applied between the metallic needle of the syringe with protein or protein/polymer solution and the collector, and then the protein solution is injected into the collector to create protein nanofiber membranes, as shown in Figure 7.1a [18]. This technique provides the possibility of creating various protein nanofibers/microfibers by using a wide range of plant and animal proteins [7]. For example, Kim et al. prepared recombinant mussel adhesive protein nanofibers, and Min et al. created multifunctional silk protein nanofibers by electrospinning [18, 22]. However, the applied high potential will cause the partly denaturation of protein molecules, affecting their biocompatibility and activity in some ways.

Solution blowing has been one of the industrially available techniques to produce nanofibers with different diameters. Protein solution jet was set coaxially with the surrounding high-pressure air jet, creating uniform protein nanofibers in the collection chamber to form blown nanofiber mats, as indicated

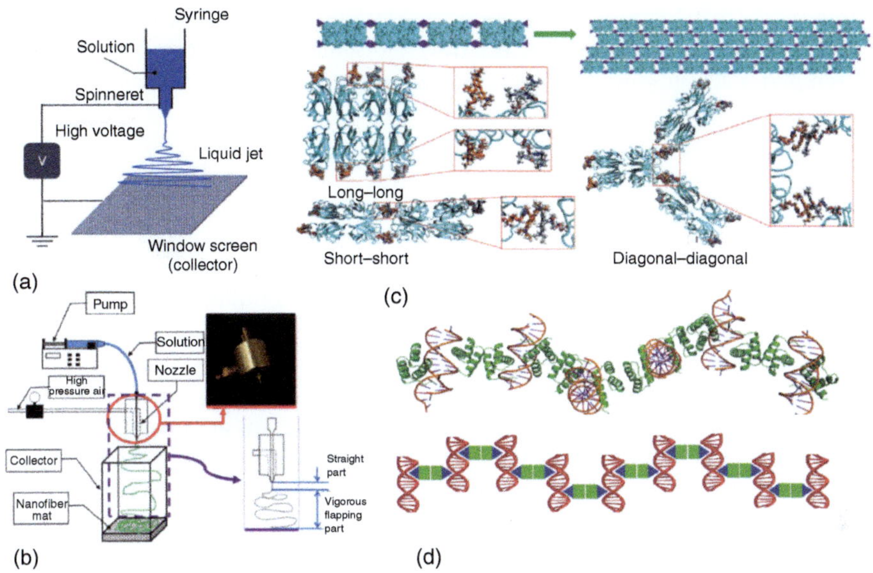

Figure 7.1 Synthesis techniques for fibrous proteins. (a) Electrospinning. *Source:* Min et al. 2018 [18]. https://www.nature.com/articles/s41598-018-27917-w#rightslink. Licensed Under CC BY 4.0. (b) Solution blowing. *Source:* Reproduced with the permission from Kolbasov et al. [19]. Copyright 2016, American Chemical Society. (c) Self-assembly. *Source:* Reproduced with the permission from Yang et al. [20]. Copyright 2017, John Wiley and Sons.
(d) Template-assisted assembly. *Source:* Reproduced with the permission from Mou et al. [21]. Copyright 2015, Springer Nature.

in Figure 7.1b [19]. Previously, nanofibers of silk protein and bovine serum albumin (BSA) have been fabricated by the solution blowing technique [23]. Recently, Kolbasov et al. demonstrated an industrial-scale solution blowing of soy protein nanofibers with a diameter of 500–600 nm, which were collected with an area on the order of 0.1–1 m^2 in 10 seconds [19].

Self-assembly of biomolecules (protein, peptide, DNA, and others) has been the widely used and most promising strategy to design and synthesize functional biomaterials with enhanced mechanical properties, novel structure, and new functions. The formation of fibrous protein materials could be mediated by adjusting several factors, including pH, temperature, ionic strength, denaturants, enzymes, and others [3]. Previously, the self-assembly technique has been utilized to create various fibrous protein materials [11, 20, 24, 25]. For instance, Yang et al. demonstrated highly ordered self-assembly of native proteins (LecA/R1G) into nanofibers through the long–long, short–short, and diagonal–diagonal packing of the proteins (Figure 7.1c) [20]. It was found that both lectin–sugar interactions and π–π stacking are crucial for the ligand-mediated protein self-assembly.

Template-assisted synthesis can also be utilized for the creation of fibrous protein materials by using nanoscale templates like carbon nanotubes (CNTs) and DNA [8, 21, 26]. For example, previously Wei et al. created water-soluble fibronectin–CNT hybrid nanofibers by covalent assembly of proteins along the functionalized CNTs [8]. In another case, Mou et al. designed and created protein–DNA hybrid nanowires by noncovalent interaction-mediated molecular co-assembly [21], in which DNA building blocks served as both binding sites and templates for guiding the assembly of proteins to nanowires, as shown in Figure 7.1d.

7.2.2 Structure

Generally, the synthesized fibrous protein materials include electrospun protein/polymer nanofibers [27, 28], self-assembled protein nanofibrils [29–31], nanotubes [32], and nanoribbons [33], as well as other kind of microscale fibers [11, 34], as shown in Figure 7.2.

For instance, Wang and Chen produced hybrid hordein/zein nanofibers by electrospinning [28], in which the electrospun nanofibers exhibited a diameter of several hundred nanometers (Figure 7.2a). Zhou et al. investigated the self-assembly of Ure2-alkaline phosphatase (Ure2-AP) to protein nanofibrils (Figure 7.2b) [31]. In general, the self-assembled amyloid nanofibrils have a

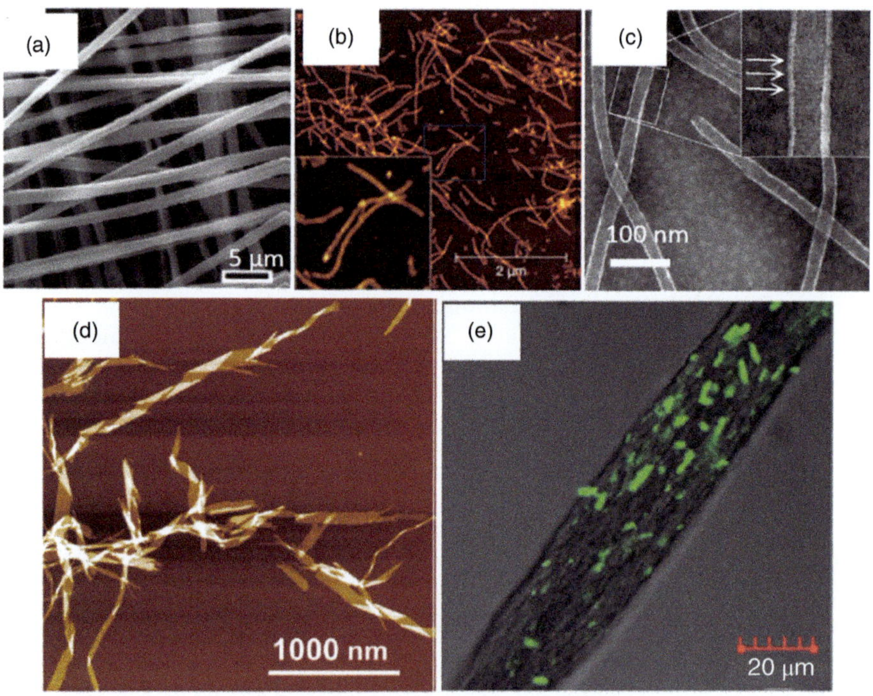

Figure 7.2 Morphological structures of several fibrous proteins. (a) Electrospun polymer/protein nanofibers. *Source:* Reproduced with the permission from Wang and Chen [28]. Copyright 2014, American Chemical Society. (b) Self-assembled protein nanofibers. *Source:* Reproduced with the permission from Zhou et al. [31]. Copyright 2015, American Chemical Society. (c) Self-assembled protein nanotubes. *Source:* Reproduced with the permission from Rao et al. [32]. Copyright 2018, American Chemical Society. (d) Self-assembled protein nanoribbons. *Source:* Reproduced with the permission from Lara et al. [33]. Copyright 2011, American Chemical Society. (e) Self-assembled protein microfibers. *Source:* Reproduced with the permission from Wang et al. [34]. Copyright 2014, John Wiley and Sons.

diameter of a few to tens of nm with a length from 100 nm to a few μm. Rao et al. demonstrated the controllable self-assembly of capsid proteins to flexible protein nanotubes (Figure 7.2c) with average inside diameter of 18 nm and outside diameter of 28 nm [32]. Lara et al. created helical giant multistranded amyloid lysozyme and β-lactoglobulin ribbons (Figure 7.2d) by controlling the self-assembly of proteins under 90 °C and 30 hours incubation time [33]. The created protein nanoribbons showed a width of about 173 nm and a length of a few μm. In another case, Wang et al. presented the design and synthesis of multicolored microfibers by controlling the self-assembly of DNA, histone proteins, and cationic conjugated polymers [34]. The diameter of the created microfibers was about 30 μm, and the length of fibers could be up to 30 cm (Figure 7.2e).

7.2.3 Characterizations

The structural and morphological characteristics of fibrous proteins can be measured by several surface analysis techniques including atomic force microscopy (AFM), scanning electron microscopy (SEM), transmission electron microscopy (TEM), and confocal laser scanning microscopy (CLSM). The mechanical and physical properties (tensile strength, Young's modulus, thermal stability, conductivity, and others) of fibrous protein materials can be obtained by using tensile tester, differential scanning calorimetry (DSC), thermal gravimetric analysis (TGA), and impedance analyzer. In addition, several other techniques, such as circular dichroism (CD) spectroscopy, small-angle X-ray scattering (SAXS), Fourier transform infrared spectroscopy (FTIR), and Raman spectroscopy, are highly useful to explore the self-assembly mechanism of protein molecules toward fibrous protein structures like nanofibrils, nanotubes, nanoribbons, and microfibers.

For instance, Beun et al. observed the unidirectional living growth of self-assembled protein nanofibrils with super-resolution microscopy [35]. In another case, Helbing et al. created hybrid BSA/hemoglobin nanofibers by controlling the co-self-assembly of two proteins in ethanol [36]. The self-assembly mechanisms of proteins to nanofibrils were explored by using AFM, TEM, CD, tip-enhanced Raman spectroscopy (TERS), and force spectroscopy.

7.3 Applications of Fibrous Protein Materials

In this section, the potential applications of fibrous protein materials in tissue engineering, biomedical engineering, sensors, nanodevices, energy, and environmental fields are presented and discussed.

7.3.1 Bone Tissue Engineering

The inorganic hydroxyapatite (HA) and the natural fibrous protein collagen are the main ingredients for bone tissue engineering. As organic scaffolds, fibrous protein materials show excellent biocompatibility, biodegradability, morphologic flexibility, permeability, and tangible mechanical properties due to the abundant functional groups on the surface of protein.

Ding et al. developed original injectable hydrogels that served as osteogenesis biomaterials based on nanoscale silk–HA hybrids [37]. In this work, the silk-coated HA nanoparticles (NPs) were firstly prepared by using silk as template to promote the nucleation and growth of HA NPs. Then the SF nanofibers were formed in water at 60 °C. Finally, the silk-coated HA NPs and SF nanofibers were mixed together to fabricate hybrid SF–HA hydrogel, which could serve as the injectable bone materials for biomedical applications. The *in vitro* and *in vivo* cell studies demonstrated that the SF–HA hybrid hydrogel has good cytocompatibility and osteogenic differentiation capacity. In a further study, they developed a multiple biomimetic scaffold based on HA and SF to improve the osteogenesis capacity [38]. In another study, Yao and coworkers fabricated a three-dimensional (3D) hydrogel that can modulate the nucleation and growth of HA crystals via a simple ion diffusion method [39].

SF nanofibers can also be combined with peptides and chitosan to form scaffold for bone tissue engineering application. For example, Aliramaji et al. fabricated the SF/chitosan-based magnetic scaffolds using different amounts of magnetite NPs by a freeze-casting method [40]. In their work, the magnetite was firstly prepared by coprecipitation under ultrasound treatment. Then, the magnetite scaffold was fabricated by adding the magnetite ferrofluid into silk and chitosan solution with a crosslinker agent. The stability of the scaffold was tested, and the results demonstrated that the fabricated scaffold has good stability with improved magnetic responsivity. In another study, Behera et al. designed a platform based on the tripeptide Arg-Gly-Asp (RGD) motif and silk fibrils for bone tissue engineering, in which the SF fibrils benefited the HA nucleation and growth and the RGD provided supports cell attachment and proliferation [41].

Amyloid protein nanofibrils are other main promising fibrous protein biomaterials for bone tissue engineering due to their extraordinary characteristics in terms of exceptional mechanical properties, rich and diverse surface functional groups, strong adhesion to both hydrophobic and hydrophilic surfaces, and excellent biocompatibility. For instance, Li et al. developed an amyloid–HA bone biomimetic composites scaffold by a "bottom-up" method for bone tissue engineering application [42]. In their work, HA platelets were firstly prepared by dissolving $CaCl_2$ and NaH_2PO_4 into CH_3COONa solution and adjusting the pH to 5.5 and followed by heating for 24 hours at 160 °C. Subsequently, amyloid–HA hybrid composites were created by mixing HA platelets with amyloid fibrils (AFs) at pH 7, which were then vacuum-filtered through a nylon filtration membrane to obtain the amyloid–HA hybrid film, as shown in Figure 7.3. The created amyloid–HA bone biomimetic composites exhibited excellent properties, leave complete flexibility on the exact final properties, and have excellent biocompatibility. For instance, the human trabecular bone-derived preosteoblastic cells could efficiently adhere and grow on these composites.

Amyloid protein nanofibrils can also be conjugated onto graphene materials for bone tissue engineering application. Previously, graphene–fibrinogen nanofibril hybrid composites on silicon substrate have been prepared by the layer-by-layer method and used for the nucleation and growth of HA in simulated body fluid at different periods [43].

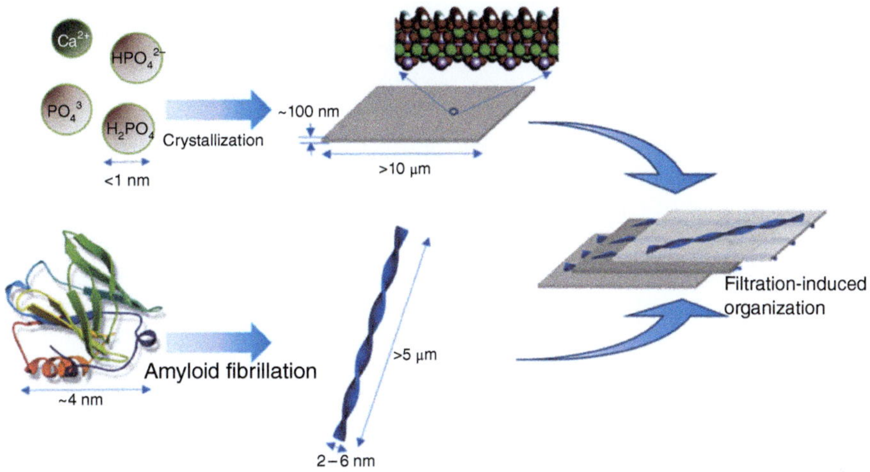

Figure 7.3 Amyloid–HA hybrid film for bone tissue engineering application. *Source:* Reproduced with the permission from Li et al. [42]. Copyright 2014, John Wiley and Sons.

7.3.2 Biomedical Engineering

Fibrous protein materials have also been widely used for biomedical engineering including drug delivery and bioimaging. In recent years, a lot of fibrous proteins, including AFs and silk fibrils, have been further fabricated to hydrogels or combined with other components like graphene, CNTs, and biomolecules to create bio-scaffolds for biomedical applications [44]. For example, Kaplan and coworkers designed a pH-responsive and injectable hydrogel based on silk nanofibers for drug delivery [45], as shown in Figure 7.4. In their work, the silk nanofibers were formed by a concentration dilution process, which were then fabricated to nanofiber hydrogels by incubating for 24 hours at 60 °C. Then doxorubicin (DOX) was directly dissolved into the silk nanofiber hydrogel to form DOX-loaded hydrogels. The obtained DOX-loaded silk nanofiber hydrogels were injectable and could release anticancer drug DOX by adjusting the

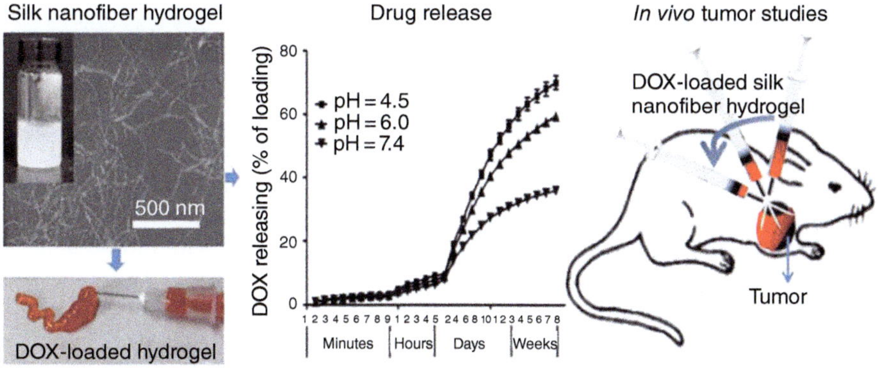

Figure 7.4 Self-assembled silk nanofiber hydrogel for drug delivery application. *Source:* Reproduced with the permission from Wu et al. [45]. Copyright 2016, American Chemical Society.

pH of solution to 4.5. Furthermore, the *in vitro* cytotoxicity test indicated that the formed DOX-loaded silk nanofiber hydrogels exhibited better therapeutic effects with the similar cytotoxicity to free DOX.

Silk fibrils can also co-assemble with biomolecules including peptide, protein, and DNA to form hydrogels or NPs for drug delivery [46]. For instance, a silk-elastin-like protein polymer (SELP) formed by silk fibrils and elastin protein have been utilized to trigger the delivery of DOX. The created SELP polymer system has a few advantages such as high biocompatibility, very low toxicity, and high feasibility in all aqueous systems, showing high potential for further applications in terms of cancer treatments.

Fibrous protein materials can also be used for bioimaging application. Previously, Vannoy et al. investigated the conjugation of amyloid lysozyme fibrils with CdSe–ZnS quantum dots for biolabeling and bioimaging [47]. In another case, Wei et al. demonstrated the creation of fibronectin nanofibers by controlling the self-assembly of proteins in highly concentrated ethanol solution, which were further conjugated with CdSe–ZnS quantum dots by covalent interaction to construct biophotonic nanohybrids [48].

7.3.3 Sensors and Biosensors

Amyloid-like proteins as promising biomaterials that can self-assemble into amyloid nanofibrils, which can interact with various ligands to fabricate sensors and biosensors for analytical and biomedical treatments [49]. For example, Yang et al. combined α-synuclein AFs and pentacosadiynoic acid (PCDA) to produce a sensor [50]. The presence of molecular freedom in the PCDA bundles made the fabricated sensor very sensitive to temperature and organic solvents. Under UV irradiation, both AF/PCDA and AF + PCDA had a color transition on paper by changing the temperature or organic solvent. By heating the sensor containing AF + PCDA on the hot plate, it can be found that the color of the sensor changed constantly as the temperature rose in the heating process, and therefore the temperature change can be monitored at any time. It should be noted that the fabricated sensors were unreversible under temperature variations, while they had certain reversibility in two environmental transitions, heating, and UV radiation. According to the experimental results, this reversibility was due to the existence of redundant PCDA monomers. This new sensor can be widely used to monitor temperature change over time as well as monitoring organic compound pollution. In addition, the synthesized AFs can be combined with graphene oxide (GO) nanosheets through electrostatic interactions, and the obtained composites can be used for Au nanocatalysts or enzyme fixation [51]. It can not only maintain the catalytic activity of the catalyst but also keep outstanding electrochemical properties. Besides, AF can also be used as a glucose sensor system in the field of electrochemistry.

Sensors containing protein fibers can also be used to detect biomolecules. For example, Kim et al. prepared mesoporous WO_3 nanofibers using hollow apoferritin protein as nanoscale biological template [52]. The addition of catalytic NPs made the gas sensor more accurate, and it had a good ability to recognize biomolecules, which can directly monitor the target biomarkers

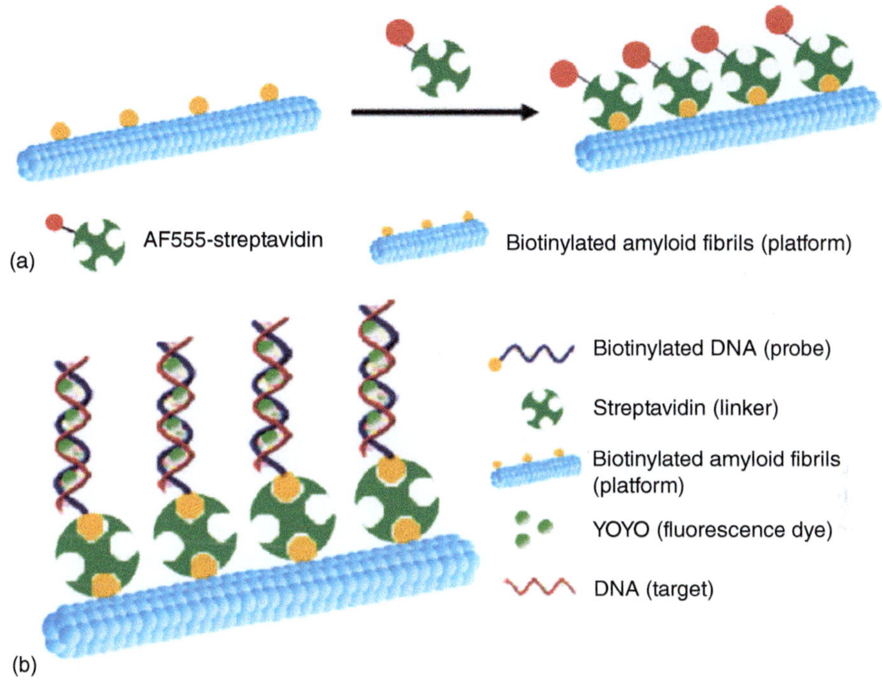

Figure 7.5 Self-assembling Aβ fibrillar platform for detecting biomolecules: (a) detecting AF555-streptavidin and (b) detecting poly$(A)_{15}$ DNA. *Source:* Reproduced with the permission from Chan and Li [53]. Copyright 2011, American Chemical Society.

in exhaled breath. In another example, Chan and Li formed functional self-assembled protein nanofibrils by using beta-amyloid (Aβ) protein [53]. It can combine different biological probes and thus form biological sensors to monitor different biomolecules including Alexa Fluor 555-labeled streptavidin (AF555-Streptavidin) and poly$(A)_{15}$ DNA, as shown in Figure 7.5. This sensor had the characteristics of high efficiency and throughput, and the coding mode used in this process made the monitoring good diversity. These sensors for detecting biological molecules provide good platforms for studying the structure and functions of biomaterials. Very recently, Nandi et al. demonstrated that self-assembled BSA amyloid nanofibrils could bioinspire the formation of Au clusters, which are possible for cysteine sensing both *in vitro* and inside live cells [54].

7.3.4 Nanodevices

The insulating protein nanofibers can be treated in different ways to have electrical properties. For example, metal polymer, a conductor coating, can be attached to the protein fibers to form an excellent biomolecular scaffold for electrochemical nanodevices. Poly(3,4-ethylenedioxythiophene) (PEDOT) has good biocompatibility and electrochemical stability, while the alkoxy-sulfonate version of PEDOT (PEDOT-S) is conductive and water soluble, which can participate in the self-assembly process of biological molecules. Therefore,

Elfwing et al. combined PEDOT-S coating with AFs to study the electrical conductivity of the coated fiber [55]. The results showed that the single AFs with metal polymer coating was more conductive, which indicated the direction of biological interface with excellent performance. In another case, Meier et al. used amyloid nanofibers as biological templates to form core–shell nanofibers by oxidizing polymerization of aniline [56]. The created nanofibers could also be used for the development of nanodevices.

In addition, microbial protein nanofibers play also important roles in electrical conductivity to enhance electron transfer through thermal activation. For example, Lampa-pastirk et al. demonstrated that the pilus protein fibers were conducive to the transfer of electric charge [57]. Due to the thermal dependence of the electrical conductivity of the pili, they proposed a multistep mechanism of thermal activation to ensure better charge transfer of the pili between intracellular and external. The obtained results contributed to the application of pili as protein nanofibers in nanoelectronics. On the other hand, microbial protein nanofibers can also be charged by changing the amino acid species. For instance, Tan et al. used genetic techniques to introduce tryptophan into PilA (the pilus monomer) of *Geobacter sulfurreducens* to produce synthetic protein nanowires (called the W51W57 pili) [58]. Modified pili nanowires (Figure 7.6a) were not only smaller in diameter, but also more conductive than native wild-type pili nanowires. The inhomogeneity of the nanowires leads to differences in the current voltage response of the pili at different nanoscales. At pH 2, the current voltage response of synthetic protein nanowires shown in Figure 7.6b was the highest. In spite of that, the minimum conductivity of synthetic protein nanowires was higher than native wild-type pili, and there was also a significant difference (Figure 7.6c) in the conductivity between them at higher pH. In addition, it can be found that the conductivity of synthetic protein nanowires was proportional to the proton doping amount. When compared with other organic materials with similar diameters, the synthetic protein nanowires exhibited similar electrical conductivity to CNTs, as shown in Figure 7.6d. The fabricated nontoxic and efficient protein nanowires produced by using the regenerative matrix are attractive candidates for fabricating functional nanodevices.

7.3.5 Energy Application

Biomimetic nanomaterials have been widely used for energy applications due to their unique capabilities for generating and storing energy in efficient ways [59]. Fibrous proteins exhibited great advantages for energy applications due to their unique structure and properties. They can serve as both templates and active components to form uniform nanostructures and hybrid materials. Previously, fibrous protein-based nanomaterials have been utilized for the applications in energy storage (fuel cells [60], batteries [61, 62], and supercapacitors [63–65]) and harvesting [27, 66, 67].

Electrospun protein nanofibers are potential materials for fuel cells. Zhou et al. demonstrated the application of electrospun spider silk nanofibers for further templated synthesis of carbon nanofibers (CNFs), which could be used as sustainable electrocatalysts for microbial fuel cells [60]. It was found that the fabricated

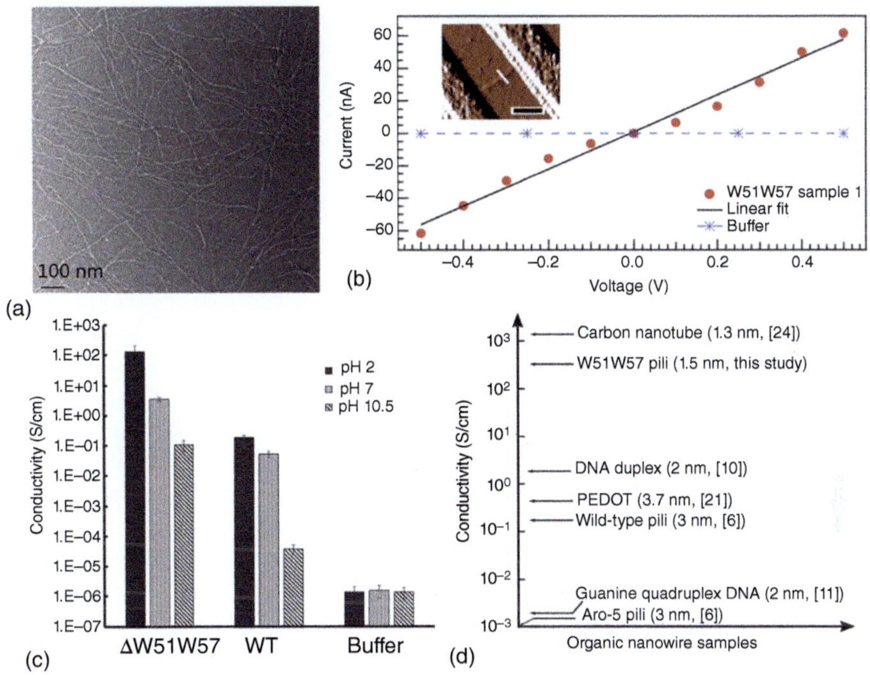

Figure 7.6 The conductivity of synthetic protein nanowires. (a) The nanowires created from *Geobacter sulfurreducens* strain W51W57 producing a genetically modified PilA. (b) Current–voltage response of protein nanowires. (c) pH effect on the conductivity of synthetic protein nanowires. (d) Conductivity of protein nanowires compared with the conductivities of other organic nanowires. *Source:* Reproduced with the permission from Tan et al. [58]. Copyright 2016, John Wiley and Sons.

silk-based CNFs exhibited higher oxygen reduction reaction (ORR) activity than the Pt/C catalyst and most reported metal-free carbon catalysts in alkaline conditions. The high catalytic activity of CNFs is ascribed to the electronegative N and S atoms, high surface area, and porous structure of silk nanofibers. The fabricated microbial fuel cell equipped with the CNF cathode revealed a maximum power density of 1800 mW/m^2, which is 1.56 time higher than that of Pt/C-based fuel cell.

The unique self-assembly and mineralization/carbonization properties of proteins make it possible to create novel hybrid materials for batteries. For instance, Guo et al. synthesized novel protein nanofibers by controlling the self-assembly of GPG-AG3 protein and further prepared Pt and Pd nanotubes by biomimetic mineralizing 3–5 nm Pt and Pd NPs on the surface of protein nanofibers [61]. The synthesized Pt and Pd nanotubes exhibited potential application for lithium–O_2 batteries. In another case, Hou et al. reported the synthesis of hierarchical porous N-doped carbon nanosheets by using the self-assembled lamellar-like layered silk structure [62], which revealed high performance for lithium-ion batteries and supercapacitors by serving as the electrode material in both devices.

Fibrous protein nanomaterials could be utilized for energy storage supercapacitors by combining with other electroactive materials such as graphene and conductive polymers. Wang et al. demonstrated the self-assembly of SF on reduced graphene oxide (RGO) for the formation of RGO/SF nanofibril composites, which were further frizzed, dried, and carbonized to RGO/carbon nanocomposites [63]. The fabricated RGO/carbon-based supercapacitors exhibited remarkable stability with a capacitance retention ratio of 96.3%, high energy density (14.4 Wh/kg), and ultrahigh power density (40 kW/kg). Pal et al. for the first time introduced the fabrication of flexible, fully organic, degradable microsupercapacitors (μSCs) by using silk proteins, conductive polymer PEDOT polystyrene sulfonate, and RGO [64]. Besides acting as electrode materials, fibrous proteins can serve as separators for batteries and supercapacitors. For instance, Tan et al. created silk nanofibrils (SNFs) through the denaturation of silk fibers in deep eutectic solvent and then fabricated freestanding SNF membranes by the vacuum filtration [65]. The fabricated nanofibril membranes showed nanoporous structure, solvent insolubility, good mechanical stability, and high thermal stability, presenting potential application as an effective, stable, and sustainable separator for supercapacitors and electrochemical batteries.

Self-assembled fibrous protein scaffolds offer an attractive way to construct light-harvesting systems through a precisely spaced arraying of chromophores, which can transport the energy to photosynthetic reaction centers and facilitate the conversion of sunlight to chemical bonds [66]. Previously, Miller et al. reported that self-assembled tobacco mosaic virus (TMV) coat protein nanorods can be used to collect light over a broad range of wavelengths after conjugating with different types of chromophores [66]. Recently, Sun et al. designed a novel light-harvesting system by using core-crosslinked micelles (CCMs) to direct the self-assembly of cricoid stable protein one (SP1), as shown in Figure 7.7a [67]. It can be found that one micelle can interact with two central SP1 to form a sandwich structure due to the structural properties of CCMs and SP1, promoting the formation of supramolecular protein nanowires. By labeling donor and acceptor chromophores to the rings of SP1 (DSP1) and micelles (ECCM), respectively, it is possible to develop an artificial light-harvesting nanowire system. The fluorescent test indicated that a significant decrease of fluorescence intensity of DSP1 in the developed DSP1/ECCM light-harvesting system (Figure 7.7b) suggesting energy can be transferred to acceptor chromophores through direct donor-to-energy (path 1) and multiple donor-to-donor (path 2) transfers (Figure 7.7c). The bioinspired light-harvesting systems will be highly valuable for developing novel light-sensitive solar cells, photocatalytic synthesis, and optical sensors. Besides the light-harvesting systems, recently Kim et al. developed a silk nanofiber network-based bio-triboelectric generator (Silk Bio-TEG) by electropinning [27]. The fabricated Silk Bio-TEG exhibited durable and reliable energy harvesting and generating properties due to the high surface-to-volume ratio and strong mechanical ability of silk nanofibers, proving high potentials for the fabrication of energy and electrical devices.

7.3.6 Environmental Application

Fibrous protein materials have exhibited great advantages for the purification of water due to the facile production, quick dispersion in water, low toxicity, and

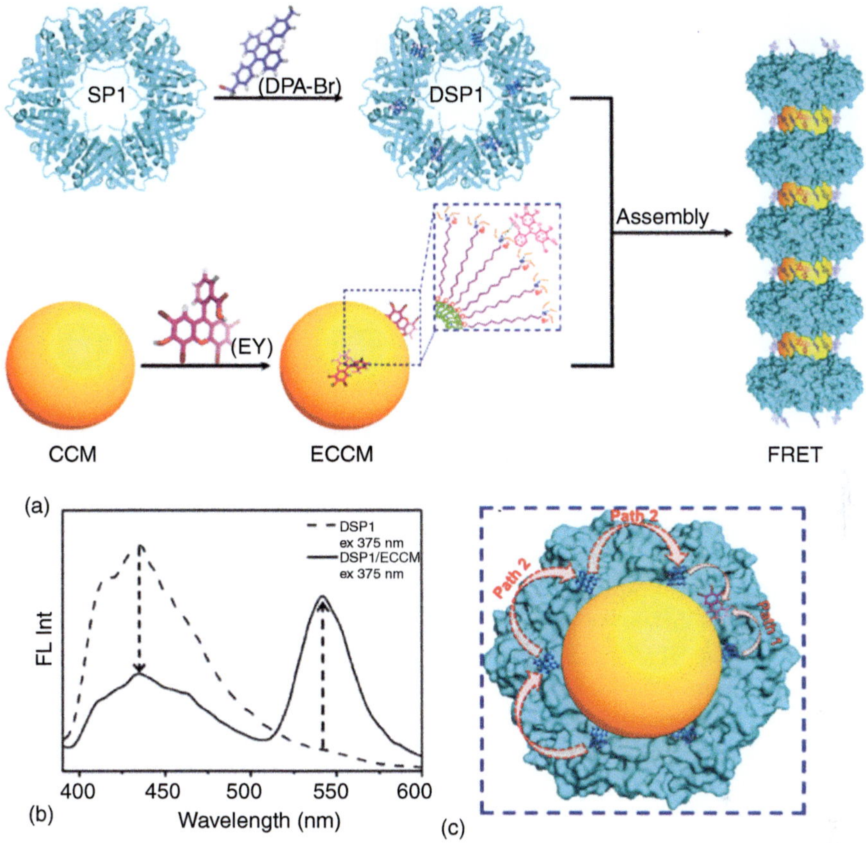

Figure 7.7 Light harvesting of protein nanowire-based hybrid nanomaterials: (a) design of donor-labeled SP1 (DSP1) and acceptor-labeled micelles (ECCM), (b) fluorescence spectrum of DSP1 and DSP1/ECCM light-harvesting system, and (c) light energy transfer mechanisms. *Source:* Reproduced with the permission from Sun et al. [67]. Copyright 2016, American Chemical Society.

high biodegradability. For instance, the electrospun SF nanofiber membranes have been utilized for the adsorption of Cu^{2+} [68–70], and the wool-derived keratin nanofiber membrane has been used for removing Cu^{2+}, Ni^{2+}, and Co^{2+} from aqueous solutions [71]. The fabricated protein nanofiber membranes are effective to remove the metal ions due to the high affinity between protein substrate and metal ions; however, the fabricated membranes revealed limited performance for removing other metallic ions, NPs, dyes, biomolecules, and others from water. One of the potential strategies to solve this problem is to use the hybrid protein nanofibrous materials. Previously, Bolisetty and Mezzenga demonstrated the fabrication of hybrid nanofibrous membranes based on amyloid β-lactoglobulin nanofibrils and activated porous carbon and their highly effective water purification [72]. The amyloid–carbon hybrid membranes exhibited high separation ability and selectivity toward heavy metal ions and radioactive waste. In another case, Ling et al. designed novel biomimetic multilayer membranes based on

SNFs and *in situ* mineralized HA [73]. The fabricated SNF/HA membranes showed ultrafast water penetration and high performance to remove some contaminants such as heavy metal ions, dyes, biomolecules, and NPs from water systems.

Nanofibrous protein materials could be utilized for removing hazardous azo dyes and NPs from aqueous solutions through the coagulation process. For instance, Morshedi et al. demonstrated that the self-assembled lysozyme nanofibrils were able to remove acid red 88, Bismarck brown R, direct violet 51, reactive black 5, and Congo red with the adsorption of 60–98% [74]. It was found that the coagulation process is affected by a few factors like the temperature, ionic strength, and pH value. Dhandayuthapani et al. prepared PVA/gluten hybrid nanofibers by electrospinning for removing gold and silver NPs from water [75]. The conjugation of gluten with polymer nanofibers increased the number and type of functional groups on the nanofiber surface, which enhances the adsorption of NPs with nanofibers. It was found that the using a small amount of protein in the electrospun nanofibers caused in high extraction efficiency of 99% toward both NPs.

Electrospun fibrous protein membranes provide potential applications for air purification to solve the worldwide air pollution problem. Wang et al. reported the electrospun fabrication of SF nanofiber-based air filter for filtering both particulate matter (PM) and submicron particles [76]. It was found that the fabricated SF membranes exhibited filtration efficiency of 98.8% and 96.2% for $PM_{2.5}$ and 300 nm particles, respectively. In addition, the produced air filter revealed several attractive features such as light weight (3.4 g/m^2, about 1/60th of traditional filter), high efficiency, low airflow resistance, high compatibility, and antibacterial activity (after binding nanoscale Ag). Recently, Min et al. demonstrated the fabrication of highly efficient, eco-friendly, optically translucent, and multifunctional air filters by electrospinning SF/PEO nanofibers (Figure 7.8) [18]. The fabricated SF/PEO membranes could be produced a window screen to block the entering of PM from outside (Figure 7.8a), and the optical transmittances of the fabricate air filters were adjustable by changing the area density of SF/PEO nanofibers (Figure 7.8b). To evaluate the filtration efficiencies, a simple "one-way" setup was designed, as shown in Figure 7.8c. Through checking the nanofibrous air filter before and after air filtration process, it can be found that PMs ($PM_{2.5}$ and PM_{10}) were effectively captured by the SF/PEO nanofibers due to the high adhesion properties of SF nanofibers (Figure 7.8d,e). In addition, the incorporation of SF nanofibers with organic dye probes made it possible to produce novel air filters as highly sensitive chemosensors for detecting hazardous and volatile vapors.

Besides silk protein, a type of abundant plant protein, soy protein, has also been utilized to fabricate nanofibrous membranes for highly efficient and multifunctional air filtration [77]. The produced nanofibrous filter exhibited high filtration efficiency for both particles with a broad size range ($PM_{2.5}$ and $PM_{10\text{-}2.5}$) and toxic gases (HCHO and CO).

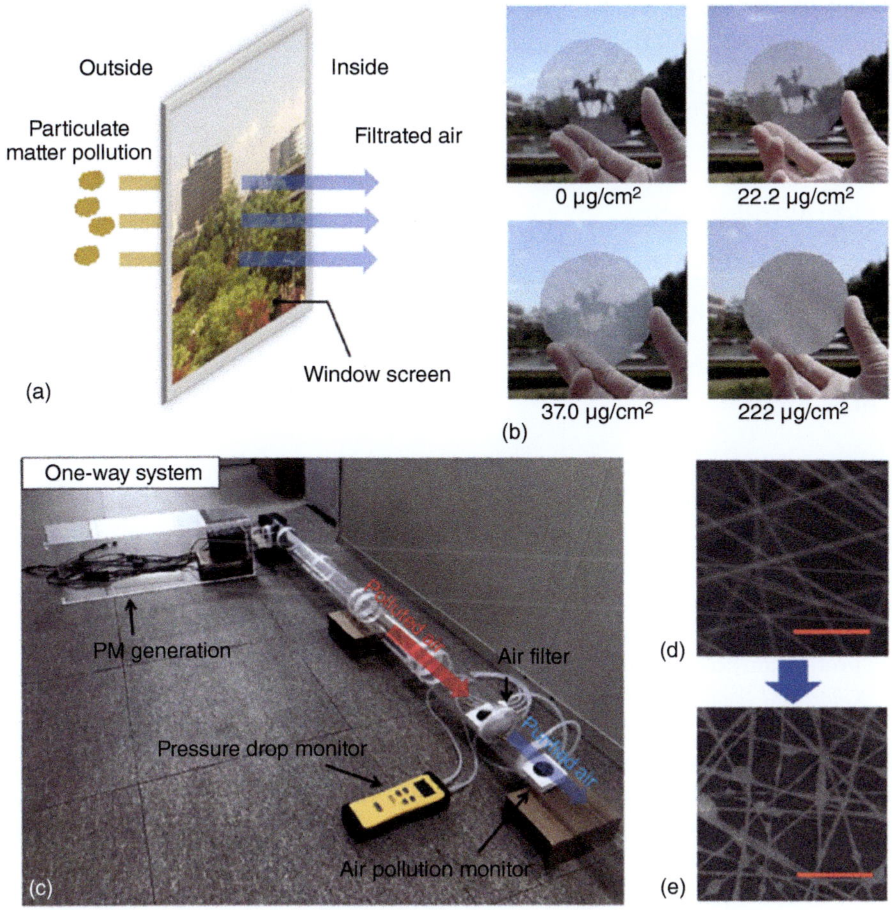

Figure 7.8 Electrospun nanofibrous NF membranes for air filtration: (a) NF membrane as window screen, (b) optical transmittances of membranes with various NF area densities, (c) one-way setup for the evaluation of filter efficiency, and (d, e) SEM images of NF filter (d) before and (e) after air capture. *Source:* Reproduced with the permission from Min et al. [18]. Copyright 2018, Springer Nature.

7.4 Conclusions

In summary, recent advances in the synthesis and related applications of fibrous protein materials were introduced and discussed in this chapter. Several techniques, such as electrospinning, solution blowing, flowing, self-assembly, template synthesis, and others, have been utilized for creating various fibrous protein materials. Fibrous proteins could be classified into nanofibers, nanofibrils, nanotubes, nanoribbons, and microfibers based on their specific fabrication methods and structures, which are measured by various characterization techniques like AFM, SEM, TEM, FTIR, and others. In addition, the synthesized fibrous protein materials have exhibited wide applications from materials science, nanotechnology, energy, and environment fields to tissue engineering and biomedical engineering.

Acknowledgments

This work was supported by the High-Grade Talents Plan of Yan'an University and the National Natural Science Foundation of China (NSFC) (no. 51873225).

References

1 Ming, J.F., Pan, F.K., and Zuo, B.Q. (2015). Structure and properties of protein-based fibrous hydrogels derived from silk fibroin and sodium alginate. *Journal of Sol-Gel Science and Technology* 74 (3): 774–782.
2 Silva, R., Fabry, B., and Boccaccini, A.R. (2014). Fibrous protein-based hydrogels for cell encapsulation. *Biomaterials* 35 (25): 6727–6738.
3 Wei, G., Su, Z.Q., Reynolds, N.P. et al. (2017). Self-assembling peptide and protein amyloids: from structure to tailored function in nanotechnology. *Chemical Society Reviews* 46 (15): 4661–4708.
4 Zhang, W.S., Yu, X.Q., Li, Y. et al. (2018). Protein-mimetic peptide nanofibers: motif design, self-assembly synthesis, and sequence-specific biomedical applications. *Progress in Polymer Science* 80: 94–124.
5 Knowles, T.P.J. and Mezzenga, R. (2016). Amyloid fibrils as building blocks for natural and artificial functional materials. *Advanced Materials* 28: 6546–6561.
6 Cherny, I. and Gazit, E. (2008). Amyloids: not only pathological agents but also ordered nanomaterials. *Angewandte Chemie International Edition* 47 (22): 4062–4069.
7 Babitha, S., Rachita, L., Karthikeyan, K. et al. (2017). Electrospun protein nanofibers in healthcare: a review. *International Journal of Pharmaceutics* 523 (1): 52–90.
8 Wei, G., Zhang, J.T., Xie, L., and Jandt, K.D. (2011). Biomimetic growth of hydroxyapatite on super water-soluble carbon nanotube-protein hybrid nanofibers. *Carbon* 49 (7): 2216–2226.
9 Ko, S. and Jang, J. (2007). Protein immobilization on aminated poly(glycidyl methacrylate) nanofibers as polymeric carriers. *Biomacromolecules* 8 (5): 1400–1403.
10 Sinha-Ray, S., Khansari, S., Yarin, A.L., and Pourdeyhimi, B. (2012). Effect of chemical and physical cross-linking on tensile characteristics of solution-blown soy protein nanofiber mats. *Industrial and Engineering Chemistry Research* 51 (46): 15109–15121.
11 Kamada, A., Mittal, N., Soderberg, L.D. et al. (2017). Flow-assisted assembly of nanostructured protein microfibers. *Proceedings of the National Academy of Sciences of the United States of America* 114 (6): 1232–1237.

12 Chung, H.J. and Park, T.G. (2009). Self-assembled and nanostructured hydrogels for drug delivery and tissue engineering. *Nano Today* 4 (5): 429–437.
13 Ling, S.J., Li, C.M., Jin, K. et al. (2016). Liquid exfoliated natural silk nanofibrils: applications in optical and electrical devices. *Advanced Materials* 28 (35): 7783–7790.
14 Sasso, L., Suei, S., Domigan, L. et al. (2014). Versatile multi-functionalization of protein nanofibrils for biosensor applications. *Nanoscale* 6 (3): 1629–1634.
15 Peralta, M.D.R., Karsai, A., Ngo, A. et al. (2015). Engineering amyloid fibrils from β-solenoid proteins for biomaterials applications. *ACS Nano* 9 (1): 449–463.
16 Wei, G., Wang, L., Sun, L.L. et al. (2007). Type I collagen-mediated synthesis and assembly of UV-photoreduced gold nanoparticles and their application in surface-enhanced Raman scattering. *Journal of Physical Chemistry C* 111 (5): 1976–1982.
17 Sun, Y.J., Wei, G., Song, Y.H. et al. (2008). Type I collagen-templated assembly of silver nanoparticles and their application in surface-enhanced Raman scattering. *Nanotechnology* 19 (11): 115604.
18 Min, K., Kim, S., and Kim, S. (2018). Silk protein nanofibers for highly efficient, eco-friendly, optically translucent, and multifunctional air filters. *Scientific Reports* 8: 9598.
19 Kolbasov, A., Sinha-Ray, S., Joijode, A. et al. (2016). Industrial-scale solution blowing of soy protein nanofibers. *Industrial and Engineering Chemistry Research* 55 (1): 323–333.
20 Yang, G., Ding, H.M., Kochovski, Z. et al. (2017). Highly ordered self-assembly of native proteins into 1D, 2D, and 3D structures modulated by the tether length of assembly-inducing ligands. *Angewandte Chemie International Edition* 56 (36): 10691–10695.
21 Mou, Y., Yu, J.Y., Wannier, T.M. et al. (2015). Computational design of co-assembling protein–DNA nanowires. *Nature* 525 (7568): 230–233.
22 Kim, B.J., Kim, S., Oh, D.X. et al. (2015). Mussel-inspired adhesive protein based electrospun nanofibers reinforced by Fe(III)-DOPA complexation. *Journal of Materials Chemistry B* 3 (1): 112–118.
23 Khansari, S., Sinha-Ray, S., Yarin, A.L., and Pourdeyhimi, B. (2013). Biopolymer-based nanofiber mats and their mechanical characterization. *Industrial and Engineering Chemistry Research* 52 (43): 15104–15113.
24 Hernandez-Garcia, A., Kraft, D.J., Janssen, A.F.J. et al. (2014). Design and self-assembly of simple coat proteins for artificial viruses. *Nature Nanotechnology* 9 (9): 698–702.
25 Brodin, J.D., Smith, S.J., Carr, J.R., and Tezcan, F.A. (2015). Designed, helical protein nanotubes with variable diameters from a single building block. *Journal of the American Chemical Society* 137 (33): 10468–10471.
26 Li, H.Y., Park, S.H., Reif, J.H. et al. (2004). DNA-templated self-assembly of protein and nanoparticle linear arrays. *Journal of the American Chemical Society* 126 (2): 418–419.
27 Kim, H.J., Kim, J.H., Jun, K.W. et al. (2016). Silk nanofiber-networked bio-triboelectric generator: silk bio-TEG. *Advanced Energy Materials* 6 (8): 1502329.

28 Wang, Y.X. and Chen, L.Y. (2014). Cellulose nanowhiskers and fiber alignment greatly improve mechanical properties of electrospun prolamin protein fibers. *ACS Applied Materials & Interfaces* 6 (3): 1709–1718.

29 Wei, G., Reichert, J., Bossert, J., and Jandt, K.D. (2008). Novel biopolymeric template for the nucleation and growth of hydroxyapatite crystals based on self-assembled fibrinogen fibrils. *Biomacromolecules* 9 (11): 3258–3267.

30 Wei, G., Reichert, J., and Jandt, K.D. (2008). Controlled self-assembly and templated metallization of fibrinogen nanofibrils. *Chemical Communications* 33: 3903–3905.

31 Zhou, X.M., Shimanovich, U., Herling, T.W. et al. (2015). Enzymatically active microgels from self-assembling protein nanofibrils for microflow chemistry. *ACS Nano* 9 (6): 5772–5781.

32 Rao, G.B., Fu, Y., Li, N. et al. (2018). Controllable assembly of flexible protein nanotubes for loading multifunctional modules. *ACS Applied Materials & Interfaces* 10 (30): 25135–25145.

33 Lara, C., Adamcik, J., Jordens, S., and Mezzenga, R. (2011). General self-assembly mechanism converting hydrolyzed globular proteins into giant multistranded amyloid ribbons. *Biomacromolecules* 12 (5): 1868–1875.

34 Wang, F.Y., Liu, Z., Wang, B. et al. (2014). Multi-colored fibers by self-assembly of DNA, histone proteins, and cationic conjugated polymers. *Angewandte Chemie International Edition* 53 (2): 424–428.

35 Beun, L.H., Albertazzi, L., van der Zwaag, D. et al. (2016). Unidirectional living growth of self assembled protein nanofibrils revealed by super-resolution microscopy. *ACS Nano* 10 (5): 4973–4980.

36 Helbing, C., Deckert-Gaudig, T., Firkowska-Boden, I. et al. (2018). Protein handshake on the nanoscale: how albumin and hemoglobin self-assemble into nanohybrid fibers. *ACS Nano* 12 (2): 1211–1219.

37 Ding, Z.Z., Han, H.Y., Fan, Z.H. et al. (2017). Nanoscale silk-hydroxyapatite hydrogels for injectable bone biomaterials. *ACS Applied Materials & Interfaces* 9 (20): 16914–16922.

38 Ding, Z.Z., Fan, Z.H., Huang, X.W. et al. (2016). Silk-hydroxyapatite nanoscale scaffolds with programmable growth factor delivery for bone repair. *ACS Applied Materials & Interfaces* 8 (37): 24463–24470.

39 Jin, Y.S., Kundu, B., Cai, Y.R. et al. (2015). Bio-inspired mineralization of hydroxyapatite in 3D silk fibroin hydrogel for bone tissue engineering. *Colloids and Surfaces B: Biointerfaces* 134: 339–345.

40 Aliramaji, S., Zamanian, A., and Mozafari, M. (2017). Super-paramagnetic responsive silk fibroin/chitosan/magnetite scaffolds with tunable pore structures for bone tissue engineering applications. *Materials Science and Engineering C* 70: 736–744.

41 Behera, S., Naskar, D., Sapru, S. et al. (2017). Hydroxyapatite reinforced inherent RGD containing silk fibroin composite scaffolds: promising platform for bone tissue engineering. *Nanomedicine: Nanotechnology, Biology and Medicine* 13 (5): 1745–1759.

42 Li, C.X., Born, A.K., Schweizer, T. et al. (2014). Amyloid-hydroxyapatite bone biomimetic composites. *Advanced Materials* 26 (20): 3207–3212.

43 Wang, J.H., Wang, H.X., Wang, Y.Z. et al. (2014). Alternate layer-by-layer assembly of graphene oxide nanosheets and fibrinogen nanofibers on a silicon substrate for a biomimetic three-dimensional hydroxyapatite scaffold. *Journal of Materials Chemistry B* 2 (42): 7360–7368.

44 Kapoor, S. and Kundu, S.C. (2016). Silk protein-based hydrogels: promising advanced materials for biomedical applications. *Acta Biomaterialia* 31: 17–32.

45 Wu, H.C., Liu, S.S., Xiao, L.Y. et al. (2016). Injectable and pH-responsive silk nanofiber hydrogels for sustained anticancer drug delivery. *ACS Applied Materials & Interfaces* 8 (27): 17118–17126.

46 Xia, X.X., Wang, M., Lin, Y.A. et al. (2014). Hydrophobic drug-triggered self-assembly of nanoparticles from silk-elastin-like protein polymers for drug delivery. *Biomacromolecules* 15 (3): 908–914.

47 Vannoy, C.H., Xu, J.M., and Leblanc, R.M. (2010). Bioimaging and self-assembly of lysozyme fibrils utilizing CdSe/ZnS quantum dots. *Journal of Physical Chemistry C* 114 (2): 766–773.

48 Wei, G., Keller, T.F., Zhang, J.T., and Jandt, K.D. (2011). Novel 1-D biophotonic nanohybrids: protein nanofibers meet quantum dots. *Soft Matter* 7 (5): 2011–2018.

49 Hauser, C.A.E., Maurer-Stroh, S., and Martins, I.C. (2014). Amyloid-based nanosensors and nanodevices. *Chemical Society Reviews* 43 (15): 5326–5345.

50 Yang, J.E., Lee, S., Bhak, G. et al. (2016). Litmus-type thermochromic and solvatochromic sensors prepared with α-synuclein amyloid fibrils and polydiacetylene. *Sensors and Actuators B: Chemical* 227: 313–319.

51 Wu, X.C., Li, M.J., Li, Z.H. et al. (2017). Amyloid-graphene oxide as immobilization platform of Au nanocatalysts and enzymes for improved glucose-sensing activity. *Journal of Colloid and Interface Science* 490: 336–342.

52 Kim, S.J., Choi, S.J., Jang, J.S. et al. (2016). Mesoporous WO_3 nanofibers with protein-templated nanoscale catalysts for detection of trace biomarkers in exhaled breath. *ACS Nano* 10 (6): 5891–5899.

53 Chan, H.M. and Li, H.W. (2011). Multifunctional encoded self-assembling protein nanofibrils as platform for high throughput and multiplexed detection of biomolecules. *Analytical Chemistry* 83 (24): 9370–9377.

54 Nandi, I., Chall, S., Chowdhury, S. et al. (2018). Protein fibril-templated biomimetic synthesis of highly fluorescent gold nanoclusters and their applications in cysteine sensing. *ACS Omega* 3 (7): 7703–7714.

55 Elfwing, A., Backlund, F.G., Musumeci, C. et al. (2015). Protein nanowires with conductive properties. *Journal of Materials Chemistry C* 3 (25): 6499–6504.

56 Meier, C., Lifincev, I., and Welland, M.E. (2015). Conducting core–shell nanowires by amyloid nanofiber templated polymerization. *Biomacromolecules* 16 (2): 558–563.

57 Lampa-Pastirk, S., Veazey, J.P., Walsh, K.A. et al. (2016). Thermally activated charge transport in microbial protein nanowires. *Scientific Reports* 6: 23517.

58 Tan, Y., Adhikari, R.Y., Malvankar, N.S. et al. (2016). Synthetic biological protein nanowires with high conductivity. *Small* 12 (33): 4481–4485.

59 Lee, J.H., Lee, J.H., Lee, Y.J., and Nam, K.T. (2013). Protein/peptide based nanomaterials for energy application. *Current Opinion in Biotechnology* 24 (4): 599–605.

60 Zhou, L.H., Fu, P., Cai, X.X. et al. (2016). Naturally derived carbon nanofibers as sustainable electrocatalysts for microbial energy harvesting: a new application of spider silk. *Applied Catalysis B: Environmental* 188: 31–38.

61 Guo, G.L., Truong, T.H.A., Tan, H.T.T. et al. (2014). Platinum and palladium nanotubes based on genetically engineered elastin-mimetic fusion protein-fiber templates: synthesis and application in lithium-O_2 batteries. *Chemistry - An Asian Journal* 9 (9): 2555–2559.

62 Hou, J.H., Cao, C.B., Idrees, F., and Ma, X.L. (2015). Hierarchical porous nitrogen-doped carbon nanosheets derived from silk for ultrahigh-capacity battery anodes and supercapacitors. *ACS Nano* 9 (3): 2556–2564.

63 Wang, Y.X., Song, Y.F., Wang, Y. et al. (2015). Graphene/silk fibroin based carbon nanocomposites for high performance supercapacitors. *Journal of Materials Chemistry A* 3 (2): 773–781.

64 Pal, R.K., Kundu, S.C., and Yadavalli, V.K. (2018). Fabrication of flexible, fully organic, degradable energy storage devices using silk proteins. *ACS Applied Materials & Interfaces* 10 (11): 9620–9628.

65 Tan, X.X., Zhao, W.C., and Mu, T.C. (2018). Controllable exfoliation of natural silk fibers into nanofibrils by protein denaturant deep eutectic solvent: nanofibrous strategy for multifunctional membranes. *Green Chemistry* 20: 3625–3633.

66 Miller, R.A., Presley, A.D., and Francis, M.B. (2007). Self-assembling light-harvesting systems from synthetically modified tobacco mosaic virus coat proteins. *Journal of the American Chemical Society* 129 (11): 3104–3109.

67 Sun, H.C., Zhang, X.Y., Miao, L. et al. (2016). Micelle-induced self-assembling protein nanowires: versatile supramolecular scaffolds for designing the light-harvesting system. *ACS Nano* 10 (1): 421–428.

68 Ki, C.S., Gang, E.H., Um, I.C., and Park, Y.H. (2007). Nanofibrous membrane of wool keratose/silk fibroin blend for heavy metal ion adsorption. *Journal of Membrane Science* 302 (1–2): 20–26.

69 Baek, D.H., Ki, C.S., Um, I.C., and Park, Y.H. (2007). Metal ion adsorbability of electrospun wool keratose/silk fibroin blend nanofiber mats. *Fibers and Polymers* 8 (3): 271–277.

70 Zhou, W.T., He, J.X., Cui, S.Z., and Gao, W.D. (2011). Preparation of electrospun silk fibroin/cellulose acetate blend nanofibers and their applications to heavy metal ions adsorption. *Fibers and Polymers* 12 (4): 431–437.

71 Aluigi, A., Corbellini, A., Rombaldoni, F., and Mazzuchetti, G. (2013). Wool-derived keratin nanofiber membranes for dynamic adsorption of heavy-metal ions from aqueous solutions. *Textile Research Journal* 83 (15): 1574–1586.

72 Bolisetty, S. and Mezzenga, R. (2016). Amyloid-carbon hybrid membranes for universal water purification. *Nature Nanotechnology* 11 (4): 365–372.

73 Ling, S.J., Qin, Z., Huang, W.W. et al. (2017). Design and function of biomimetic multilayer water purification membranes. *Science Advances* 3 (4): e1601939.

74 Morshedi, D., Mohammadi, Z., Boojar, M.M.A., and Aliakbari, F. (2013). Using protein nanofibrils to remove azo dyes from aqueous solution by the coagulation process. *Colloids and Surfaces B: Biointerfaces* 112: 245–254.
75 Dhandayuthapani, B., Mallampati, R., Sriramulu, D. et al. (2014). PVA/gluten hybrid nanofibers for removal of nanoparticles from water. *ACS Sustainable Chemistry & Engineering* 2 (4): 1014–1021.
76 Wang, C.Y., Wu, S.Y., Jian, M.Q. et al. (2016). Silk nanofibers as high efficient and lightweight air filter. *Nano Research* 9 (9): 2590–2597.
77 Souzandeh, H., Johnson, K.S., Wang, Y. et al. (2016). Soy-protein-based nanofabrics for highly efficient and multifunctional air filtration. *ACS Applied Materials & Interfaces* 8 (31): 20023–20031.

8

Microbial Polyhydroxyalkanoates (PHAs): From Synthetic Biology to Industrialization

Yuki Miyahara[1], *Ayaka Hiroe*[1,4], *Shunsuke Sato*[2], *Takeharu Tsuge*[3,4], *and Seiichi Taguchi*[1,4]

[1] *Tokyo University of Agriculture, Department of Chemistry for Life Sciences and Agriculture, 1-1-1 Sakuragaoka, Setagaya, Tokyo 156-8502, Japan*
[2] *Kaneka Corporation, Biotechnology Laboratories, 1-8 Miyamae-cho, Takasago-cho, Takasago, Hyogo 676-8688, Japan*
[3] *Tokyo Institute of Technology, Department of Materials Science and Engineering, 4259 Nagatsuta, Midori-ku, Yokohama 226-8502, Japan*
[4] *Japan Science and Technology agency, CREST, 4-1-8 Honcho, Kawaguchi Saitama 332-0012, Japan*

8.1 Introduction

The potential environmental and economic benefits of biotechnology should be considered by creating a sustainable production system for wide range of biobased products. The industrial application of biotechnology could lead to have the establishment of "bioeconomy," where a substantial share of economic output is partly dependent on the use of biological materials. The development of biobased polymeric materials has attracted much attention as well as various biofuels and biobased chemicals. The paradigm shift from petroleum-based polymers to biobased polymers has been motivated from the two viewpoints, renewable feedstock utilization and biodegradation of used end products. These points shed light on biobased polymers that are recognized as carbon-neutral materials in order to realize sustainable society and industrial systems. Polylactide (PLA) is the representative biobased polyester manufactured via chemo-bio process starting from biomass-derived resources. Production scale of PLA currently counts 250 000 ton/yr. However, biodegradability of PLA is limited to be the compostable conditions with fermentation-induced temperature increase. On the other hand, petrochemical polycaprolactone and polyglycolic acid can be biodegraded in natural environments. Like these examples, biomass-derived and biodegradable polymers should be promising ideal targets.

Natural polyesters, polyhydroxyalkanoates (PHAs), satisfy aforementioned both requirements. PHAs can be intrinsically synthesized and accumulated as energy storage inside the microbial cells from renewable biomass-derived carbons sources [1]. It should be notable that PHAs have large amount of capacity of monomeric constituents over 160 available for building blocks (Figure 8.1). Dependent on the chain length of the monomeric unit, PHAs have

Biopolymers for Biomedical and Biotechnological Applications, First Edition.
Edited by Bernd H. A. Rehm and M. Fata Moradali.

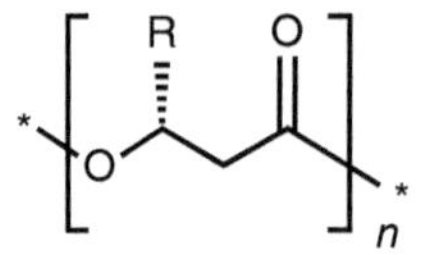

Figure 8.1 Chemical structure of PHAs; R = methyl (C1) to tridecyl (C13).

been categorized into two classes: short-chain-length PHAs (scl-PHAs) that consist of 3–5 carbon atoms and are obtained from several bacteria such as *Cupriavidus necator* (formerly *Ralstonia eutropha*) and *Aeromonas caviae* [1]. The prototype PHA is a homopolymer of 3-hydroxybutyrate (3HB), P(3HB) that was discovered from the Pasteur Institute. P(3HB) has similar thermal property with that of polypropylene (PP), but it is stiff and brittle. In principle, the copolymerization with any other monomeric constituents is effective to overcoming this mechanical defect. In fact, incorporation of 3-hydroxyvalerate (HV) into P(3HB) to form a copolymer; poly(3-hydroxybutyrate-*co*-3-hydroxyvalerate) (P(3HB-*co*-3HV)) makes the material tough and more flexible and acquires broader thermal processing properties [1]. This copolymer was manufactured as a brand name of Biopol by ICI Co. Ltd, United Kingdom. It is compostable, degrading harmlessly in soil, after a few months. The second class of PHA includes medium-chain-length polyhydroxyalkanoates (mcl-PHAs) that consist of 6–14 carbon atoms in the chain. The mcl-PHAs are obtained from several bacteria such as *Pseudomonas putida* [2, 3]. The effectiveness of copolymerization in gaining mechanical benefits aforementioned was further justified by using 3-hydroxyhexanoate (3HHx) with larger side-chain length [4]. The copolymer of P(3HB-*co*-3HHx) has been industrially produced in the scale of over one thousand ton per year by Japanese leading company, Kaneka Co. Ltd. Taken together, natural polyester PHA can provide us with the satisfactory requirements, alternative polymer properties of petroleum-based plastics with biodegradation-involved carbon-neutral process. There should be concerns about both input and output of manufactured products in harmonic manner to maintain the social and environmental balances. Considering the political targets for bioeconomy products is therefore useful to direct the technical design of sustainable PHA production, for example, in integrated lignocellulose biorefineries.

In this chapter, synthetic biology and industrial production of PHAs will be described with the three representative topics. First case is the practical production process of P(3HB-*co*-3HHx), Kaneka Biodegradable Polymer™ PHBH, developed by industry. Second one is also the creation of new homopolymers, mcl-PHAs with homogeneous side-chain lengths. Third one is the creation of novel type polyesters, lactate (LA)-based PHAs such as P(LA-*co*-3HB). These case studies can be achieved in the engineered microbial platforms by utilizing oil-based and sugar-based renewable biomass. Considering the chemical structure, properties, and biodegradability/biocompatibility, most of the polymers have been continuously biosynthesized by effective approaches, enzyme-directed evolution, metabolic engineering, and fermentation technology (Figure 8.2). From the aspect of bioeconomy, these polymeric products are expected to be adapted in the desirable applications contributing to biomedical, food industry, and so on.

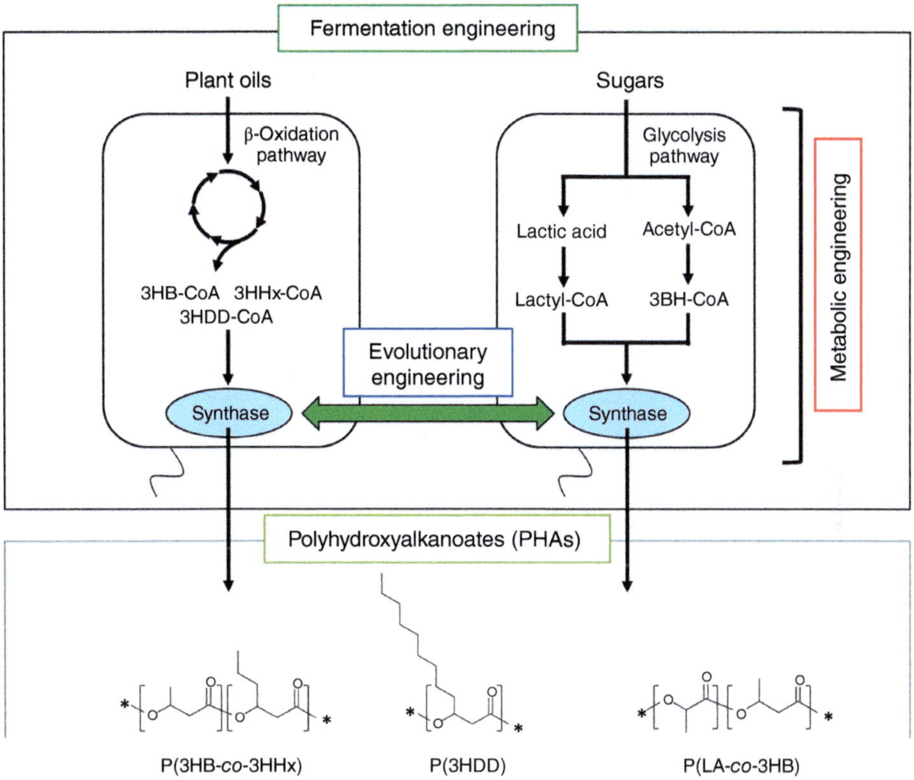

Figure 8.2 Microbial production system for three polyesters based on biotechnological approaches.

8.2 Synthetic Biology for Production of Kaneka PHBH

8.2.1 Isolation of Bacterium Producing Poly(3-hydroxybutyrate-*co*-3-hydroxyhexanoate)

As mentioned in introduction, homopolymer of (*R*)-3HB, P(3HB) is the most abundant PHA in nature environment and prototype PHA. The melting point and tensile strength of P(3HB) are close to polypropylene, and it was expected as an environmentally friendly material. However, the melting point of P(3HB) (175 °C) is close to its pyrolysis temperature, and the process window is too narrow, which has a fatal drawback as a thermoplastic. Moreover, as aforementioned, P(3HB) is highly crystalline, hard, and brittle; its practical applications are limited. For this reason, in order to improve physical property of P(3HB), some groups have developed PHAs with a second monomer constituent. For example, Tianjin GreenBio Corp. develops poly(3-hydroxybutyrate-*co*-4-hydroxybutyrate) (P(3HB-*co*-4HB)), and TianAn Biologic Materials Corp. develops P(3HB-*co*-3HV), respectively. Among them, P(3HB-*co*-3HHx) (PHBH) (Figure 8.3) is consist of (*R*)-3HB and (*R*)-3HHx.

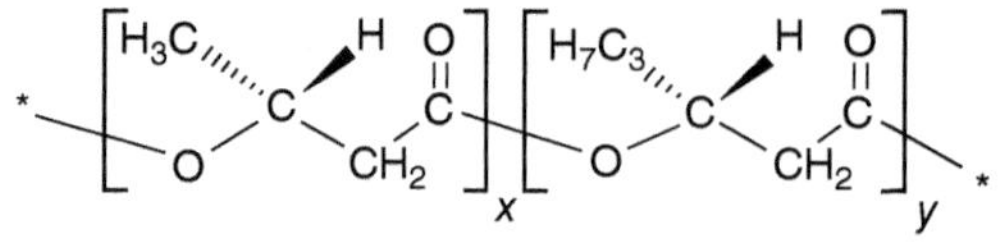

Figure 8.3 Chemical structure of PHBH.

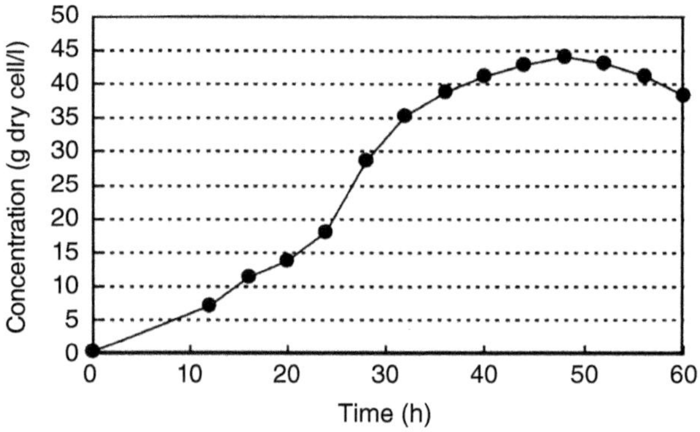

Figure 8.4 Fermentative production of PHBH by *Aeromonas caviae* FA440.

This kind of PHA is accumulated as energy storage in bacterial cell and is expected as an alternative to petroleum-based plastics since it can be synthesized from renewable feedstocks such as plant oils, sugars, or gas. In 1993, PHBH-producing bacteria "*Aeromonas caviae* FA440" was isolated from soil at Takasago plant of Kaneka Co. Ltd, and its physical properties and biodegradability were evaluated [5]. However, according to our advance study with 2l fermentor, PHBH productivity in *A. caviae* FA440 was only 8.3 g_{PHBH}/l/60 h (21.3 wt% of PHBH content in the cell), and this productivity was too low for an industrial production for cost (Figure 8.4).

8.2.2 Material Properties of PHBH

PHBH is a crystalline resin that is thought to be crystallized when the lamella structure formed in the part of 3HB chain is bound by hydrogen bond. In the case of PHBH, the propylene group of the (*R*)-3HHx monomer units is sterically hindered and cannot enter the crystalline structure phase formed by the 3HB chain. Therefore, the higher the 3HHx fraction, the lower the crystallinity, and as result PHBH become flexible depending on the molar ratio of 3HHx monomer. For example, when 3HHx fraction is 10.8 mol%, elongation at break improved to 300%, and melting temperature (T_m) is dropped to 113.5 °C. In this way, by increase of 3HHx fraction, PHBH can take physical property such as polypropylene (PP) or low-density polyethylene (LDPE) (Figure 8.5 and Table 8.1). Also, introduction of 3HHx monomer lowers the T_m, allows molding in a temperature range sufficiently lower than the thermal pyrolysis temperature, and has the advantage in that the process window widens widely. From the above viewpoint, it is very important to control the 3HHx monomer ratio that affects the physical properties and processability of PHBH.

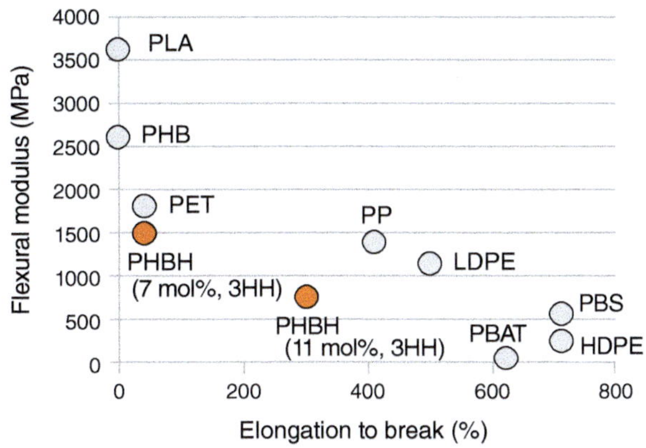

Figure 8.5 Plots for modulus vs. elongation of various polymers.

Table 8.1 DSC analysis of PHBH with various 3HHx fractions.

3HHx fraction (mol%)	First heating process			Cooling process			Second heating process	
	T_m (°C)	ΔH_m (J/g)	X_c (%)	T_g (°C)	T_c (°C)	ΔH_c (J/g)	T_{cc} (°C)	ΔH_{cc} (J/g)
0 (PHB)[b)]	175.1	103	70.5	0.4	77.4	80.2	—	—
2.7[a)]	150.8	72.9	49.9	−2.3	50.2	25	50.3	21.7
5.2	142	70.8	48.5	−2.9	41	3.6	58.8	47.1
5.9	138.9	69.5	47.6	−2.4	—	—	68.2	43.3
7.1	133.5	66.3	45.4	−2.1	—	—	74.5	27.2
7.9	131.1	67.5	46.2	−2.2	—	—	76.7	20.2
9.1	116.4	65.9	45.1	−3.4	—	—	77	12.2
10.8	113.5	55.8	38.2	−4.7	—	—	79.9	1.2

a) T_m, melting temperature; ΔH_m, melting enthalpy; X_c, crystallinity; T_g, glass transition temperature; T_c, crystallization temperature; ΔH_c, crystallization enthalpy; T_{cc}, cold crystallization temperature; ΔH_{cc}, cold crystallization enthalpy.
b) Purchased from Sigma-Aldrich.

8.2.3 Industrial PHBH Production Process

As shown in Figure 8.6, PHBH production process consists of fermentation and purification process. In fermentation process, an aeration/agitation parameters and culture medium, which consists of minerals, nitrogen, and phosphate as nutrients, should be optimized to achieve highest yield (Figure 8.7). In purification process, PHBH granule accumulated inside the bacterial cells is recovered and purified, especially to remove a color and odor from PHBH granules. The purification process should be very important to be used as a practical plastic in these points of view. Many methodologies have been developed so far [6, 7]. Depending on the material properties of PHAs and production strain, different recovery techniques should be selected. As it is mentioned in Figure 8.6, our

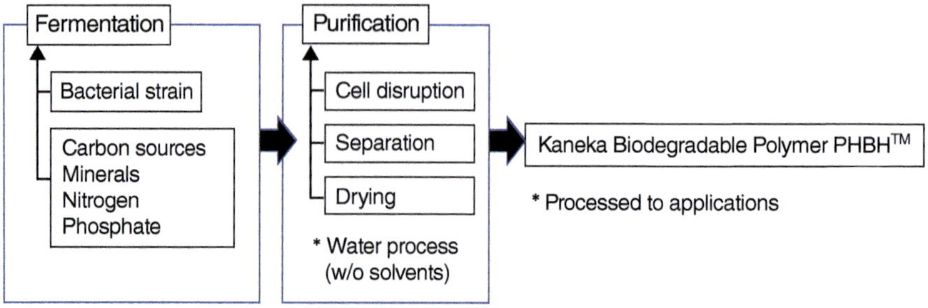

Figure 8.6 Industrial production process of PHBH.

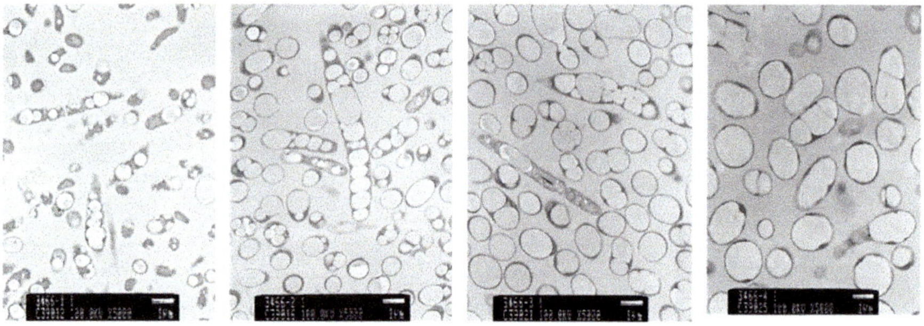

Figure 8.7 Achievements in PHBH accumulating cells. The transmission electron morphological images showing the fermentation ability of PHBH: from the beginning (left) to the present status (right).

purification process of PHBH does not require any extraction solvent such as chloroform or acetone and thus is environmentally friendly purification process operated in the water system. Next, a PHBH synthesis pathway in recombinant *R. eutropha* (presently *C. necator*) H16 strain from plant oil is shown in Figure 8.8. Acetyl-CoA is the starting substance of the TCA cycle and supplied in large amount via a β-oxidation pathway. Acetoacetyl-CoA is synthesized by dimerization of two molecules of acetyl-CoA with thiolase (PhaA); after that (*R*)-3HB-CoA is generated by reduction of acetoacetyl-CoA by (*R*)-specific reductase (PhaB). On the other hand, (*R*)-3HHx-CoA is mainly supplied by hydrolyzation of 2-hexenoyl-CoA with (*R*)-specific enoyl-CoA hydratase (PhaJ). The regulation of the molar ratio of two comonomers should be very important for the production of PHBH with required material property.

8.2.4 Molecular Breeding of PHBH-Producing Bacteria

Several processes for PHBH production using recombinant *R. eutropha* H16 have been reported [8–15] since *R. eutropha* has relatively high-level productivity of P(3HB) [16, 17]. Moreover, as *R. eutropha* was accessed to be used for single-cell proteins like animal feed, it is very safe bacteria; we selected as a host strain

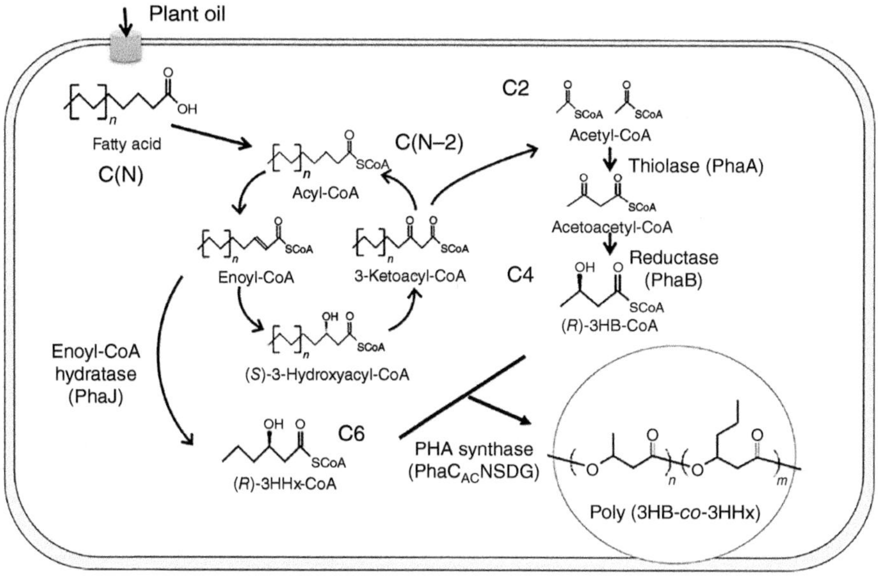

Figure 8.8 PHBH synthesis pathway from plant oil.

for industrial production [18, 19]. However, *R. eutropha* originally produces P(3HB), not PHBH. This is because the original PHA synthase (PhaC_{Re}) from *R. eutropha* cannot recognize (*R*)-3HHx-CoA as a substrate to polymerize. Therefore, to breed a PHBH-producing mutant strain of *R. eutropha*, *phaC1*$_{Re}$ gene was replaced by a mutant of PHBH synthase gene (*phaC*$_{Ac}$NSDG) that is derived from *A. caviae* FA440 (KNK005) (Figure 8.9) [20, 21]. The strain KNK005 produces and accumulates high amount of PHBH, but the 3HHx fraction was still low (7.2 mol%). For broad application deployment, it is necessary to control the 3HHx fraction precisely and produce PHBH separately as necessary in industrial scale. For example, some researchers constructed the recombinant *R. eutropha* strains lacking the multiple *phaB* genes (*phaB*, *B2*, and *B3*). These mutant strains produced PHBH containing more than 10 mol% of 3HHx, but at low levels [22, 23]. PhaB is responsible for reduction of acetoacetyl-CoA to produce (*R*)-3HB-CoA, and it was considered that the decrease in (*R*)-3HB-CoA production rate due to the inactivation of *phaB* genes resulted in the decrease in PHBH productivity. Therefore, from an economic perspective, deletion of *phaB* was counterproductive for industrial production. To overcome this problem, we supplied butyrate as a precursor of 3-ketohexanoyl-CoA to increase the 3HHx content because 3-ketothiolase encoded by the *bktB* is expected to enhance the biosynthesis of 3-ketohexanoyl-CoA as precursor of (*R*)-3HHx-CoA from acetyl-CoA and butyryl-CoA (Figure 8.10) [24]. In this system, the 3HHx fraction was successfully increased from 7.2 to 13.0 mol% by changing the amount of butyrate to be added as a co-substrate [25]. But further studies to reduce the requirement of additional butyrate are warranted for cost reduction.

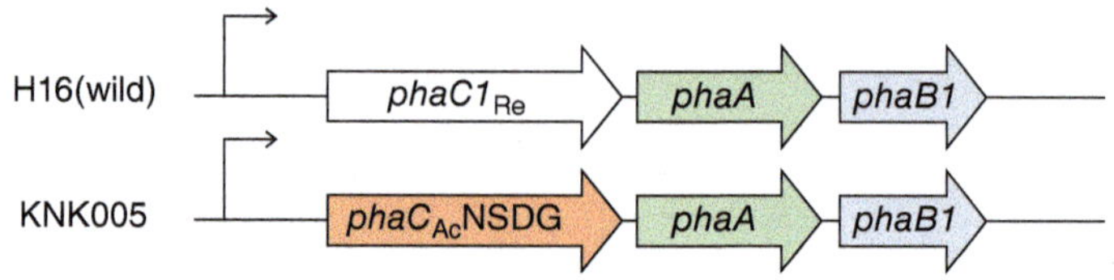

Figure 8.9 Genotype of KNK-005.

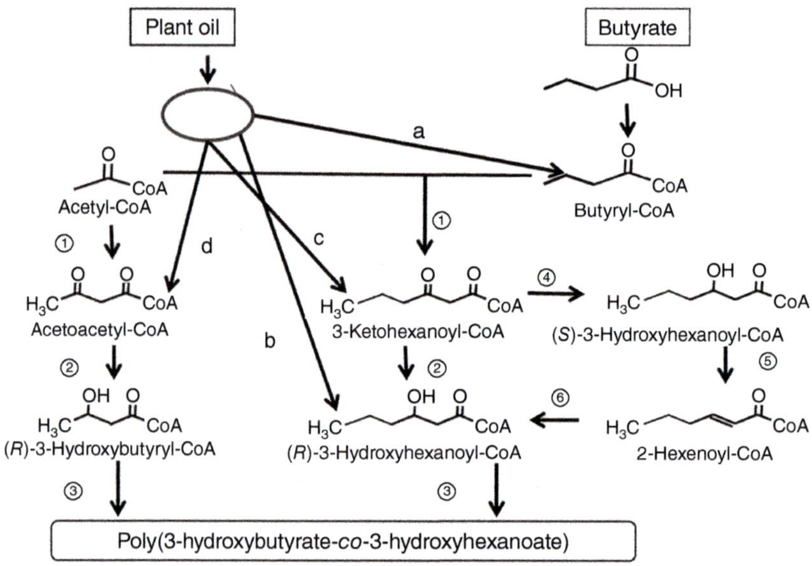

Figure 8.10 Pathway for conversion of acetyl-CoA and butyryl-CoA to (*R*)-3HHx-CoA. 1: 3-ketothiolase, 2: (*R*)-specific reductase, 3: PHA synthase, 4: hydroxyacyl-CoA dehydrogenase, 5: enoyl-CoA hydratase. a, c, d: Intermediates of the β-oxidation pathway, butyryl-CoA, 3-ketohexanoyl-CoA, and acetoacetyl-CoA, respectively. b: (*R*)-3HHx-CoA is produced from 2-hexeoyl-CoA by PhaJ.

8.2.5 Precise Control of 3HHx Fraction by Genetic Modification of *Ralstonia eutropha*

Ultimately it was necessary to control the 3HHx fraction without adding butyrate. For that purpose, precise expression control of *phaJ* that is a key enzyme of 3HHx monomer biosynthesis from plant oil via β-oxidation pathway was necessary (Figure 8.8). The *phaJ* encodes (*R*)-specific enoyl-CoA hydratase that catalyzes the addition of a water molecule to the C—C double bond in enoyl-CoA, which generates the corresponding (*R*)-3-hydroxyacyl-CoA. In this case, it is expected to generate (*R*)-3HHx-CoA from 2-hexenoyl-CoA that is the intermediate of β-oxidation pathway when plant oil is used as raw material. *R. eutropha* possess three *phaJ* genes that can generate (*R*)-3HHx-CoA. Then, two genes, *phaJ4a* and *phaJ4b*, were chosen for precise control of 3HHx fraction [26]. Constructed strains were able to produce PHBHs of various 3HHx fractions (Table 8.2). These results demonstrated that it is possible to precisely control

Table 8.2 PHBH production by *phaJ* expression-regulated strains.

Strain[a)]	Dry cell weight (mg/ml)	PHA content (% dry cell weight)	PHA (mg/ml)	Real cell mass (mg/ml)	3HHx fraction (mol%)
Control	19.1 ± 0.1	84.2 ± 0.2	16.1 ± 0.1	3.0 ± 0.1	2.8 ± 0.1
PphaC1-phaJ4a	19.3 ± 0.2	84.7 ± 2.0	16.4 ± 0.2	2.9 ± 0.4	5.3 ± 0.1
Ptrp-dRBS-phaJ4a	19.1 ± 0.6	84.2 ± 2.7	16.1 ± 0.0	3.0 ± 0.6	7.8 ± 0.0
Ptrp-phaJ4a	19.1 ± ± 0.3	82.3 ± 1.5	15.8 ± 0.0	3.4 ± 0.3	8.6 ± 0.1
PlacUV5-dRBS-phaJ4a	19.2 ± 0.5	83.4 ± 0.9	16.0 ± 0.3	3.2 ± 0.3	9.1 ± 9.9
PlacUV5-phaJ4a	18.7 ± 0.8	84.0 ± 1.1	15.7 ± 0.4	3.0 ± 0.3	9.9 ± 0.2
Ptrc-phaJ4b	18.2 ± 0.4	84.0 ± 3.3	15.3 ± 1.0	2.9 ± 0.5	10.7 ± 0.1

a) *R. eutropha* H16 strains to produce PHBH with various 3HHx compositions. 3HHx fraction in the polymers was regulated by the combination of 3HHx-CoA supplying genes (*phaJ4a, phaJ4b* derived from *R. eutropha*), their promoters (Ptrc, PlacUV5, and Ptrp derived from *E. coli* and PphaC1 derived from *R. eutropha*), and their ribosomal binding sequences (RBS and dRBS).

the metabolic flux according to the expression level of *phaJ* [27]. Moreover, it is confirmed that these strains are possible to produce PHBH of various 3HHx compositions without reduced productivity even under high-density cultivation process similar to actual industrial production conditions.

8.2.6 Business Plan for Kaneka PHBH Industrialization

Despite the fact that more than 90 years have passed since PHA discovered, PHA is still in the process of practical use. Considering that PHAs including PHBH are natural polyesters that microorganisms originally synthesize and accumulate as an energy and carbon storage compound such as starch, oils, and fats, we have to meet the issue that PHBH has low physical and mechanical properties as a plastic so far. However, in recent years, PHBH formulating and molding processing technology has greatly advanced; processing such as injection molding, extrusion molding, and blown-film molding become possible to produce sheets, films, and bottles; and moreover, foaming and spinning are becoming possible as well (Figure 8.11). As a major differentiation point of PHBH over the conventional plastics, it has excellent biodegradability derived from naturally occurring structure, especially in real environment and marine. When a PHBH film is immersed in seawater, it has been found that a large amount of biofilm is formed on the surface by seawater-derived microbial flora and is efficiently biodegraded by microorganisms in the biofilm [28, 29]. PHBH has already obtained "OK biodegradable MARINE" certification from TÜV AUSTRIA Belgium NV (former name Vincotte), an international certification body (Figure 8.12). Making use of the characteristic that PHBH biodegrades even in seawater and freshwater environments should be ongoing project, which consequently will contribute to the conservation of the global environment and realization of a sustainable society.

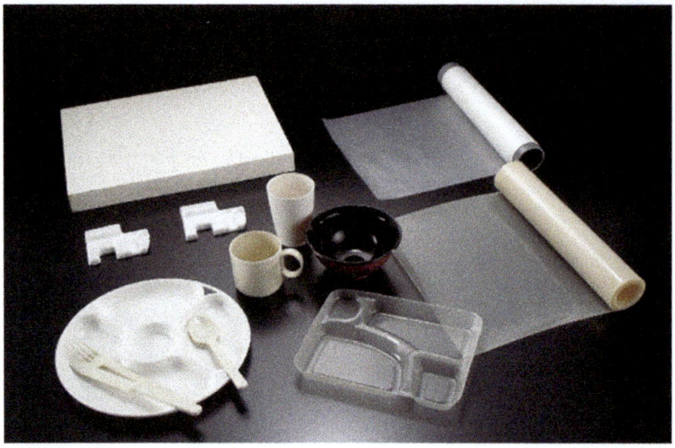

Figure 8.11 PHBH molded samples.

Figure 8.12 Logo mark of OK biodegradable MARINE certification.

8.3 Synthetic Biology for Production of Medium-Chain-Length PHAs with Homogeneous Side-Chain Lengths (Homo-PHAs)

8.3.1 Copolymers Based on Medium-Chain-Length PHA Monomeric Constituents

As mentioned in Introduction, there are mainly two types of PHAs in general, which has been categorized depending upon the length of the monomeric carbon chain: scl-PHAs with 2–5 carbons and mcl-PHAs with 6–14 carbons or more 15 carbons (Figure 8.13). Among those, mcl-PHAs are the most promising and possess superior features that can meet the requirement for broader applications. The pseudomonads including *Pseudomonas putida*, *Pseudomonas oleovorans*, and *Pseudomonas aeruginosa* are native mcl-PHA producers and able to utilize structurally related and unrelated carbon substrates such as sugars and fatty acids [2, 3]. Typically, the mcl-PHAs produced exist as copolymers containing predominantly 3-hydroxyoctanoate (3HO) and 3-hydroxydecanoate (3HD) monomers. Depending on the carbon source selection, (*R*)-3HA-CoAs are supplied as mcl-3HA precursors either via the fatty acid degradation (β-oxidation)

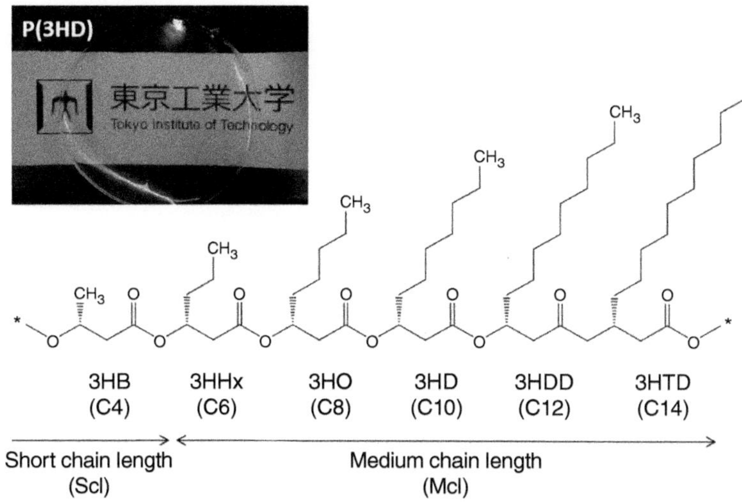

Figure 8.13 Medium-chain-length (mcl) 3-hydroxyalkanoates as building blocks of PHA. 3HB: 3-hydroxybutyrate, 3HHx: 3-hydroxyhexanoate, 3HO: 3-hydroxyoctanoate, 3HD: 3-hydroxydecanoate, 3HDD: 3-hydroxydodecanoate, 3HTD: 3-hydroxytetradecanoate. High transparency of P(3HD) solvent cast film (inset).

or fatty acid biosynthesis pathway. These (*R*)-3HA-CoAs, which vary in their chain length, are later polymerized by PHA synthase (PhaC) with broad substrate specificity. In the mcl-PHA biosynthesis pathways, some specific enzymes play important role for conversion of the intermediates to (*R*)-3HA-CoA monomers. As depicted in Figure 8.14, (*R*)-specific enoyl-CoA hydratase (PhaJ) and (*R*)-3-hydroxyacyl-acyl-carrier-protein (ACP)-CoA transferase (PhaG) are highly effective for supplying (*R*)-3HA-CoA from *trans*-2-enoyl-CoA and (*R*)-3HA-ACP, respectively. These enzymes were naturally found in some bacteria, which are involved in PHA biosynthesis [30, 31]. In addition, there are also several other enzymes that have been reported with higher ability to supply the PHA monomers [32, 33]. Of these, 3-ketoacyl-ACP reductase (FabG) is inherently involved in the fatty acid biosynthesis pathway. What is more interesting is that FabG also has an ability to convert other intermediate from β-oxidation, whereby acyl-CoA generated is channeled toward the formation of (*R*)-mcl-3HA-CoA as well [33].

The physicochemical and thermal properties of PHAs depend on the structure of the repeating units. For instance, the highly flexible mcl-PHAs are due to its amorphous structure, while the rigid and brittle properties of scl-PHAs are caused by their high crystallinity. The exceptional characteristic of mcl-PHAs provides a wide range of applicability, with the most notable one being widely used in biomedical field [34–36]. Despite these, mcl-PHAs are naturally existed as sticky elastomers, which bring difficulties for processing them to be films and fibers. Therefore, it is necessary to alter the mcl-PHA composition, which will suit for variety of applications. Like these, by using *Pseudomonas* spp., various PHAs can be undoubtedly obtained with different monomer structures, applicable to the various application targets [37, 38].

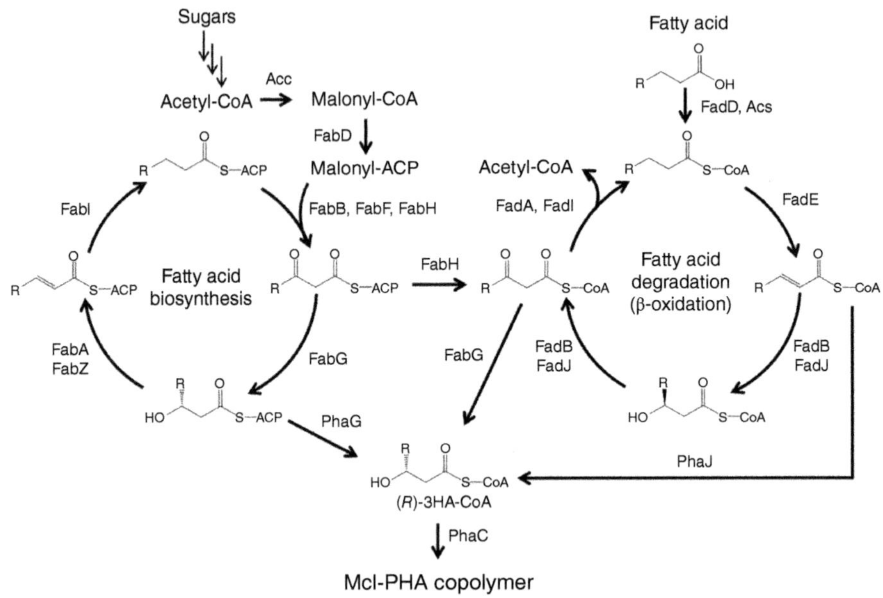

Figure 8.14 Metabolic pathways for mcl-PHA. The enzymes involved in fatty acid biosynthesis and β-oxidation represent FabA–Z and FadA–J, respectively. PhaC: PHA synthase, PhaG: (*R*)-3-hydroxydecanoyl-ACP:CoA transacylase, PhaJ: (*R*)-specific enoyl-CoA hydratase, Acc: acetyl-CoA carboxylase, Acs: acyl-CoA synthetase.

8.3.2 Pathway Engineering for Homo-PHA Production

Mcl-PHA homopolymers (homo-PHA) are ideally composed of single 3HA unit with 6–14 carbon atoms and have been reported as new types of bioplastics synthesized by metabolically engineered bacteria [39, 40]. Compared with the conventional mcl-PHA copolymers obtained from *Pseudomonas* species, the homo-PHAs acquired from *Escherichia coli* are easily processed into stretchable and highly transparent films (Figure 8.13). The unique physical properties of homo-PHAs provide a bright prospect to be developed as stretchable films for new biomaterial applications. Since the past, large numbers of works focused on the studies of the mcl-PHA copolymers with heterogeneous side-chain lengths [37, 38], yet merely few studies reported the development of homo-PHAs [36, 41]. The homo-PHAs are unnaturally produced and therefore require the gene alteration for constructing the artificial PHA biosynthetic pathway in bacteria. To date, homo-PHAs can be synthesized directly from fatty acid as carbon sources by strains in *Pseudomonas* [36, 41–45] and *E. coli* [39, 40, 46–49] with mutants blocking the intermediate steps of β-oxidation pathway. Nevertheless, the genetic modification strategy for synthesizing homo-PHA was different in these two bacteria. Since *Pseudomonas* strains are the natural mcl-PHA producers, the homo-PHA synthesis can merely be attained by deleting the key genes related to β-oxidation. Meanwhile, *E. coli* strains are not natural PHA producers and therefore require heterologous expression of PHA biosynthetic enzymes, such as (*R*)-specific enoyl-CoA hydratase (PhaJ) and PHA synthase

(PhaC) in addition to host's gene deletion. A β-oxidation defective strain (*E. coli* LSBJ) was successfully demonstrated as an effective platform for homo-PHA production, together with the overproduction of PhaJ and PhaC after all successfully maintaining the uniformity of the repeating monomers [49]. The uniformity of monomers relies on the fatty acid substrate used, as pure fatty acids are required for channeling toward the PHA precursor during the biosynthesis steps [39]. Since the key genes responsible for β-oxidation pathway were eliminated from the chromosomal DNA of *E. coli* strains, fatty acids supplemented are directly converted into 3HA monomers, having similar carbon numbers to those corresponding fatty acids [40, 49]. The homo-PHA production system based on *E. coli* offers a wide range of homo-PHAs, synthesized with 6–12 carbons from feeding of the corresponding fatty acid [40, 49].

In yet another interesting approach, it has been reported the synthesis of mcl-PHA copolymers containing higher 3-hydroxydecanoate (3HDD) and 3-hydroxydodecanoate (3HTD) fractions via a hybrid metabolic pathway, involving both fatty acid biosynthesis and β-oxidation [50]. Construction of the complex artificial pathway was performed by deleting the β-oxidation genes (*fadRABIJ*) along with the heterologous expression of (*R*)-specific enoyl-CoA hydratase (PhaJ3), PhaC2, acyl-ACP thioesterase, and acyl-CoA synthetase to allow glucose conversion into 3HA monomers. The resulting strategy enables to produce PHA containing up to 78 mol% 3HDD and 22 mol% 3HTD; however the homo-PHA synthesis has not been successfully achieved.

8.3.3 Improved Microbial Production of Homo-PHAs

Previously, various homo-PHAs containing 6–12 carbons have been reported [36, 39–49]. Of these, poly(3-hydroxydecanoate) (P(3HD)) and poly(3-hydroxydodecanoate) (P(3HDD)) with monomer repeating units of C10 and C12, respectively, are attractive because they are easy to handle for polymer processing [39]. Hence, this section will mainly focus on the development studies for P(3HD) and P(3HDD). In the synthetic biology, early research focuses on the inhibition of fatty acid degradation (β-oxidation) pathway in the host strain. Defective β-oxidation *E. coli* strain CAG18496 (*fadA* and *fadB* deletion mutant) was demonstrated for the biosynthesis of homo-PHAs by utilizing *trans*-2-alkenoic acids [40] as PHA precursors. This further complements the artificial pathway introduced in the host strain by expression of mcl-PHA biosynthetic enzymes from DNA plasmid that carry genes from *Pseudomonas* strains. The strategy employed able to produce near-homopolymeric mcl-PHAs with 6–12 carbons. After all, the amount of P(3HD) acquired was at 0.14 g/l consisting of 99.6 mol% of the dominant monomer unit under 2-decenoic acid supplementation. Further, Tappel et al. developed a strategy for effective control of the mcl-PHA monomers by deletion of two key genes (*fadB* and *fadJ*) responsible for β-oxidation in *E. coli* LS5218 [*fadR*601, *atoC*(Con)], hence named as *E. coli* LSBJ [49]. The precise control strategy enabled to produce a tailor-made mcl-PHAs with similar chain length to those fatty acids supplemented. Together with expression of mcl-specific (*R*)-hydratase (PhaJ4$_{Pp}$ from *P. putida*) and a mutant PHA synthase (PhaC1$_{Ps}$ STQK mutant) from the introduced plasmid, the system allowed a direct conversion of the corresponding fatty acids toward the mcl-PHA biosynthetic pathway. This strategy resulted in producing 0.26 g/l of P(3HD) when grown on the related fatty acid [49]. Following that, the P(3HD) production was further enhanced to 0.60 g/l by removal of transcriptional regulatory gene (*arcA*) that inhibits the β-oxidation-related genes from the LSBJ strain [48]. Afterward, optimization of culture conditions led to enhance the near P(3HD) homopolymer production up to 1.47 g/l when grown on a mineral-based medium enriched with yeast extract [39].

In more recent study, an engineered PHA synthase ($PhaC1_{Pp}$) is obtained through evolutionary engineering (expressed in *E. coli* LSBJ strain) capable of increasing P(3HDD) yield to 1.6-fold [46] (Figure 8.15). Despite all these, the main challenge of the current homo-PHA production system is still related closely to the fatty acid substrates itself. The fatty acid has long been known to provide toxicity effects to cells. Hence, another study demonstrated a feeding strategy for producing higher amount of homo-PHAs through intermittent carbon feeding, which aimed to support high biomass production and reduce fatty acid toxicity toward *E. coli* cells. By employing the intermittent carbon feeding strategy, P(3HD) production increased up to 5.44 g/l [47], which is the highest P(3HD) titer ever reported. Similarly, the high P(3HDD) amount reported at 3.50 g/l was also achieved with intermittent feeding of dodecanoic acid as a co-carbon source at low concentration [47]. Besides, the highest P(3HDD)

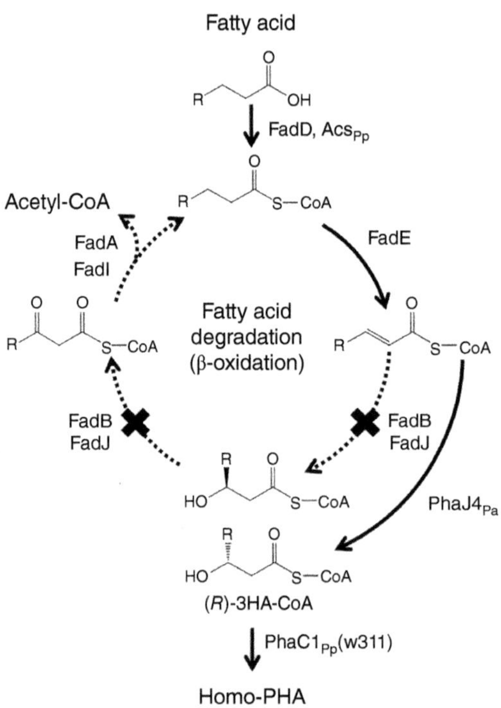

Figure 8.15 Metabolic pathway for enhanced synthesis of homo-PHA constructed in *E. coli* LSBJ (Δ*fadB*, Δ*fadJ*) [46]. $PhaC1_{Pp}$(w311): *P. putida* PHA synthase with two site-specific mutations, Glu358→Gly (E358G) and Asn398→Ser (N398S), $PhaJ4_{Pa}$: (*R*)-specific enoyl-CoA hydratase from *P. aeruginosa*, Acs_{Pp}: acyl-CoA synthetase from *P. putida*.

titer has been reported at 7.3 g/l, which achieved by growing recombinant *Pseudomonas entomophila* LAC23 (Δ*fadAB*, ΔPSEEN0664) cultures in 6-l jar fermenter [42].

8.3.4 Material Properties of Homo-PHAs

Homo-PHAs, specifically P(3HD) and P(3HDD), are previously described as stretchable materials with glossy and transparent features [39]. In the same study, the melting temperatures (T_m) for P(3HD) and P(3HDD) were observed at 70 and 82 °C, respectively. Meanwhile, the glass transition temperature (T_g) for P(3HD) was noticeable at −46 °C, but for P(3HDD), the T_g value from differential scanning calorimetry (DSC) analysis is indistinct. Two cold crystallization temperatures (T_{cc}) and two melting temperatures (T_m) of both P(3HD) and P(3HDD) during the second heating scan of DSC have been observed consistently in other reports as well [39, 40, 45]. The two types of crystalline structures formed according to the distinctive thermograms were temperature dependence. It has been proposed that a smectic liquid-crystalline structure is formed at lower temperature, caused by the intramolecular and/or intermolecular interactions between branched alkyl chains and low ordered main chains. In other studies, the crystallization behaviors of homo-PHAs were previously examined [51]. As the finding, increasing the temperature presumably breaks the smectic aggregates between the side chains and eventually formed the crystal structure through placement of both main and side chains. Besides, the results of wide-angle X-ray diffraction (WAXD) revealed that the crystallinity degree of the P(3HD) was only 25.9%. This is relatively low when compared with P(3HB) with more than 60% [52]. As perceived, the lower crystallinity is said to be partly responsible for the flexibility and transparency features of the P(3HD) cast film.

The mechanical properties of homo-PHAs are shown in Table 8.3. Generally, the mechanical properties of P(3HDD) were similar to those of LDPE [39]. Young's modulus values for P(3HD) and P(3HDD) reported by Hiroe's research group were three- to sixfold higher than the other studies [12, 13], considering that there were differences in crystallinity of the polymer samples caused by different aging conditions during solvent casting. When subjected to a drawing treatment at draw ratio of 3-4, the crystalline structure experienced an orientation that led to enhancement of the tensile strengths of P(3HD) and P(3HDD) samples by 5.6-fold (from 8 to 45 MPa) and 2.5-fold (from 11 to 27 MPa), respectively. The cast films' transparency index (shown by the respective haze values) of P(3HB), P(3HO), P(3HD), and P(3HDD) samples with thickness of 57–72 μm is also presented in Table 8.3. Among all PHAs examined, P(3HDD) exhibited the lowest haze value (9%), which is far less than that of LDPE (31%). Although homo-PHAs are crystalline polymers, they are highly transparent because of its low crystallinity with finer crystal particle size. Indeed, after drawing for several times, the transparency of P(3HD) and P(3HDD) films remained high. This property gives homo-PHAs as highly valuable materials to be used as blend polymers that are applicable for colorless application with non-deceptive appearance.

Table 8.3 Material properties of homo-PHAs.

Polymer	T_g (°C)	T_m (°C)	ΔH_m[a)] (J/g)	Young's modulus (MPa)	Tensile strength (MPa)	Elongation to break (%)	Haze[b)] (%)
P(3HHx) (C6)	−28	59	34	—[c)]	—[c)]	—[c)]	—[c)]
P(3HO) (C8)	−47	51	29	24	9	470	20
P(3HD) (C10)	−53	69	40	118	8	226	13
P(3HDD) (C12)	n.d.	82	51	180	11	270	9

n.d., not detected.
a) Enthalpy of fusion.
b) Haze value of solvent cast film.
c) Sticky film.
Source: From Abe et al. 2012 [51]; Hiroe et al. 2016 [39].

8.3.5 Integrated Production Process of Homo-PHAs from Renewable Feedstock

A practical and industrially feasible process for production of homo-PHAs is desirable to keep PHAs highly competitive. This can be achieved through inexpensive use of non-edible and crude fatty acids obtained as by-products from plant oil processing for the feedstock materials. Nevertheless, homo-PHAs are only produced by using pure fatty acids as a carbon source for PHA production. Thus, the crude fatty acids require a pretreatment prior to utilization. For this reason, the dominant fraction of fatty acid with desired chain length can be concentrated by a urea treatment. This has been demonstrated through utilization of the palm kernel fatty acid distillate (PKFAD), which is a by-product of palm kernel oil refining and further treated with urea before added to the culture as a co-substrate. The urea-treated PKFAD containing 29–50% dodecanoate fraction was used for 3HDD-rich polymer production [39]. The best performing condition for 3HDD-rich polymer was achieved with urea-treated PKFAD, which produced up to 85 mol% 3HDD fraction. The thermal and mechanical properties of 3HDD-rich polymers were reported to behave as in between the range of P(3HD) and P(3HDD) [39]. The transparency of the cast film of 3HDD-rich polymer (85 mol% 3HDD) is virtually the same as the P(3HDD) homopolymer. Currently, the production of homo-PHAs from fatty acid mixture including PKFAD seems to be quite challenging. It is more likely toward the increasing of dominant monomers in mcl-PHA through a pretreatment step of fatty acid mixtures. Despite this, the ideal homo-PHA production definitely requires further development for controlling the metabolism by using the current engineered strain even using the mixture of fatty acids. To implement this, a basic study on PHA metabolism and redesigning the existing biosynthetic pathway are highly important. Additionally, the balance between utilization of substrate and conserving energy is necessary to ensure an effective sustainable resource management through a chemical recycling system. As previously reported, thermal degradation of PHAs suggests a feasible approach for efficient recovery of monomeric chemicals [40]. It was found that the product of the pyrolysis of homo-PHAs, 2-alkenoic acid, exists as a main component in volatile products as well as other scl-PHAs including P(3HB). Besides, the 2-alkenoic acids also served as starting materials for PHA biosynthesis and have protective effects toward fatty acid inhibition [40]. With integration of thermal decomposition with aid of synthetic biology for mcl-PHA production, an innovative recycling route for homo-PHAs can be established that will eventually create more sustainable framework of the target PHAs.

8.4 Synthetic Biology for Production of Lactate-Based Polymers

8.4.1 Creation of Lactate-Polymerizing Enzyme (LPE)

PLA, which can be synthesized via chemo-bio process from renewable resources such as corn and sugarcane, is the currently most practically used biobased plastic. The stereocomplex of polylactide (sc-PLA) formed by blending the enantiomeric poly(L-lactide) (PLLA) and poly(D-lactide) (PDLA) has melting point of 230 °C and superior mechanical properties compared with solitary PLLA (Table 8.4) [53, 55]. Industrial production of PLA is mostly carried out as follows (Figure 8.16): (i) Lactic acid (LA) is generated from biomass-derived sugars by microbial fermentation [56]. (ii) LA is converted into lactides (cyclic dimer of lactic acid) via polycondensation and depolymerization steps. (iii) Lactides are polymerized to produce PLA by a ring-opening polymerization (ROP) with heavy metal catalysts. This chemo-bio process for PLA production is still

Table 8.4 Comparison of material properties of biobased plastics and petroleum plastics.

	Biobased plastics					Petroleum plastics		
	PLLA	Sc-PLA	P(3HB)	PBS	P(3HB-*co*-10 mol% 3HHx)	PET	PBT	PP
T_m (°C)[a]	160–170	220–240	177	114	127	260	220	164
T_g (°C)[b]	60	65–72	4	−32	−1	80	50	5
Tensile strength (MPa)	68	90	43	57	21	57	62	32
Elongation at break (%)	4	30	5	700	400	300	10	500

PLLA, poly(L)-lactic acid; Sc-PLA, stereocomplex of PLA; P(3HB), poly[(*R*)-3-hydroxybutyrate]; PBS, poly(butylene succinate); P(3HB-*co*-10 mol% 3HHx), poly(3-hydroxybutyrate-*co*-10 mol% 3-hydroxyhexanoate); PET, poly(ethylene terephthalate); PBT, poly(butylene terephthalate); PP, polypropylene.

a) Melting temperature.

b) Glass transition temperature.

Source: From Kakuta et al. 2009 [53]; Tsuge 2002 [54].

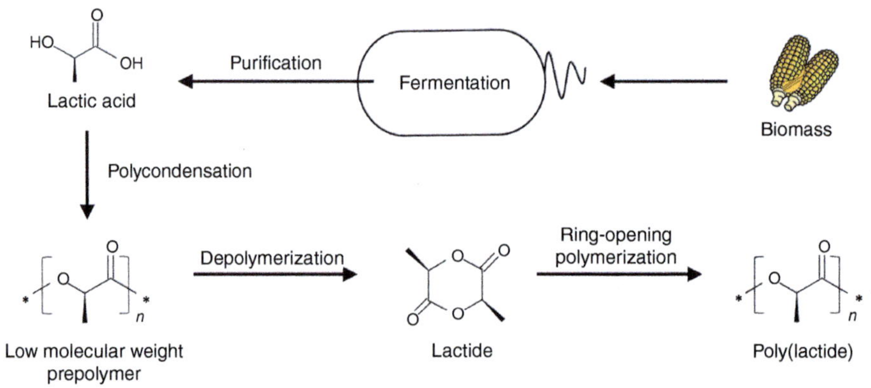

Figure 8.16 Industrial production of poly(lactide) via chemo-bio process.

expensive relative to that of petroleum-based polymers, owing to the multistep process including especially cost/energy-consuming LA purification step [57]. Is there any solution to overcome these problems? As already described, PHAs can be synthesized via one-pot bioprocess by nontoxic enzymatic catalysts. If PLA could be produced according to the microbial PHA production system, it would be advantageous in terms of process shortcut and human body friendly over conventional chemical process.

When looking at chemical structures of PLA and P(3HB), individual monomers LA and 3HB are similar to each other in chemical structure. PHA synthase (PhaC) is a key enzyme that polymerizes various 3HA monomers including 3HB. If LA monomer (lactyl-CoA) could be recognized by any PHA synthase, PLA could be synthesized by means of one-pot bioprocess. To demonstrate this proof of concept, the most important point should be finding or development of lactate-polymerizing enzyme (LPE). LPE has not been identified so far until then [58–60]. Taguchi's group attempted to explore LPE candidate(s) from a set of mutant libraries of PHA synthases prepared through long-term evolutionary engineering studies [61–67]. PHA synthase from *Pseudomonas* sp. 61-3 ($PhaC1_{Ps}$) originally has broad substrate specificity and can polymerize both short-chain-length (scl) and medium-chain-length (mcl) monomers [68]. Takase et al. expanded its substrate specificity toward SCL monomer (3HB monomer) by evolutionary engineering [66, 67]. Namely, four beneficial mutation points (Glu130, Ser325, Ser477, and Gln481) were identified as the beneficial sites for enzyme functions . An enzymatic activity assay revealed that Glu130, Ser325, and Ser477 contribute to the enzyme activity and Ser477 and Gln481 contribute to the substrate specificity[61–63, 67, 104]. By using water–organic solvent two phase *in vitro* system [69], in 2008, Taguchi et al. demonstrated the proof of concept that $PhaC1_{STQK}$ ($PhaC1_{Ps}$ with two mutations of Ser325Thr and Gln481Lys) was capable of polymerizing LA monomers together with 3HB monomers [70]. Therefore, $PhaC1_{STQK}$ was formally established as LPE. Beneficial mutation effects of STQK in LPE have been finely interpreted on the currently solved tertiary structures of class I PHA synthases from *R. eutropha* and *Chromobacterium* sp. USM2 [71].

8.4.2 Biosynthesis of Lactate-Based Polymers

The discovery of LPE opens the door to incorporate the unusual/unnatural monomers like LA into the polymer chain. In this part, LPE-mediated microbial platform for production of LA-based polymers will be described with respect to the regulation of LA fraction as well as the increased polymer production. To biosynthesize LA-based polymer, P(LA-*co*-3HB), four genes encoding 3-ketothiolase (PhaA), acetoacetyl-CoA reductase (PhaB), propionyl-CoA transferase (PCT), and LPE ($PhaC1_{STQK}$) were introduced into *E. coli* (Figure 8.17). The first LA-based polymer with 6 mol% of LA unit was biosynthesized in the recombinant *E. coli* [70]. In order to enrich LA fraction in the copolymer, various approaches were applied, as shown in Figure 8.18.

First of all, the generation of starting substance LA was reinforced by metabolic engineering and change in aeration conditions [72–74]. Yamada et al. chose the mutant strain of *E. coli* with the disrupted gene encoding pyruvate formate lyase-activating enzyme PflA. The recombinant Δ*pflA* mutant strain actually enhanced the LA fraction by 4.3-fold (26 mol% LA) compared with wild-type strain. Generally anaerobic cultivation causes the enhanced LA flux in the cells through shifting from pyruvate to LA. In fact, the LA fraction was efficiently increased up to 47 mol% [74]. However, at the same time, the polymer yield was drastically decreased since the cell growth was repressed under anaerobic conditions. This is a kind of trade-off relationship between LA fraction increase and polymer production decrease. Then, the aerobic condition has been examined to maintain the cell growth in the following studies.

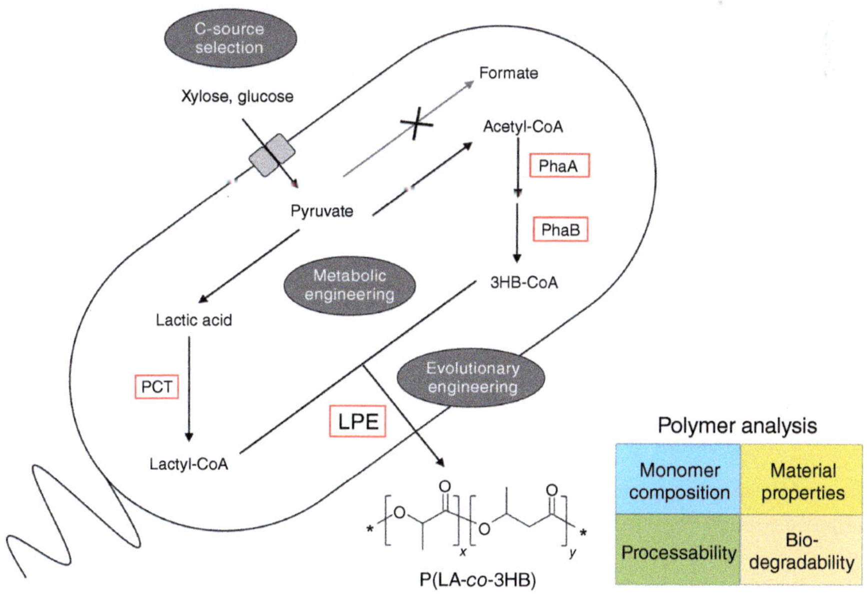

Figure 8.17 LA-based polymer biosynthesis pathway from sugars. PCT: propionyl-CoA transferase from *Megasphaera elsdenii*, LPE: lactate-polymerizing enzyme, PhaA: 3-ketothiolase, PhaB: acetoacetyl-CoA reductase.

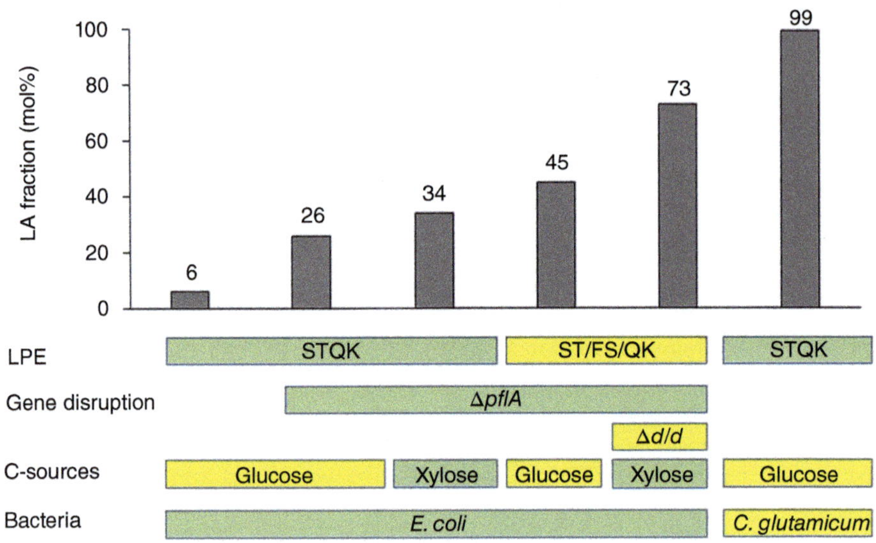

Figure 8.18 The enhancement of LA fraction in LA-based polymer. STQK: $PhaC1_{STQK}$, ST/FS/QK: $PhaC1_{ST/FS/QK}$, *pflA*: pyruvate formate lyase-activating gene, *dld*: NAD^+-independent lactate dehydrogenase gene.

Next, as another approach, an evolved variant of LPE was created by evolutionary engineering using $PhaC1_{STQK}$ as a prototype template. Then, the beneficial mutation of F420S gained in the PHA synthase from *R. eutropha* [75, 76] was introduced into the corresponding position of $PhaC1_{STQK}$. The resultant modified LPE ($PhaC1_{ST/FS/QK}$) also successfully enhanced the LA fraction in P(LA-*co*-3HB) polymer with maintaining the polymer accumulation level [74]. Thus, $PhaC1_{ST/FS/QK}$ was established as an evolved LPE. This strongly suggests the functional compatibility between differently categorized both enzymes, $PhaC1_{Ps}$ and $PhaC_{Re}$. In fact, this hypothesis can be supported by structure–function relationship study on the 3D structures of PHA synthases recently solved [71]. This third new mutation occurred at position 392 in $PhaC1_{Ps}$ (corresponding to position 420 in $PhaC_{Re}$) is located at interfacial region upon the dimer formation of $PhaC_{Re}$. After these achievements, similar studies have been followed by several research groups using the LPE homologues [77].

Regarding the carbon source, xylose was effective for giving a higher LA fraction in the P(LA-*co*-3HB) from 26 to 34 mol% instead of glucose[72]. This finding is supposed to be specifically suited for the case of LA-based polymer synthesis. The beneficial result caused by the carbon source change is not still feasibly interpreted. Utrilla et al. reported that the overexpression of GatC, a novel xylose transporter, enhanced LA production [78]. As described, xylose was also effective biomass-derived sugar for P(LA-*co*-3HB) production [72, 73]. By overproducing the GatC gene in the recombinant *E. coli*, the uptake of xylose was efficiently performed to produce 14.4 g/l of polymer with 66 mol% LA fraction [73] on 40 g/l of xylose. This polymer yield was the highest so far recorded, reaching near 75% of the level performed by Kaneka company at shaking flask level.

As shown in Figure 8.18, the regulation of LA fraction can be achieved in the range from 6 to 73 mol% in the recombinant *E. coli* by combination of LPEs, LA enrichment caused mutant strains, and carbon source selection [73]. Surprisingly, when *Corynebacterium glutamicum* was used as a host strain for production of LA-based polymer by using the same expression constructs for *E. coli*, the recombinant strain could produce the polymer with over 99 mol% of LA fraction (almost the same as PDLA), the highest LA fraction so far recorded [79, 80]. Such a big difference in LA-enrichment ability in both microbial platforms is attributed to the difference in concentration/stability of the intracellular LA-CoA in individual bacterial cells [81]. Extremely high LA fraction (96 mol%) was unexpectedly also obtained as tercopolymers, P(LA-*co*-3HB-*co*-3HV), for the recombinant *E. coli* via an inverse reaction of β-oxidation pathway, grown on glucose with a feeding of valerate as the monomer precursor [82]. This suggests a potential for synthesis of highly LA-enriched polymer even *E. coli* recombinant system if proper conditions could be employed.

As a separated approach, modifications of microbial platform were carried out in genome-wide. Kadoya et al. applied a transposon mutagenesis to obtain mutants of *E. coli* representing high polymer accumulation [83, 84]. Several beneficial mutations were isolated in the genome-wide region. Most of them had apparently indirect functions with synthetic pathways for production of the target polymers. For example, the strain bearing a disruption of *mtgA*, which is associated with a formation of the peptidoglycan strand, enhanced accumulation of the LA-based polymer (7.0 g/l) compared with the control wild-type strain (5.2 g/l). Interestingly, the cell size of Δ*mtgA* strain was enlarged like a balloon associated with the polymer accumulation inside the cells, and this unique phenotype consequently contributed to the increase in polymer production [84]. Similar studies have been carried out by Chen's group. However, most of the mutant strains exhibited extraordinary changes in morphology of the cells like filamentous phenotype formation, causing negatively affecting toward cell growth [85].

8.4.3 Integrated Production Process of Lactate-Based Polymers from Renewable Feedstock

The tailor-made production of P(LA-*co*-3HB) from commercially available pure carbon sources has been achieved by metabolic engineering, evolutionary engineering, and cultivation conditioning. In this section, the establishment of integrated production system for LA-based polymers by utilization of renewable feedstock will be described.

Figure 8.19 illustrates the integrated production process of LA-based polymer from renewable feedstock. As for the case studies of PHBH and homo-PHAs, oil-based biomass was useful for cost-effective production of the target polymers. In case of LA-based polymer, sugar-based biomass was used throughout the works, as described below. In order to exploit lignocellulose (the most

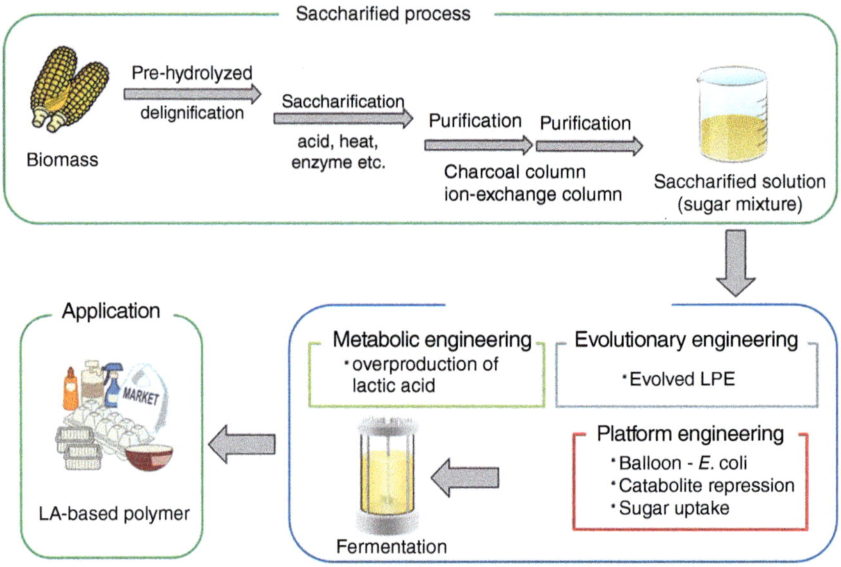

Figure 8.19 Integrated production process of LA-based polymer from renewable feedstock.

abundant sugar-based biomass) as a feedstock, the current ways of bio-refinery are generally carried out in the following three steps: (i) pretreatment of biomass (e.g. hydrothermal technique, mechanochemical processing), (ii) releasing of monosaccharide (simple sugars) from polysaccharide by enzymatic or chemical hydrolysis reactions, and (iii) microbial fermentation of saccharified solution (various sugars including glucose, xylose, galactose, and so on) for the production of target polymers [86, 87]. By using saccharified solution of *eucalyptus* woodchip and hybrid *Miscanthus*, P(LA-*co*-3HB) was successfully biosynthesized in the yields of 5.4 and 5.3 g/l of polymer with 5.5 and 16.9 mol% of LA fraction, respectively [88, 89]. To improve the polymer production by utilizing saccharified solution from hybrid *Miscanthus*, Kadoya et al. modified sugar uptake system of *E. coli*. When various sugars coexisted, it was known that glucose tends to be preferentially consumed and xylose is remained in the culture medium. To avoid this unbalanced sugar assimilation and convert both sugars to polymer effectively, overexpression of Mlc, a multiple regulator of glucose and xylose uptake, was attempted. As a result, simultaneous consumption of glucose and xylose was observed, and P(LA-*co*-3HB) production was increased up to 10.6 g/l with 10.9 mol% LA fraction [90].

As a typical integrated process, Salamanca-Cardona et al. attempted the direct utilization of polysaccharide, xylan, as a feedstock for P(LA-*co*-3HB) production [91–93]. The recombinant *E. coli* harboring the genes (*xylB* and *xynB*) encoding two xylan-degrading enzymes, endoxylanase and xylosidase, synthesized the copolymer containing 16 mol% of LA fraction with 0.44 g/l yield from commercial beechwood xylan[93].

8.4.4 Biosynthesized Lactate-Based Polymer Shows Superior Properties

Considering the copolymer P(LA-*co*-3HB), it is of interest to compare again the material properties of homopolymers: PLA and P(3HB). PLA is rigid, transparent, and low biodegradable material. On the other hand, P(3HB) is rigid and opaque but exhibits high biodegradability. In such a sense, P(LA-*co*-3HB) has both advantages of PLA and P(3HB): superior transparency and biodegradability. In addition, P(LA-*co*-3HB) exhibits varied flexibilities depending on the LA fraction. Table 8.5 shows the summary of mechanical and thermal properties of P(LA-*co*-3HB)s with various LA fractions (6–66 mol%) [94, 95]. P(LA-*co*-3HB) exhibited flexibilities with the ranges of 122–283% of elongation at break, depending on LA fraction. In addition, thermal properties were improved; especially process window could be expanded due to the reduction in between melting temperature (T_m) and degradation temperature. Glass transition temperature (T_g), which affects polymer processability, was also changed in the range from 4 to 60 °C, depending on LA fraction. P(LA-*co*-3HB) represents good processability, and various molding products such as film, compact, and (nano)fiber were already produced by way of trial. On a side note, molecular weight is also one of the basic indexes affecting the material strength and processability. In case of P(LA-*co*-3HB), molecular weight was decreased with increasing LA fraction. This inverse relationship between molecular weight and LA fraction has been investigated from several viewpoints [81]. The bio-production of LA-based polymer with higher molecular weight is one of big challenges in the future. PDLA synthesized by the recombinant *E. coli* described in Section 8.4.2 exhibited identical thermal properties to those of the chemically synthesized PLLA [82]. In addition, the blended polymer of

Table 8.5 Material properties of P(3HB), P(LA-*co*-3HB), and PLA [94–96].

LA fraction (mol %)	Molecular weight		Mechanical properties[a)]			Thermal properties	
	M_w (×10^4)	M_w/M_n	Young's modulus (MPa)	Tensile strength (MPa)	Elongation at break (%)	T_g (°C)	T_m (°C)
0			2259 ± 21	28 ± 2	3 ± 1	4	170
6	13	2.3	769 ± 144	11 ± 4	3 ± 1	7	160
21	23	2.9	556 ± 196	14 ± 1	283 ± 79	13	157
33	18	1.9	417 ± 74	9 ± 1	244 ± 58	22	148
48	8	2.3	386 ± 69	9 ± 2	203 ± 12	26	125
66	6	2.5	498 ± 46	13 ± 0	122 ± 15	25	n.d.
100[a)]	20		1020	52 ± 2	2	60	154

n.d., not detected.
a) Data from [96]. PLA was chemically synthesized.

biologically synthesized PDLA and chemically synthesized PLLA could form a stereocomplex associated with causing a higher melting temperature [82].

A good biodegradability of P(3HB) is contrasted with an undegradable PLA in natural environments. As for P(LA-*co*-3HB), an existence of 3HB unit contributes to well biodegradability of the polymer. P(3HB) is a most typical microbial polyester; therefore P(3HB) (3HB-3HB linkage) can be easily degraded by microbial degrading enzymes (PhaZs). In a biodegradation assay using a specific depolymerase isolated from *Variovorax* sp. C34 ($PhaZ_{Vs}$), degradation rate of P(67 mol% LA-*co*-3HB) powders was near the same with that of P(3HB) under the liquid state [97]. Analyses of degradation products indicated that not only 3HB-3HB linkage but also LA–LA linkage were hydrolyzed using a series of synthetic trimers with LA and 3HB sequences as substrates [98]. It should be noted that PLLA and PDLA could not be degraded by $PhaZ_{Vs}$ at all. Then, enzymatic hydrolysis took place for oligomers generated from alkali-hydrolyzed PDLA by using the same depolymerase [98]. This implies that enzymatic hydrolysis of P(LA-*co*-3HB) can be accounted for by the crystallinity of polymer not monomer sequence-dependent substrate specificity. These findings will allow us to apply this type of polymer to the environmental sites.

Taken together, P(LA-*co*-3HB) is a promising new polyester. LPE has two beneficial properties: extremely high enantiomeric selectivity and broad substrate specificity of the monomer substrates. By using LPE, various types of chiral polyesters incorporating the varied unusual/unnatural monomers [99, 100] can be biosynthesized via one-pot process from renewable feedstocks. In the near future, it is expected that LA-based polymers would be used as one of commercial biobased plastics that are different from PLA.

8.5 Outlook

In this chapter, three kinds of polymers are presented from academic and industrial missions. Industrial production of PHBH Kaneka polymer should be a symbolic milestone of PHA research works that we hope to reach. At the same time, ongoing academic achievements should be continuously updated as an important key piece for further progressing the field of biobased polymers in order to realize sustainable low carbon society. At the present, several promising findings have been already reported:

(i) *Tertiary structure of PHA synthases*: Structure-based enzyme engineering will allow us to create tailor-made PHA synthases to satisfy the requirements. To this end, enzyme-substrate complex form should be solved for fine-tuned enzyme engineering.

(ii) *New type of polymers incorporating new monomers*: Associated with development of evolved PHA synthases, a structural variety of monomeric constituents other than LA has been already expanded [99, 100], as described in this chapter.

(iii) *Functionalization of PHAs*: Taking advantage of various monomer incorporation capacity, chemical reactions like click reaction will be extensively applied to modify the PHA functions.

(iv) *Sequence-regulated polymers*: A pioneering study on biosynthesis of block copolymer has been published under special conditions [81]. This breakthrough technology is expected to be ubiquitous for varied type of PHA.
(v) *Polyhydroxyalkanoate oligomer (PHAO) secretion*: This has been currently reported [101–103]. PHAO is recognized as building blocks for making wide range of bioactive compounds, platform chemicals, and polymeric materials.
(vi) *Biocompatibility and biodegradability of PHA*: These polyesters should intrinsically have attractive properties for biomedical and environmental applications. Especially, it should be noted that Kaneka acquired the "OK biodegradable MARINE" certificate in September 2017 from TÜV AUSTRIA Belgium NV (formerly termed Vincotte), an international certification organization based in Belgium. It is a good kickoff for global standardization of biobased biodegradable polymers.

Environmental assessment tools like life cycle analysis and the product environmental footprint imply that PHAs can be good targets contributing to greenhouse gas emission reduction and waste reduction as well as innovation in the biotechnology sector. So far PHA research works have been limited to be mostly biosynthesis of the target polymers. However, this is not a final goal; of course the feasibility of polymer biosynthesis should be justified by characterization of polymer products in order to grow up the manufacturing material polymers from the primitive level. Industrial production of PHAs grew rapidly in the past years; their total market share is still marginal, owing to their higher price. In such a sense, the recent advances described in this chapter will encourage the progress of PHA studies from now on.

References

1 Madison, L.L. and Huisman, G.W. (1999). Metabolic engineering of poly(3-hydroxyalkanoates): from DNA to plastic. *Microbiology and Molecular Biology Reviews* 63 (1): 21–53.
2 Huisman, G.W., de Leeuw, O., Eggink, G. et al. (1989). Synthesis of poly-3-hydroxyalkanoates is a common feature of fluorescent pseudomonads. *Applied and Environmental Microbiology* 55 (8): 1949–1954.
3 Witholt, B. and Kessler, B. (1999). Perspectives of medium chain length poly(hydroxyalkanoates), a versatile set of bacterial bioplastics. *Current Opinion in Biotechnology* 10 (3): 279–285. https://doi.org/10.1016/S0958-1669(99)80049-4.
4 Abe, H. and Doi, Y. (2002). Side-chain effect of second monomer units on crystalline morphology, thermal properties, and enzymatic degradability for random copolyesters of (*R*)-3-hydroxybutyric acid with (*R*)-3-hydroxyalkanoic acids. *Biomacromolecules* 3: 133–138. https://doi.org/10.1021/bm0155975.
5 Shimamura, E., Kasuya, K., Kobayashi, G. et al. (1994). Physical properties and biodegradability of microbial

poly(3-hydroxybutyrate-*co*-3-hydroxyhexanoate). *Macromolecules* 27 (3): 878–880. https://doi.org/10.1021/ma00081a041.

6 Gumel, A.M., Annuar, M.S.M., and Chisti, Y. (2013). Recent advances in the production, recovery and applications of polyhydroxyalkanoates. *Journal of Polymers and the Environment* 21 (2): 580–605. https://doi.org/10.1007/s10924-012-0527-1.

7 Madkour, M.H., Heinrich, D., Alghamdi, M.A. et al. (2013). PHA recovery from biomass. *Biomacromolecules* 14 (9): 2963–2972. https://doi.org/10.1021/bm4010244.

8 Fukui, T. and Doi, Y. (1998). Efficient production of polyhydroxyalkanoates from plant oils by *Alcaligenes eutrophus* and its recombinant strain. *Applied Microbiology and Biotechnology* 49 (3): 333–336. https://doi.org/10.1007/s002530051178.

9 Fukui, T., Abe, H., and Doi, Y. (2002). Engineering of *Ralstonia eutropha* for production of poly(3-hydroxybutyrate-*co*-3-hydroxyhexanoate) from fructose and solid-state properties of the copolymer. *Biomacromolecules* 3 (3): 618–624. https://doi.org/10.1021/bm0255084.

10 Loo, C.Y., Lee, W.H., Tsuge, T. et al. (2005). Biosynthesis and characterization of poly(3-hydroxybutyrate-*co*-3-hydroxyhexanoate) from palm oil products in a *Wautersia eutropha* mutant. *Biotechnology Letters* 27 (18): 1405–1410. https://doi.org/10.1007/s10529-005-0690-8.

11 Mifune, J., Nakamura, S., and Fukui, T. (2008). Targeted engineering of *Cupriavidus necator* chromosome for biosynthesis of poly(3-hydroxybutyrate-*co*-3-hydroxyhexanoate) from vegetable oil. *Canadian Journal of Chemistry* 86 (6): 621–627. https://doi.org/10.1111/j.1365-2648.2008.04878.x.

12 Riedel, S.L., Bader, J., Brigham, C.J. et al. (2012). Production of poly(3-hydroxybutyrate-*co*-3-hydroxyhexanoate) by *Ralstonia eutropha* in high cell density palm oil fermentations. *Biotechnology and Bioengineering* 109 (1): 74–83. https://doi.org/10.1002/bit.23283.

13 Tsuge, T., Saito, Y., Kikkawa, Y. et al. (2004). Biosynthesis and compositional regulation of poly[(3-hydroxybutyrate)-*co*-(3-hydroxyhexanoate)] in recombinant *Ralstonia eutropha* expressing mutated polyhydroxyalkanoate synthase genes. *Macromolecular Bioscience* 4 (3): 238–242. https://doi.org/10.1002/mabi.200300077.

14 Tsuge, T., Watanabe, S., Sato, S. et al. (2007a). Variation in copolymer composition and molecular weight of polyhydroxyalkanoate generated by saturation mutagenesis of *Aeromonas caviae* PHA synthase. *Macromolecular Bioscience* 7 (6): 846–854. https://doi.org/10.1002/mabi.200700023.

15 Wong, Y.M., Brigham, C.J., Rha, C.K. et al. (2012). Biosynthesis and characterization of polyhydroxyalkanoate containing high 3-hydroxyhexanoate monomer fraction from crude palm kernel oil by recombinant *Cupriavidus necator*. *Bioresource Technology* 121: 320–327. https://doi.org/10.1016/j.biortech.2012.07.015.

16 Oliver, P.P. and Sinskey, A.J. (1989). Poly-β-hydroxybutyrate (PHB) biosynthesis in *Alcaligenes eutrophus* H16. *Journal of Biological Chemistry* 264 (26): 15298–15303.

17 Pohlmann, A., Fricke, W.F., Reinecke, F. et al. (2006). Genome sequence of the bioplastic-producing "Knallgas" bacterium *Ralstonia eutropha* H16. *Nature Biotechnology* 24: 1257–1262. https://doi.org/10.1038/nbt1244.
18 Calloway, D.H. and Kumar, A.M. (1969). Protein quality of the bacterium *Hydrogenomonas eutropha. Applied Microbiology* 17 (1): 176–178.
19 Waslien, C.I. and Calloway, D.H. (1969). Nutritional value of lipids in *Hydrogenomonas eutropha* as measured in the rat. *Applied Microbiology* 18 (2): 152–155.
20 Kichise, T., Taguchi, S., and Doi, Y. (2002). Enhanced accumulation and change monomer composition in polyhydroxyalkanoate (PHA) copolyester by *in vitro* evolution of *Aeromonas caviae* PHA synthase. *Appliel and Environmental Microbiology* 68 (5): 2411–2419. https://doi.org/10.1128/AEM.68.5.2411-2419.2002.
21 Tsuge, T., Watanabe, S., Shimada, D. et al. (2007b). Combination of N149S and D171G mutations in *Aeromonas caviae* polyhydroxyalkanoate synthase and impact on polyhydroxyalkanoate biosynthesis. *FEMS Microbiology Letters* 277 (2): 217–222. https://doi.org/10.1111/j.1574-6968.2007.00958.x.
22 Budde, C.F., Riedel, S.L., Willis, L.B. et al. (2011). Production of poly(3-hydroxybutyrate-*co*-3-hydroxyhexanoate) from plant oil by engineered *Ralstonia eutropha* strains. *Applied and Environmental Microbiology* 77 (9): 2847–2854. https://doi.org/10.1128/AEM.02429-10.
23 Jeon, J.M., Brigham, C.J., Kim, Y.H. et al. (2014). Biosynthesis of poly(3-hydroxybutyrate-*co*-3-hydroxyhexanoate) (P[HB-*co*-HHx]) from butyrate using engineered *Ralstonia eutropha. Applied Microbiology and Biotechnology* 98 (12): 5461–5469. https://doi.org/10.1007/s00253-014-5617-7.
24 Slater, S., Houmiel, K.L., Tran, M. et al. (1998). Multiple β-ketothiolases mediate poly (β-Hydroxyalkanoate) copolymer synthesis in *Ralstonia eutropha. Journal of Bacteriology* 180 (8): 1979–1987.
25 Sato, S., Maruyama, H., Fujiki, T. et al. (2015). Regulation of 3-hydroxyhexanoate composition in PHBH synthesized by recombinant *Cupriavidus necator* H16 from plant oil by using butyrate as a co-substrate. *Journal of Bioscience and Bioengineering* 120 (3): 246–251. https://doi.org/10.1016/j.jbiosc.2015.01.016.
26 Kawashima, Y., Cheng, W., Mifune, J. et al. (2012). Characterization and functional analyses of *R*-specific enoyl coenzyme a hydratases in polyhydroxyalkanoate-producing *Ralstonia eutropha. Appliel and Environmental Microbiology* 78 (2): 493–502. https://doi.org/10.1128/AEM.06937-11.
27 Arikawa, H. and Matsumoto, K. (2016). Evaluation of gene expression cassettes and production of poly(3-hydroxybutyrate-*co*-3-hydroxyhexanoate) with a fine modulated monomer composition by using it in *Cupriavidus necator. Microbial Cell Factories* 15 (1): 1–11. https://doi.org/10.1186/s12934-016-0583-7.
28 Morohoshi, T., Ogata, K., Okura, T. et al. (2018a). Molecular characterization of the bacterial community in biofilms for degradation of

poly(3-hydroxybutyrate-*co*-3-hydroxyhexanoate) films in seawater. *Microbes and Environments* 33 (1): 19–25. https://doi.org/10.1264/jsme2.ME17052.

29 Morohoshi, T., Oi, T., Aiso, H. et al. (2018b). Biofilm formation and degradation of commercially available biodegradable plastic films by bacterial consortiums in freshwater environments. *Microbes and Environments* 33 (3): 332–335. https://doi.org/10.1264/jsme2.ME18033.

30 Sato, S., Kanazawa, H., and Tsuge, T. (2011). Expression and characterization of (*R*)-specific enoyl coenzyme a hydratases making a channeling route to polyhydroxyalkanoate biosynthesis in *Pseudomonas putida*. *Applied Microbiology and Biotechnology* 90 (3): 951–959. https://doi.org/10.1007/s00253-011-3150-5.

31 Tsuge, T., Taguchi, K., Taguchi, S. et al. (2003). Molecular characterization and properties of (*R*)-specific enoyl-CoA hydratases from *Pseudomonas aeruginosa*: metabolic tools for synthesis of polyhydroxyalkanoates *via* fatty acid β-oxidation. *International Journal of Biological Macromolecules* 31 (4–5): 195–205. https://doi.org/10.1016/S0141-8130(02)00082-X.

32 Nomura, C.T., Taguchi, K., Taguchi, S. et al. (2004). Coexpression of genetically engineered 3-ketoacyl-ACP synthase III (*fabH*) and polyhydroxyalkanoate synthase (*phaC*) genes leads to short-chain-length-medium-chain-length polyhydroxyalkanoate copolymer production from glucose in *Escherichia coli* JM109. *Applied and Environmental Microbiology* 70 (2): 999–1007. https://doi.org/10.1128/AEM.70.2.999-1007.2004.

33 Nomura, C.T., Tanaka, T., Eguen, T.E. et al. (2008). FabG mediates polyhydroxyalkanoate production from both related and nonrelated carbon sources in recombinant *Escherichia coli* LS5218. *Biotechnology Progress* 24 (2): 342–351. https://doi.org/10.1021/bp070303y.

34 Philip, S., Keshavarz, T., and Roy, I. (2007). Polyhydroxyalkanoates: biodegradable polymers with a range of applications. *Journal of Chemical Technology and Biotechnology* 82 (3): 233–247. https://doi.org/10.1002/jctb.1667.

35 Rai, R., Keshavarz, T., Roether, J.A. et al. (2011a). Medium chain length polyhydroxyalkanoates, promising new biomedical materials for the future. *Materials Science and Engineering R: Reports* 72 (3): 29–47. https://doi.org/10.1016/j.mser.2010.11.002.

36 Rai, R., Yunos, D.M., Boccaccini, A.R. et al. (2011b). Poly-3-hydroxyoctanoate P(3HO), a medium chain length polyhydroxyalkanoate homopolymer from *Pseudomonas mendocina*. *Biomacromolecules* 12 (6): 2126–2136. https://doi.org/10.1021/bm2001999.

37 Abe, H., Doi, Y., Fukushima, T. et al. (1994). Biosynthesis from gluconate of a random copolyester consisting of 3-hydroxybutyrate and medium-chain-length 3-hydroxyalkanoates by *Pseudomonas* sp. 61-3. *International Journal of Biological Macromolecules* 16 (3): 115–119. https://doi.org/10.1016/0141-8130(94)90036-1.

38 Huang, P., Okoshi, T., Mizuno, S. et al. (2018). Gas chromatography-mass spectrometry-based monomer composition analysis of medium-chain-length polyhydroxyalkanoates biosynthesized by *Pseudomonas* spp. *Bioscience,*

Biotechnology, and Biochemistry 82 (9): 1615–1623. https://doi.org/10.1080/09168451.2018.1473027.

39 Hiroe, A., Ishii, N., Ishii, D. et al. (2016). Uniformity of monomer composition and material properties of medium-chain-length polyhydroxyalkanoates biosynthesized from pure and crude fatty acids. *ACS Sustainable Chemistry & Engineering* 4 (12): 6905–6911. https://doi.org/10.1021/acssuschemeng.6b01851.

40 Sato, S., Ishii, N., Hamada, Y. et al. (2012). Utilization of 2-alkenoic acids for biosynthesis of medium-chain-length polyhydroxyalkanoates in metabolically engineered *Escherichia coli* to construct a novel chemical recycling system. *Polymer Degradation and Stability* 97 (3): 329–336. https://doi.org/10.1016/j.polymdegradstab.2011.12.007.

41 Wang, H.H., Li, X.T., Chen, G.Q. et al. (2009). Production and characterization of homopolymer polyhydroxyheptanoate (P3HHp) by a *fadBA* knockout mutant *Pseudomonas putida* KTOY06 derived from *P. putida* KT2442. *Process Biochemistry* 44 (1): 106–111. https://doi.org/10.1016/j.procbio.2008.09.014.

42 Chung, A.L., Jin, H.L., Huang, L.J. et al. (2011). Biosynthesis and characterization of poly(3-hydroxydodecanoate) by β-oxidation inhibited mutant of *Pseudomonas entomophila* L48. *Biomacromolecules* 12 (10): 3559–3566. https://doi.org/10.1021/bm200770m.

43 Liu, Q., Luo, G., Zhou, X.R. et al. (2011). Biosynthesis of poly(3-hydroxydecanoate) and 3-hydroxydodecanoate dominating polyhydroxyalkanoates by β-oxidation pathway inhibited *Pseudomonas putida*. *Metabolic Engineering* 13 (1): 11–17. https://doi.org/10.1016/j.ymben.2010.10.004.

44 Wang, H.H., Zhou, X.R., Liu, Q. et al. (2011). Biosynthesis of polyhydroxyalkanoate homopolymers by *Pseudomonas putida*. *Applied Microbiology and Biotechnology* 89 (5): 1497–1507. https://doi.org/10.1007/s00253-010-2964-x.

45 Wang, Y., Chung, A., and Chen, G.Q. (2017). Synthesis of medium-chain-length polyhydroxyalkanoate homopolymers, random copolymers, and block copolymers by an engineered strain of *Pseudomonas entomophila*. *Advanced Healthcare Materials* 6 (7): 1–10. https://doi.org/10.1002/adhm.201601017.

46 Hiroe, A., Watanabe, S., Kobayashi, M. et al. (2018). Increased synthesis of poly(3-hydroxydodecanoate) by random mutagenesis of polyhydroxyalkanoate synthase. *Applied Microbiology and Biotechnology* 102 (18): 7927–7934. https://doi.org/10.1007/s00253-018-9230-z.

47 Mohd Fadzil, F.I., Mizuno, S., Hiroe, A. et al. (2018). Low carbon concentration feeding improves medium-chain-length polyhydroxyalkanoate production in *Escherichia coli* strains with defective β-oxidation. *Frontiers in Bioengineering and Biotechnology* 6: 1–8. https://doi.org/10.3389/fbioe.2018.00178.

48 Scheel, R.A., Ji, L., Lundgren, B.R. et al. (2016). Enhancing poly(3-hydroxyalkanoate) production in *Escherichia coli* by the removal of the regulatory gene *arcA*. *AMB Express* 6 (120): 1–12. https://doi.org/10.1186/s13568-016-0291-z.

49 Tappel, R.C., Wang, Q., and Nomura, C.T. (2012). Precise control of repeating unit composition in biodegradable poly(3-hydroxyalkanoate) polymers synthesized by *Escherichia coli*. *Journal of Bioscience and Bioengineering* 113 (4): 480–486. https://doi.org/10.1016/j.jbiosc.2011.12.004.

50 Agnew, D.E., Stevermer, A.K., Youngquist, J.T. et al. (2012). Engineering *Escherichia coli* for production of C12–C14 polyhydroxyalkanoate from glucose. *Metabolic Engineering* 14 (6): 705–713. https://doi.org/10.1016/j.ymben.2012.08.003.

51 Abe, H., Ishii, N., Sato, S. et al. (2012). Thermal properties and crystallization behaviors of medium-chain-length poly(3-hydroxyalkanoate)s. *Polymer* 53 (14): 3026–3034. https://doi.org/10.1016/j.polymer.2012.04.043.

52 Kabe, T., Tsuge, T., Kasuya, K.I. et al. (2012). Physical and structural effects of adding ultrahigh-molecular-weight poly[(*R*)-3-hydroxybutyrate] to wild-type poly[(*R*)-3-hydroxybutyrate]. *Macromolecules* 45 (4): 1858–1865. https://doi.org/10.1021/ma202285c.

53 Kakuta, M., Hirata, M., and Kimura, Y. (2009). Stereoblock polylactides as high-performance bio-based polymers. *Polymer Reviews* 49 (2): 107–140. https://doi.org/10.1080/15583720902834825.

54 Tsuge, T. (2002). Metabolic improvements and use of inexpensive carbon sources in microbial production of polyhydroxyalkanoates. *Journal of Bioscience and Bioengineering* 94 (6): 579–584. https://doi.org/10.1016/S1389-1723(02)80198-0.

55 Ikada, Y., Jamshidi, K., Tsuji, H. et al. (1987). Stereocomplex formation between enantiomeric poly(lactides). *Macromolecules* 20 (4): 904–906. https://doi.org/10.1021/ma00170a034.

56 Yu, R.C. and Hang, Y.D. (1989). Kinetics of direct fermentation of agricultural commodities to L(+) lactic acid by *Rhizopus oryzae*. *Biotechnology Letters* 11 (8): 597–600. https://doi.org/10.1007/BF01040043.

57 Lasprilla, A.J.R., Martinez, G.A.R., Lunelli, B.H. et al. (2012). Poly-lactic acid synthesis for application in biomedical devices – a review. *Biotechnology Advances* 30 (1): 321–328. https://doi.org/10.1016/j.biotechadv.2011.06.019.

58 Valentin, H.E. and Steinbiichel, A. (1994). Application of enzymatically synthesized short-chain-length hydroxy fatty acid coenzyme A thioesters for assay of polyhydroxyalkanoic acid synthases. *Applied Microbiology and Biotechnology* 40 (5): 699–709. https://doi.org/10.1007/BF00173332.

59 Yuan, W., Jia, Y., Tian, J. et al. (2001). Class I and III polyhydroxyalkanoate synthases from *Ralstonia eutropha* and *Allochromatium vinosum*: characterization and substrate specificity studies. *Archives of Biochemistry and Biophysics* 394 (1): 87–98. https://doi.org/10.1006/abbi.2001.2522.

60 Zhang, S., Kamachi, M., Takagi, Y. et al. (2001). Comparative study of the relationship between monomer structure and reactivity for two polyhydroxyalkanoate synthases. *Applied Microbiology and Biotechnology* 56: 131–136. https://doi.org/10.1007/s002530000562.

61 Matsumoto, K., Takase, K., Aoki, E. et al. (2005). Synergistic effects of Glu130Asp substitution in the type II polyhydroxyalkanoate (PHA) synthase:

enhancement of PHA production and alteration of polymer molecular weight. *Biomacromolecules* 6 (1): 99–104. https://doi.org/10.1021/bm049650b.

62 Matsumoto, K., Aoki, E., Takase, K. et al. (2006). *In vivo* and *in vitro* characterization of Ser477X mutations polyhydroxyalkanoate (PHA) synthase 1 from *Pseudomonas* sp. 61-3: effects of beneficial mutations on enzymatic activity, substrate specificity, and molecular weight of PHA. *Biomacromolecules* 7 (8): 2436–2442. https://doi.org/10.1021/bm0602029.

63 Matsumoto, K., Shozui, F., Satoh, Y. et al. (2009). Kinetic analysis of engineered polyhydroxyalkanoate synthases with broad substrate specificity. *Polymer Journal* 41 (3): 237–240. https://doi.org/10.1295/polymj.PJ2008253.

64 Nomura, C.T. and Taguchi, S. (2007). PHA synthase engineering toward superbiocatalysts for custom-made biopolymers. *Applied Microbiology and Biotechnology* 73 (5): 969–979. https://doi.org/10.1007/s00253-006-0566-4.

65 Taguchi, S. and Doi, Y. (2004). Evolution of polyhydroxyalkanoate (PHA) production system by "enzyme evolution": successful case studies of directed evolution. *Macromolecular Bioscience* 4 (3): 146–156. https://doi.org/10.1002/mabi.200300111.

66 Takase, K., Taguchi, S., and Doi, Y. (2003). Enhanced synthesis of poly(3-hydroxybutyrate) in recombinant *Escherichi coli* by means of error-prone PCR mutagenesis, saturation mutagenesis, and *iv vitro* recombinantion of the type II polyhydroxyalkanoate synthase gene. *Journal of Biochemistry* 133 (1): 139–145. https://doi.org/10.1093/jb/mvg015.

67 Takase, K., Matsumoto, K., Taguchi, S. et al. (2004). Alteration of substrate chain-length specificity of type II synthase for polyhydroxyalkanoate biosynthesis by *in vitro* evolution: *in vivo* and *in vitro* enzyme assays. *Biomacromolecules* 5 (2): 480–485. https://doi.org/10.1021/bm034323+.

68 Kato, M., Bao, H.J., Kang, C.K. et al. (1996). Production of a novel copolyester of 3-hydroxybutyric acid and medium-chain-length 3-hydroxyalkanoic acids by *Pseudomonas* sp. 61-3 from sugars. *Applied Microbiology and Biotechnology* 45 (3): 363–370. https://doi.org/10.1007/s002530050697.

69 Tajima, K., Satoh, Y., Nakazawa, K. et al. (2004). Chemoenzymatic synthesis of poly(3-hydroxybutyrate) in a water-organic solvent two-phase system. *Macromolecules* 37 (12): 4544–4546. https://doi.org/10.1021/ma049828m.

70 Taguchi, S., Yamada, M., Matsumoto, K. et al. (2008). A microbial factory for lactate-based polyesters using a lactate-polymerizing enzyme. *Proceedings of the National Academy of Sciences of the United States of America* 105 (45): 17323–17327. https://doi.org/10.1073/pnas.0805653105.

71 Chek, M.F., Hiroe, A., Hakoshima, T. et al. (2018). PHA synthase (PhaC): interpreting the functions of bioplastic-producing enzyme from a structural perspective. *Applied Microbiology and Biotechnology* 103 (3): 1131–1141. https://doi.org/10.1007/s00253-018-9538-8.

72 Nduko, J.M., Matsumoto, K., Ooi, T. et al. (2013). Effectiveness of xylose utilization for high yield production of lactate-enriched P(lactate-*co*-3-hydroxybutyrate) using a lactate-overproducing strain of

Escherichia coli and an evolved lactate-polymerizing enzyme. *Metabolic Engineering* 15: 159–166. https://doi.org/10.1016/j.ymben.2012.11.007.

73 Nduko, J.M., Matsumoto, K., Ooi, T. et al. (2014). Enhanced production of poly(lactate-*co*-3-hydroxybutyrate) from xylose in engineered *Escherichia coli* overexpressing a galactitol transporter. *Applied Microbiology and Biotechnology* 98 (6): 2453–2460. https://doi.org/10.1007/s00253-013-5401-0.

74 Yamada, M., Matsumoto, K., Shimizu, K. et al. (2010). Adjustable mutations in lactate (LA)-polymerizing enzyme for the microbial production of LA-based polyesters with tailor-made monomer composition. *Biomacromolecules* 11 (3): 815–819. https://doi.org/10.1021/bm901437z.

75 Normi, Y.M., Hiraishi, T., Taguchi, S. et al. (2005). Site-directed saturation mutagenesis at residue F420 and recombination with another beneficial mutation of *Ralstonia eutropha* polyhydroxyalkanoate synthase. *Biotechnology Letters* 27 (10): 705–712. https://doi.org/10.1007/s10529-005-5186-z.

76 Taguchi, S., Nakamura, H., Hiraishi, T. et al. (2002). *In vitro* evolution of a polyhydroxybutyrate synthase by intragenic suppression-type mutagenesis. *Journal of Biochemistry* 131 (6): 801–806. https://doi.org/10.1093/oxfordjournals.jbchem.a003168.

77 Park, S.J., Kim, T.W., Kim, M.K. et al. (2012). Advanced bacterial polyhydroxyalkanoates: towards a versatile and sustainable platform for unnatural tailor-made polyester. *Biotechnology Advances* 30 (6): 1196–1206. https://doi.org/10.1016/j.biotechadv.2011.11.007.

78 Utrilla, J., Licona-Cassani, C., Marcellin, E. et al. (2012). Engineering and adaptive evolution of *Escherichia coli* for D-lactate fermentation reveals GatC as a xylose transporter. *Metabolic Engineering* 14 (5): 469–476. https://doi.org/10.1016/j.ymben.2012.07.007.

79 Matsumoto, K., Tobitani, K., Aoki, S. et al. (2014). Improved production of poly(lactic acid)-like polyester based on metabolite analysis to address the rate-limiting step. *AMB Express* 4 (83): 1–5. https://doi.org/10.1186/s13568-014-0083-2.

80 Song, Y., Matsumoto, K., Yamada, M. et al. (2012). Engineered *Corynebacterium glutamicum* as an endotoxin-free platform strain for lactate-based polyester production. *Applied Microbiology and Biotechnology* 93 (5): 1917–1925. https://doi.org/10.1007/s00253-011-3718-0.

81 Matsumoto, K., Iijima, M., Hori, C. et al. (2018). *In vitro* analysis of D-lactyl-CoA-polymerizing polyhydroxyalkanoate synthase in polylactate and poly(lactate-*co*-3-hydroxybutyrate) syntheses. *Biomacromolecules* 19 (7): 2889–2895. https://doi.org/10.1021/acs.biomac.8b00454.

82 Shozui, F., Matsumoto, K., Motohashi, R. et al. (2011). Biosynthesis of a lactate (LA)-based polyester with a 96 Mol% LA fraction and its application to stereocomplex formation. *Polymer Degradation and Stability* 96 (4): 499–504. https://doi.org/10.1016/j.polymdegradstab.2011.01.007.

83 Kadoya, R., Kodama, Y., Matsumoto, K. et al. (2015a). Indirect positive effects of a sigma factor RpoN deletion on the lactate-based polymer production in *Escherichia coli*. *Bioengineered* 6 (5): 307–311. https://doi.org/10.1080/21655979.2015.1069449.

84 Kadoya, R., Matsumoto, K., Ooi, T. et al. (2015b). MtgA deletion-triggered cell enlargement of *Escherichia coli* for enhanced intracellular polyester accumulation. *PLoS One* 10 (6): 1–11. https://doi.org/10.1371/journal.pone.0125163.

85 Chen, G. and Jiang, X. (2018). Engineering microorganisms for improving polyhydroxyalkanoate biosynthesis. *Current Opinion in Biotechnology* 53: 20–25. https://doi.org/10.1016/j.copbio.2017.10.008.

86 Mosier, N., Wyman, C., Dale, B. et al. (2005). Features of promising technologies for pretreatment of lignocellulosic biomass. *Bioresource Technology* 96 (6): 673–686. https://doi.org/10.1016/j.biortech.2004.06.025.

87 Wyman, C.E., Dale, B.E., Elander, R.T. et al. (2005). Coordinated development of leading biomass pretreatment technologies. *Bioresource Technology* 96 (18): 1959–1966. https://doi.org/10.1016/j.biortech.2005.01.010.

88 Sun, J., Utsunomia, C., Sasaki, S. et al. (2016). Microbial production of poly(lactate-*co*-3-hydroxybutyrate) from hybrid *Miscanthus*-derived sugars. *Bioscience, Biotechnology, and Biochemistry* 80 (4): 818–820. https://doi.org/10.1080/09168451.2015.1116933.

89 Takisawa, K., Ooi, T., Matsumoto, K. et al. (2017). Xylose-based hydrolysate from *eucalyptus* extract as feedstock for poly(lactate-*co*-3-hydroxybutyrate) production in engineered *Escherichia coli*. *Process Biochemistry* 54: 102–105. https://doi.org/10.1016/j.procbio.2016.12.019.

90 Kadoya, R., Matsumoto, K., Takisawa, K. et al. (2018). Enhanced production of lactate-based polyesters in *Escherichia coli* from a mixture of glucose and xylose by Mlc-mediated catabolite derepression. *Journal of Bioscience and Bioengineering* 125 (4): 365–370. https://doi.org/10.1016/j.jbiosc.2017.11.003.

91 Salamanca-cardona, L., Ashe, C.S., Stipanovic, A.J. et al. (2014a). Enhanced production of polyhydroxyalkanoates (PHAs) from beechwood xylan by recombinant *Escherichia coli*. *Applied Microbiology and Biotechnology* 98 (2): 831–842. https://doi.org/10.1007/s00253-013-5398-4.

92 Salamanca-cardona, L., Scheel, R.A., Lundgren, B.R. et al. (2014b). Deletion of the *pflA* gene in *Escherichia coli* LS5218 and its effects on the production of polyhydroxyalkanoates using beechwood xylan as a feedstock. *Bioengineered* 5 (5): 284–287. https://doi.org/10.4161/bioe.29595.

93 Salamanca-Cardona, L., Scheel, R.A., Bergey, N.S. et al. (2016). Consolidated bioprocessing of poly(lactate-*co*-3-hydroxybutyrate) from xylan as a sole feedstock by genetically-engineered *Escherichia coli*. *Journal of Bioscience and Bioengineering* 122 (4): 406–414. https://doi.org/10.1016/j.jbiosc.2016.03.009.

94 Ishii, D., Takisawa, K., Matsumoto, K. et al. (2017). Effect of monomeric composition on the thermal, mechanical and crystalline properties of poly[(*R*)-lactate-*co*-(*R*)-3-hydroxybutyrate]. *Polymer* 122 (28): 169–173. https://doi.org/10.1016/j.polymer.2017.06.039.

95 Yamada, M., Matsumoto, K., Uramoto, S. et al. (2011). Lactate fraction dependent mechanical properties of semitransparent poly(lactate-*co*-3-hydroxybutyrate)s produced by control of lactyl-CoA monomer fluxes in recombinant *Escherichia coli*. *Journal of Biotechnology* 154 (4): 255–260. https://doi.org/10.1016/j.jbiotec.2011.05.011.

96 Zaman, H.U., Song, J.C., Park, L.S. et al. (2011). Poly(lactic acid) blends with desired end-use properties by addition of thermoplastic polyester elastomer and MDI. *Polymer Bulletin* 67 (1): 187–198. https://doi.org/10.1007/s00289-011-0448-5.

97 Sun, J., Matsumoto, K., Nduko, J.M. et al. (2014). Enzymatic characterization of a depolymerase from the isolated bacterium *Variovorax* sp. C34 that degrades poly(enriched lactate-*co*-3-hydroxybutyrate). *Polymer Degradation and Stability* 110: 44–49. https://doi.org/10.1016/j.polymdegradstab.2014.08.013.

98 Sun, J., Matsumoto, K., Tabata, Y. et al. (2015). Molecular weight-dependent degradation of D-lactate-containing polyesters by polyhydroxyalkanoate depolymerases from *Variovorax* sp. C34 and *Alcaligenes faecalis* T1. *Applied Microbiology and Biotechnology* 99 (22): 9555–9563. https://doi.org/10.1007/s00253-015-6756-1.

99 Matsumoto, K. and Taguchi, S. (2013a). Biosynthetic polyesters consisting of 2-hydroxyalkanoic acids: current challenges and unresolved questions. *Applied Microbiology and Biotechnology* 97 (18): 8011–8021. https://doi.org/10.1007/s00253-013-5120-6.

100 Matsumoto, K. and Taguchi, S. (2013b). Enzyme and metabolic engineering for the production of novel biopolymers: crossover of biological and chemical processes. *Current Opinion in Biotechnology* 24 (6): 1054–1060. https://doi.org/10.1016/j.copbio.2013.02.021.

101 Utsunomia, C. and Taguchi, S. (2018). *Green Polymer Chemistry: New Products, Processes, and Applications*, Chapter 4 (eds. H.N. Cheng, R.A. Gross and P.B. Smith). Washington, DC: American Chemical Society.

102 Utsunomia, C., Matsumoto, K., Date, S. et al. (2017a). Microbial secretion of lactate-enriched oligomers for efficient conversion into lactide: a biological shortcut to polylactide. *Journal of Bioscience and Bioengineering* 124 (2): 204–208. https://doi.org/10.1016/j.jbiosc.2017.03.002.

103 Utsunomia, C., Matsumoto, K., and Taguchi, S. (2017b). Microbial secretion of D-lactate-based oligomers. *ACS Sustainable Chemistry & Engineering* 5 (3): 2360–2367. https://doi.org/10.1021/acssuschemeng.6b02679.

104 Shozui, F., Matsumoto, K., Sasaki, T. et al. (2009). Engineering of polyhydroxyalkanoate synthase by Ser477X/Gln481X saturation mutagenesis for efficient production of 3-hydroxybutyrate-based copolyesters. *Applied Microbiology and Biotechnology* **84** (6): 1117–1124. https://doi.org/10.1007/s00253-009-2052-2.

9

Natural and Synthetic Biopolymers in Drug Delivery and Tissue Engineering

John D. Schneible[1], *Michael A. Daniele*[2], *and Stefano Menegatti*[1]

[1] *NC State University, Department of Chemical and Biomolecular Engineering, 911 Partners Way, Raleigh, 27695, USA*
[2] *NC State University and UNC – Chapel Hill, Joint Department of Biomedical Engineering, Chapel Hill, 27599, USA*

9.1 Introduction

Through the last three decades, hydrogels for biomedical applications have grown into a vast class of materials, featuring a large variety of ingredients, fabrication techniques, and uses [1–3]. Together with a massive body of literature, the value of hydrogels is illustrated by myriad commercial products, ranging from drug delivery to tissue engineering and regenerative medicine. Key to their success is the high biocompatibility and physicochemical diversity, which enable functional adaptation to many different physiological microenvironments [4, 5]. Herein, we survey the reports on peptides and polysaccharides, in particular focusing on chemical constituency, structure–function relations, and their impact on drug delivery and tissue engineering [6–12].

Peptide and polysaccharide hydrogels have gained popularity as biocompatible and biodegradable drug delivery vectors [13–18], as micro-/nanoparticles [19–21], micelles [22–24], and bulk hydrogels (Figure 9.1) [25–27]. Peptides have gained popularity for their ability to tune drug loading and release the payload in a stimuli-/environment-controlled manner [15, 17, 18]. Peptides can also provide accumulation and internalization of bioactive ingredients at target sites by acting as site-selective ligands and cell-penetrating vectors [28–34]. Similarly, polysaccharides have been extensively studied as drug delivery systems [35], owing to their wide availability [36–49], physicochemical diversity, and high number of reactive groups per chain, which enable facile chemical modification for tuning drug loading and release [35, 50, 51], stability [52, 53], and tissue adhesion [54].

These materials have also been utilized to engineer scaffolds for tissue engineering and wound healing [55–60]. A wide variety of fabrication techniques (e.g. phase separation [61–64], 3D printing [65, 66], supercritical fluid technology [67, 68], porogen leaching [64, 69–72], and micro-patterning techniques [64, 73–75]) have been combined with a host of crosslinking and functionalization chemistries (e.g. mechanical robustness, display and storage of growth

Biopolymers for Biomedical and Biotechnological Applications, First Edition.
Edited by Bernd H. A. Rehm and M. Fata Moradali.

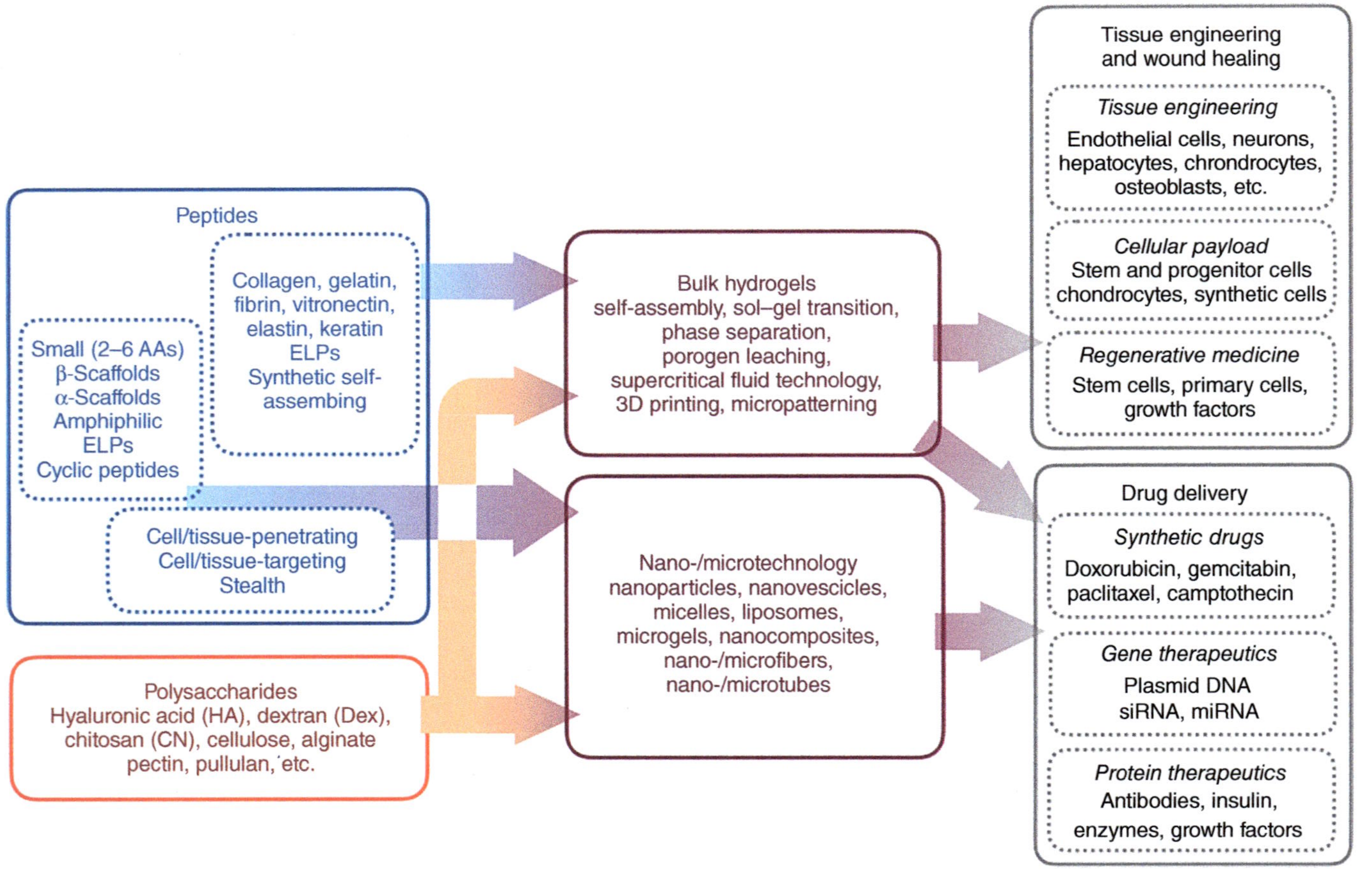

Figure 9.1 Overview of the substrate families, fabrication technologies, and applications of peptides, proteins, and polysaccharides in biomedical engineering.

factors, and stimuli responsiveness [76]) to generate hydrogels with biomimetic morphology, mechanical properties, and biochemical functionality. These systems have been demonstrated for tissue engineering and repair *in vivo*, as well as in surgical applications as filling materials and implants [77–82]. While an exhaustive treatment of these topics is not permitted by the vastness of the field, we will nonetheless attempt to provide a portrait of the technologies and applications centered around peptides and polysaccharides that hold promise to improve human health.

9.2 Synthetic and Natural Substrates

Polysaccharides and polypeptides can be derived from renewable resources or be obtained synthetically [83, 84]. The polysaccharides of bioengineering value include plant exudates and algal derivatives (alginate, galactans, carrageenan), animal (chitin, chitosan [CN], glycosaminoglycan, hyaluronic acid [HA]), and microbial (dextran [Dex], gellan gum, pullulan, xanthan gum, and cellulose) polysaccharides (Figure 9.2) [85–87]. Their peptide counterparts include collagen, elastin, silk fibroin, fibrin, fibrinogen, fibronectin, gelatin, laminin, vitronectin, keratin, silk, and adhesive proteins (Figure 9.3) [88, 89]. The introduction of synthetic routes to produce peptides and oligosaccharides has impressed considerable acceleration to the science and engineering of biomedical materials. Using commercially available reagents and equipment, peptides of up to 50 residues can be routinely prepared with high yield and reproducibility [90]. *De novo* synthesis enables a superior control of chemical identity compared with natural materials, which often lack chemical definition and can contain harmful contaminants that are difficult to define and remove. Further, synthetic biopolymers can introduce non-natural moieties that provide advantageous functionalities, such as conjugation to other biomolecules, adhesion to inorganic materials (e.g. metal and silica), and radioactivity or fluorescence [91]. The synthesis of polysaccharides [92–94], while conceptually similar to that of peptides, is considerably more challenging, due to the stereospecific linkage of sugar monomers. The introduction of novel orthogonal protection/deprotection schemes [95–98] has enabled the synthesis of oligosaccharides and condensed polysaccharides, which have been successfully utilized in as diagnostics [99, 100], vectors for drug delivery [101, 102] and gene therapy [103–105], and tissue engineering [14].

9.3 Applications of Natural and Synthetic Polypeptides

9.3.1 Drug Delivery Vehicles

Hydrogels for drug delivery generally comprise one or more active ingredients either dispersed into or covalently bound to a polymer [25, 106, 107] whose properties determine the pharmacokinetics of the therapeutic payload [108–113]. The drug is released either by diffusion through the polymer [27], by

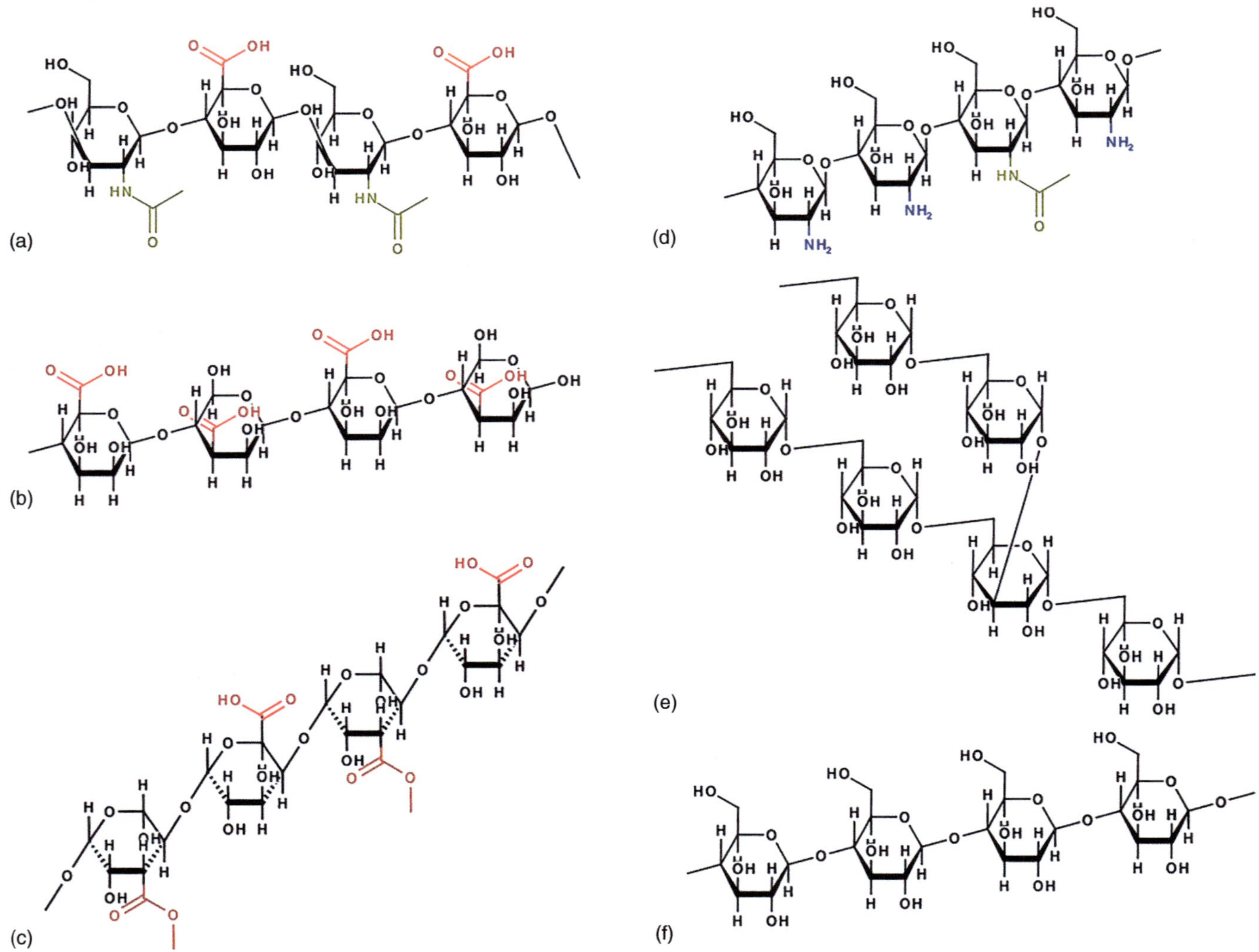

Figure 9.2 Structures of polysaccharides utilized as substrates in drug delivery, tissue engineering, and regenerative medicine: (a) hyaluronic acid, (b) alginic acid, (c) pectin, (d) chitosan, (e) dextran, and (f) cellulose. Native functional groups used for modification are highlighted, including carboxyl and primary amine groups.

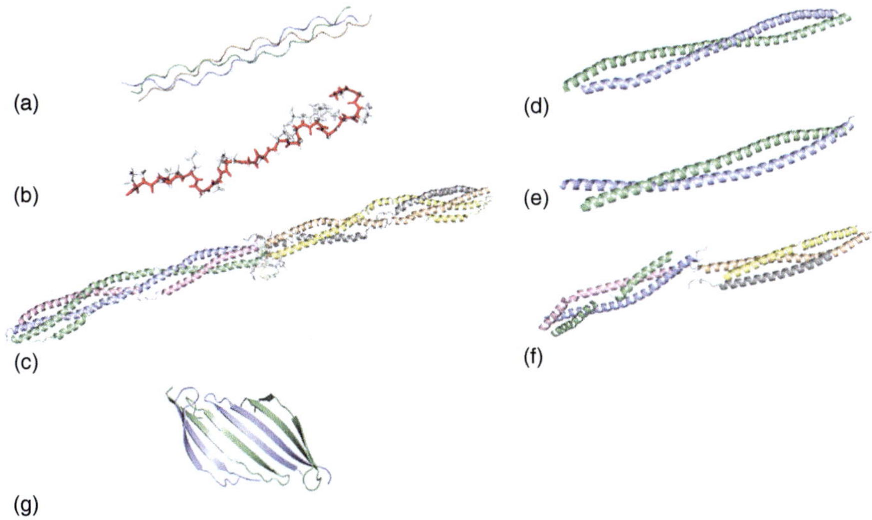

Figure 9.3 Structures of natural polypeptides utilized as substrates in drug delivery, tissue engineering, and regenerative medicine: (a) collagen (Protein Data Bank, PDB ID: 1BKV), (b) gelatin, (c) fibrinogen (PDB ID: 3GHG), (d) keratin 1/10 (PDB ID: 4ZRY), (e) keratin 5/14 (PDB ID: 3TNU), (f) laminin (PDB ID: 5XAU), and (g) silk fibroin (PDB ID: 3UA0). The higher-order morphology directly impact functionality, e.g. tertiary helical structure of collagen, especially in tissue growth and regeneration.

degradation of the polymer [30, 31], or by loss of the supramolecular structure of the carrier [114, 115]. Peptides have been extensively utilized as building blocks for self-assembling 3D nanostructures (e.g. nanovesicles, nanofibers, nanotubes, nanoribbons) and hydrogels for drug delivery [12, 15, 17, 18, 116–119]. Notably, peptide-based systems are superior to traditional vehicles (e.g. liposomes), as they confer higher drug loading and stabilization, lower leakage, sustained release, and specific targeting and permeation to cells [120]. Further, the behavior of self-assembled peptide structures can be fine-tuned through the amino acid makeup [117, 121–123] or adding polyelectrolytes or templates [124, 125] to become responsive to pH [126, 127], ionic strength [128, 129], and temperature [130, 131] of the environments. Peptide-based drug delivery vehicles can be classified into (Figure 9.4):

1. *Ultra-small peptides*: These inexpensive short peptide sequences are widely utilized to form self-assembled hydrogels. For example, KLD and EAK motifs have been used to construct hydrogels capable of controlled release of anticancer drugs such as doxorubicin (DOX) [132] and ellipticine (EPT) [133]. Environment-responsive small peptides, such as V_6K_2 and L_6K_2, are of particular interest for their ability to self-assemble in structures capable of morphing from nanotubes to nanovesicles [134]. These have been utilized for delivering anticancer drugs (DOX and paclitaxel [PTX]) [135] and gene therapeutics [134, 136, 137]. Similar nanovesicles assembled with dipeptides VE, LE, and $G_{n\text{-p}}$E ($G_{n\text{-p}}$: *n*-propyl glycine) were utilized to encapsulate DOX and other biomolecules to be released in response to calcium ions within the

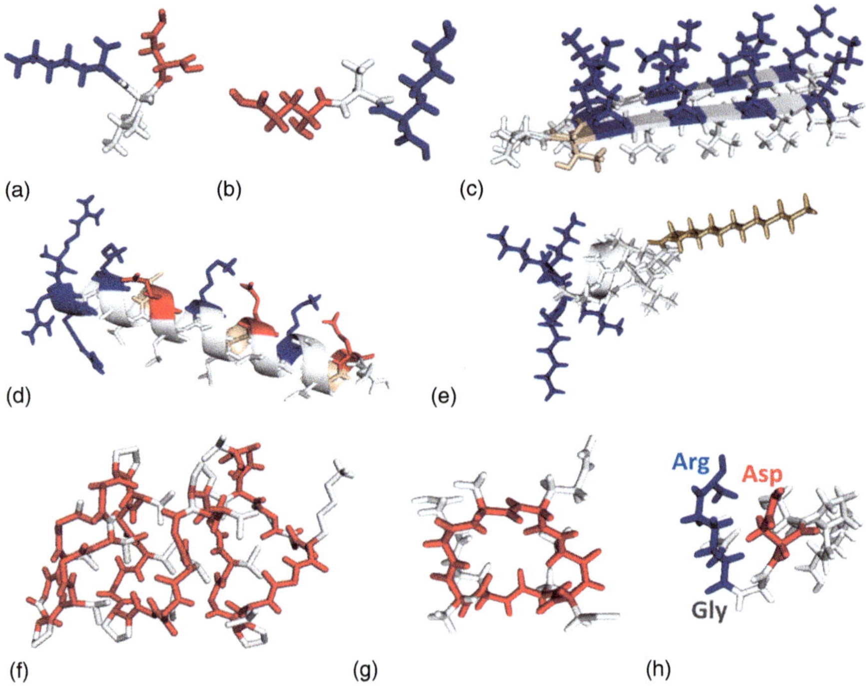

Figure 9.4 Structures of synthetic polypeptides utilized as substrates in drug delivery, tissue engineering, and regenerative medicine: (a) KLD (Lys-Leu-Asp), (b) EAK (Glu-Ala-Lys), (c) MAX1 (PDB ID: 2N1E), (d) $R_3(AKALTEL)_3$, (e) peptide amphiphile $[C_{16}]$-VVVAAAKKKK, (f) ELP (VKV_6F), (g) cyclic peptide $Q^DAE^DAQ^DAC^DA$, (h) G-RGD-S.

target cells [138]. Tetrapeptides GAIL and GFIL served as building blocks to construct hydrogels with thermo- and pH-responsive behavior, which were utilized to encapsulate DOX and afforded controlled release with 90% release of DOX after 45 hours [139], showing good therapeutic effect together and non-detectable side effects.

2. β *scaffolds*: β-Sheet structures originating from repeated peptide domains have been observed in several natural proteins such as silk fibroin [140–142], eggshell proteins [143], and adenovirus fiber protein [144]. Synthetic peptides comprising alternated hydrophobic and charged amino acids (e.g. EAK16-II, AEAEAKAKAEAEAKAK) [133] also assemble into β-sheet structures whose properties can be controlled by tuning the ratio of enantiomeric peptides. These materials have been extensively utilized for drug delivery [145, 146], owing to their high loading capacity for hydrophobic drugs and their tissue penetration ability [147–152]. Notable examples include β-sheet fibrils of FEFQFNFK for sustained release of EPT to MCF-7 and A549 cells [153]; hydrogel-forming β-sheet peptide KRRMKWKKK for sustained release of DOX and Smac – (second mitochondria-derived activator of caspases)-derived proapoptotic peptide (SDPP) [132]; β-hairpin peptide $VKVKVKVKV^DPPTKVEVKVKV$ [154], which self-assembles in response to an increase in the ionic strength of the medium [155], to deliver poorly

soluble anticancer drugs [156]; or Nap-GFFYGRGD, which was used to encapsulate and release curcumin to HepG2 liver carcinoma cells in a mouse model [157]. Among β-sheet-forming synthetic peptides, the most widely utilized sequences are RADA16-I16 and RADA16-II16. These peptides form nanofibers of 10–60 nm in size and have been utilized for the sustained release of cytotoxic drugs [158, 159], in particular anticancer therapeutics [146, 160, 161], protein therapeutics, and growth factors [10, 162–168]. Prolonged release of growth factors makes RADA16-based substrates ideal for cell culture [164, 169], tissue repair [170], and tissue engineering [171, 172], as will be discussed later.

3. α *scaffolds*: α-Helical peptides consist of repeats of the heptad $X_1X_2X_3X_4X_5X_6X_7$, wherein X_1 and X_4 are hydrophobic amino acids and X_5 and X_7 are charged amino acids [173–175]. The hydrophobic interaction between X_1 and X_4 and the electrostatic interaction between X_5 and X_7 confer these peptides with the coiled α-helical structure [176–178]. Positions X_2 and X_3 can be replaced with amino acids that provide weaker interactions, resulting in more flexible α-helices that form physical hydrogels, featuring smaller and softer bundles of fibers [179]. This rationale is exemplified by numerous examples, such as the peptides A_3(LETLAKA)$_3$ [180], αFFP(QQLAREL)$_4$ [181], and R_3(AKALTEL)$_3$ [182], the hydrogelating self-assembling fiber (hSAF) peptides [179], the 30-mer GALA peptide [183], and the coiled-coil pentamers (e.g. QLAREL(QQLAREL)$_4$) [184–187]. α-Helical peptides have been extensively characterized for their ability to form supramolecular [18] structures (e.g. nanofibers and nanofilaments) [188–191] but have only minimally been applied for drug delivery with most applications focusing on gene delivery, showing efficient gene loading and transfection in target cells [183, 192, 193].
4. *Peptide amphiphiles* (*APs*): These diblock biopolymers comprise a hydrophobic peptide tail, rich in A, G, L, and F amino acids, and a hydrophilic bioactive peptide head, rich in D, E, H, and R amino acids. These sequences self-assemble into nanostructured hydrogels that demonstrated great potential in drug delivery [12, 194]. For example, the lipid-like peptides G_4DD, G_6DD, G_8DD, A_6D, A_6K, and KA_4 sequences self-assemble into nanostructures at their critical aggregation concentration and have been utilized as drug delivery vectors [134, 194, 195]. When the hydrophobic tail is replaced by one or multiple alkyl chains or an amphiphilic surfactant [12, 196], the resulting APs self-assemble into cylindrical fibrils whose surface displays a high density of peptide antigens [197]. Palmitoyl-GGGAAAR and palmitoyl-GGGAAAKRK, for example, assemble into nanofibers that have been utilized for drug delivery to the brain [198]; while capable of penetrating the neurons, these nanofibers were found to be degraded by plasma enzymes, which limits accumulation in the brain and reduces potential side toxicity; in another example, fibrils of palmitoyl-YDAGFLR encapsulating a dalargin prodrug were utilized for crossing the blood–brain barrier and controlling the uptake and release into brain tissue of mice models [199]. Other examples include nanofiber-based hydrogel of palmitic acid-GTAGGLIGQRGDS for cisplatin release, [200] [C_{12}]-PPPPRRRR cell-penetrating nanofibers

for delivery of DOX and PTX in MCF-7 cells [201], or core–shell micelles of [C_{16}]-VVVAAAKKKK-folate for folate targeted delivery of DOX to MDA-MB-231 cells [202]. Notably, a number of AP-based vectors have been utilized for controlled delivery of biotherapeutics, such as growth factors and gene therapy products [203–206]. Among the latter, notable examples include a cholesterol–peptide conjugate for DNA delivery into HepG2 and HEK293 cells [207], capable of amino acid sequence-dependent gene expression efficiency [208]; a multiblock peptide comprising a hydrophobic cell membrane-binding sequence linked to a cell-permeating sequence (KKKRKV) and capable of high-efficiency complexing and intracellular trafficking of therapeutic small interfering RNA (siRNA) to a variety of cell lines [209–211]; and the peptide Ac-$(AF)_6$-H_5K_{15}, which self-assembles into a micelle carrier that co-delivered a synergistic combination of DOX and p53 genes into HepG2 human liver carcinoma cells [212]. Another notable application of peptide amphiphiles (PAs) is drug delivery to bones. For example, oligo-aspartates that bind hydroxyapatite have in fact been shown to promote accumulation of small drugs in bones *in vivo* [213, 214]; similarly, peptide nanofibers displaying phosphorylated serine and tyrosine ($P(S_{PO4}F)_5S_{PO4}P$ and $P(Y_{PO4}F)_5Y_{PO4}P$) were used for surface nucleation of hydroxyapatite [215–217] to improve bone mineralization [218–220]. Finally, PAs have also been utilized to engineer bio-responsive delivery systems. For example, palmitoyl-GTAGLIGQRGDS were used to construct a cisplatin-loaded nanofiber gel decorated with RGD peptides for cancer targeting; the gel matrix was degraded by matrix metalloproteinase-2 (MMP2), thus releasing cisplatin at the tumor site [200]. Similarly, a KRRASVAGK-[C_{12}]-based hydrogel containing a substrate for protein kinase A (PKA) was developed for PKA-responsive delivery of DOX to MDA-MB-231 cancer cells [221].

5. *Elastin-like polypeptides* (*ELPs*): ELPs are a renowned class of natural polymers that exhibit thermo-responsive behavior. ELPs are constituted by repeats of VPGXG, wherein X is a variable amino acid that dictates the phase transition temperature and the behavior of the resulting peptide-rich phase [222, 223]. This characteristic has been extensively utilized to engineer injectable formulations that, upon exposure to the body temperature, form coacervate structures for sustained local delivery [224–226]. Of particular interest is the ability of ELPs to efficiently encapsulate drugs that present high hydrophobicity or abundant hydrogen bonding groups and are traditionally challenging to formulate [223]. Combinations of ELPs and anticancer drugs (e.g. DOX, PTX, camptothecin [CPT], gemcitabine [GEM]) [226–228] form micellar structures that have been demonstrated for chemotherapeutic treatment both *in vitro* and *in vivo* systems [229, 230]. To facilitate cellular uptake of the therapeutic payload, fusions of ELPs with cell-penetrating peptides (e.g. CPPs-Bac, Tat, and SynB1) have been genetically engineered [226, 230]. Conjugates or aggregates of these peptides with anticancer drugs have shown superior accumulation and penetration into cancer tissues compared with non-targeted systems and free drugs, with remarkable reduction of side effects [231, 232].

6. *Cyclic peptides* (*CPs*): Finally, CPs that form self-assembled nanotubes via hydrogen bonding have been also utilized as drug delivery vehicles [233–235]. This is owed to their higher mechanical and biochemical stability and higher tissue permeability compared with structures formed by their linear counterparts [233–235]. Among the numerous examples, cyclic peptide nanotubes of $(WDL)_4QDL$ afforded a high level of 5-fluorouracil (FU) penetration in tumor cells, resulting in potent anticancer activity [236]; self-assembled cyclic peptide nanotubes with the sequence $Q^DAE^DAQ^DAC^DA$ loaded with DOX have been shown to increase drug uptake and cytotoxic effect in human breast cancer cells, demonstrating potential to overcome multidrug resistance [237].

9.3.2 Targeting Agents

Targeted drug delivery have been shown to impact pharmacokinetics and improve therapeutic outcomes [238–242]. Peptides are playing a growing role as targeting agents displayed onto a drug carriers for the treatments of lung [243–250], colon [251–255], skin [256–260], and breast cancer [261–268]. Of growing interest are tumor-targeting peptides (TTPs) [269, 270], which provide superior targeting activity compared with naked drug delivery systems by recognizing the proteins expressed by tumor-associated endothelium while showing no affinity for healthy endothelium [243]. Liposomes are a popular drug delivery modality that have benefited from the incorporation of TTPs, and the list of examples is endless. PEGylated liposomes functionalized with PIVO-8 and PIVO-24 afforded a twofold higher accumulation of DOX in subcutaneous murine lung cancer compared with non-targeted liposomes [243]. Liposomes decorated with (immunoglobulin heavy-chain binding protein) targeting peptide WIFPWIQL showed a fourfold higher accumulation within neovascular cells compared with prostate and colon cancer cells, indicating potential for blocking tumor invasion by attacking its vascularization [251]. Similarly, liposomes functionalized with peptides GNGRG and APRPG attacked tumor vascularization by targeting the aminopeptidase N (APN) onto tumor-associated endothelial cells [252, 271, 272]; functionalization with GNGRG and APRPG not only affords stronger antiproliferative effect, but also reduced drug accumulation in the spleen, liver, kidneys, lung, and heart. The disulfide-cyclic variant CNGRCGV and the linear peptide CPRECESA, which target aminopeptidase A (APA), have also been utilized to functionalize liposomes to target the tumor-associated vasculature [273–275]. Liposomes decorated with peptides targeting APN and APA have been utilized to deliver DOX to endothelial and perivascular cells that support neuroblastoma tumor in a mouse model, showing therapeutic efficacy [276]. Peptides that target hormones and growth factors are also widely utilized TTPs. Among the former, we recognize the somatostatin-binding peptide FCFWKTCT [247], the luteinizing hormone-releasing hormone (LHRH)-binding peptide Pyr-HWSYGLRPG [263], cholecystokinin-binding GDYMGWMDF [260], and neurotensin-targeting RRPYIL [254]. FCYWKTCT-functionalized liposomes incorporating metal chelating lipids were used to deliver radioactive $^{111}InCl_3$ to AR42J pancreatic cancer cells [250]. Growth factor-binding peptides include VEGF-targeting WHSDMEWWYLLG, neuropilin-1-targeting ATWLPPR, αV integrin-targeting GARYCRGDCFDG, and PDGF receptor-targeting GCR-GRRST [258, 259]. Liposomes functionalized with peptide KRFKQDGGWSHW, which binds and activates tumor growth factor β (TGF-β) [253], have been shown to halt tumor progression in mice compared with non-targeting liposomes. In another example, the peptide TDSILRSYDWTY, which specifically targets non-small cell lung cancer (NSCLC) cells [244], was used to functionalize DOX-loaded liposomes for treating xenografted mice, affording a fivefold higher DOX delivery than free drug and twofold higher delivery than traditional liposomes. DOX-loaded liposomes functionalized with a similar NSCLC-targeting peptide, CSNIDARAC, were shown to halt tumor growth and produce minimal accumulation in the liver, lung, spleen, and heart in comparison with free DOX and non-targeting DOX-loaded liposomes [248].

A special place is occupied by integrin-binding peptides, to date the most utilized targeting moieties in the history of drug delivery, first and foremost RGD-based sequences. Innumerable RGD-functionalized systems have been developed as anticancer nanomedicines, carrying either traditional chemotherapeutics [277, 278] or gene drugs [279]. Lately, cyclic variants of RGD have been employed to functionalize liposomes for drug delivery against breast cancer [280–282], colon cancer [261, 283, 284], and melanoma [285], as well as diagnostics against lung tumors [286]. A notable RGD-derivative targeting α5β1 integrin, KSSPHSRNSGSGSGSGSGRGDSP [287–290], has utilized to confer liposomes and polymersomes with both cell targeting and penetrating activities; these formulations were effective in delivering diverse drugs (5-FU, DOX, tumor necrosis factor-α, siRNA, and plasmid DNA) to colon cancer cells [261, 267, 289], prostate cancer cells [291, 292], and porcine Langerhans islets [293]. It is worth noting that linear and cyclic RGD peptides have been extensively utilized for developing synthetic mimetics of cells, especially platelets. In an early work, RGD-decorated nanoparticles were shown to reduce bleeding time by half after IV administration in a rat model [294]; later, liposomes, albumin particles, latex particles, and poly-L-lysine-poly(lactide-*co*-glycolide) (PLL-PLGA) nanoparticles functionalized with RGD peptides were also developed [294–303]. Recently, liposomes functionalized with von Willebrand factor-binding peptides (VBPs), collagen-binding peptides (CBPs), and GRGDS peptides were shown to be highly efficacious in both microfluidic and murine wound models [304]. Lastly, layer-by-layer, discoidal, soft nanoparticles coated with poly(amidoamine) (PAMAM) dendrimers functionalized with VBP, CBP, or GRGDS peptides have demonstrated a remarkable 65% reduction in bleeding time relative to natural platelets, the highest reported value to date [305].

9.3.3 Cell-Permeating Peptides

Together with cell-targeting peptides, cell-permeating peptides (CPPs) have been extensively integrated with drug delivery systems [34, 238, 306, 307] to promote the internalization of cargoes such as siRNA [308–310], nucleic acids [311, 312], small molecule therapeutic agents [313, 314], proteins [315, 316], peptides [317, 318], and contrast agents [319]. Recently, novel CPPs have also been developed for nuclear localization [320–324], organelle and cytoplasmic protein targeting [325–331], and stimuli-responsive delivery [227, 232, 332]. Protein-derived sequences were among first CPPs developed. Among these, the most notable are the Tat-derived sequence Tat_{49-57} [333, 334], $penetratin_{43-58}$ [335, 336], the herpes virus-derived VP22 [337], and the Kaposi fibroblast growth factor (FGF) signal sequences [338–340], which derive from viral capsid proteins and have been utilized for intracellular delivery of antibodies [341, 342], peptides [343–348], proteins [349–353], nucleic acids [354–356], peptide nucleic acids (PNAs) [357], and drugs [358, 359]. A plethora of synthetic sequences have also been developed [32, 34], including cationic [360], amphiphilic, and hydrophobic CPPs and skin-permeating peptides (SPPs). Cationic sequences, such as poly-R and $(WR)_n$ $(n > 4)$ [361, 362], are among the first synthetic CPPs [363–367]; these CPPs have been reported to induce a wide variety of side effects, chiefly cell membrane disruption, which comprises cell viability and leads to severe cell toxicity and tissue irritation [368–370]. Amphiphilic CPPs contain both polar and nonpolar residues [371–373]. Some are derived from proteins, such as pVEC, ARF(1–22), BPrPr(1–28), MPG, penetratin, and CADY, while others are synthetic, like proline-rich peptides, and the model amphipathic peptide (MAP) and its derivatives MAP17, MAP12, GALA, and KALA [374, 375]. Amphipathic CPPs can be α-helical, with the coil featuring a hydrophobic side and a charged or polar side, or β-sheet, comprising one hydrophobic and one hydrophilic string of amino acids. Conjugates of Tat, penetratin, CADY, and poly-R peptides with DOX have been developed for treating different cell lines *in vitro* and *in vivo* [358, 376–385]. In other studies, peptides YTA2 and YTA4 have been conjugated to methotrexate (MTX) and fluorescent probes for treatment and imaging of breast cancer cells [313, 386, 387]. Amphiphilic CPPs MPG8 and PEP3 have been utilized for delivering short regulatory oligonucleotides in both cancer and healthy cell lines to modulate gene expression [320, 388, 389]. Finally, hydrophobic CPPs form a smaller class that includes stapled peptides, prenylated peptides, and pepducins [390–392], the signal sequence for integrin β3 (VTVLALGALAGVGVG), and Kaposi FGF (AAVALLPAVLLALLAP) [393], whose potential for drug delivery has been marginally studied. Finally, SPPs have been developed, the majority of which has been identified by screening phage-display libraries specifically to achieve transdermal drug delivery. Examples include peptide DRTTLTN, IMT-P8, and SPACE™. Peptide DRTTLTN reaches the epidermis via fluidization of the lipids and extension of the keratin fibers in the *stratum corneum* (SC) and has been applied for transdermal delivery of unfractionated heparin (UFH), yielding a remarkable increase of UFH permeation compared with UFH alone [394]. IMT-P8, an arginine-rich SPP [395, 396], penetrates the SC, permeates the epidermis, and accumulates hair follicles; fusions of this peptide with green fluorescent protein (GFP) and proapoptotic peptide KLA were evaluated on murine skin *in vivo* [397] and found to achieve higher skin penetration compared with Tat, inducing localization in mitochondria and cell death.

Finally, the disulfide-cyclic "skin permeating and cell entering" (SPACE) peptide ACHSALTKHCG has been extensively characterized for dermal delivery of siRNA, small molecules, proteins, and biopolymers. Conjugates of SPACE and siRNA targeting interleukin-10 and GAPDH have been shown to absorb in skin porcine with a 10-fold increase compared with siRNA solution, achieving effective knockdown (60–80%) of the corresponding proteins both *in vitro* and *in vivo*. Ethosomal carriers constructed with SPACE-phospholipid conjugates have been utilized to deliver HA across both human and porcine skin *in vitro* and murine skin *in vivo*, with significant increments compared with aqueous HA [398]. SPACE was also found effective in permeating small drugs without covalent conjugation. In one example, a hydroalcoholic solution of SPACE and corticosterone enhanced epidermal permeation compared with Cortisone™ cream, both *in vitro* on porcine skin and in hairless mice models [399]. In another study, SPACE was utilized to promote *in vitro* dermal absorption of cyclosporine A (CsA), achieving a ninefold increase compared with a hydroalcoholic solution of CsA [400]. Recently, a computational algorithm was developed for the *de novo* identification of disulfide-cyclic peptides that enhance transdermal delivery of CsA. Selected heptameric sequences ACSATLQHSCG, ACSLTVNWNCG, ACTSTGRNACG, and ACSASTNHNCG and octameric peptide ACNAHQARSTCG were tested *in vitro* and found to afford a CsA permeation comparable with or higher than SPACE, without triggering any toxic effect on skin keratinocytes, even at high concentration [401].

9.3.4 Peptides in Tissue Engineering and Regenerative Medicine

Hydrogels produced with self-assembling proteins and peptides have been extensively applied as scaffolds for tissue engineering and regenerative medicine (Figure 9.5) [10, 162, 402–414]. Among the many building blocks utilized in these efforts, we must mention:

1. *Collagen, gelatin, and fibrin*: These proteins, either produced recombinantly [415–423] or obtained from mammalian sources [424–428], are the most widely utilized biologically derived substrates for tissue engineering [89]. Collagen, the main component of connective tissue, tendon, skin, and cartilage, is known for its biocompatibility, bioactivity, and wide availability. Collagen is often modified by covalent conjugation or physical adsorption of growth factors to promote regenerative outcomes, such as neovascularization and angiogenesis [429–433] or cartilage repair [434, 435], osteogenesis [436–438], and infiltration of human fibroblasts to promote cellularization and vascularization of implants [439–441]. Gelatin, i.e. hydrolyzed and denatured collagen, is also frequently functionalized with growth factors (e.g. TGF-β, bone morphogenetic proteins [BMPs], and VEGF) and utilized as substrate for cartilage regeneration [442–445], angiogenesis, and osteogenesis. [446, 447] Fibrin has been frequently utilized as matrix for the immobilization of growth factors FGF and BMP and the conditioning of platelets, leukocytes, fibroblasts, and endothelial cells for bone tissue engineering and regeneration [448–451].
2. *Elastin-like polypeptides (ELPs)*: ELPs have gained great popularity in tissue engineering [452–455], owing to their biocompatibility and facile physicochemical customization. Their sequence can in fact be tailored to enable crosslinking [456–460] and functionalization with biomolecular cues

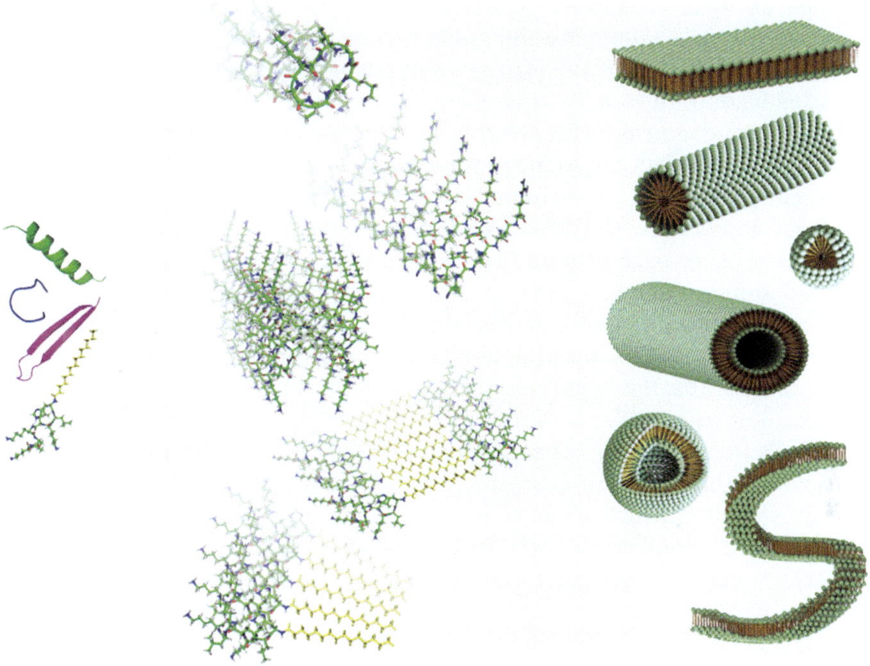

Figure 9.5 Self-assembly of monomer peptides (random coil, α-helix, β-sheet, and peptide amphiphile) into ordered assemblies and subsequently into nanostructures (nanosheet, nanotube, micelle, and nanofilament).

[461–463] and to adjust their thermomechanical properties for easy processing into gels [459, 460, 464–467], films [458, 463], foams [467], and fibers [468, 469]. Non-crosslinked ELPs have been shown to be ideal substrates for cartilage engineering [470], as they promote biosynthesis and retention of chondrocyte-associated extracellular matrix (ECM) and long-term phenotype maintenance of chondrocytes and human adipose-derived adult stem (hADAS) cells [471, 472]. Soft, non-crosslinked ELPs are not ideal for load-bearing materials to promote cartilage regeneration. Crosslinked ELPs, on the other hand, form mechanically robust hydrogels that are capable of hosting primary chondrocytes and promote formation of cartilage matrix [459] and have been used as injectable 3D scaffolds for cartilage regeneration in animal models [457, 460, 462, 473–475]. Crosslinked ELP hydrogels have also been employed in engineering intervertebral discs that were successfully implanted in animal models [476, 477], where they promoted formation of new cartilage without significant inflammation and local degradation; analogous ELP hydrogels have been developed to engineer load-bearing cartilage tissue for defect filling [475]. Notably, the mechanical properties of ELPs can also be controlled thermally by taking advantage of sequence-dependent thermo-responsive behavior. Upon temperature-induced gelation, in fact, ELP hydrogels present a rheological behavior similar to that of strong collagen and hyaluronan substrates, thereby serving as ideal substrates for cartilage engineering [471, 478, 479] and implants for *in vivo* applications [480]. These ELP hydrogels exhibit high durability and biostability *in vivo*, without triggering inflammatory responses or calcification [481–483]. Being produced by recombinant techniques, ELPs can also be fused with proteins, such as growth factors and cytokines, to better direct cell adhesion, proliferation, and differentiation [484, 485]. For example, an ELP-interleukin-1 receptor antagonist (ELP-IL-1Ra) fusion was developed for intra-articular treatment of osteoarthritis [486, 487]; numerous other examples of ELP-based fusions with growth factors (e.g. soluble tumor necrosis factor receptor II [sTNFRII], epidermal growth factor, and oncogene inhibitors) have been developed for anti-inflammatory use, cellularization of surgical implants, or halting the progress of solid tumors [227, 345, 346, 488]. ELPs can also be fused with cell-binding small peptide domains [489]. Numerous examples are reported of crosslinked ELP hydrogels displaying fibronectin-derived REDV, laminin-derived IKVAV, or integrin-binding RGD motifs [463, 485, 489, 490], utilized as mimetics of the ECM. These hydrogels have been utilized as substrates for fibroblasts, platelets, and endothelial cells to engineer small diameter vascular grafts that effectively integrate with host tissues [458, 461–463, 484, 485, 489–492], as well as stem cells and primary chondrocytes for cartilage and intervertebral disc engineering [457–459, 462, 466, 483–485, 489, 490, 493, 494]. Notably, ELPs can also be mixed with synthetic polymers and formed into processed materials. For example, ELP-based nonwoven nanofiber substrates have been developed for vascular engineering [481, 483] and ocular tissue engineering [495, 496]. Hybrids of ELPs and polyacrylic acid (PAA) and polyethylenimine (PEI) have been processed into substrates for culturing and controlling the differentiation of primary hepatocytes [497]; notably, ELP-PAA and ELP-PEI trigger high production of both urea and albumin in hepatocytes, proving to be ideal substrates to culture hepatocytes for liver regeneration [497, 498].

3. *Synthetic peptides*: [499] Synthetic peptides have gathered significant momentum as affordable and tailorable building blocks for biomimetic hydrogels for tissue engineering and reconstitution [475, 494, 500–503]. A plethora of sequences are known that (i) self-assemble into structures with customized physiochemical properties, most notably stimuli responsiveness to environmental conditions; (ii) can be crosslinked *in vitro* and *in vivo* by chemical, enzymatic, or electromagnetic triggers; or (iii) can be combined with biopolymers, resulting in ECM-mimetic matrices with controlled morphology, biocompatibility, and biodegradability [459, 466, 494, 499, 500, 502, 504–510]. Among the most studied synthetic peptides are RAD16-I and RAD16-II, which self-assemble into ribbons and fibrils [511] that form highly porous hydrogels with high water content (99–99.5%) [162, 512, 513]. Their high porosity favors migration of nutrients, growth factors, and oxygen [403, 514], making these materials ideal substrates to sustain growth and differentiation of a variety of mammalian cells [166, 170, 402, 514–527]. Primarily, these peptides have been used in cardiovascular engineering applications [166, 514, 519, 522, 526]. Examples include RAD16-I hydrogels functionalized with laminin-derived peptides YIGSR and RYVVLPR

to culture aortic endothelial cells into tubules [528]; a RAD16-I hydrogel functionalized with a VEGF agonist peptide and RGD peptides to grow and differentiate human umbilical vein endothelial cells (HUVECs) toward tubulogenic phenotype [529, 530]; and a RAD16-II scaffold to engineer murine neonatal cardiomyocytes, in combination with endothelial cells, into ventricular tissue [514, 531]. Numerous studies have also demonstrated the potential of these materials for *in vivo* applications. For example, RAD16-II nanofibers loaded with mesenchymal stem cells (MSCs) were injected into a rat model of myocardial infarction, where they promoted proliferation and differentiation of MSCs into cardiomyocytes to reduce ischemic heart tissue and restore cardiac function [519]. In another example, a RAD16-I hydrogel hosting a combination of skeletal myoblasts and adipose-derived MSCs achieved remarkable reduction of necrotic tissue and reformation of functional heart muscle in a murine model of infarcted myocardium [522]. A RAD16-II hydrogel loaded with insulin-like growth factor 1 (IGF-1) [163] was injected in the left ventricle of mouse models, where it became rapidly populated by capillary-forming endothelial cells and smooth muscle cells that formed functional vessel-like structures and arterioles [514], resulting in improved heart functionality within 30 days upon injection. More recently, the RAD16-I and RAD16-II have demonstrated applicability in neural tissue engineering. RAD16-I hydrogels mixed with fibronectin, laminin, collagen, and bone marrow-homing peptides have been utilized as 3D scaffolds for adhesion and differentiation of neural stem cells (NSCs) [513, 532] and engineering neurite outgrowth and synaptic brain tissue [515]. A RAD16-I hydrogel with brain tissue-like softness was functionalized with laminin-derived peptide IKVAV to guide the migration, adhesion, and neuronal differentiation of NSCs to reconstruct injured brain tissues [533]. Further, several reports have illustrated the potential use of injectable cell-loaded RAD-16-based hydrogels for neuronal and brain tissue repair [514, 530, 534, 535]. For example, an NSC-loaded hydrogel implanted in a rat model of damaged cerebral neocortex yielded a notable regeneration in brain tissue [533]. Similar RAD16-II microporous hydrogels have been utilized to achieve homogeneous encapsulation and growth of neuronal cells for repairing damaged optic nerves [534], spinal cords [517], and brain injuries *in vivo* [155, 402, 534–536]. Many other fibril-forming β-sheet peptides and α-helical peptides have been extensively utilized to engineer 3D substrates for mammalian cells [155, 402, 405, 407, 410, 537]. Among hydrogel-forming β-sheet peptides, P11 sequences are noted for their biocompatibility and non-immunogenicity and ability to self-assemble under desired conditions of pH and ionic strength [538–541]. For example, a RGD-decorated hydrogel constructed with P11-2 (Ac-QQKFQFQFEQQ) was developed as substrate with adjustable stiffness for HUVECs [542, 543] and human dermal fibroblasts [544]; the peptide/water and RGD/non-RGD peptide ratios were used as tuning parameters, showing correlation between mechanical properties and phenotypic response. The MAX peptide family is another notable example of β-hairpin peptides, known for cytocompatibility and antibacterial properties [545, 546] and their ability to self-assemble when combined with cell suspensions in cell culture media, thereby encapsulating the cells without damage [155, 547]. Many of these nanofiber-forming β-sheet peptides have also been assembled into multi-domain peptides (MDPs) in combination with cell-binding and cell-signaling peptides and sequences, enabling enzymatic crosslinking or digestion. These MDPs have been engineered into scaffolds for vascular grafts [548] and injectable gels for tissue repair [549, 550]. Similarly, many α-helical peptides have been extensively utilized to engineer 3D hydrogel substrates for endothelial cells, neuronal cells, hepatocytes, chondrocytes, and osteoblasts [194, 403, 515, 528, 529] in combination with proteins for cell differentiation and formation of vascular or organ structures [9, 410, 513, 530, 549, 551, 552]. PAs have also been utilized as cell culture and tissue engineering substrates, chiefly for their ability to undergo remodeling by resident cell populations [553, 554]. This is particularly useful in stem cell differentiation, where matrix remodeling is necessary to elicit the desired differentiation response and support the new fully differentiated daughter cell population. Hydrogels obtained by combining PAs with synthetic biopolymers (e.g. polyglycolic acid) were used as 3D scaffolds in cell culture bioreactors to obtain bone and vascular grafts from MSCs [59, 555–558]. *In vivo*, IKVAV-AP gels supporting neural progenitor cells promoted neuronal differentiation and were applied for treating spinal cord injuries in mice models, resulting in axon growth and functional recovery [405, 559]. Analogous PA-based hydrogels loaded with growth factors and cytokines (e.g. TGF-β and BMPs) have been developed for cell growth and differentiation of human mesenchymal stem cells (hMSCs) toward chondrogenic [560], osteogenic [557, 561, 562], myogenic, adipogenic, and neurogenic lineages [405, 563].

9.4 Applications of Polysaccharides

9.4.1 Drug Delivery

Like proteins and polypeptides, polysaccharides have been widely utilized as drug delivery vehicles [564–574], owing to their excellent biocompatibility and biodegradability, ease of chemical functionalization and processing, commercial abundance, and low cost. Their versatility enables a multitude of applications, including drug conjugates and drug-loaded self-assembled structures to increase the solubility and circulation time of otherwise insoluble or rapidly cleared drugs and particle coating for targeted delivery or as alternatives to PEGylation for stealth function. Research on polysaccharide–drug conjugates has grown hand in hand with bioconjugation technology [575–583], resulting in new linkers for site-specific conjugation [584, 585], pH-responsive release [582, 586, 587], especially useful in the mildly acidic environment of tumor cells [575, 588], or enzyme-directed cleavage for intracellular drug release [575, 581, 589–591]. Of recent interest is sustained intestinal release of traditional injectable therapeutics (e.g. insulin and immunotherapies) and drugs for fighting colon cancer, ulcers, Crohn's disease, and gastrointestinal infections. Polysaccharides are ideal for colon targeting, as they resist the gastrointestinal environment but are degraded by the bacterial enzymes in the colon, resulting in release of the therapeutic payload and bioresorbable oligosaccharides [592–595].

1. *Hyaluronic acid (HA)*: [596] HA consists of D-gluconic acid–D-*N*-acetylglucosamine dimer repeats whose hydroxyl and carboxyl groups have been extensively utilized for drug conjugation [597]. HA has gained great popularity as drug delivery vector for its ability to specifically target cell surface marker CD44 [598–604], which is overexpressed in many cancer and cancer stem cells [599, 605]. A great deal of studies have been devoted to optimizing HA–drug conjugates to prevent aggregation [588, 600], improve conjugation [602, 606–608], and understand cell internalization [561, 609–611]. In this role, HA has served as carrier for (i) butyric acid to treat Lewis lung, melanoma, and leukemia [612–614]; (ii) PTX against ovarian, bladder, breast, and brain cancer [600, 610, 615–619]; (iii) CPT for peritoneal cancer [620, 621]; (iv) DOX to treat breast cancer [622]; and (v) cisplatin for breast cancer, squamous cell cancer of the head and neck (SCCHN), melanoma, sarcoma, and breast cancer [623–628]. In a recent study, DOX and CPT were conjugated to HA at a 4.5 : 1 molar ratio CPT:DOX, which is synergistic against HER2+ BT-474 breast cancer cells [629]. Analogous formulations were prepared by conjugating DOX to HA via a pH-labile hydrazone bond [630] and PTX via ester bond; the former provided higher anticancer efficacy *in vivo* compared with free DOX [622], while the latter showed targeted toxicity *in vitro* and strong anticancer activity *in vivo* in human ovarian carcinoma xenografts [600] and bladder carcinoma [631, 632]. HA micelles, liposomes, and nanoparticles have also been developed for the delivery of insoluble anticancer drugs. Their PEGylation and functionalization with proteins and peptides result in prolonged circulation and active targeting, while incorporation of thermo-/pH-responsive building blocks enables site-selective release. Examples of HA-based nanoparticulate formulations include:
   (i) Micelles carrying DOX for treatment of breast and prostate cancer [633, 634] or PTX to treat breast cancer and SSCHN [635–638].
   (ii) Liposomes loaded with mitomycin C (MMC) to treat primary, metastatic cancers, and Lewis lung carcinoma in mice [639] or GEM against pancreatic cancer [640, 641]; DOX-loaded liposomes against breast cancer [642, 643] were also successfully tested on human breast cancer xenografts in mice [622]; analogous PTX liposomes were developed against hepatocellular carcinoma [644] and CD44+ human ovarian carcinoma, showing antitumor activity *in vivo* [600].
   (iii) Core–shell nanoparticles wherein synthetic polymers (e.g. polylactic acid [PLA] and PLGA) and HA, respectively, populate the inner and outer layers have been developed to carry chemotherapeutic drugs against various breast cancer cell lines; [645–649] Ehrlich ascites tumor, breast cancer, and SSCHN; [650–654] and sarcoma [655]. Similar lipid–HA nanoparticles have been developed to deliver chemotherapeutics to treat breast cancer and carcinoma [650–654, 656–662], colon cancer

[663], melanoma [655, 664–666], and lung cancer [667]. Great effort has been devoted into studying the biodistribution, side toxicity, tumor accumulation, and antitumor efficacy *in vitro* and *in vivo* of these lipid–HA systems [645, 647, 648].

2. *Dextran (Dex)*: Dex is also widely employed as drug carrier by formation of non-covalent complexes [668–671], as well as conjugates based on either traditional chemistries [672] or innovative linkers enabling environment-controlled release [673–676]. Conjugates with chemotherapeutics are naturally the most widely studied. Among the examples of traditional conjugates, we find a Dex–cytarabine system that greatly improved survival in leukemia murine models [677] and Dex–MTX conjugates that showed therapeutic efficacy in mouse models of brain tumor [678] and leukemia [679, 680]. Short peptide linkers have then gained popularity for drug conjugation to Dex carriers. For example, an oligo-glycine linkers were used to construct Dex–CPT conjugates that showed significant antitumor activity *in vivo* against an ensemble of human tumor xenografts [673, 674, 676, 681, 682] and colon cancer [587, 683]; using the same GGGF linker, a conjugate of Dex and exatecan, a CPT analog, was prepared, which exhibited strong therapeutic activity against a panel of human xenografts and murine solid tumors [675, 684, 685]. The linker Jeffamine-PVGLIG, which is cleaved by matrix metalloproteinase enzymes, was used to link MTX to Dex; the resulting conjugate showed improved efficacy against human tumor xenograft models compared with the free drug [586, 590, 591]. To facilitate the conjugation of the drug payload through cleavable bonds, modified versions of Dex, such as oxidized (Ox) or carboxymethyl (CM) Dex, have been developed. PTX–CM–Dex ester conjugates showed significant antitumor activity against multiple colon cancer cell lines *in vivo* [587, 683]; similarly, an imine conjugate of DOX to Ox–Dex showed higher therapeutic activity and lower heart accumulation and acute toxicity compared with free DOX in a Lewis lung carcinoma rat model [686]. Notably, a CM–Dex conjugate with delimotecan, a CPT analog, has entered clinical trial; two cancer patients responded to therapy but showed adverse effects [676]. Other clinical studies were performed with a CM–Dex–CPT conjugate, which showed stabilization of the disease in 50% of patients affected by advanced solid tumors [685], and an Ox–Dex–DOX conjugate administered through IV infusion, which led to stabilization of the disease in 90% of patients, although toxic side effects were reported [687]. Worthy of notice are also the PEGylated Dex nanogels utilized for gene therapy, in particular for IV administration of small interfering RNA (siRNA) and microRNA [688–691], which have successfully achieved gene knockdown in human hepatoma and glioblastoma [692, 693]. Among others, it is worth reporting a series of studies employing Dex nanogels for delivering siRNA that silences the genes responsible for multidrug resistance in drug-resistant cancer cell lines [694–696]; the cells were sensitized to drug treatment, which resulted in effective suppression of cell growth *in vitro* and tumor growth *in vivo*.
3. *Chitosan (CN)*: CN is the most widely utilized cationic biopolymer as drug carrier [697–699]. CN owes its success to the amino groups displayed on the

glucosamine monomers, which enable fine-tuning of its gelation behavior and mechanical properties by chemical modification and conjugation of anticancer drugs. Direct drug conjugates with CN have been developed as injectable macromolecular drug carriers. A number of conjugates with MMC, for example, have showed good antitumor activity against murine leukemias, melanoma, sarcoma, metastatic liver cancer, and hepatic cell carcinoma *in vitro* [700] and comparable efficacy but reduced systemic toxicity against sarcoma *in vivo* [700–702]. DOX–CN conjugates, which formed micelles in solution comprising a DOX-rich core and a CN-rich shell, were shown to suppress tumor growth without systemic toxicity *in vivo* against breast cancer [703], melanoma [699], and mesothelioma cells [704]. Other micelle-forming PTX–CN conjugates, developed for oral administration, showed significant inhibition of murine melanoma [601], and similar docetaxel (DTX)–CN conjugates provided increased bioavailability, lower acute toxicity, and effective antitumor efficacy *in vivo* [705]. Conjugates with insulin for oral administration have also been developed, which released the payload in the intestine, increasing its bioavailability and controlling blood glucose levels effectively for several hours in diabetic rats [706]. Because of its insolubility at physiological pH, CN has been mostly utilized in the form of bulk and particulate hydrogels [670, 671, 707–710]. Numerous methods have been developed to fabricate CN micro- and nanoparticles for sustained drug release, including crosslinking, solvent exchange, ionic gelation, spray drying, emulsion polymerization, and precipitation/reconstitution [35, 711–728]. Notably, DOX-loaded nanoparticles were shown to circulate systemically without drug leaching and accumulate into tumor tissue, where they delivered the payload intracellularly by pH-triggered release [704]; analogous DOX-loaded nanospheres functionalized with HER2-targeting antibodies showed great promise for target delivery against HER2+ breast and ovarian cancers [671]. Being positively charged, NC nanoparticles make attractive vectors for oligonucleotide conjugations and gene delivery [729–731]. In several studies, CN particles crosslinked with deoxycholic acid and loaded with plasmid DNA were shown to effectively transfect COS-1 cells [732–737]. CN nanoparticles co-delivering tumor-suppressing IL17RB siRNA and DOX induced apoptosis and inhibited migration of breast cancer cells, owing to siRNA-induced sensitization of cells to DOX [738–744]. PEGylated RGD-functionalized CN nanoparticles loaded with siRNA targeting the lactate transporter MCT1 and the glutamine transporter ASCT2 showed strong antiproliferative effect both *in vitro* and *in vivo*. In other studies, CN nanoparticles were developed to deliver siRNA that silences galectin-1, a lectin associated with brain tumor progression; the particles were delivered to the central nervous system upon intranasal administration and showed significant reduction of Gal-1 expression in tumor-bearing mice. Like nanoparticles, CN nanofibers also show great potential for drug and gene delivery [745–748]. Their high specific surface and porosity improve drug loading and cell binding [749] and enable high integration of optically and magnetically active materials for cancer imaging [750–774]. Examples include CN thin films loaded with mesoporous aluminosilicate to fine-tune the delivery rate of metformin [774] or DOX [756, 768]; CN nanoparticles modified with magnetic iron oxide nanoparticles to control the delivery of DOX while concurrently enabling cell imaging [773]; particles loaded with magnetic nanocrystals for magnetic resonance (MR) cancer imaging and pH-controlled release of DOX [760]; and iron oxide nanoparticles coating CN for concurrent cancer imaging and treatment by delivery of MTX [765, 769] or MTX and GEM [759, 765, 769]. To complete the list of uses revolving around CN's cationic nature, it is worth mentioning complexation with negatively charged polyelectrolytes, which has been utilized as a method for constructing robust drug delivery systems such as particles, tablets, gels, and films [775]. CN-alginate constructs, for example, have been utilized to deliver anticancer [776–778], ocular [779], asthma and pulmonary [780], and anti-inflammatory drugs [781]. Similarly, CN-HA systems have been developed for ocular applications [782–784] and nonviral gene delivery for treating asthma and osteoarthritis [785, 786]. Tablets of CN and xanthan gum also show great promise for sustained drug release in the intestine [787, 788].

9.4.2 Tissue Engineering and Regenerative Medicine

The qualities that make polysaccharides ideal as drug delivery vectors – hydrophilicity, biocompatibility, controlled water uptake and degradation kinetics, and tailorable biochemical functionalization – make them attractive substrates for tissue engineering [8, 789–800] and regeneration [54, 570, 694, 801–808]. A multitude of methods have been implemented to adapt polysaccharides for tissue engineering applications, including enzymatic and photoinduced crosslinking, chemical crosslinking by Schiff base reaction, Michaels addition, and click chemistry and physical methods like ion-, pH-, and temperature-induced gelation (Figure 9.6) [809–817]. Let us peruse some examples:

1. *Plant and bacterial cellulose (BC)*: A plethora of 3D porous structures, such as sponges, hydrogels, and nanofiber matrices, have been developed using plant cellulose and its blends with other polymers for tissue engineering [818–820]. Particular attention has been devoted toward osteogenic and chondrogenic tissue scaffolds, since the porosity and mechanical strength of cellulose are ideal for osteoblast and chondrocyte infiltration and proliferation. In a few studies, cellulose microparticles were sintered into a 3D scaffold with mechanical properties similar to those of native bone and functionalized with collagen to support adhesion, spreading, and maturation of human osteoblasts, with consequent mineralization of the substrate [803, 821]. Similar scaffolds have been utilized to produce porous bone grafts and nonporous screws and plates for orthopedic applications [822, 823]. Cellulose has also been electrospun into fiber matrices and coated with BMP proteins for osteogenic differentiation of bone marrow stromal cells (BMSCs) into bone tissue after implantation [824]. Cellulose scaffolds coated with calcium phosphates have also been developed to sustain proliferation and phenotype development of human chondrocytes cells and have been widely utilized for

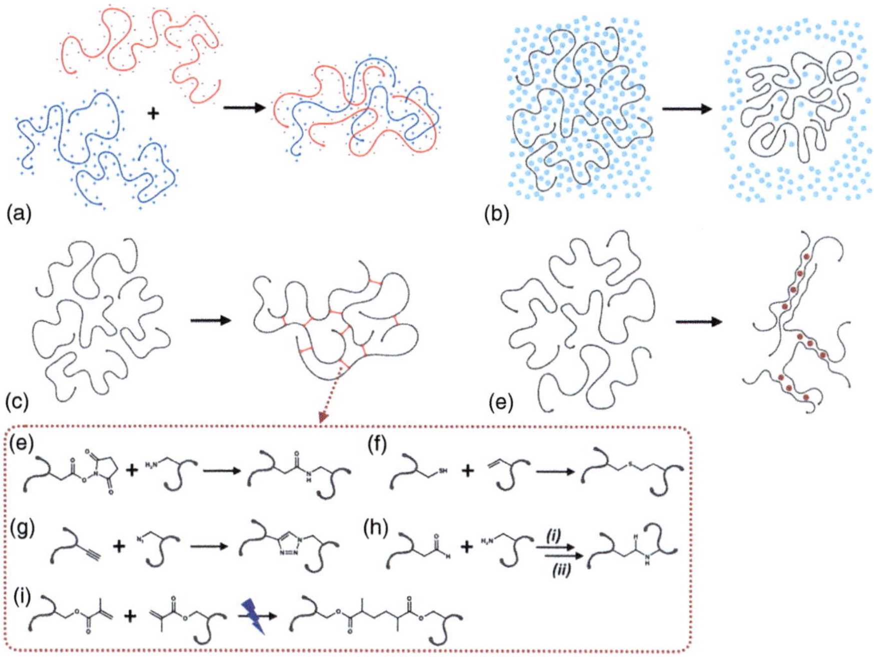

Figure 9.6 Fabrication techniques to convert polysaccharides into hydrogel substrates for tissue engineering and regenerative medicine: (a) electrostatic interaction, (b) solvent removal, (c) chemical crosslinking, (d) salt-mediated non-covalent crosslinking, (e) reaction of a *N*-hydroxysuccinimide (NHS) ester with a primary amine forming an amide bond, (f) thiol–ene Michael addition forming a thioether bond, (g) alkyne–azide "click" reaction forming a triazole bond, (h) reaction of an aldehyde and a primary amine forming a Schiff base (imine bond) followed by "reductive amination" forming a secondary amine bond, and (i) UV-catalyzed reaction of acrylic groups resulting in an aliphatic linkage.

cartilage engineering [818]. More recently, BC has been introduced, which combines the biocompatibility of its plant cognate with higher water content and tensile strength, thus making a better substrate for cartilage and bone tissues [825–828], as has been shown in clinical trials [829]. BC is obtained from bacterial cultures of *Acetobacter xylinum* as a highly crystalline nanofibrous material similar to collagen [825–827, 830] and, like all other polysaccharide substrates, can be functionalized with biochemical cues to be employed as a cell culture substrate. In one study, BC functionalized with osteogenic growth peptide supported osteoblast development *in vitro*, promoting high levels of alkaline phosphatase expression and mineralization, while showing no cytotoxicity and genotoxicity *in vitro* [831]. In another example, BC scaffolds coated with BMP-2 were utilized for adhesion and proliferation of fibroblast-like cells and were subsequently implanted subcutaneously in mice, resulting in ectopic bone formation [832].

2. *Chitosan (CN)*: Crosslinked CN hydrogels and composites play a leading role as scaffolds for soft and hard tissue engineering and injectable systems for tissue regeneration. The amine groups on CN enable fine-tuning of the elastic

modulus [833] and conjugation of ECM proteins and bioactive ingredients to support a variety of resident cell populations [834, 835]. To sample the versatility of CN composites, one can consider case studies where CN has been combined with other biopolymers (e.g. HA, alginate, aloe vera, etc.) to produce fibers and films for engineering ligaments [836], muscle tissue, skin [837], cartilage [838, 839], and even salivary glands [840]. Electrospinning of nanofiber composites has gained popularity as a processing technology for CN in combination with natural and synthetic biopolymers. For example, polycaprolactone (PCL)-CN coated nanofibers were electrospun with collagen to serve as osteogenic and chondrogenic cell scaffolds. Similarly, nanofibrous CN membranes were developed to support the adhesion and proliferation of Schwann cells, which play a crucial role in axonal regeneration in the peripheral nervous system [841], as well as cells from mouse neuroblastoma that produce the nerve growth factors (NGFs) implicated in the reconstruction of peripheral nerve lesions [842]. Among synthetic complements to CN, a notable role is played by stimuli-responsive polymers, which have been integrated with CN to enable on-demand cell entrapment or detachment. A multilayer ultraviolet (UV)-curable CN methacrylate was developed to encapsulate cells upon UV exposure and showed cell protection after crosslinking [843]. A thermo-responsive CN-pNIPAAm hydrogel [844] promoted cell adhesion and spreading upon heating, where pNIPAAm hydrophobically collapses, and allowed detachment of cell sheets upon cooling, where pNIPAAm regains hydrophilic character. Other composites based on CN and biopolymers such as gelatin, collagen, and starch have also been developed as 3D porous scaffolds for cartilage engineering [798, 845–847], tubular scaffolds for vascular grafts [846, 848, 849], asymmetric porous scaffolds for skin grafts [850–852], and membranes for constructing adipose tissue grafts [853]. Ceramic materials have also been integrated with CN to create composites with fine-tuned 3D morphology. Examples include CN crosslinked with silica particles for skin engineering and regeneration [854] and a CN-nanohydroxyapatite framework for growing bone tissue using stem cell cells extracted from human periodontal ligament tissues [855]. By combining their bioactive properties and tunable mechanics, CN composites have shown promise as injectable systems to promote tissue regeneration. An illustrative list of case studies includes coated CN tubules for engineering adipose stem cells into neurosphere cells to repair transected sciatic nerve in Sprague-Dawley rats [856]; a rigid CN hydrogel to promote skin regeneration in third-degree burns inflicted to female minipigs [857]; a thin film of CN and silk fibroin to repair ventral hernia in guinea pigs [858]; a blended hydrogel of CN and hydroxyethyl cellulose to repair of articular cartilage defects in sheep [859]; an *in situ* gelling mixture of CN and sodium alginate dialdehyde loaded with corneal endothelial cells for corneal regeneration [860]; and CN microspheres dispersed in a calcium phosphate paste for bone regeneration of damaged femur in rabbits [861]. In all these examples, CN scaffolds promote 3D tissue remodeling, with integration of the new tissue with the native one, deposition of new ECM, uniform vascularization, and cellular infiltration in repair sites, which contributed to ensuring mechanical and functional stability of the regenerated tissue. Owing to their ability to rapidly exchange gas and adsorb fluids, CN-based porous constructs are ideal materials for wound dressing [862]. CN hemostatic bandages and fillings have been reported to stop bleeding from deadly wounds in large animals much more rapidly than gauze dressings and increase the rate of survival [863]. The positive charges on CN chains, in fact, bridge erythrocytes by binding their negatively charged cell membrane and stimulates platelet intervention and rapid clotting. CN hemostats are frequently blended with biodegradable organic acids (e.g. succinic or lactic acid) to increase sponginess and make these materials bioresorbable. A number of hemostatic dressings are now commercial, such as the HemCon® bandage [862, 864], and are utilized by the United States and European armed forces. The abundance of amine groups on CN has also been utilized to develop materials for treating peptic ulcers by neutralizing gastric acids and form a protective layer in the stomach [865, 866]. CN-based composites have also been developed to retain moisture and enable easy peeling for treatment of skin injuries and have been loaded with antibiotics to ensure a sterile environment for the healing of burns. Examples include blends of CN, poly(vinyl alcohol), and alginate, loaded with broad-spectrum antibacterials (e.g. ciprofloxacin, minocycline, and ornidazole) and cast into moist films [867, 868] that form flexible and elastic membranes on wounds and provide superior healing activity compared with conventional gauzes. To further facilitate dressing removal from critical wounds, thermo-responsive pNIPAAm has been combined with CN into films that can be easily peeled from the wound upon cooling [869]. Further, in place of common antibiotics, silver and zinc oxide nanoparticles have been integrated CN films to avoid microorganism proliferation in infection-prone wounds [869–876].

3. *Hyaluronic acid (HA)*: As a major component of the ECM in soft connective tissues, HA regulates lubrication, inter- and intracellular communication, exchange of nutrients and metabolic waste, and rheological properties and shock absorption [877, 878]. Its biocompatibility, biodegradability, and ease of functionalization make HA an ideal material for tissue engineering [879, 880]. Echoing its biological role, HA is mostly applied as a substrate for cartilage and bone tissue engineering and regeneration. HA, in fact, induces cartilage stem and progenitor cells to differentiate into chondrocytes [881, 882] and stimulate the deposition of ECM [883–887], thus promoting a microenvironment that is conducive to cartilage production and *in situ* regeneration [885, 888–890]. Similarly, HA also supports the adhesion, differentiation, and growth of MSCs and BMSCs for osteochondral tissue formation [891, 892] and bone formation and regeneration [893]. Thus, countless HA blends with natural and synthetic biopolymers [885–887, 894] and chemically modified, photopolymerizable, and thermo-responsive [895–901] HA gels have been developed to encapsulate chondrocytes, supporting excellent cellular viability and proliferation and secretion of cartilaginous ECM compounds (aggrecan, collagen type I and type II) [885, 902]. These materials have been utilized to construct and regenerate cartilage [902, 903], intervertebral disc [904] and nucleus pulposus [905], human meniscus [906], and vocal cords [907]. Analogous HA ingredients, loaded with growth factors (e.g. BMP-2) and functionalized with

complementary reactive groups or UV-reactive moieties (e.g. acrylate and benzophenone) for *in situ* formation of strong hydrogels, have been developed for bone tissue engineering and regeneration [893, 908–915]. Notably, HA hydrogels are degraded by hyaluronidase at a rate dictated by the type and degree of modification, crosslinking, and blending with other materials. HA degradation enables the diffusion of growth factors, migration of cells, and distribution of ECM, ultimately promoting integration with the native tissues [916]. While high molecular weight HA is a known anti-inflammatory and inhibitor of cell proliferation [917, 918], low molecular weight HA has been found to promote cell migration and angiogenesis [917, 919, 920]. Functionalized hyaluronic acid oligomers (HAOs), *per se* [917, 921], or displayed on other biomaterials [922–924] promote vessel sprouting and outgrowth by driving cell migration through ECM and have been utilized to develop scaffolds for vascular grafts. HAO release by hyaluronidases from a gellan gum-HA hydrogel has been shown to promote the proliferation of HUVECs [925]. In another study, conductive thin films of HA were used to develop electrically responsive systems that induce angiogenesis on demand *in vivo* [926]. HA porous hydrogels have also proven effective in the engineering and regeneration of neuronal tissue in animal models [927, 928]. Early work conducted with localized injections of HA in animal models of nerve damage showed axonal regeneration and improved conduction, together with a decrease of perineural scar tissue [929, 930]. Soft HA hydrogels, in fact, possess brain tissue-like modulus (<1 kPa) and are ideal substrates for differentiating MSCs into neurogenic lineages and NSCs into neuronal lineages [931–933]. As biochemical functionalization became recognized as critical to ensure high regenerative outcomes, HA hydrogels functionalized with RGD peptides and growth factors multiplied. A RGD-decorated HA matrix with tunable stiffness and peptide density was developed to guide neural progenitor cells and promote neurite outgrowth [934]; while neurite extension was noted in HA matrices without cell-binding peptides, RGD-HA with optimized peptide density showed superior neurite outgrowth; similarly, the elasticity of the HA matrix showed a strong influence over the behavior of hippocampal neural progenitor cells, with an optimum of outgrowth and density of neurites at a storage modulus of 400 Pa. Porous RGD-HA hydrogels were effectively employed in healing of cortex defects [935] and spinal cord injury (SCI) *in vivo* [936] by improving vascularization and neurite extension while reducing scar formation in the defect area. A set of studies focused on HA–ephrinB2 conjugates that promote *in vivo* formation and differentiation of new neurons in areas of the brain that are in a prolonged state of inactivity through sustained delivery of ephrinB2 [937–939]. More recently, structural anisotropy (e.g. uniaxial fibrils and nano-/microtubular constructs, matrices with gradient porosity) has been combined with biochemical functionalization (e.g. NGF, neurotrophin-3 [NT-3], brain-derived neurotrophic factor [BDNF], glial cell line-derived neurotrophic factor [GDNF], and ciliary neurotrophic factor [CNTF]) to better guide the migration of cells toward lesion sites and promote growth and functional regeneration [940–944]. A number of studies have focused on the use of HA prefilled with pluripotent stem cells to sustain survival, differentiation, and proliferation of stem cells into neuroblasts, thereby achieving better repair of nerve injuries [945–947]. This work has led to biomimetic cell-filled HA gels being utilized as inks for 3D bioprinting of neural replacement tissues [948]. The HA component of the bio-ink contains the biomolecular cues that drive cell differentiation into specific neural types upon printing. Bioprinted neuronal constructs feature precise control of the 3D spatial distribution of cell types and scaffold properties and therefore superior biomimetic and restorative behavior. Other studies on HA scaffolds have been focused on constructing adipose grafts or promoting proliferation and retention of adipose tissue *in vivo* [949]. HYAFF-11® scaffolds, which consist of porous bioresorbable foams of HA esters [950, 951], possess the physical and biochemical properties that are suitable for the development of functional adipose tissue. *In vitro* tests with HYAFF-11 scaffolds have demonstrated adhesion and proliferation of adipocytes through the 3D scaffolds; furthermore, *in vivo* studies with patients with small tissue deficit showed that implants produced by incubating adipose stromal cells into HYAFF-11 scaffolds and implanted as subcutaneous pocket promoted the reconstruction of an adipose layer without triggering any toxicity or inflammation [952, 953]. Notably, these HA scaffolds have been also utilized for inducing human adipose stem cells into forming cornea-like structures [954].

4. *Alginate*: Easy to form, non-immunogenic, and nontoxic, alginate has become widely applied as injectable hydrogel, sponge, and fiber [955–960]. Like HA, alginate has been utilized to support vascular grafts and cardiac tissue repair, as well as cartilage, bone, neuronal, and adipose tissues. Examples of injectable hydrogels of alginate blends with HA and gelatin have shown excellent performance in encapsulating and maintaining the viability of primary chondrocytes, supporting their proliferation and migration within the polymer matrix, and integrating well with the ECM and the resulting cartilage with negligible inflammatory and oxidative stress responses [961, 962]. On the front of bone tissue engineering, porous structures of alginate and CN are produced by coacervation, which supported adhesion, proliferation, and production of calcified matrix by osteoblasts, resulting in excellent *in vivo* biocompatibility [963]. Recent studies have presented blends of alginate and calcium salts [964], micro- and nanoparticles [965–969], and photo-crosslinkable polymers [970, 971] in the form of injectable pastes and cements preloaded with MSCs and biochemical cues for stimulating differentiation toward the osteogenic lineage; these formulations exhibited osteo-differentiation and secretion of bone minerals both *in vitro* and *in vivo* and promoted bone repair and in small and large animal models [972, 973]. Alginate patches functionalized with biochemical cues (e.g. RGD peptides and growth factors) [974] have also been utilized as scaffolds for cardiac tissue engineering. In several studies, functionalized alginate patches supported adhesion and proliferation of neonatal rat cardiac cells and increased cardiac tissue regeneration compared with control alginate patches [974]; shortly after the application, micro-fibrous bundles of cardiomyocytes were noted, which showed integration with native tissue and regeneration of functional tissue in scar or necrotic areas [960, 975, 976].

9.5 Conclusions and Future Outlook

Three decades of research and preclinical development of peptide- and polysaccharide-based biomaterials for drug delivery and tissue engineering have brought a prodigious wealth of fundamental knowledge and technological capabilities to fruition of the scientific, pharmaceutical, and medical community. Having established the correlations between the physicochemical and biomolecular properties of these materials with their biological outcome *in vitro* and *in vivo*, a new epoch has recently begun, centered on the integration of these materials with modern design and fabrication techniques, devices for real-time monitoring, and cutting-edge biochemical and biological discoveries that can morph the human body into its own therapy manufacturer. The latest advancements in computational tools, in fact, enable accurate *in silico* modeling of complex supramolecular systems for extended simulation times that were unthinkable a handful of years ago and will soon extend modeling to the interactions between these systems and cells. In parallel, the development of new fabrication technologies, such as high-throughput 3D printing, and optical instrumentation enables precise assembly of truly biomimetic systems and accurate spatiotemporal control of their performance. Biopolymer materials are also undergoing integration with miniaturized implantable systems that collect biochemical data and control in real time the behavior of the implanted material and the release of the therapeutic or cellular payload and communicate data with medical personnel. Finally, capitalizing on their ability as delivery agents, these biomaterials will enable next-generation therapeutic approaches, such as gene therapy, that harness the bio-machinery of the body to produce target-specific therapeutics.

References

1 Van Tomme, S.R., Storm, G., and Hennink, W.E. (2008). In situ gelling hydrogels for pharmaceutical and biomedical applications. *International Journal of Pharmaceutics* 355 (1): 1–18.
2 Huang, R., Qi, W., Fan, L. et al. (2011). Self-assembling peptide–polysaccharide hybrid hydrogel as a potential carrier for drug delivery. *Soft Matter* 7 (13): 6222–6230.
3 Hoffman, A.S. (2012). Hydrogels for biomedical applications. *Advanced Drug Delivery Reviews* 64: 18–23.
4 Drury, J.L. and Mooney, D.J. (2003). Hydrogels for tissue engineering: scaffold design variables and applications. *Biomaterials* 24 (24): 4337–4351.
5 Seliktar, D. (2012). Designing cell-compatible hydrogels for biomedical applications. *Science* 336 (6085): 1124–1128.
6 Mahler, A., Reches, M., Rechter, M. et al. (2006). Rigid, self-assembled hydrogel composed of a modified aromatic dipeptide. *Advanced Materials* 18 (11): 1365–1370.
7 Sui, Z., King, W.J., and Murphy, W.L. (2008). Protein-based hydrogels with tunable dynamic responses. *Advanced Functional Materials* 18 (12): 1824–1831.

8 Jin, R., Teixeira, L.S.M., Dijkstra, P.J. et al. (2009). Injectable chitosan-based hydrogels for cartilage tissue engineering. *Biomaterials* 30 (13): 2544–2551.
9 Koutsopoulos, S., Unsworth, L.D., Nagai, Y., and Zhang, S. (2009). Controlled release of functional proteins through designer self-assembling peptide nanofiber hydrogel scaffold. *Proceedings of the National Academy of Sciences of the United States of America* 106 (12): 4623–4628.
10 Yanlian, Y., Ulung, K., Xiumei, W. et al. (2009). Designer self-assembling peptide nanomaterials. *Nano Today* 4 (2): 193–210.
11 Zhou, M., Smith, A.M., Das, A.K. et al. (2009). Self-assembled peptide-based hydrogels as scaffolds for anchorage-dependent cells. *Biomaterials* 30 (13): 2523–2530.
12 Cui, H., Webber, M.J., and Stupp, S.I. (2010). Self-assembly of peptide amphiphiles: from molecules to nanostructures to biomaterials. *Peptide Science* 94 (1): 1–18.
13 Coviello, T., Matricardi, P., Marianecci, C., and Alhaique, F. (2007). Polysaccharide hydrogels for modified release formulations. *Journal of Controlled Release* 119 (1): 5–24.
14 Chawla, K., Yu, T.-B., Liao, S.W., and Guan, Z. (2011). Biodegradable and biocompatible synthetic saccharide-peptide hydrogels for three-dimensional stem cell culture. *Biomacromolecules* 12 (3): 560–567.
15 Altunbas, A. and Pochan, D.J. (2012). Peptide-based and polypeptide-based hydrogels for drug delivery and tissue engineering. *Topics in Current Chemistry* 310: 135–167.
16 Jonker, A.M., Löwik, D.W.P.M., and van Hest, J.C.M. (2012). Peptide- and protein-based hydrogels. *Chemistry of Materials* 24 (5): 759–773.
17 Panda, J.J. and Chauhan, V.S. (2014). Short peptide based self-assembled nanostructures: implications in drug delivery and tissue engineering. *Polymer Chemistry* 5 (15): 4418–4436.
18 Habibi, N., Kamaly, N., Memic, A., and Shafiee, H. (2016). Self-assembled peptide-based nanostructures: smart nanomaterials toward targeted drug delivery. *Nano Today* 11 (1): 41–60.
19 De Jong, W.H. and Borm, P.J.A. (2008). Drug delivery and nanoparticles: applications and hazards. *International Journal of Nanomedicine* 3 (2): 133–149.
20 Hu, Q., Sun, W., Wang, C., and Gu, Z. (2016). Recent advances of cocktail chemotherapy by combination drug delivery systems. *Advanced Drug Delivery Reviews* 98: 19–34.
21 Rizvi, S.A.A. and Saleh, A.M. (2018). Applications of nanoparticle systems in drug delivery technology. *Saudi Pharmaceutical Journal* 26 (1): 64–70.
22 Croy, S.R. and Kwon, G.S. (2006). Polymeric micelles for drug delivery. *Current Pharmaceutical Design* 12 (36): 4669–4684.
23 Kedar, U., Phutane, P., Shidhaye, S., and Kadam, V. (2010). Advances in polymeric micelles for drug delivery and tumor targeting. *Nanomedicine: Nanotechnology, Biology and Medicine* 6 (6): 714–729.
24 Zhang, Y., Huang, Y., and Li, S. (2014). Polymeric micelles: nanocarriers for cancer-targeted drug delivery. *AAPS PharmSciTech* 15 (4): 862–871.

25 Hoare, T.R. and Kohane, D.S. (2008). Hydrogels in drug delivery: progress and challenges. *Polymer* 49 (8): 1993–2007.
26 Park, H., Park, K., and Shalaby, W.S.W. (2011). *Biodegradable Hydrogels for Drug Delivery*, 272. CRC Press.
27 Li, J. and Mooney, D.J. (2016). Designing hydrogels for controlled drug delivery. *Nature Reviews Materials* 1 (12): 1–17.
28 Snyder, E.L. and Dowdy, S.F. (2004). Cell penetrating peptides in drug delivery. *Pharmaceutical Research* 21 (3): 389–393.
29 Temsamani, J. and Vidal, P. (2004). The use of cell-penetrating peptides for drug delivery. *Drug Discovery Today* 9 (23): 1012–1019.
30 Tréhin, R. and Merkle, H.P. (2004). Chances and pitfalls of cell penetrating peptides for cellular drug delivery. *European Journal of Pharmaceutics and Biopharmaceutics* 58 (2): 209–223.
31 Foged, C. and Nielsen, H.M. (2008). Cell-penetrating peptides for drug delivery across membrane barriers. *Expert Opinion on Drug Delivery* 5 (1): 105–117.
32 Vivès, E., Schmidt, J., and Pèlegrin, A. (2008). Cell-penetrating and cell-targeting peptides in drug delivery. *Biochimica et Biophysica Acta (BBA) - Reviews on Cancer* 1786 (2): 126–138.
33 Regberg, J., Srimanee, A., and Langel, Ü. (2012). Applications of cell-penetrating peptides for tumor targeting and future cancer therapies. *Pharmaceuticals* 5 (9): 991–1007.
34 Wang, F., Wang, Y., Zhang, X. et al. (2014). Recent progress of cell-penetrating peptides as new carriers for intracellular cargo delivery. *Journal of Controlled Release* 174: 126–136.
35 Liu, Z., Jiao, Y., Wang, Y. et al. (2008). Polysaccharides-based nanoparticles as drug delivery systems. *Advanced Drug Delivery Reviews* 60 (15): 1650–1662.
36 Maia, J., Ferreira, L., Carvalho, R. et al. (2005). Synthesis and characterization of new injectable and degradable dextran-based hydrogels. *Polymer* 46 (23): 9604–9614.
37 Augst, A.D., Kong, H.J., and Mooney, D.J. (2006). Alginate hydrogels as biomaterials. *Macromolecular Bioscience* 6 (8): 623–633.
38 Dhaneshwar, S.S., Kandpal, M., Gairola, N., and Kadam, S.S. (2006). Dextran: a promising macromolecular drug carrier. *Indian Journal of Pharmaceutical Sciences* 68 (6): 705.
39 Bhattarai, N., Gunn, J., and Zhang, M. (2010). Chitosan-based hydrogels for controlled, localized drug delivery. *Advanced Drug Delivery Reviews* 62 (1): 83–99.
40 Kar, R., Mohapatra, S., Bhanja, S. et al. (2010). Formulation and in vitro characterization of xanthan gum-based sustained release matrix tables of isosorbide-5- mononitrate. *Iranian Journal of Pharmaceutical Research : IJPR* 9 (1): 13–19.
41 Prabaharan, M. (2011). Prospective of guar gum and its derivatives as controlled drug delivery systems. *International Journal of Biological Macromolecules* 49 (2): 117–124.

42 Lee, K.Y. and Mooney, D.J. (2012). Alginate: properties and biomedical applications. *Progress in Polymer Science* 37 (1): 106–126.

43 Oprea, A.-M., Profire, L., Lupusoru, C.E. et al. (2012). Synthesis and characterization of some cellulose/chondroitin sulphate hydrogels and their evaluation as carriers for drug delivery. *Carbohydrate Polymers* 87 (1): 721–729.

44 Varshosaz, J. (2012). Dextran conjugates in drug delivery. *Expert Opinion on Drug Delivery* 9 (5): 509–523.

45 Badwaik, H.R., Giri, T.K., Nakhate, K.T. et al. (2013). Xanthan gum and its derivatives as a potential bio-polymeric carrier for drug delivery system. *Current Drug Delivery* 10 (5): 587–600.

46 Kono, H., Otaka, F., and Ozaki, M. (2014). Preparation and characterization of guar gum hydrogels as carrier materials for controlled protein drug delivery. *Carbohydrate Polymers* 111: 830–840.

47 Ahmadi, F., Oveisi, Z., Mohammadi Samani, S., and Amoozgar, Z. (2015). Chitosan based hydrogels: characteristics and pharmaceutical applications. *Research in Pharmaceutical Sciences* 10 (1): 1–16.

48 Neves, S.C., Gomes, D.B., Sousa, A. et al. (2015). Biofunctionalized pectin hydrogels as 3D cellular microenvironments. *Journal of Materials Chemistry B* 3 (10): 2096–2108.

49 Hamedi, H., Moradi, S., Hudson, S.M., and Tonelli, A.E. (2018). Chitosan based hydrogels and their applications for drug delivery in wound dressings: a review. *Carbohydrate Polymers* 199: 445–460.

50 Matricardi, P., Di Meo, C., Coviello, T. et al. (2013). Interpenetrating polymer networks polysaccharide hydrogels for drug delivery and tissue engineering. *Advanced Drug Delivery Reviews* 65 (9): 1172–1187.

51 Li, S., Xiong, Q., Lai, X. et al. (2016). Molecular modification of polysaccharides and resulting bioactivities. *Comprehensive Reviews in Food Science and Food Safety* 15 (2): 237–250.

52 Ahmed, E.M. (2015). Hydrogel: preparation, characterization, and applications: a review. *Journal of Advanced Research* 6 (2): 105–121.

53 Kumar, D., Pandey, J., Raj, V., and Kumar, P. (2017). A review on the modification of polysaccharide through graft copolymerization for various potential applications. *The Open Medicinal Chemistry Journal* 11: 109–126.

54 Lee, J.W., Park, J.H., and Robinson, J.R. (2000). Bioadhesive-based dosage forms: the next generation. *Journal of Pharmaceutical Sciences* 89 (7): 850–866.

55 Ma, T., Li, Y., Yang, S.T., and Kniss, D.A. (2000). Effects of pore size in 3-D fibrous matrix on human trophoblast tissue development. *Biotechnology and Bioengineering* 70 (6): 606–618.

56 Xie, Y., Yang, S.T., and Kniss, D.A. (2001). Three-dimensional cell-scaffold constructs promote efficient gene transfection: implications for cell-based gene therapy. *Tissue Engineering* 7 (5): 585–598.

57 Shimaoka, H., Dohi, Y., Ohgushi, H. et al. (2004). Recombinant growth/differentiation factor-5 (GDF-5) stimulates osteogenic differentiation of marrow mesenchymal stem cells in porous hydroxyapatite ceramic. *Journal of Biomedical Materials Research Part A* 68A (1): 168–176.

58 Hosseinkhani, H., Azzam, T., Kobayashi, H. et al. (2006). Combination of 3D tissue engineered scaffold and non-viral gene carrier enhance in vitro DNA expression of mesenchymal stem cells. *Biomaterials* 27 (23): 4269–4278.
59 Hosseinkhani, H., Yamamoto, M., Inatsugu, Y. et al. (2006). Enhanced ectopic bone formation using a combination of plasmid DNA impregnation into 3-D scaffold and bioreactor perfusion culture. *Biomaterials* 27 (8): 1387–1398.
60 Collins, M.N. and Birkinshaw, C. (2013). Hyaluronic acid based scaffolds for tissue engineering—a review. *Carbohydrate Polymers* 92 (2): 1262–1279.
61 Schugens, C., Maquet, V., Grandfils, C. et al. (1996). Polylactide macroporous biodegradable implants for cell transplantation. II. Preparation of polylactide foams by liquid-liquid phase separation. *Journal of Biomedical Materials Research* 30 (4): 449–461.
62 Schugens, C., Maquet, V., Grandfils, C. et al. (1996). Biodegradable and macroporous polylactide implants for cell transplantation: 1. Preparation of macroporous polylactide supports by solid-liquid phase separation. *Polymer* 37 (6): 1027–1038.
63 Nam, Y.S. and Park, T.G. (1999). Porous biodegradable polymeric scaffolds prepared by thermally induced phase separation. *Journal of Biomedical Materials Research* 47 (1): 8–17.
64 Taboas, J.M., Maddox, R.D., Krebsbach, P.H., and Hollister, S.J. (2003). Indirect solid free form fabrication of local and global porous, biomimetic and composite 3D polymer-ceramic scaffolds. *Biomaterials* 24 (1): 181–194.
65 Zein, I., Hutmacher, D.W., Tan, K.C., and Teoh, S.H. (2002). Fused deposition modeling of novel scaffold architectures for tissue engineering applications. *Biomaterials* 23 (4): 1169–1185.
66 Hutmacher, D.W., Ng, K.W., Kaps, C. et al. (2003). Elastic cartilage engineering using novel scaffold architectures in combination with a biomimetic cell carrier. *Biomaterials* 24 (24): 4445–4458.
67 Mooney, D.J., Baldwin, D.F., Suh, N.P. et al. (1996). Novel approach to fabricate porous sponges of poly(D,L-lactic-co-glycolic acid) without the use of organic solvents. *Biomaterials* 17 (14): 1417–1422.
68 Harris, L.D., Kim, B.S., and Mooney, D.J. (1998). Open pore biodegradable matrices formed with gas foaming. *Journal of Biomedical Materials Research* 42 (3): 396–402.
69 Oxley, H.R., Corkhill, P.H., Fitton, J.H., and Tighe, B.J. (1993). Macroporous hydrogels for biomedical applications: methodology and morphology. *Biomaterials* 14 (14): 1064–1072.
70 Nikpour, M., Chaouk, H., Mau, A. et al. (1999). Porous conducting membranes based on polypyrrole–PMMA composites. *Synthetic Metals* 99 (2): 121–126.
71 Chen, G., Ushida, T., and Tateishi, T. (2001). Preparation of poly(L-lactic acid) and poly(DL-lactic-co-glycolic acid) foams by use of ice microparticulates. *Biomaterials* 22 (18): 2563–2567.
72 Lin, A.S.P., Barrows, T.H., Cartmell, S.H., and Guldberg, R.E. (2003). Microarchitectural and mechanical characterization of oriented porous polymer scaffolds. *Biomaterials* 24 (3): 481–489.

73 Chirila, T.V., Constable, I.J., Crawford, G.J. et al. (1993). Poly(2-hydroxyethyl methacrylate) sponges as implant materials: in vivo and in vitro evaluation of cellular invasion. *Biomaterials* 14 (1): 26–38.
74 Kane, R.S., Takayama, S., Ostuni, E. et al. (1999). Patterning proteins and cells using soft lithography. *Biomaterials* 20 (23): 2363–2376.
75 Zhang, X., Jiang, X.N., and Sun, C. (1999). Micro-stereolithography of polymeric and ceramic microstructures. *Sensors and Actuators A: Physical* 77 (2): 149–156.
76 Balakrishnan, B. and Banerjee, R. (2011). Biopolymer-based hydrogels for cartilage tissue engineering. *Chemical Reviews* 111 (8): 4453–4474.
77 Bent, A.E., Foote, J., Siegel, S. et al. (2001). Collagen implant for treating stress urinary incontinence in women with urethral hypermobility. *The Journal of Urology* 166 (4): 1354–1357.
78 Kogan, G., Soltés, L., Stern, R., and Gemeiner, P. (2007). Hyaluronic acid: a natural biopolymer with a broad range of biomedical and industrial applications. *Biotechnology Letters* 29 (1): 17–25.
79 Kim, I.L., Mauck, R.L., and Burdick, J.A. (2011). Hydrogel design for cartilage tissue engineering: a case study with hyaluronic acid. *Biomaterials* 32 (34): 8771–8782.
80 Chang, S.J., Huang, Y.-T., Yang, S.-C. et al. (2012). In vitro properties of gellan gum sponge as the dental filling to maintain alveolar space. *Carbohydrate Polymers* 88 (2): 684–689.
81 Fagien, S. and Cassuto, D. (2012). Reconstituted injectable hyaluronic acid: expanded applications in facial aesthetics and additional thoughts on the mechanism of action in cosmetic medicine. *Plastic and Reconstructive Surgery* 130 (1): 208–217.
82 Mossaad, B.M. and Frame, J.D. (2012). Correction of breast contour deformities using polyurethane breast implant capsule in revisional breast surgery. *Journal of Plastic, Reconstructive & Aesthetic Surgery : JPRAS* 65 (10): 1425–1429.
83 Chan, W. and White, P. (1999). *Fmoc Solid Phase Peptide Synthesis: A Practical Approach*, 371. Oxford University Press.
84 Xiao, R. and Grinstaff, M.W. (2017). Chemical synthesis of polysaccharides and polysaccharide mimetics. *Progress in Polymer Science* 74: 78–116.
85 Mano, J.F., Silva, G.A., Azevedo, H.S. et al. (2007). Natural origin biodegradable systems in tissue engineering and regenerative medicine: present status and some moving trends. *Journal of the Royal Society, Interface* 4 (17): 999–1030.
86 Pawar, R., Jadhav, W., Bhusare, S.R. et al. (2008). 1 - Polysaccharides as carriers of bioactive agents for medical applications. In: *Natural-Based Polymers for Biomedical Applications* (eds. R.L. Reis et al.), 3–53. Woodhead Publishing.
87 Senni, K., Pereira, J., Gueniche, F. et al. (2011). Marine polysaccharides: a source of bioactive molecules for cell therapy and tissue engineering. *Marine Drugs* 9 (9): 1664–1681.

88 Malafaya, P.B., Silva, G.A., and Reis, R.L. (2007). Natural–origin polymers as carriers and scaffolds for biomolecules and cell delivery in tissue engineering applications. *Advanced Drug Delivery Reviews* 59 (4): 207–233.
89 Gomes, S., Leonor, I.B., Mano, J.F. et al. (2012). Natural and genetically engineered proteins for tissue engineering. *Progress in Polymer Science* 37 (1): 1–17.
90 Cemazar, M. and Craik, D.J. (2008). Microwave-assisted Boc-solid phase peptide synthesis of cyclic cysteine-rich peptides. *Journal of Peptide Science* 14 (6): 683–689.
91 Collier, J.H. and Segura, T. (2011). Evolving the use of peptides as biomaterials components. *Biomaterials* 32 (18): 4198–4204.
92 Ito, Y. and Manabe, S. (1998). Solid-phase oligosaccharide synthesis and related technologies. *Current Opinion in Chemical Biology* 2 (6): 701–708.
93 Andrade, R.B., Plante, O.J., Melean, L.G., and Seeberger, P.H. (1999). Solid-phase oligosaccharide synthesis: preparation of complex structures using a novel linker and different glycosylating agents. *Organic Letters* 1 (11): 1811–1814.
94 Seeberger, P.H. and Haase, W.-C. (2000). Solid-phase oligosaccharide synthesis and combinatorial carbohydrate libraries. *Chemical Reviews* 100 (12): 4349–4394.
95 Zhu, T. and Boons, G.-J. (2000). A new set of orthogonal-protecting groups for oligosaccharide synthesis on a polymeric support. *Tetrahedron: Asymmetry* 11 (1): 199–205.
96 Wang, Q., Dordick, J.S., and Linhardt, R.J. (2002). Synthesis and application of carbohydrate-containing polymers. *Chemistry of Materials* 14 (8): 3232–3244.
97 Rauter, A.P., Lindhorst, T., and Queneau, Y. (2014). *Carbohydrate Chemistry*, 671. Royal Society of Chemistry.
98 Polyakova, S.M., Nizovtsev, A.V., Kunetskiy, R.A., and Bovin, N.V. (2015). New protecting groups in the synthesis of oligosaccharides. *Russian Chemical Bulletin* 64 (5): 973–989.
99 Bouchet-Spinelli, A., Reuillard, B., Coche-Guérente, L. et al. (2013). Oligosaccharide biosensor for direct monitoring of enzymatic activities using QCM-D. *Biosensors and Bioelectronics* 49: 290–296.
100 Sun, Y.-S. and Zhu, X.D. (2017). Real-time, label-free characterization of oligosaccharide-binding proteins using carbohydrate microarrays and an ellipsometry-based biosensor. *Instrumentation Science and Technology* 45 (5): 506–524.
101 Boltje, T.J., Buskas, T., and Boons, G.-J. (2009). Opportunities and challenges in synthetic oligosaccharide and glycoconjugate research. *Nature Chemistry* 1 (8): 611–622.
102 Ravichandran, V., Kothandaraman, G.P., Bories, C. et al. (2018). Synthetic polysaccharides as drug carriers: synthesis of polyglucose-amphotericin B conjugates and in vitro evaluation of their anti-fungal and anti-leishmanial activities. *Journal of Nanoscience and Nanotechnology* 18 (4): 2405–2414.
103 Nair, J.K., Willoughby, J.L., Chan, A. et al. (2014). Multivalent N-acetylgalactosamine-conjugated siRNA localizes in hepatocytes and elicits

robust RNAi-mediated gene silencing. *Journal of the American Chemical Society* 136 (49): 16958–16961.
104 Huang, Y. (2017). Preclinical and clinical advances of GalNAc-decorated nucleic acid therapeutics. *Molecular Therapy--Nucleic Acids* 6: 116–132.
105 Springer, A.D. and Dowdy, S.F. (2018). GalNAc-siRNA conjugates: leading the way for delivery of RNAi therapeutics. *Nucleic Acid Therapeutics* 28 (3): 109–118.
106 Bouhadir, K.H. (2000). Sustained and controlled release of daunomycin from cross-linked poly(aldehyde guluronate) hydrogels. *Journal of Pharmaceutical Sciences* 89 (7): 10.
107 Rodrı´ guez, R.A., Alvarez-Lorenzo, C., and Concheiro, A. (2003). Interactions of ibuprofen with cationic polysaccharides in aqueous dispersions and hydrogels. *European Journal of Pharmaceutical Sciences* 20 (4–5): 429–438.
108 Zhang, L., Radovic-Moreno, A.F., Alexis, F. et al. (2007). Co-delivery of hydrophobic and hydrophilic drugs from nanoparticle-aptamer bioconjugates. *ChemMedChem* 2 (9): 1268–1271.
109 Li, S.-D. and Huang, L. (2008). Pharmacokinetics and biodistribution of nanoparticles. *Molecular Pharmaceutics* 5 (4): 496–504.
110 Farokhzad, O.C. and Langer, R. (2009). Impact of nanotechnology on drug delivery. *ACS Nano* 3 (1): 16–20.
111 Li, S.-D. and Huang, L. (2010). Stealth nanoparticles: high density but sheddable PEG is a key for tumor targeting. *Journal of Controlled Release* 145 (3): 178–181.
112 Liu, D., Bimbo, L.M., Mäkilä, E. et al. (2013). Co-delivery of a hydrophobic small molecule and a hydrophilic peptide by porous silicon nanoparticles. *Journal of Controlled Release* 170 (2): 268–278.
113 Guo, S. and Huang, L. (2014). Nanoparticles containing insoluble drug for cancer therapy. *Biotechnology Advances* 32 (4): 778–788.
114 Yui, N. (2002). *Supramolecular Design for Biological Applications*, 424. CRC Press.
115 Webber, M.J. and Langer, R. (2017). Drug delivery by supramolecular design. *Chemical Society Reviews* 46 (21): 6600–6620.
116 de Bruyn Ouboter, D. (2011). *Rational Design of Purely Peptidic Amphiphiles for Drug Delivery Applications*, 133 S. University of Basel.
117 Mandal, D., Shirazi, A.N., and Parang, K. (2014). Self-assembly of peptides to nanostructures. *Organic & Biomolecular Chemistry* 12 (22): 3544–3561.
118 Eskandari, S., Guerin, T., Toth, I., and Stephenson, R.J. (2017). Recent advances in self-assembled peptides: implications for targeted drug delivery and vaccine engineering. *Advanced Drug Delivery Reviews* 110–111: 169–187.
119 Fan, T., Yu, X., Shen, B., and Sun, L. (2017). Peptide self-assembled nanostructures for drug delivery applications. *Journal of Nanomaterials* 9: 1–16.
120 Kumar, P., Pillay, V., Modi, G. et al. (2011). Self-assembling peptides: implications for patenting in drug delivery and tissue engineering. *Recent Patents on Drug Delivery & Formulation* 5 (1): 24–51.

121 Hong, Y., Legge, R.L., Zhang, S., and Chen, P. (2003). Effect of amino acid sequence and pH on nanofiber formation of self-assembling peptides EAK16-II and EAK16-IV. *Biomacromolecules* 4 (5): 1433–1442.

122 Larsen, T.H., Branco, M.C., Rajagopal, K. et al. (2009). Sequence-dependent gelation kinetics of β-hairpin peptide hydrogels. *Macromolecules* 42 (21): 8443–8450.

123 Bowerman, C.J., Liyanage, W., Federation, A.J., and Nilsson, B.L. (2011). Tuning β-sheet peptide self-assembly and hydrogelation behavior by modification of sequence hydrophobicity and aromaticity. *Biomacromolecules* 12 (7): 2735–2745.

124 Pu, K.-Y., Li, K., and Liu, B. (2010). Multicolor conjugate poly-electrolyte/peptide complexes as self-assembled nanoparticles for receptor-targeted cellular imaging. *Chemistry of Materials* 22 (24): 6736–6741.

125 Wang, W. and Chau, Y. (2012). Self-assembly mediated platform for rapid and facile preparation of peptide-functionalized nanoparticles with high stability. *Chemistry of Materials* 24 (5): 946–953.

126 Jin, Y., Huang, Y., Yang, H. et al. (2015). A peptide-based pH-sensitive drug delivery system for targeted ablation of cancer cells. *Chemical Communications* 51 (77): 14454–14457.

127 Chang, C., Liang, P., Chen, L. et al. (2017). pH-responsive nanoparticle assembly from peptide amphiphiles for tumor targeting drug delivery. *Journal of Biomaterials Science Polymer Edition* 28 (13): 1338–1350.

128 Feng, Y., Taraban, M., and Yu, Y.B. (2012). The effect of ionic strength on the mechanical, structural and transport properties of peptide hydrogels. *Soft Matter* 8 (46): 11723–11731.

129 Li, R., Horgan, C.C., Long, B. et al. (2015). Tuning the mechanical and morphological properties of self-assembled peptide hydrogels via control over the gelation mechanism through regulation of ionic strength and the rate of pH change. *RSC Advances* 5 (1): 301–307.

130 Mackay, J.A. and Chilkoti, A. (2008). Temperature sensitive peptides: engineering hyperthermia-directed therapeutics. *International Journal of Hyperthermia* 24 (6): 483–495.

131 Aladini, F., Araman, C., and Becker, C.F.W. (2016). Chemical synthesis and characterization of elastin-like polypeptides (ELPs) with variable guest residues. *Journal of Peptide Science* 22 (5): 334–342.

132 Yishay-Safranchik, E., Golan, M., and David, A. (2014). Controlled release of doxorubicin and Smac-derived pro-apoptotic peptide from self-assembled KLD-based peptide hydrogels. *Polymers for Advanced Technologies* 25 (5): 539–544.

133 Lu, S., Wang, H., Sheng, Y. et al. (2012). Molecular binding of self-assembling peptide EAK16-II with anticancer agent EPT and its implication in cancer cell inhibition. *Journal of Controlled Release* 160 (1): 33–40.

134 von Maltzahn, G., Vauthey, S., Santoso, S., and Zhang, S. (2003). Positively charged surfactant-like peptides self-assemble into nanostructures. *Langmuir* 19 (10): 4332–4337.

135 Jabbari, E., Yang, X., Moeinzadeh, S., and He, X. (2013). Drug release kinetics, cell uptake, and tumor toxicity of hybrid VVVVVVKK peptide-assembled polylactide nanoparticles. *European Journal of Pharmaceutics and Biopharmaceutics* 84 (1): 49–62.

136 Yan, X., He, Q., Wang, K. et al. (2007). Transition of cationic dipeptide nanotubes into vesicles and oligonucleotide delivery. *Angewandte Chemie International Edition* 46 (14): 2431–2434.

137 McCarthy, H.O., McCafferty, J., McCrudden, C.M. et al. (2014). Development and characterization of self-assembling nanoparticles using a bio-inspired amphipathic peptide for gene delivery. *Journal of Controlled Release* 189: 141–149.

138 Naskar, J., Roy, S., Joardar, A. et al. (2011). Self-assembling dipeptide-based nontoxic vesicles as carriers for drugs and other biologically important molecules. *Organic & Biomolecular Chemistry* 9 (19): 6610–6615.

139 Naskar, J., Palui, G., and Banerjee, A. (2009). Tetrapeptide-based hydrogels: for encapsulation and slow release of an anticancer drug at physiological pH. *The Journal of Physical Chemistry B* 113 (35): 11787–11792.

140 Hayashi, C.Y., Shipley, N.H., and Lewis, R.V. (1999). Hypotheses that correlate the sequence, structure, and mechanical properties of spider silk proteins. *International Journal of Biological Macromolecules* 24 (2–3): 271–275.

141 Matsumoto, A., Chen, J., Collette, A.L. et al. (2006). Mechanisms of silk fibroin sol–gel transitions. *The Journal of Physical Chemistry B* 110 (43): 21630–21638.

142 Slotta, U., Hess, S., Spiess, K. et al. (2007). Spider silk and amyloid fibrils: a structural comparison. *Macromolecular Bioscience* 7 (2): 183–188.

143 Aggeli, A., Hamodrakas, S.J., Komitopoulou, K., and Konsolaki, M. (1991). Tandemly repeating peptide motifs and their secondary structure in *Ceratitis capitata* eggshell proteins Ccs36 and Ccs38. *International Journal of Biological Macromolecules* 13 (5): 307–315.

144 Green, N.M., Wrigley, N.G., Russell, W.C. et al. (1983). Evidence for a repeating cross-beta sheet structure in the adenovirus fibre. *The EMBO Journal* 2 (8): 1357–1365.

145 Gelain, F., Horii, A., and Zhang, S. (2007). Designer self-assembling peptide scaffolds for 3-D tissue cell cultures and regenerative medicine. *Macromolecular Bioscience* 7 (5): 544–551.

146 Fung, S.Y., Yang, H., and Chen, P. (2008). Sequence effect of self-assembling peptides on the complexation and in vitro delivery of the hydrophobic anticancer drug ellipticine. *PLoS One* 3 (4): e1956.

147 Mastrobattista, E., van der Aa, M.A., Hennink, W.E., and Crommelin, D.J. (2006). Artificial viruses: a nanotechnological approach to gene delivery. *Nature Reviews Drug Discovery* 5 (2): 115–121.

148 Henricus, M.M., Johnson, K.T., and Banerjee, I.A. (2008). Investigation of insulin loaded self-assembled microtubules for drug release. *Bioconjugate Chemistry* 19 (12): 2394–2400.

149 Lim, Y.-B., Moon, K.-S., and Lee, M. (2009). Recent advances in functional supramolecular nanostructures assembled from bioactive building blocks. *Chemical Society Reviews* 38 (4): 925–934.

150 Chung, E.K., Lee, E., Kim, Y.B., and Lee, M. (2010). Cyclic peptide facial amphiphile preprogrammed to self-assemble into bioactive peptide capsules. *Chemistry A European Journal* 16 (18): 5305–5309.

151 Zhang, J., Hao, R., Huang, L. et al. (2011). Self-assembly of a peptide amphiphile based on hydrolysed *Bombyx mori* silk fibroin. *Chemical Communications* 47 (37): 10296–10298.

152 Wang, Q., Zhang, X., Zheng, J., and Liu, D. (2014). Self-assembled peptide nanotubes as potential nanocarriers for drug delivery. *RSC Advances* 4 (48): 25461–25469.

153 Sadatmousavi, P. (2009). Peptide-mediated anticancer drug delivery. Master's thesis. University of Waterloo.

154 Leonard, S.R., Cormier, A.R., Pang, X. et al. (2013). Solid-state NMR evidence for β-hairpin structure within MAX8 designer peptide nanofibers. *Biophysical Journal* 105 (1): 222–230.

155 Haines-Butterick, L., Rajagopal, K., Branco, M. et al. (2007). Controlling hydrogelation kinetics by peptide design for three-dimensional encapsulation and injectable delivery of cells. *Proceedings of the National Academy of Sciences of the United States of America* 104 (19): 7791–7796.

156 Altunbas, A., Lee, S.J., Rajasekeran, S.A. et al. (2011). Encapsulation of curcumin in self-assembling peptide hydrogels as injectable drug delivery vehicles. *Biomaterials* 32 (25): 5906–5914.

157 Liu, J., Liu, J., Xu, H. et al. (2014). Novel tumor-targeting, self-assembling peptide nanofiber as a carrier for effective curcumin delivery. *International Journal of Nanomedicine* 9: 197–207.

158 Zarzhitsky, S. and Rapaport, H. (2011). The interactions between doxorubicin and amphiphilic and acidic β-sheet peptides towards drug delivery hydrogels. *Journal of Colloid and Interface Science* 360 (2): 525–531.

159 Wu, H., Zhou, T., Tian, L. et al. (2017). Self-assembling RADA16-I peptide hydrogel scaffold loaded with tamoxifen for breast reconstruction. *BioMed Research International* 3656193: 1–10.

160 Fung, S.Y., Yang, H., and Chen, P. (2007). Formation of colloidal suspension of hydrophobic compounds with an amphiphilic self-assembling peptide. *Colloids and Surfaces B: Biointerfaces* 55 (2): 200–211.

161 Fung, S.Y., Yang, H., Bhola, P.T. et al. (2009). Self-assembling peptide as a potential carrier for hydrophobic anticancer drug ellipticine: complexation, release and in vitro delivery. *Advanced Functional Materials* 19 (1): 74–83.

162 Zhang, S. (2002). Emerging biological materials through molecular self-assembly. *Biotechnology Advances* 20 (5): 321–339.

163 Davis, M.E., Hsieh, P.C.H., Takahashi, T. et al. (2006). Local myocardial insulin-like growth factor 1 (IGF-1) delivery with biotinylated peptide nanofibers improves cell therapy for myocardial infarction. *Proceedings of the National Academy of Sciences of the United States of America* 103 (21): 8155–8160.

164 Hsieh, P.C.H., MacGillivray, C., Gannon, J. et al. (2006). Local controlled intramyocardial delivery of platelet-derived growth factor improves postinfarction ventricular function without pulmonary toxicity. *Circulation* 114 (7): 637–644.

165 Segers, V.F.M., Tokunou, T., Higgins, L.J. et al. (2007). Local delivery of protease-resistant stromal cell derived factor-1 for stem cell recruitment after myocardial infarction. *Circulation* 116 (15): 1683–1692.

166 Padin-Iruegas, M.E., Misao, Y., Davis, M.E. et al. (2009). Cardiac progenitor cells and biotinylated insulin-like growth factor-1 nanofibers improve endogenous and exogenous myocardial regeneration after infarction. *Circulation* 120 (10): 876–887.

167 Guo, H.-D., Cui, G.-H., Yang, J.-J. et al. (2012). Sustained delivery of VEGF from designer self-assembling peptides improves cardiac function after myocardial infarction. *Biochemical and Biophysical Research Communications* 424 (1): 105–111.

168 Zhou, A., Chen, S., He, B. et al. (2016). Controlled release of TGF-beta 1 from RADA self-assembling peptide hydrogel scaffolds. *Drug Design, Development and Therapy* 10: 3043–3051.

169 Hsieh, P.C.H., Davis, M.E., Gannon, J. et al. (2006). Controlled delivery of PDGF-BB for myocardial protection using injectable self-assembling peptide nanofibers. *Journal of Clinical Investigation* 116 (1): 237–248.

170 Kim, J.H., Jung, Y., Kim, S.-H. et al. (2011). The enhancement of mature vessel formation and cardiac function in infarcted hearts using dual growth factor delivery with self-assembling peptides. *Biomaterials* 32 (26): 6080–6088.

171 Xu, F.-F., Wang, Y.-C., Sun, S. et al. (2015). Comparison between self-assembling peptide nanofiber scaffold (SAPNS) and fibrin sealant in neurosurgical hemostasis. *Clinical and Translational Science* 8 (5): 490–494.

172 Wu, X., He, L., Li, W. et al. (2017). Functional self-assembling peptide nanofiber hydrogel for peripheral nerve regeneration. *Regenerative Biomaterials* 4 (1): 21–30.

173 Liu, W., Jawerth, L.M., Sparks, E.A. et al. (2006). Fibrin fibers have extraordinary extensibility and elasticity. *Science (New York, NY)* 313 (5787): 634.

174 Brown, A.E.X., Litvinov, R.I., Discher, D.E., and Weisel, J.W. (2007). Forced unfolding of coiled-coils in fibrinogen by single-molecule AFM. *Biophysical Journal* 92 (5): L39–L41.

175 Piechocka, I.K., Bacabac, R.G., Potters, M. et al. (2010). Structural hierarchy governs fibrin gel mechanics. *Biophysical Journal* 98 (10): 2281–2289.

176 Yu, Y.B. (2002). Coiled-coils: stability, specificity, and drug delivery potential. *Advanced Drug Delivery Reviews* 54 (8): 1113–1129.

177 Ulijn, R.V. and Smith, A.M. (2008). Designing peptide based nanomaterials. *Chemical Society Reviews* 37 (4): 664–675.

178 Fletcher, N.L., Lockett, C.V., and Dexter, A.F. (2011). A pH-responsive coiled-coil peptide hydrogel. *Soft Matter* 7 (21): 10210–10218.

179 Banwell, E.F., Abelardo, E.S., Adams, D.J. et al. (2009). Rational design and application of responsive α-helical peptide hydrogels. *Nature Materials* 8 (7): 596–600.

180 Kojima, S., Kuriki, Y., Yoshida, T. et al. (1997). Fibril formation by an amphipathic α-helix-forming polypeptide produced by gene engineering. *Proceedings of the Japan Academy, Series B* 73 (1): 7–11.

181 Zimenkov, Y., Conticello, V.P., Guo, L., and Thiyagarajan, P. (2004). Rational design of a nanoscale helical scaffold derived from self-assembly of a dimeric coiled coil motif. *Tetrahedron* 60 (34): 7237–7246.

182 Kojima, S., Kuriki, Y., Yazaki, K., and Miura, K. (2005). Stabilization of the fibrous structure of an α-helix-forming peptide by sequence reversal. *Biochemical and Biophysical Research Communications* 331 (2): 577–582.

183 Li, W., Nicol, F., and Szoka, F.C. (2004). GALA: a designed synthetic pH-responsive amphipathic peptide with applications in drug and gene delivery. *Advanced Drug Delivery Reviews* 56 (7): 967–985.

184 Melnik, T.N., Villard, V., Vasiliev, V. et al. (2003). Shift of fibril-forming ability of the designed α-helical coiled-coil peptides into the physiological pH region. *Protein Engineering, Design & Selection* 16 (12): 1125–1130.

185 Kajava, A.V., Potekhin, S.A., Corradin, G., and Leapman, R.D. (2004). Organization of designed nanofibrils assembled from α-helical peptides as determined by electron microscopy. *Journal of Peptide Science* 10 (5): 291–297.

186 Villard, V., Kalyuzhniy, O., Riccio, O. et al. (2006). Synthetic RGD-containing α-helical coiled coil peptides promote integrin-dependent cell adhesion. *Journal of Peptide Science* 12 (3): 206–212.

187 Zimenkov, Y., Dublin, S.N., Ni, R. et al. (2006). Rational design of a reversible pH-responsive switch for peptide self-assembly. *Journal of the American Chemical Society* 128 (21): 6770–6771.

188 Wagner, D.E., Phillips, C.L., Ali, W.M. et al. (2005). Toward the development of peptide nanofilaments and nanoropes as smart materials. *Proceedings of the National Academy of Sciences of the United States of America* 102 (36): 12656.

189 Gribbon, C., Channon, K.J., Zhang, W. et al. (2008). MagicWand: a single, designed peptide that assembles to stable, ordered α-helical fibers. *Biochemistry* 47 (39): 10365–10371.

190 Yan, C. and Pochan, D.J. (2010). Rheological properties of peptide-based hydrogels for biomedical and other applications. *Chemical Society Reviews* 39 (9): 3528–3540.

191 Wu, Y. and Collier, J.H. (2017). α-Helical coiled-coil peptide materials for biomedical applications. *Wiley Interdisciplinary Reviews: Nanomedicine and Nanobiotechnology* 9 (2): e1424.

192 Gabrielson, N.P., Lu, H., Yin, L. et al. (2012). Reactive and bioactive cationic α-helical polypeptide template for nonviral gene delivery. *Angewandte Chemie International Edition* 51 (5): 1143–1147.

193 Lee, D., Li, Y.M., Kim, J. et al. (2015). Enhanced tumor-targeted gene delivery by bioreducible polyethylenimine tethering EGFR divalent ligands. *Biomaterials Science* 3 (7): 1096–1104.

194 Caplan, M.R., Schwartzfarb, E.M., Zhang, S. et al. (2002). Control of self-assembling oligopeptide matrix formation through systematic variation of amino acid sequence. *Biomaterials* 23 (1): 219–227.

195 Vauthey, S., Santoso, S., Gong, H. et al. (2002). Molecular self-assembly of surfactant-like peptides to form nanotubes and nanovesicles. *Proceedings of the National Academy of Sciences of the United States of America* 99 (8): 5355–5360.

196 Hamley, I.W. (2011). Self-assembly of amphiphilic peptides. *Soft Matter* 7 (9): 4122–4138.

197 Krogstad, D.V. (2012). *Investigating the Structure-Property Relationships of Aqueous Self-Assembled Materials*, 262. Santa Barbara, CA: University of California.

198 Mazza, M., Patel, A., Pons, R. et al. (2014). Peptide nanofibres as molecular transporters: from self-assembly to in vivo degradation. *Faraday Discussions* 166: 181–194.

199 Mazza, M., Norman, R., Anwar, J. et al. (2013). Nanofiber-based delivery of therapeutic peptides to the brain. *ACS Nano* 7 (2): 1016–1026.

200 Kim, J.-K., Anderson, J., Jun, H.-W. et al. (2009). Self-assembling peptide amphiphile-based nanofiber gel for bioresponsive cisplatin delivery. *Molecular Pharmaceutics* 6 (3): 978–985.

201 Sardan, M., Kilinc, M., Genc, R. et al. (2014). Cell penetrating peptide amphiphile integrated liposomal systems for enhanced delivery of anticancer drugs to tumor cells. *Faraday Discussions* 166: 269–283.

202 Boekhoven, J., Zha, R.H., Tantakitti, F. et al. (2015). Alginate–peptide amphiphile core–shell microparticles as a targeted drug delivery system. *RSC Advances* 5 (12): 8753–8756.

203 Ghanaati, S., Webber, M.J., Unger, R.E. et al. (2009). Dynamic in vivo biocompatibility of angiogenic peptide amphiphile nanofibers. *Biomaterials* 30 (31): 6202–6212.

204 Chow, L.W., Wang, L.J., Kaufman, D.B., and Stupp, S.I. (2010). Self-assembling nanostructures to deliver angiogenic factors to pancreatic islets. *Biomaterials* 31 (24): 6154–6161.

205 Angeloni, N.L., Bond, C.W., Tang, Y. et al. (2011). Regeneration of the cavernous nerve by sonic hedgehog using aligned peptide amphiphile nanofibers. *Biomaterials* 32 (4): 1091–1101.

206 Preslar, A.T., Parigi, G., McClendon, M.T. et al. (2014). Gd(III)-labeled peptide nanofibers for reporting on biomaterial localization in vivo. *ACS Nano* 8 (7): 7325–7332.

207 Kim, H.-J., Kim, T., and Lee, M. (2011). Responsive nanostructures from aqueous assembly of rigid–flexible block molecules. *Accounts of Chemical Research* 44 (1): 72–82.

208 Guo, X.D., Tandiono, F., Wiradharma, N. et al. (2008). Cationic micelles self-assembled from cholesterol-conjugated oligopeptides as an efficient gene delivery vector. *Biomaterials* 29 (36): 4838–4846.

209 Morris, M.C., Vidal, P., Chaloin, L. et al. (1997). A new peptide vector for efficient delivery of oligonucleotides into mammalian cells. *Nucleic Acids Research* 25 (14): 2730–2736.

210 Simeoni, F., Morris, M.C., Heitz, F., and Divita, G. (2003). Insight into the mechanism of the peptide-based gene delivery system MPG: implications for delivery of siRNA into mammalian cells. *Nucleic Acids Research* 31 (11): 2717–2724.

211 Crombez, L., Morris, M.C., Deshayes, S. et al. (2008). Peptide-based nanoparticle for ex vivo and in vivo drug delivery. *Current Pharmaceutical Design* 14 (34): 3656–3665.

212 Wiradharma, N., Tong, Y.W., and Yang, Y.-Y. (2009). Self-assembled oligopeptide nanostructures for co-delivery of drug and gene with synergistic therapeutic effect. *Biomaterials* 30 (17): 3100–3109.

213 Ouyang, L., Huang, W., He, G. et al. (2009). Bone targeting prodrugs based on peptide dendrimers, synthesis and hydroxyapatite binding in vitro. 6 (4): 272–277.

214 Jiang, T., Yu, X., Carbone, E.J. et al. (2014). Poly aspartic acid peptide-linked PLGA based nanoscale particles: potential for bone-targeting drug delivery applications. *International Journal of Pharmaceutics* 475 (1–2): 547–557.

215 Sone, E.D. and Stupp, S.I. (2004). Semiconductor-encapsulated peptide–amphiphile nanofibers. *Journal of the American Chemical Society* 126 (40): 12756–12757.

216 Palmer, L.C., Newcomb, C.J., Kaltz, S.R. et al. (2008). Biomimetic systems for hydroxyapatite mineralization inspired by bone and enamel. *Chemical Reviews* 108 (11): 4754–4783.

217 Spoerke, E.D., Anthony, S.G., and Stupp, S. (2009). Enzyme directed templating of artificial bone mineral. *Advanced Materials* 21 (4): 425–430.

218 Sutton, S., Campbell, N.L., Cooper, A.I. et al. (2009). Controlled release from modified amino acid hydrogels governed by molecular size or network dynamics. *Langmuir* 25 (17): 10285–10291.

219 Polo-Corrales, L., Latorre-Esteves, M., and Ramirez-Vick, J.E. (2014). Scaffold design for bone regeneration. *Journal of Nanoscience and Nanotechnology* 14 (1): 15–56.

220 Rapaport, H. (2014). Amphiphilic peptides and hydrogel matrices thereof for bone repair. Ben Gurion University of Negev Research and Development Authority Ltd. United States patent US 8,658,763.

221 Webber, M.J., Newcomb, C.J., Bitton, R., and Stupp, S.I. (2011). Switching of self-assembly in a peptide nanostructure with a specific enzyme. *Soft Matter* 7 (20): 9665–9672.

222 Deming, T. (ed.) (2012). *Peptide-Based Materials*, Topics in Current Chemistry. Berlin, Heidelberg: Springer-Verlag.

223 Shi, P., Gustafson, J.A., and MacKay, J.A. (2014). Genetically engineered nanocarriers for drug delivery. *International Journal of Nanomedicine* 9: 1617–1626.

224 McDaniel, J.R., Callahan, D.J., and Chilkoti, A. (2010). Drug delivery to solid tumors by elastin-like polypeptides. *Advanced Drug Delivery Reviews* 62 (15): 1456–1467.

225 Amiram, M., Luginbuhl, K.M., Li, X. et al. (2013). Injectable protease-operated depots of glucagon-like peptide-1 provide extended and tunable glucose control. *Proceedings of the National Academy of Sciences of the United States of America* 110 (8): 2792–2797.

226 MacEwan, S.R. and Chilkoti, A. (2014). Applications of elastin-like polypeptides in drug delivery. *Journal of Controlled Release* 190: 314–330.
227 Bidwell, G.L. and Raucher, D. (2010). Cell penetrating elastin-like polypeptides for therapeutic peptide delivery. *Advanced Drug Delivery Reviews* 62 (15): 1486–1496.
228 Ryu, J.S. and Raucher, D. (2014). Anti-tumor efficacy of a therapeutic peptide based on thermo-responsive elastin-like polypeptide in combination with gemcitabine. *Cancer Letters* 348 (1): 177–184.
229 McDaniel, J.R., Macewan, S.R., Dewhirst, M., and Chilkoti, A. (2012). Doxorubicin-conjugated chimeric polypeptide nanoparticles that respond to mild hyperthermia. *Journal of Controlled Release* 159 (3): 362–367.
230 Moktan, S., Perkins, E., Kratz, F., and Raucher, D. (2012). Thermal targeting of an acid-sensitive doxorubicin conjugate of elastin-like polypeptide enhances the therapeutic efficacy compared with the parent compound in vivo. *Molecular Cancer Therapeutics* 11 (7): 1547–1556.
231 Bidwell, G.L. III, Fokt, I., Priebe, W., and Raucher, D. (2007). Development of elastin-like polypeptide for thermally targeted delivery of doxorubicin. *Biochemical Pharmacology* 73 (5): 620–631.
232 Walker, L., Perkins, E., Kratz, F., and Raucher, D. (2012). Cell penetrating peptides fused to a thermally targeted biopolymer drug carrier improve the delivery and antitumor efficacy of an acid-sensitive doxorubicin derivative. *International Journal of Pharmaceutics* 436 (1–2): 825–832.
233 Horne, W.S., Stout, C.D., and Ghadiri, M.R. (2003). A heterocyclic peptide nanotube. *Journal of the American Chemical Society* 125 (31): 9372–9376.
234 Ashkenasy, N., Horne, W.S., and Ghadiri, M.R. (2006). Design of self-assembling peptide nanotubes with delocalized electronic states. *Small* 2 (1): 99–102.
235 Brea, R.J., Reiriz, C., and Granja, J.R. (2010). Towards functional bionanomaterials based on self-assembling cyclic peptide nanotubes. *Chemical Society Reviews* 39 (5): 1448–1456.
236 Liu, H., Chen, J., Shen, Q. et al. (2010). Molecular insights on the cyclic peptide nanotube-mediated transportation of antitumor drug 5-fluorouracil. *Molecular Pharmaceutics* 7 (6): 1985–1994.
237 Wang, Y., Yi, S., Sun, L. et al. (2014). Doxorubicin-loaded cyclic peptide nanotube bundles overcome chemoresistance in breast cancer cells. *Journal of Biomedical Nanotechnology* 10 (3): 445–454.
238 Heitz, F., Morris, M.C., and Divita, G. (2009). Twenty years of cell-penetrating peptides: from molecular mechanisms to therapeutics. *British Journal of Pharmacology* 157 (2): 195–206.
239 Pangburn, T.O., Petersen, M.A., Waybrant, B. et al. (2009). Peptide- and aptamer-functionalized nanovectors for targeted delivery of therapeutics. *Journal of Biomechanical Engineering* 131 (7): 074005.
240 Blanco, E., Hsiao, A., Mann, A.P. et al. (2011). Nanomedicine in cancer therapy: innovative trends and prospects. *Cancer Science* 102 (7): 1247–1252.
241 Muro, S. (2012). Challenges in design and characterization of ligand-targeted drug delivery systems. *Journal of Controlled Release* 164 (2): 125–137.

242 Pearce, T.R., Shroff, K., and Kokkoli, E. (2012). Peptide targeted lipid nanoparticles for anticancer drug delivery. *Advanced Materials* 24 (28): 3803–3822.
243 Chang, D.-K., Chiu, C.-Y., Kuo, S.-Y. et al. (2009). Antiangiogenic targeting liposomes increase therapeutic efficacy for solid tumors. *The Journal of Biological Chemistry* 284 (19): 12905–12916.
244 Chang, D.-K., Lin, C.-T., Wu, C.-H., and Wu, H.-C. (2009). A novel peptide enhances therapeutic efficacy of liposomal anti-cancer drugs in mice models of human lung cancer. *PLoS One* 4 (1): e4171.
245 Mai, J., Song, S., Rui, M. et al. (2009). A synthetic peptide mediated active targeting of cisplatin liposomes to Tie2 expressing cells. *Journal of Controlled Release* 139 (3): 174–181.
246 Meng, S., Su, B., Li, W. et al. (2010). Enhanced antitumor effect of novel dual-targeted paclitaxel liposomes. *Nanotechnology* 21 (41): 415103.
247 Zhang, J., Jin, W., Wang, X. et al. (2010). A novel octreotide modified lipid vesicle improved the anticancer efficacy of doxorubicin in somatostatin receptor 2 positive tumor models. *Molecular Pharmaceutics* 7 (4): 1159–1168.
248 He, X., Na, M.-H., Kim, J.-S. et al. (2011). A novel peptide probe for imaging and targeted delivery of liposomal doxorubicin to lung tumor. *Molecular Pharmaceutics* 8 (2): 430–438.
249 Lowery, A., Onishko, H., Hallahan, D.E., and Han, Z. (2011). Tumor-targeted delivery of liposome-encapsulated doxorubicin by use of a peptide that selectively binds to irradiated tumors. *Journal of Controlled Release* 150 (1): 117–124.
250 Helbok, A., Rangger, C., von Guggenberg, E. et al. (2012). Targeting properties of peptide-modified radiolabeled liposomal nanoparticles. *Nanomedicine: Nanotechnology, Biology and Medicine* 8 (1): 112–118.
251 Katanasaka, Y., Ishii, T., Asai, T. et al. (2010). Cancer antineovascular therapy with liposome drug delivery systems targeted to BiP/GRP78. *International Journal of Cancer* 127 (11): 2685–2698.
252 Murase, Y., Asai, T., Katanasaka, Y. et al. (2010). A novel DDS strategy, "dual-targeting", and its application for antineovascular therapy. *Cancer Letters* 287 (2): 165–171.
253 Rivera-Fillat, M.P., Reig, F., Martínez, E.M., and Grau-Oliete, M.R. (2010). Improved therapeutic responses for liposomal doxorubicin targeted via thrombospondin peptidomimetics versus untargeted doxorubicin. *Journal of Peptide Science* 16 (7): 315–321.
254 Falciani, C., Accardo, A., Brunetti, J. et al. (2011). Target-selective drug delivery through liposomes labeled with oligobranched neurotensin peptides. *ChemMedChem* 6 (4): 678–685.
255 Kuai, R., Yuan, W., Li, W. et al. (2011). Targeted delivery of cargoes into a murine solid tumor by a cell-penetrating peptide and cleavable poly(ethylene glycol) comodified liposomal delivery system via systemic administration. *Molecular Pharmaceutics* 8 (6): 2151–2161.

256 Dubey, P.K., Singodia, D., and Vyas, S.P. (2010). Liposomes modified with YIGSR peptide for tumor targeting. *Journal of Drug Targeting* 18 (5): 373–380.

257 Takara, K., Hatakeyama, H., Ohga, N. et al. (2010). Design of a dual-ligand system using a specific ligand and cell penetrating peptide, resulting in a synergistic effect on selectivity and cellular uptake. *International Journal of Pharmaceutics* 396 (1–2): 143–148.

258 Herringson, T.P. and Altin, J.G. (2011). Effective tumor targeting and enhanced anti-tumor effect of liposomes engrafted with peptides specific for tumor lymphatics and vasculature. *International Journal of Pharmaceutics* 411 (1–2): 206–214.

259 Herringson, T.P. and Altin, J.G. (2011). Increasing the antitumor efficacy of doxorubicin-loaded liposomes with peptides anchored via a chelator lipid. *Journal of Drug Targeting* 19 (8): 681–689.

260 Morisco, A., Accardo, A., Tesauro, D. et al. (2011). Peptide-labeled supramolecular aggregates as selective doxorubicin carriers for delivery to tumor cells. *Biopolymers* 96 (1): 88–96.

261 Garg, A., Tisdale, A.W., Haidari, E., and Kokkoli, E. (2009). Targeting colon cancer cells using PEGylated liposomes modified with a fibronectin-mimetic peptide. *International Journal of Pharmaceutics* 366 (1–2): 201–210.

262 Dai, W., Yang, T., Wang, X. et al. (2010). PHSCNK-modified and doxorubicin-loaded liposomes as a dual targeting system to integrin-overexpressing tumor neovasculature and tumor cells. *Journal of Drug Targeting* 18 (4): 254–263.

263 He, Y., Zhang, L., and Song, C. (2010). Luteinizing hormone-releasing hormone receptor-mediated delivery of mitoxantrone using LHRH analogs modified with PEGylated liposomes. *International Journal of Nanomedicine* 5: 697–705.

264 Negussie, A.H., Miller, J.L., Reddy, G. et al. (2010). Synthesis and in vitro evaluation of cyclic NGR peptide targeted thermally sensitive liposome. *Journal of Controlled Release* 143 (2): 265–273.

265 Wang, T., D'Souza, G.G., Bedi, D. et al. (2010). Enhanced binding and killing of target tumor cells by drug-loaded liposomes modified with tumor-specific phage fusion coat protein. *Nanomedicine (London, England)* 5 (4): 563–574.

266 Wang, T., Yang, S., Petrenko, V.A., and Torchilin, V.P. (2010). Cytoplasmic delivery of liposomes into MCF-7 breast cancer cells mediated by cell-specific phage fusion coat protein. *Molecular Pharmaceutics* 7 (4): 1149–1158.

267 Garg, A. and Kokkoli, E. (2011). pH-sensitive PEGylated liposomes functionalized with a fibronectin-mimetic peptide show enhanced intracellular delivery to colon cancer cell. *Current Pharmaceutical Biotechnology* 12 (8): 1135–1143.

268 Kang, M.J., Lee, S., Kim, B.K. et al. (2011). Pep-1 peptide-modified liposomal carriers for intracellular delivery of gold nanoparticles. *Chemical and Pharmaceutical Bulletin* 59 (1): 109–112.

269 Kuil, J., Velders, A.H., and van Leeuwen, F.W.B. (2010). Multimodal tumor-targeting peptides functionalized with both a radio- and a fluorescent label. *Bioconjugate Chemistry* 21 (10): 1709–1719.

270 Dissanayake, S., Denny, W.A., Gamage, S., and Sarojini, V. (2017). Recent developments in anticancer drug delivery using cell penetrating and tumor targeting peptides. *Journal of Controlled Release* 250: 62–76.

271 Pasqualini, R., Koivunen, E., Kain, R. et al. (2000). Aminopeptidase N is a receptor for tumor-homing peptides and a target for inhibiting angiogenesis. *Cancer Research* 60 (3): 722–727.

272 Oku, N., Asai, T., Watanabe, K. et al. (2002). Anti-neovascular therapy using novel peptides homing to angiogenic vessels. *Oncogene* 21 (17): 2662–2669.

273 Colombo, G., Curnis, F., De Mori, G.M.S. et al. (2002). Structure-activity relationships of linear and cyclic peptides containing the NGR tumor-homing motif. *The Journal of Biological Chemistry* 277 (49): 47891–47897.

274 Marchiò, S., Lahdenranta, J., Schlingemann, R.O. et al. (2004). Aminopeptidase A is a functional target in angiogenic blood vessels. *Cancer Cell* 5 (2): 151–162.

275 Loi, M., Di Paolo, D., Soster, M. et al. (2013). Novel phage display-derived neuroblastoma-targeting peptides potentiate the effect of drug nanocarriers in preclinical settings. *Journal of Controlled Release* 170 (2): 233–241.

276 Loi, M., Marchiò, S., Becherini, P. et al. (2010). Combined targeting of perivascular and endothelial tumor cells enhances anti-tumor efficacy of liposomal chemotherapy in neuroblastoma. *Journal of Controlled Release* 145 (1): 66–73.

277 Zhao, H., Wang, J.-C., Sun, Q.-S. et al. (2009). RGD-based strategies for improving antitumor activity of paclitaxel-loaded liposomes in nude mice xenografted with human ovarian cancer. *Journal of Drug Targeting* 17 (1): 10–18.

278 Du, H., Cui, C., Wang, L. et al. (2011). Novel tetrapeptide, RGDF, mediated tumor specific liposomal doxorubicin (DOX) preparations. *Molecular Pharmaceutics* 8 (4): 1224–1232.

279 Jiang, J., Yang, S.-J., Wang, J.-C. et al. (2010). Sequential treatment of drug-resistant tumors with RGD-modified liposomes containing siRNA or doxorubicin. *European Journal of Pharmaceutics and Biopharmaceutics* 76 (2): 170–178.

280 Naik, S., Patel, D., Chuttani, K. et al. (2012). In vitro mechanistic study of cell death and in vivo performance evaluation of RGD grafted PEGylated docetaxel liposomes in breast cancer. *Nanomedicine: Nanotechnology, Biology and Medicine* 8 (6): 951–962.

281 Veneti, E., Tu, R.S., and Auguste, D.T. (2016). RGD-targeted liposome binding and uptake on breast cancer cells is dependent on elastin linker secondary structure. *Bioconjugate Chemistry* 27 (8): 1813–1821.

282 Wen, X., Li, J., Cai, D. et al. (2018). Anticancer efficacy of targeted shikonin liposomes modified with RGD in breast cancer cells. *Molecules (Basel, Switzerland)* 23 (2): 268.

283 Amin, M., Mansourian, M., Koning, G.A. et al. (2015). Development of a novel cyclic RGD peptide for multiple targeting approaches of liposomes to tumor region. *Journal of Controlled Release* 220: 308–315.
284 Cheng, Y. and Ji, Y. (2019). RGD-modified polymer and liposome nanovehicles: recent research progress for drug delivery in cancer therapeutics. *European Journal of Pharmaceutical Sciences* 128: 8–17.
285 Liu, X.-Y., Ruan, L.-M., Mao, W.-W. et al. (2010). Preparation of RGD-modified long circulating liposome loading matrine, and its in vitro anti-cancer effects. *International Journal of Medical Sciences* 7 (4): 197–208.
286 Li, W., Su, B., Meng, S. et al. (2011). RGD-targeted paramagnetic liposomes for early detection of tumor: in vitro and in vivo studies. *European Journal of Radiology* 80 (2): 598–606.
287 Kokkoli, E., Mardilovich, A., Wedekind, A. et al. (2006). Self-assembly and applications of biomimetic and bioactive peptide-amphiphiles. *Soft Matter* 2 (12): 1015–1024.
288 Mardilovich, A., Craig, J.A., McCammon, M.Q. et al. (2006). Design of a novel fibronectin-mimetic peptide-amphiphile for functionalized biomaterials. *Langmuir* 22 (7): 3259–3264.
289 Pangburn, T.O., Bates, F.S., and Kokkoli, E. (2012). Polymersomes functionalized via "click" chemistry with the fibronectin mimetic peptides PR_b and GRGDSP for targeted delivery to cells with different levels of $\alpha_5\beta_1$ expression. *Soft Matter* 8 (16): 4449–4461.
290 Shroff, K., Pearce, T.R., and Kokkoli, E. (2012). Enhanced integrin mediated signaling and cell cycle progression on fibronectin mimetic peptide amphiphile monolayers. *Langmuir* 28 (3): 1858–1865.
291 Demirgöz, D., Garg, A., and Kokkoli, E. (2008). PR_b-targeted PEGylated liposomes for prostate cancer therapy. *Langmuir* 24 (23): 13518–13524.
292 Demirgöz, D., Pangburn, T.O., Davis, K.P. et al. (2009). PR_b-targeted delivery of tumor necrosis factor-α by polymersomes for the treatment of prostate cancer. *Soft Matter* 5 (10): 2011–2019.
293 Atchison, N.A., Fan, W., Papas, K.K. et al. (2010). Binding of the fibronectin-mimetic peptide, PR_b, to $\alpha_5\beta_1$ on pig islet cells increases fibronectin production and facilitates internalization of PR_b functionalized liposomes. *Langmuir* 26 (17): 14081–14088.
294 Bertram, J.P., Williams, C.A., Robinson, R. et al. (2009). Intravenous hemostat: nanotechnology to halt bleeding. *Science Translational Medicine* 1 (11): 11ra22.
295 Nishiya, T., Kainoh, M., Murata, M. et al. (2001). Platelet interactions with liposomes carrying recombinant platelet membrane glycoproteins or fibrinogen: approach to platelet substitutes. *Artificial Cells Blood Substitutes and Immobilization Biotechnology* 29 (6): 453–464.
296 Nishiya, T., Kainoh, M., Murata, M. et al. (2002). Reconstitution of adhesive properties of human platelets in liposomes carrying both recombinant glycoproteins Ia/IIa and Ibα under flow conditions: specific synergy of receptor-ligand interactions. *Blood* 100 (1): 136–142.

297 Takeoka, S., Teramura, Y., Okamura, Y. et al. (2002). Rolling properties of rGPIbα-conjugated phospholipid vesicles with different membrane flexibilities on vWf surface under flow conditions. *Biochemical and Biophysical Research Communications* 296 (3): 765–770.
298 Takeoka, S., Okamura, Y., Teramura, Y. et al. (2003). Function of fibrinogen gamma-chain dodecapeptide-conjugated latex beads under flow. *Biochemical and Biophysical Research Communications* 312 (3): 773–779.
299 Okamura, Y., Maekawa, I., Teramura, Y. et al. (2005). Hemostatic effects of phospholipid vesicles carrying fibrinogen γ chain dodecapeptide in vitro and in vivo. *Bioconjugate Chemistry* 16 (6): 1589–1596.
300 Okamura, Y., Takeoka, S., Teramura, Y. et al. (2005). Hemostatic effects of fibrinogen gamma-chain dodecapeptide-conjugated polymerized albumin particles in vitro and in vivo. *Transfusion* 45 (7): 1221–1228.
301 Okamura, Y., Fujie, T., Maruyama, H. et al. (2007). Prolonged hemostatic ability of polyethylene glycol-modified polymerized albumin particles carrying fibrinogen gamma-chain dodecapeptide. *Transfusion* 47 (7): 1254–1262.
302 Okamura, Y., Fujie, T., Nogawa, M. et al. (2008). Haemostatic effects of polymerized albumin particles carrying fibrinogen gamma-chain dodecapeptide as platelet substitutes in severely thrombocytopenic rabbits. *Transfusion Medicine (Oxford, England)* 18 (3): 158–166.
303 Okamura, Y., Takeoka, S., Eto, K. et al. (2009). Development of fibrinogen gamma-chain peptide-coated, adenosine diphosphate-encapsulated liposomes as a synthetic platelet substitute. *Journal of Thrombosis and Haemostasis : JTH* 7 (3): 470–477.
304 Ravikumar, M., Modery, C.L., Wong, T.L., and Gupta, A.S. (2012). Peptide-decorated liposomes promote arrest and aggregation of activated platelets under flow on vascular injury relevant protein surfaces in vitro. *Biomacromolecules* 13 (5): 1495–1502.
305 Anselmo, A.C., Modery-Pawloski, C.L., Menegatti, S. et al. (2014). Platelet-like nanoparticles: mimicking shape, flexibility, and surface biology of platelets to target vascular injuries. *ACS Nano* 8 (11): 11243–11253.
306 Snyder, E.L. and Dowdy, S.F. (2005). Recent advances in the use of protein transduction domains for the delivery of peptides, proteins and nucleic acids in vivo. *Expert Opinion on Drug Delivery* 2 (1): 43–51.
307 Munyendo, W.L.L., Lv, H., Benza-Ingoula, H. et al. (2012). Cell penetrating peptides in the delivery of biopharmaceuticals. *Biomolecules* 2 (2): 187–202.
308 Cheng, C.J. and Saltzman, W.M. (2011). Enhanced siRNA delivery into cells by exploiting the synergy between targeting ligands and cell-penetrating peptides. *Biomaterials* 32 (26): 6194–6203.
309 Ezzat, K., Zaghloul, E.M., El Andaloussi, S. et al. (2012). Solid formulation of cell-penetrating peptide nanocomplexes with siRNA and their stability in simulated gastric conditions. *Journal of Controlled Release* 162 (1): 1–8.
310 Nakase, I., Tanaka, G., and Futaki, S. (2013). Cell-penetrating peptides (CPPs) as a vector for the delivery of siRNAs into cells. *Molecular Biosystems* 9 (5): 855–861.

311 Bai, H., You, Y., Yan, H. et al. (2012). Antisense inhibition of gene expression and growth in gram-negative bacteria by cell-penetrating peptide conjugates of peptide nucleic acids targeted to rpoD gene. *Biomaterials* 33 (2): 659–667.
312 Nakase, I., Akita, H., Kogure, K. et al. (2012). Efficient intracellular delivery of nucleic acid pharmaceuticals using cell-penetrating peptides. *Accounts of Chemical Research* 45 (7): 1132–1139.
313 Lindgren, M., Rosenthal-Aizman, K., Saar, K. et al. (2006). Overcoming methotrexate resistance in breast cancer tumour cells by the use of a new cell-penetrating peptide. *Biochemical Pharmacology* 71 (4): 416–425.
314 Lee, J.Y., Bae, K.H., Kim, J.S. et al. (2011). Intracellular delivery of paclitaxel using oil-free, shell cross-linked HSA-multi-armed PEG nanocapsules. *Biomaterials* 32 (33): 8635–8644.
315 Liu, B.R., Huang, Y.-W., and Lee, H.-J. (2013). Mechanistic studies of intracellular delivery of proteins by cell-penetrating peptides in cyanobacteria. *BMC Microbiology* 13: 57.
316 Liu, B.R., Liou, J.-S., Chen, Y.-J. et al. (2013). Delivery of nucleic acids, proteins, and nanoparticles by arginine-rich cell-penetrating peptides in rotifers. *Marine Biotechnology (New York, NY)* 15 (5): 584–595.
317 Mäe, M. and Langel, Ü. (2006). Cell-penetrating peptides as vectors for peptide, protein and oligonucleotide delivery. *Current Opinion in Pharmacology* 6 (5): 509–514.
318 Järver, P., Mäger, I., and Langel, Ü. (2010). In vivo biodistribution and efficacy of peptide mediated delivery. *Trends in Pharmacological Sciences* 31 (11): 528–535.
319 Lewin, M., Carlesso, N., Tung, C.H. et al. (2000). Tat peptide-derivatized magnetic nanoparticles allow in vivo tracking and recovery of progenitor cells. *Nature Biotechnology* 18 (4): 410–414.
320 Bolhassani, A. (2011). Potential efficacy of cell-penetrating peptides for nucleic acid and drug delivery in cancer. *Biochimica et Biophysica Acta (BBA) - Reviews on Cancer* 1816 (2): 232–246.
321 Chen, H.-C.G., Chiou, S.-T., Zheng, J.-Y. et al. (2011). The nuclear localization signal sequence of porcine circovirus type 2 ORF2 enhances intracellular delivery of plasmid DNA. *Archives of Virology* 156 (5): 803–815.
322 Wang, H.-Y., Chen, J.-X., Sun, Y.-X. et al. (2011). Construction of cell penetrating peptide vectors with N-terminal stearylated nuclear localization signal for targeted delivery of DNA into the cell nuclei. *Journal of Controlled Release* 155 (1): 26–33.
323 Kim, B.-K., Kang, H., Doh, K.-O. et al. (2012). Homodimeric SV40 NLS peptide formed by disulfide bond as enhancer for gene delivery. *Bioorganic and Medicinal Chemistry Letters* 22 (17): 5415–5418.
324 Yin, J., Meng, X., Zhang, S. et al. (2012). The effect of a nuclear localization sequence on transfection efficacy of genes delivered by cobalt(II)-polybenzimidazole complexes. *Biomaterials* 33 (31): 7884–7894.
325 Barnett, E.M., Zhang, X., Maxwell, D. et al. (2009). Single-cell imaging of retinal ganglion cell apoptosis with a cell-penetrating, activatable peptide

probe in an in vivo glaucoma model. *Proceedings of the National Academy of Sciences of the United States of America* 106 (23): 9391–9396.

326 Kamei, N., Morishita, M., Kanayama, Y. et al. (2010). Molecular imaging analysis of intestinal insulin absorption boosted by cell-penetrating peptides by using positron emission tomography. *Journal of Controlled Release* 146 (1): 16–22.

327 Nguyen, Q.T., Olson, E.S., Aguilera, T.A. et al. (2010). Surgery with molecular fluorescence imaging using activatable cell-penetrating peptides decreases residual cancer and improves survival. *Proceedings of the National Academy of Sciences of the United States of America* 107 (9): 4317–4322.

328 Olson, E.S., Jiang, T., Aguilera, T.A. et al. (2010). Activatable cell penetrating peptides linked to nanoparticles as dual probes for in vivo fluorescence and MR imaging of proteases. *Proceedings of the National Academy of Sciences of the United States of America* 107 (9): 4311–4316.

329 Huang, C.-W., Li, Z., and Conti, P.S. (2011). In vivo near-infrared fluorescence imaging of integrin $\alpha_2\beta_1$ in prostate cancer with cell-penetrating-peptide-conjugated DGEA probe. *Journal of Nuclear Medicine* 52 (12): 1979–1986.

330 Zhai, X.-H., Liu, M., Guo, X.-J. et al. (2011). SKOV-3 cell imaging by paramagnetic particles labeled with hairpin cell-penetrating peptides. *Chinese Medical Journal* 124 (1): 111–117.

331 Olson, E.S., Whitney, M.A., Friedman, B. et al. (2012). In vivo fluorescence imaging of atherosclerotic plaques with activatable cell-penetrating peptides targeting thrombin activity. *Integrative Biology: Quantitative Biosciences from Nano to Macro* 4 (6): 595–605.

332 Koren, E., Apte, A., Jani, A., and Torchilin, V.P. (2012). Multifunctional PEGylated 2C5-immunoliposomes containing pH-sensitive bonds and TAT peptide for enhanced tumor cell internalization and cytotoxicity. *Journal of Controlled Release* 160 (2): 264–273.

333 Virès, E., Granier, C., Prevot, P., and Lebleu, B. (1997). Structure-activity relationship study of the plasma membrane translocating potential of a short peptide from HIV-1 Tat protein. *Letters in Peptide Science* 4 (4): 429–436.

334 Koren, E., Apte, A., Sawant, R.R. et al. (2011). Cell-penetrating TAT peptide in drug delivery systems: proteolytic stability requirements. *Drug Delivery* 18 (5): 377–384.

335 Derossi, D., Calvet, S., Trembleau, A. et al. (1996). Cell internalization of the third helix of the antennapedia homeodomain is receptor-independent. *The Journal of Biological Chemistry* 271 (30): 18188–18193.

336 Fischer, P.M., Zhelev, N.Z., Wang, S. et al. (2000). Structure-activity relationship of truncated and substituted analogues of the intracellular delivery vector Penetratin. *The Journal of Peptide Research* 55 (2): 163–172.

337 Elliott, G. and O'Hare, P. (1997). Intercellular trafficking and protein delivery by a herpesvirus structural protein. *Cell* 88 (2): 223–233.

338 Lin, Y.Z., Yao, S.Y., Veach, R.A. et al. (1995). Inhibition of nuclear translocation of transcription factor NF-κB by a synthetic peptide containing a cell membrane-permeable motif and nuclear localization sequence. *The Journal of Biological Chemistry* 270 (24): 14255–14258.

339 Lin, Y.Z., Yao, S.Y., and Hawiger, J. (1996). Role of the nuclear localization sequence in fibroblast growth factor-1-stimulated mitogenic pathways. *The Journal of Biological Chemistry* 271 (10): 5305–5308.

340 Rojas, M., Donahue, J.P., Tan, Z., and Lin, Y.-Z. (1998). Genetic engineering of proteins with cell membrane permeability. *Nature Biotechnology* 16 (4): 370–375.

341 Anderson, D.C., Manger, R., Schroeder, J. et al. (1993). Enhanced in vitro tumor cell retention and internalization of antibody derivatized with synthetic peptides. *Bioconjugate Chemistry* 4 (1): 10–18.

342 Anderson, D.C., Nichols, E., Manger, R. et al. (1993). Tumor cell retention of antibody Fab fragments is enhanced by an attached HIV TAT protein-derived peptide. *Biochemical and Biophysical Research Communications* 194 (2): 876–884.

343 Vives, E., Charneau, P., van Rietschoten, J. et al. (1994). Effects of the Tat basic domain on human immunodeficiency virus type 1 transactivation, using chemically synthesized Tat protein and Tat peptides. *Journal of Virology* 68 (5): 3343–3353.

344 Bidwell, G.L. and Raucher, D. (2005). Application of thermally responsive polypeptides directed against c-Myc transcriptional function for cancer therapy. *Molecular Cancer Therapeutics* 4 (7): 1076–1085.

345 Massodi, I., Bidwell, G.L., and Raucher, D. (2005). Evaluation of cell penetrating peptides fused to elastin-like polypeptide for drug delivery. *Journal of Controlled Release* 108 (2–3): 396–408.

346 Bidwell, G.L., Davis, A.N., and Raucher, D. (2009). Targeting a c-Myc inhibitory polypeptide to specific intracellular compartments using cell penetrating peptides. *Journal of Controlled Release* 135 (1): 2–10.

347 Massodi, I., Thomas, E., and Raucher, D. (2009). Application of thermally responsive elastin-like polypeptide fused to a lactoferrin-derived peptide for treatment of pancreatic cancer. *Molecules (Basel, Switzerland)* 14 (6): 1999–2015.

348 Massodi, I., Moktan, S., Rawat, A. et al. (2010). Inhibition of ovarian cancer cell proliferation by a cell cycle inhibitory peptide fused to a thermally responsive polypeptide carrier. *International Journal of Cancer* 126 (2): 533–544.

349 Stauber, R.H. and Pavlakis, G.N. (1998). Intracellular trafficking and interactions of the HIV-1 Tat protein. *Virology* 252 (1): 126–136.

350 Schwarze, S.R., Ho, A., Vocero-Akbani, A., and Dowdy, S.F. (1999). In vivo protein transduction: delivery of a biologically active protein into the mouse. *Science (New York, NY)* 285 (5433): 1569–1572.

351 Vocero-Akbani, A.M., Vader Heyden, N., Lissy, N.A. et al. (1999). Killing HIV-infected cells by transduction with an HIV protease-activated caspase-3 protein. *Nature Medicine* 5 (1): 29–33.

352 Caron, N.J., Torrente, Y., Camirand, G. et al. (2001). Intracellular delivery of a Tat-eGFP fusion protein into muscle cells. *Molecular Therapy* 3 (3): 310–318.

353 Shamji, M.F., Chen, J., Friedman, A.H. et al. (2008). Synthesis and characterization of a thermally-responsive tumor necrosis factor antagonist. *Journal of Controlled Release* 129 (3): 179–186.

354 Astriab-Fisher, A., Sergueev, D.S., Fisher, M. et al. (2000). Antisense inhibition of P-glycoprotein expression using peptide–oligonucleotide conjugates. *Biochemical Pharmacology* 60 (1): 83–90.

355 Tung, C.-H., Mueller, S., and Weissleder, R. (2002). Novel branching membrane translocational peptide as gene delivery vector. *Bioorganic and Medicinal Chemistry* 10 (11): 3609–3614.

356 Chen, T.-H.H., Bae, Y., and Furgeson, D.Y. (2008). Intelligent biosynthetic nanobiomaterials (IBNs) for hyperthermic gene delivery. *Pharmaceutical Research* 25 (3): 683–691.

357 Pooga, M., Soomets, U., Hällbrink, M. et al. (1998). Cell penetrating PNA constructs regulate galanin receptor levels and modify pain transmission in vivo. *Nature Biotechnology* 16 (9): 857–861.

358 Rousselle, C., Clair, P., Lefauconnier, J.M. et al. (2000). New advances in the transport of doxorubicin through the blood-brain barrier by a peptide vector-mediated strategy. *Molecular Pharmacology* 57 (4): 679–686.

359 Mazel, M., Clair, P., Rousselle, C. et al. (2001). Doxorubicin-peptide conjugates overcome multidrug resistance. *Anti-Cancer Drugs* 12 (2): 107–116.

360 El-Sayed, A., Futaki, S., and Harashima, H. (2009). Delivery of macromolecules using arginine-rich cell-penetrating peptides: ways to overcome endosomal entrapment. *The AAPS Journal* 11 (1): 13–22.

361 Mitchell, D.J., Kim, D.T., Steinman, L. et al. (2000). Polyarginine enters cells more efficiently than other polycationic homopolymers. *The Journal of Peptide Research* 56 (5): 318–325.

362 Futaki, S., Suzuki, T., Ohashi, W. et al. (2001). Arginine-rich peptides. An abundant source of membrane-permeable peptides having potential as carriers for intracellular protein delivery. *The Journal of Biological Chemistry* 276 (8): 5836–5840.

363 Maiolo, J.R., Ferrer, M., and Ottinger, E.A. (2005). Effects of cargo molecules on the cellular uptake of arginine-rich cell-penetrating peptides. *Biochimica et Biophysica Acta (BBA) - Biomembranes* 1712 (2): 161–172.

364 Kale, A.A. and Torchilin, V.P. (2007). Enhanced transfection of tumor cells in vivo using "smart" pH-sensitive TAT-modified pegylated liposomes. *Journal of Drug Targeting* 15 (7–8): 538–545.

365 Kale, A.A. and Torchilin, V.P. (2007). "Smart" drug carriers: PEGylated TATp-modified pH-sensitive liposomes. *Journal of Liposome Research* 17 (3–4): 197–203.

366 Abes, R., Moulton, H.M., Clair, P. et al. (2008). Delivery of steric block morpholino oligomers by $(R\text{-}X\text{-}R)_4$ peptides: structure–activity studies. *Nucleic Acids Research* 36 (20): 6343–6354.

367 Verdurmen, W.P.R. and Brock, R. (2011). Biological responses towards cationic peptides and drug carriers. *Trends in Pharmacological Sciences* 32 (2): 116–124.

368 Pooga, M., Elmquist, A., and Langel, Ü. (2002). Toxicity and side effects of cell-penetrating peptides. *Cell-Penetrating Peptides*: 245.

369 Saar, K., Lindgren, M., Hansen, M. et al. (2005). Cell-penetrating peptides: a comparative membrane toxicity study. *Analytical Biochemistry* 345 (1): 55–65.

370 Nasrollahi, S.A., Taghibiglou, C., Azizi, E., and Farboud, E.S. (2012). Cell-penetrating peptides as a novel transdermal drug delivery system. *Chemical Biology and Drug Design* 80 (5): 639–646.

371 Lundberg, P., Magzoub, M., Lindberg, M. et al. (2002). Cell membrane translocation of the N-terminal (1–28) part of the prion protein. *Biochemical and Biophysical Research Communications* 299 (1): 85–90.

372 Deshayes, S., Plénat, T., Aldrian-Herrada, G. et al. (2004). Primary amphipathic cell-penetrating peptides: structural requirements and interactions with model membranes. *Biochemistry* 43 (24): 7698–7706.

373 Rydström, A., Deshayes, S., Konate, K. et al. (2011). Direct translocation as major cellular uptake for CADY self-assembling peptide-based nanoparticles. *PLoS One* 6 (10): e25924.

374 Oehlke, J., Scheller, A., Wiesner, B. et al. (1998). Cellular uptake of an α-helical amphipathic model peptide with the potential to deliver polar compounds into the cell interior non-endocytically. *Biochimica et Biophysica Acta (BBA) - Biomembranes* 1414 (1): 127–139.

375 Milletti, F. (2012). Cell-penetrating peptides: classes, origin, and current landscape. *Drug Discovery Today* 17 (15): 850–860.

376 Wender, P.A., Mitchell, D.J., Pattabiraman, K. et al. (2000). The design, synthesis, and evaluation of molecules that enable or enhance cellular uptake: peptoid molecular transporters. *Proceedings of the National Academy of Sciences of the United States of America* 97 (24): 13003.

377 Rousselle, C., Smirnova, M., Clair, P. et al. (2001). Enhanced delivery of doxorubicin into the brain via a peptide-vector-mediated strategy: saturation kinetics and specificity. *The Journal of Pharmacology and Experimental Therapeutics* 296 (1): 124–131.

378 Aroui, S., Ram, N., Appaix, F. et al. (2009). Maurocalcine as a non toxic drug carrier overcomes doxorubicin resistance in the cancer cell line MDA-MB 231. *Pharmaceutical Research* 26 (4): 836–845.

379 Chen, J., Li, S., and Shen, Q. (2012). Folic acid and cell-penetrating peptide conjugated PLGA–PEG bifunctional nanoparticles for vincristine sulfate delivery. *European Journal of Pharmaceutical Sciences* 47 (2): 430–443.

380 Kondo, E., Saito, K., Tashiro, Y. et al. (2012). Tumour lineage-homing cell-penetrating peptides as anticancer molecular delivery systems. *Nature Communications* 3: 951.

381 Li, Y., Zheng, X., Cao, Z. et al. (2012). Self-assembled peptide (CADY-1) improved the clinical application of doxorubicin. *International Journal of Pharmaceutics* 434 (1–2): 209–214.

382 Nakase, I., Konishi, Y., Ueda, M. et al. (2012). Accumulation of arginine-rich cell-penetrating peptides in tumors and the potential for anticancer drug delivery in vivo. *Journal of Controlled Release* 159 (2): 181–188.

383 Shi, N.-Q., Gao, W., Xiang, B., and Qi, X.-R. (2012). Enhancing cellular uptake of activable cell-penetrating peptide–doxorubicin conjugate by enzymatic cleavage. *International Journal of Nanomedicine* 7: 1613–1621.

384 Uchida, Y., Raina, D., Kharbanda, S., and Kufe, D. (2013). Inhibition of the MUC1-C oncoprotein is synergistic with cytotoxic agents in the treatment of breast cancer cells. *Cancer Biology & Therapy* 14 (2): 127–134.

385 Komin, A., Russell, L.M., Hristova, K.A., and Searson, P.C. (2017). Peptide-based strategies for enhanced cell uptake, transcellular transport, and circulation: mechanisms and challenges. *Advanced Drug Delivery Reviews* 110–111: 52–64.

386 Patel, L.N., Zaro, J.L., and Shen, W.-C. (2007). Cell penetrating peptides: intracellular pathways and pharmaceutical perspectives. *Pharmaceutical Research* 24 (11): 1977–1992.

387 Mäe, M., Rautsi, O., Enbäck, J. et al. (2012). Tumour targeting with rationally modified cell-penetrating peptides. *International Journal of Peptide Research and Therapeutics* 18 (4): 361–371.

388 Crombez, L., Morris, M.C., Heitz, F., and Divita, G. (2011). A non-covalent peptide-based strategy for ex vivo and in vivo oligonucleotide delivery. *Methods in Molecular Biology (Clifton, NJ)* 764: 59–73.

389 Margus, H., Padari, K., and Pooga, M. (2012). Cell-penetrating peptides as versatile vehicles for oligonucleotide delivery. *Molecular Therapy* 20 (3): 525–533.

390 Schafmeister, C.E., Po, J., and Verdine, G.L. (2000). An all-hydrocarbon cross-linking system for enhancing the helicity and metabolic stability of peptides. *Journal of the American Chemical Society* 122 (24): 5891–5892.

391 Covic, L., Gresser, A.L., Talavera, J. et al. (2002). Activation and inhibition of G protein-coupled receptors by cell-penetrating membrane-tethered peptides. *Proceedings of the National Academy of Sciences of the United States of America* 99 (2): 643–648.

392 Ochocki, J.D., Mullen, D.G., Wattenberg, E.V., and Distefano, M.D. (2011). Evaluation of a cell penetrating prenylated peptide lacking an intrinsic fluorophore via in situ click reaction. *Bioorganic and Medicinal Chemistry Letters* 21 (17): 4998–5001.

393 Watkins, C.L., Brennan, P., Fegan, C. et al. (2009). Cellular uptake, distribution and cytotoxicity of the hydrophobic cell penetrating peptide sequence PFVYLI linked to the proapoptotic domain peptide PAD. *Journal of Controlled Release* 140 (3): 237–244.

394 Gennari, C.G.M., Franzè, S., Pellegrino, S. et al. (2016). Skin penetrating peptide as a tool to enhance the permeation of heparin through human epidermis. *Biomacromolecules* 17 (1): 46–55.

395 Gautam, A., Chaudhary, K., Kumar, R. et al. (2013). In silico approaches for designing highly effective cell penetrating peptides. *Journal of Translational Medicine* 11: 74.

396 Gautam, A., Sharma, M., Vir, P. et al. (2015). Identification and characterization of novel protein-derived arginine-rich cell-penetrating peptides. *European Journal of Pharmaceutics and Biopharmaceutics* 89: 93–106.

397 Gautam, A., Nanda, J.S., Samuel, J.S. et al. (2016). Topical delivery of protein and peptide using novel cell penetrating peptide IMT-P8. *Scientific Reports* 6: 26278.

398 Chen, M., Gupta, V., Anselmo, A.C. et al. (2014). Topical delivery of hyaluronic acid into skin using SPACE-peptide carriers. *Journal of Controlled Release* 173: 67–74.

399 Kumar, S., Chen, M., Anselmo, A.C. et al. (2015). Enhanced epidermal localization of topically applied steroids using SPACE™ peptide. *Drug Delivery and Translational Research* 5 (5): 523–530.

400 Chen, M., Kumar, S., Anselmo, A.C. et al. (2015). Topical delivery of cyclosporine A into the skin using SPACE-peptide. *Journal of Controlled Release* 199: 190–197.

401 Menegatti, S., Zakrewsky, M., Kumar, S. et al. (2016). De novo design of skin-penetrating peptides for enhanced transdermal delivery of peptide drugs. *Advanced Healthcare Materials* 5 (5): 602–609.

402 Zhang, S., Holmes, T.C., DiPersio, C.M. et al. (1995). Self-complementary oligopeptide matrices support mammalian cell attachment. *Biomaterials* 16 (18): 1385–1393.

403 Kisiday, J., Jin, M., Kurz, B. et al. (2002). Self-assembling peptide hydrogel fosters chondrocyte extracellular matrix production and cell division: implications for cartilage tissue repair. *Proceedings of the National Academy of Sciences of the United States of America* 99 (15): 9996–10001.

404 Zhang, S. (2003). Fabrication of novel biomaterials through molecular self-assembly. *Nature Biotechnology* 21 (10): 1171–1178.

405 Silva, G.A., Czeisler, C., Niece, K.L. et al. (2004). Selective differentiation of neural progenitor cells by high-epitope density nanofibers. *Science (New York, NY)* 303 (5662): 1352–1355.

406 Law, B., Weissleder, R., and Tung, C.-H. (2006). Peptide-based biomaterials for protease-enhanced drug delivery. *Biomacromolecules* 7 (4): 1261–1265.

407 Ramachandran, S. and Yu, Y.B. (2006). Peptide-based viscoelastic matrices for drug delivery and tissue repair. *BioDrugs* 20 (5): 263–269.

408 Zhao, X. and Zhang, S. (2006). Molecular designer self-assembling peptides. *Chemical Society Reviews* 35 (11): 1105–1110.

409 Zhao, X. and Zhang, S. (2006). Self-assembling nanopeptides become a new type of biomaterial. In: *Polymers for Regenerative Medicine* (ed. C. Werner), 145–170. Berlin, Heidelberg: Springer-Verlag.

410 Horii, A., Wang, X., Gelain, F., and Zhang, S. (2007). Biological designer self-assembling peptide nanofiber scaffolds significantly enhance osteoblast proliferation, differentiation and 3-D migration. *PLoS One* 2 (2): 1–9.

411 Segers, V.F.M. and Lee, R.T. (2007). Local delivery of proteins and the use of self-assembling peptides. *Drug Discovery Today* 12 (13–14): 561–568.

412 Branco, M.C. and Schneider, J.P. (2009). Self-assembling materials for therapeutic delivery. *Acta Biomaterialia* 5 (3): 817–831.

413 Loo, Y., Zhang, S., and Hauser, C.A.E. (2012). From short peptides to nanofibers to macromolecular assemblies in biomedicine. *Biotechnology Advances* 30 (3): 593–603.

414 Luo, Z. and Zhang, S. (2012). Designer nanomaterials using chiral self-assembling peptide systems and their emerging benefit for society. *Chemical Society Reviews* 41 (13): 4736–4754.

415 Vuorela, A., Myllyharju, J., Nissi, R. et al. (1997). Assembly of human prolyl 4-hydroxylase and type III collagen in the yeast *Pichia pastoris*: formation of a stable enzyme tetramer requires coexpression with collagen and assembly of a stable collagen requires coexpression with prolyl 4-hydroxylase. *The EMBO Journal* 16 (22): 6702–6712.
416 Vaughn, P.R., Galanis, M., Richards, K.M. et al. (1998). Production of recombinant hydroxylated human type III collagen fragment in *Saccharomyces cerevisiae. DNA and Cell Biology* 17 (6): 511–518.
417 Myllyharju, J., Nokelainen, M., Vuorela, A., and Kivirikko, K.I. (2000). Expression of recombinant human type I-III collagens in the yeast *Pichia pastoris. Biochemical Society Transactions* 28 (4): 353–357.
418 Toman, P.D., Chisholm, G., McMullin, H. et al. (2000). Production of recombinant human type I procollagen trimers using a four-gene expression system in the yeast *Saccharomyces cerevisiae. The Journal of Biological Chemistry* 275 (30): 23303–23309.
419 Nokelainen, M., Tu, H., Vuorela, A. et al. (2001). High-level production of human type I collagen in the yeast *Pichia pastoris. Yeast (Chichester, England)* 18 (9): 797–806.
420 Olsen, D.R., Leigh, S.D., Chang, R. et al. (2001). Production of human type I collagen in yeast reveals unexpected new insights into the molecular assembly of collagen trimers. *The Journal of Biological Chemistry* 276 (26): 24038–24043.
421 Liu, W., Merrett, K., Griffith, M. et al. (2008). Recombinant human collagen for tissue engineered corneal substitutes. *Biomaterials* 29 (9): 1147–1158.
422 Tojo, N., Miyagi, I., Miura, M., and Ohi, H. (2008). Recombinant human fibrinogen expressed in the yeast *Pichia pastoris* was assembled and biologically active. *Protein Expression and Purification* 59 (2): 289–296.
423 Sallach, R.E., Conticello, V.P., and Chaikof, E.L. (2009). Expression of a recombinant elastin-like protein in *Pichia pastoris. Biotechnology Progress* 25 (6): 1810–1818.
424 Tillet, E., Wiedemann, H., Golbik, R. et al. (1994). Recombinant expression and structural and binding properties of α1(VI) and α2(VI) chains of human collagen type VI. *European Journal of Biochemistry* 221 (1): 177–187.
425 Fichard, A., Tillet, E., Delacoux, F. et al. (1997). Human recombinant α1(V) collagen chain HOMOTRIMERIC ASSEMBLY AND SUBSEQUENT PROCESSING. *Journal of Biological Chemistry* 272 (48): 30083–30087.
426 Frischholz, S., Beier, F., Girkontaite, I. et al. (1998). Characterization of human type X procollagen and its NC-1 domain expressed as recombinant proteins in HEK293 cells. *The Journal of Biological Chemistry* 273 (8): 4547–4555.
427 Chen, M., Costa, F.K., Lindvay, C.R. et al. (2002). The recombinant expression of full-length type VII collagen and characterization of molecular mechanisms underlying dystrophic epidermolysis bullosa. *The Journal of Biological Chemistry* 277 (3): 2118–2124.
428 Butler, S.P., O'Sickey, T.K., Lord, S.T. et al. (2004). Secretion of recombinant human fibrinogen by the murine mammary gland. *Transgenic Research* 13 (5): 437–450.

429 Nillesen, S.T.M., Geutjes, P.J., Wismans, R. et al. (2007). Increased angiogenesis and blood vessel maturation in acellular collagen-heparin scaffolds containing both FGF2 and VEGF. *Biomaterials* 28 (6): 1123–1131.

430 Shen, Y.H., Shoichet, M.S., and Radisic, M. (2008). Vascular endothelial growth factor immobilized in collagen scaffold promotes penetration and proliferation of endothelial cells. *Acta Biomaterialia* 4 (3): 477–489.

431 Borselli, C., Ungaro, F., Oliviero, O. et al. (2010). Bioactivation of collagen matrices through sustained VEGF release from PLGA microspheres. *Journal of Biomedical Materials Research Part A* 92 (1): 94–102.

432 He, Q., Zhao, Y., Chen, B. et al. (2011). Improved cellularization and angiogenesis using collagen scaffolds chemically conjugated with vascular endothelial growth factor. *Acta Biomaterialia* 7 (3): 1084–1093.

433 Miyagi, Y., Chiu, L.L.Y., Cimini, M. et al. (2011). Biodegradable collagen patch with covalently immobilized VEGF for myocardial repair. *Biomaterials* 32 (5): 1280–1290.

434 Côté, M.-F., Laroche, G., Gagnon, E. et al. (2004). Denatured collagen as support for a FGF-2 delivery system: physicochemical characterizations and in vitro release kinetics and bioactivity. *Biomaterials* 25 (17): 3761–3772.

435 Maehara, H., Sotome, S., Yoshii, T. et al. (2010). Repair of large osteochondral defects in rabbits using porous hydroxyapatite/collagen (HAp/Col) and fibroblast growth factor-2 (FGF-2). *Journal of Orthopaedic Research* 28 (5): 677–686.

436 Takeda, Y., Tsujigiwa, H., Nagatsuka, H. et al. (2005). Regeneration of rat auditory ossicles using recombinant human BMP-2/collagen composites. *Journal of Biomedical Materials Research Part A* 73 (2): 133–141.

437 Gavénis, K., Klee, D., Pereira-Paz, R.M. et al. (2007). BMP-7 loaded microspheres as a new delivery system for the cultivation of human chondrocytes in a collagen type-I gel. *Journal of Biomedical Materials Research Part B, Applied Biomaterials* 82 (2): 275–283.

438 Zhao, Y., Zhang, J., Wang, X. et al. (2010). The osteogenic effect of bone morphogenetic protein-2 on the collagen scaffold conjugated with antibodies. *Journal of Controlled Release* 141 (1): 30–37.

439 Sun, B., Chen, B., Zhao, Y. et al. (2009). Crosslinking heparin to collagen scaffolds for the delivery of human platelet-derived growth factor. *Journal of Biomedical Materials Research Part B, Applied Biomaterials* 91 (1): 366–372.

440 Teixeira, S., Yang, L., Dijkstra, P.J. et al. (2010). Heparinized hydroxyapatite/collagen three-dimensional scaffolds for tissue engineering. *Journal of Materials Science Materials in Medicine* 21 (8): 2385–2392.

441 Wu, J.M., Xu, Y.Y., Li, Z.H. et al. (2011). Heparin-functionalized collagen matrices with controlled release of basic fibroblast growth factor. *Journal of Materials Science Materials in Medicine* 22 (1): 107–114.

442 Srouji, S., Rachmiel, A., Blumenfeld, I., and Livne, E. (2005). Mandibular defect repair by TGF-beta and IGF-1 released from a biodegradable osteoconductive hydrogel. *Journal of Cranio-Maxillo-Facial Surgery* 33 (2): 79–84.

443 Chou, C.-H., Cheng, W.T.K., Lin, C.-C. et al. (2006). TGF-beta1 immobilized tri-co-polymer for articular cartilage tissue engineering. *Journal of Biomedical Materials Research Part B, Applied Biomaterials* 77 (2): 338–348.
444 Fan, H., Hu, Y., Li, X. et al. (2006). Ectopic cartilage formation induced by mesenchymal stem cells on porous gelatin-chondroitin-hyaluronate scaffold containing microspheres loaded with TGF-beta1. *The International Journal of Artificial Organs* 29 (6): 602–611.
445 Ogawa, T., Akazawa, T., and Tabata, Y. (2010). In vitro proliferation and chondrogenic differentiation of rat bone marrow stem cells cultured with gelatin hydrogel microspheres for TGF-beta1 release. *Journal of Biomaterials Science Polymer Edition* 21 (5): 609–621.
446 Kempen, D.H.R., Lu, L., Hefferan, T.E. et al. (2008). Retention of in vitro and in vivo BMP-2 bioactivities in sustained delivery vehicles for bone tissue engineering. *Biomaterials* 29 (22): 3245–3252.
447 Kempen, D.H.R., Lu, L., Heijink, A. et al. (2009). Effect of local sequential VEGF and BMP-2 delivery on ectopic and orthotopic bone regeneration. *Biomaterials* 30 (14): 2816–2825.
448 Royce, S.M., Askari, M., and Marra, K.G. (2004). Incorporation of polymer microspheres within fibrin scaffolds for the controlled delivery of FGF-1. *Journal of Biomaterials Science Polymer Edition* 15 (10): 1327–1336.
449 Campbell, P.G., Miller, E.D., Fisher, G.W. et al. (2005). Engineered spatial patterns of FGF-2 immobilized on fibrin direct cell organization. *Biomaterials* 26 (33): 6762–6770.
450 Chung, Y.-I., Ahn, K.-M., Jeon, S.-H. et al. (2007). Enhanced bone regeneration with BMP-2 loaded functional nanoparticle-hydrogel complex. *Journal of Controlled Release* 121 (1–2): 91–99.
451 Park, K.-H., Kim, H., Moon, S., and Na, K. (2009). Bone morphogenic protein-2 (BMP-2) loaded nanoparticles mixed with human mesenchymal stem cell in fibrin hydrogel for bone tissue engineering. *Journal of Bioscience and Bioengineering* 108 (6): 530–537.
452 Urry, D.W. (1997). Physical chemistry of biological free energy transduction as demonstrated by elastic protein-based polymers. *The Journal of Physical Chemistry B* 101 (51): 11007–11028.
453 Urry, D.W. (1998). Five axioms for the functional design of peptide-based polymers as molecular machines and materials: principle for macromolecular assemblies. *Peptide Science* 47 (2): 167–178.
454 Meyer, D.E. and Chilkoti, A. (2004). Quantification of the effects of chain length and concentration on the thermal behavior of elastin-like polypeptides. *Biomacromolecules* 5 (3): 846–851.
455 Lim, D.W., Trabbic-Carlson, K., Mackay, J.A., and Chilkoti, A. (2007). Improved non-chromatographic purification of a recombinant protein by cationic elastin-like polypeptides. *Biomacromolecules* 8 (5): 1417–1424.
456 Lee, J., Macosko, C.W., and Urry, D.W. (2001). Mechanical properties of cross-linked synthetic elastomeric polypentapeptides. *Macromolecules* 34 (17): 5968–5974.

457 Trabbic-Carlson, K., Setton, L.A., and Chilkoti, A. (2003). Swelling and mechanical behaviors of chemically cross-linked hydrogels of elastin-like polypeptides. *Biomacromolecules* 4 (3): 572–580.
458 Nowatzki, P.J. and Tirrell, D.A. (2004). Physical properties of artificial extracellular matrix protein films prepared by isocyanate crosslinking. *Biomaterials* 25 (7): 1261–1267.
459 McHale, M.K., Setton, L.A., and Chilkoti, A. (2005). Synthesis and in vitro evaluation of enzymatically cross-linked elastin-like polypeptide gels for cartilaginous tissue repair. *Tissue Engineering* 11 (11–12): 1768–1779.
460 Lim, D.W., Nettles, D.L., Setton, L.A., and Chilkoti, A. (2007). Rapid cross-linking of elastin-like polypeptides with (hydroxymethyl)phosphines in aqueous solution. *Biomacromolecules* 8 (5): 1463–1470.
461 Nicol, A., Gowda, D.C., and Urry, D.W. (1992). Cell adhesion and growth on synthetic elastomeric matrices containing Arg-Gly-Asp-Ser-^{3}. *Journal of Biomedical Materials Research* 26 (3): 393–413.
462 Welsh, E.R. and Tirrell, D.A. (2000). Engineering the extracellular matrix: a novel approach to polymeric biomaterials. I. Control of the physical properties of artificial protein matrices designed to support adhesion of vascular endothelial cells. *Biomacromolecules* 1 (1): 23–30.
463 Liu, J.C. and Tirrell, D.A. (2008). Cell response to RGD density in cross-linked artificial extracellular matrix protein films. *Biomacromolecules* 9 (11): 2984–2988.
464 Urry, D.W., Parker, T.M., Reid, M.C., and Gowda, C. (1991). Biocompatibility of the bioelastic materials, poly(GVGVP) and its γ-irradiation cross-linked matrix: summary of generic biological test results. *Journal of Bioactive and Compatible Polymers* 6 (3): 263–282.
465 Lee, J., Macosko, C.W., and Urry, D.W. (2001). Swelling behavior of γ-irradiation cross-linked elastomeric polypentapeptide-based hydrogels. *Macromolecules* 34 (12): 4114–4123.
466 Nagapudi, K., Brinkman, W.T., Leisen, J.E. et al. (2002). Photomediated solid-state cross-linking of an elastin–mimetic recombinant protein polymer. *Macromolecules* 35 (5): 1730–1737.
467 Srokowski, E.M. and Woodhouse, K.A. (2008). Development and characterisation of novel cross-linked bio-elastomeric materials. *Journal of Biomaterials Science Polymer Edition* 19 (6): 785–799.
468 Huang, L., McMillan, R.A., Apkarian, R.P. et al. (2000). Generation of synthetic elastin-mimetic small diameter fibers and fiber networks. *Macromolecules* 33 (8): 2989–2997.
469 Lee, J., Macosko, C.W., and Urry, D.W. (2001). Elastomeric polypentapeptides cross-linked into matrixes and fibers. *Biomacromolecules* 2 (1): 170–179.
470 Betre, H., Chilkoti, A., and Setton, L.A. (2002). A two-step chondrocyte recovery system based on thermally sensitive elastin-like polypeptide scaffolds for cartilage tissue engineering. *Proceedings of the 2nd Joint 24th Annual Conference and the Annual Fall Meeting of the Biomedical Engineering Society] [Engineering in Medicine and Biology.*

471 Betre, H., Setton, L.A., Meyer, D.E., and Chilkoti, A. (2002). Characterization of a genetically engineered elastin-like polypeptide for cartilaginous tissue repair. *Biomacromolecules* 3 (5): 910–916.

472 Betre, H., Ong, S.R., Guilak, F. et al. (2006). Chondrocytic differentiation of human adipose-derived adult stem cells in elastin-like polypeptide. *Biomaterials* 27 (1): 91–99.

473 Nettles, D.L., Kitaoka, K., Hanson, N.A. et al. (2008). In situ crosslinking elastin-like polypeptide gels for application to articular cartilage repair in a goat osteochondral defect model. *Tissue Engineering Part A* 14 (7): 1133–1140.

474 Hrabchak, C., Rouleau, J., Moss, I. et al. (2010). Assessment of biocompatibility and initial evaluation of genipin cross-linked elastin-like polypeptides in the treatment of an osteochondral knee defect in rabbits. *Acta Biomaterialia* 6 (6): 2108–2115.

475 Nettles, D.L., Chilkoti, A., and Setton, L.A. (2010). Applications of elastin-like polypeptides in tissue engineering. *Advanced Drug Delivery Reviews* 62 (15): 1479–1485.

476 Moss, I.L., Gordon, L., Woodhouse, K.A. et al. (2011). A novel thiol-modified hyaluronan and elastin-like polypetide composite material for tissue engineering of the nucleus pulposus of the intervertebral disc. *Spine* 36 (13): 1022–1029.

477 Muiznieks, L.D. (2018). Genipin cross-linking of elastin and elastin-based proteins. In: *Protein Scaffolds: Design, Synthesis, and Applications* (ed. A.K. Udit), 213–221. New York, NY: Springer.

478 Nicol, A., Gowda, C., and Urry, D.W. (1991). Elastic protein-based polymers as cell attachment matrices. *Journal of Vascular Surgery* 13 (5): 746–748.

479 Urry, D.W., Pattanaik, A., Xu, J. et al. (1998). Elastic protein-based polymers in soft tissue augmentation and generation. *Journal of Biomaterials Science Polymer Edition* 9 (10): 1015–1048.

480 Betre, H., Liu, W., Zalutsky, M.R. et al. (2006). A thermally responsive biopolymer for intra-articular drug delivery. *Journal of Controlled Release* 115 (2): 175–182.

481 Caves, J.M., Kumar, V.A., Martinez, A.W. et al. (2010). The use of microfiber composites of elastin-like protein matrix reinforced with synthetic collagen in the design of vascular grafts. *Biomaterials* 31 (27): 7175–7182.

482 Ravi, S. and Chaikof, E.L. (2010). Biomaterials for vascular tissue engineering. *Regenerative Medicine* 5 (1): 107–120.

483 Sallach, R.E., Cui, W., Balderrama, F. et al. (2010). Long-term biostability of self-assembling protein polymers in the absence of covalent crosslinking. *Biomaterials* 31 (4): 779–791.

484 Di Zio, K. and Tirrell, D.A. (2003). Mechanical properties of artificial protein matrices engineered for control of cell and tissue behavior. *Macromolecules* 36 (5): 1553–1558.

485 Heilshorn, S.C., DiZio, K.A., Welsh, E.R., and Tirrell, D.A. (2003). Endothelial cell adhesion to the fibronectin CS5 domain in artificial extracellular matrix proteins. *Biomaterials* 24 (23): 4245–4252.

486 Trabbic-Carlson, K., Liu, L., Kim, B., and Chilkoti, A. (2004). Expression and purification of recombinant proteins from *Escherichia coli*: comparison of an elastin-like polypeptide fusion with an oligohistidine fusion. *Protein Science* 13 (12): 3274–3284.

487 Shamji, M.F., Whitlatch, L., Friedman, A.H. et al. (2008). An injectable and in situ-gelling biopolymer for sustained drug release following perineural administration. *Spine* 33 (7): 748–754.

488 Massodi, I., Bidwell, G.L. III, Davis, A. et al. (2009). Inhibition of ovarian cancer cell metastasis by a fusion polypeptide Tat-ELP. *Clinical & Experimental Metastasis* 26 (3): 251–260.

489 Heilshorn, S.C., Liu, J.C., and Tirrell, D.A. (2005). Cell-binding domain context affects cell behavior on engineered proteins. *Biomacromolecules* 6 (1): 318–323.

490 Liu, J.C., Heilshorn, S.C., and Tirrell, D.A. (2004). Comparative cell response to artificial extracellular matrix proteins containing the RGD and CS5 cell-binding domains. *Biomacromolecules* 5 (2): 497–504.

491 Lin, H.-B., García-Echeverría, C., Asakura, S. et al. (1992). Endothelial cell adhesion on polyurethanes containing covalently attached RGD-peptides. *Biomaterials* 13 (13): 905–914.

492 Panitch, A., Yamaoka, T., Fournier, M.J. et al. (1999). Design and biosynthesis of elastin-like artificial extracellular matrix proteins containing periodically spaced fibronectin CS5 domains. *Macromolecules* 32 (5): 1701–1703.

493 Almine, J.F., Bax, D.V., Mithieux, S.M. et al. (2010). Elastin-based materials. *Chemical Society Reviews* 39 (9): 3371–3379.

494 Kim, W. and Chaikof, E.L. (2010). Recombinant elastin-mimetic biomaterials: emerging applications in medicine. *Advanced Drug Delivery Reviews* 62 (15): 1468–1478.

495 Girotti, A., Reguera, J., Rodríguez-Cabello, J.C. et al. (2004). Design and bioproduction of a recombinant multi(bio)functional elastin-like protein polymer containing cell adhesion sequences for tissue engineering purposes. *Journal of Materials Science Materials in Medicine* 15 (4): 479–484.

496 Martínez-Osorio, H., Juárez-Campo, M., Diebold, Y. et al. (2009). Genetically engineered elastin-like polymer as a substratum to culture cells from the ocular surface. *Current Eye Research* 34 (1): 48–56.

497 Janorkar, A.V., Rajagopalan, P., Yarmush, M.L., and Megeed, Z. (2008). The use of elastin-like polypeptide-polyelectrolyte complexes to control hepatocyte morphology and function in vitro. *Biomaterials* 29 (6): 625–632.

498 Swierczewska, M., Hajicharalambous, C.S., Janorkar, A.V. et al. (2008). Cellular response to nanoscale elastin-like polypeptide polyelectrolyte multilayers. *Acta Biomaterialia* 4 (4): 827–837.

499 Maude, S., Ingham, E., and Aggeli, A. (2013). Biomimetic self-assembling peptides as scaffolds for soft tissue engineering. *Nanomedicine* 8 (5): 823–847.

500 Nagapudi, K., Brinkman, W.T., Thomas, B.S. et al. (2005). Viscoelastic and mechanical behavior of recombinant protein elastomers. *Biomaterials* 26 (23): 4695–4706.

501 Chow, D., Nunalee, M.L., Lim, D.W. et al. (2008). Peptide-based biopolymers in biomedicine and biotechnology. *Materials Science and Engineering: R: Reports* 62 (4): 125–155.
502 Sallach, R.E., Cui, W., Wen, J. et al. (2009). Elastin-mimetic protein polymers capable of physical and chemical crosslinking. *Biomaterials* 30 (3): 409–422.
503 MacEwan, S.R. and Chilkoti, A. (2010). Elastin-like polypeptides: biomedical applications of tunable biopolymers. *Biopolymers* 94 (1): 60–77.
504 Hecht, M.H. (1994). De novo design of beta-sheet proteins. *Proceedings of the National Academy of Sciences of the United States of America* 91 (19): 8729–8730.
505 Xiong, H., Buckwater, B.L., Shieh, H.M., and Hecht, M.H. (1995). Periodicity of polar and nonpolar amino acids is the major determinant of secondary structure in self-assembling oligomeric peptides. *Proceedings of the National Academy of Sciences of the United States of America* 92 (14): 6349–6353.
506 Dobson, C.M. (1999). Protein misfolding, evolution and disease. *Trends in Biochemical Sciences* 24 (9): 329–332.
507 MacPhee, C.E. and Woolfson, D.N. (2004). Engineered and designed peptide-based fibrous biomaterials. *Current Opinion in Solid State and Materials Science* 8 (2): 141–149.
508 Wu, X., Sallach, R.E., Caves, J.M. et al. (2008). Deformation responses of a physically crosslinked high molecular weight elastin-like protein polymer. *Biomacromolecules* 9 (7): 1787–1794.
509 Gustafson, J.A. and Ghandehari, H. (2010). Silk-elastinlike protein polymers for matrix-mediated cancer gene therapy. *Advanced Drug Delivery Reviews* 62 (15): 1509–1523.
510 Numata, K. and Kaplan, D.L. (2010). Silk-based delivery systems of bioactive molecules. *Advanced Drug Delivery Reviews* 62 (15): 1497–1508.
511 Zhang, S., Holmes, T., Lockshin, C., and Rich, A. (1993). Spontaneous assembly of a self-complementary oligopeptide to form a stable macroscopic membrane. *Proceedings of the National Academy of Sciences of the United States of America* 90 (8): 3334–3338.
512 Yokoi, H., Kinoshita, T., and Zhang, S. (2005). Dynamic reassembly of peptide RADA16 nanofiber scaffold. *Proceedings of the National Academy of Sciences of the United States of America* 102 (24): 8414–8419.
513 Gelain, F., Bottai, D., Vescovi, A. et al. (2006). Designer self-assembling peptide nanofiber scaffolds for adult mouse neural stem cell 3-dimensional cultures. *PLoS One* 1: e119.
514 Davis, M.E., Motion, J.P., Narmoneva, D.A. et al. (2005). Injectable self-assembling peptide nanofibers create intramyocardial microenvironments for endothelial cells. *Circulation* 111 (4): 442–450.
515 Holmes, T.C., de Lacalle, S., Su, X. et al. (2000). Extensive neurite outgrowth and active synapse formation on self-assembling peptide scaffolds. *Proceedings of the National Academy of Sciences of the United States of America* 97 (12): 6728–6733.
516 Engel, F.B., Hsieh, P.C.H., Lee, R.T., and Keating, M.T. (2006). FGF1/p38 MAP kinase inhibitor therapy induces cardiomyocyte mitosis, reduces scarring, and rescues function after myocardial infarction. *Proceedings of the*

National Academy of Sciences of the United States of America 103 (42): 15546–15551.

517 Guo, J., Su, H., Zeng, Y. et al. (2007). Reknitting the injured spinal cord by self-assembling peptide nanofiber scaffold. *Nanomedicine: Nanotechnology, Biology and Medicine* 3 (4): 311–321.

518 Sieminski, A.L., Semino, C.E., Gong, H., and Kamm, R.D. (2008). Primary sequence of ionic self-assembling peptide gels affects endothelial cell adhesion and capillary morphogenesis. *Journal of Biomedical Materials Research Part A* 87 (2): 494–504.

519 Cui, X.-J., Xie, H., Wang, H.-J. et al. (2010). Transplantation of mesenchymal stem cells with self-assembling polypeptide scaffolds is conducive to treating myocardial infarction in rats. *The Tohoku Journal of Experimental Medicine* 222 (4): 281–289.

520 Guo, H.-D., Cui, G.-H., Wang, H.-J., and Tan, Y.-Z. (2010). Transplantation of marrow-derived cardiac stem cells carried in designer self-assembling peptide nanofibers improves cardiac function after myocardial infarction. *Biochemical and Biophysical Research Communications* 399 (1): 42–48.

521 Lin, Y.-D., Yeh, M.-L., Yang, Y.-J. et al. (2010). Intramyocardial peptide nanofiber injection improves postinfarction ventricular remodeling and efficacy of bone marrow cell therapy in pigs. *Circulation* 122 (11 Suppl 1): S132–S141.

522 Tokunaga, M., Liu, M.-L., Nagai, T. et al. (2010). Implantation of cardiac progenitor cells using self-assembling peptide improves cardiac function after myocardial infarction. *Journal of Molecular and Cellular Cardiology* 49 (6): 972–983.

523 Cho, H., Balaji, S., Sheikh, A.Q. et al. (2012). Regulation of endothelial cell activation and angiogenesis by injectable peptide nanofibers. *Acta Biomaterialia* 8 (1): 154–164.

524 Lin, Y.-D., Luo, C.-Y., Hu, Y.-N. et al. (2012). Instructive nanofiber scaffolds with VEGF create a microenvironment for arteriogenesis and cardiac repair. *Science Translational Medicine* 4 (146): 146ra109.

525 Kim, J.H., Jung, Y., Kim, B.-S., and Kim, S.H. (2013). Stem cell recruitment and angiogenesis of neuropeptide substance P coupled with self-assembling peptide nanofiber in a mouse hind limb ischemia model. *Biomaterials* 34 (6): 1657–1668.

526 Boopathy, A.V., Che, P.L., Somasuntharam, I. et al. (2014). The modulation of cardiac progenitor cell function by hydrogel-dependent Notch1 activation. *Biomaterials* 35 (28): 8103–8112.

527 Prat-Vidal, C., Gálvez-Montón, C., Puig-Sanvicens, V. et al. (2014). Online monitoring of myocardial bioprosthesis for cardiac repair. *International Journal of Cardiology* 174 (3): 654–661.

528 Genové, E., Shen, C., Zhang, S., and Semino, C.E. (2005). The effect of functionalized self-assembling peptide scaffolds on human aortic endothelial cell function. *Biomaterials* 26 (16): 3341–3351.

529 Narmoneva, D.A., Oni, O., Sieminski, A.L. et al. (2005). Self-assembling short oligopeptides and the promotion of angiogenesis. *Biomaterials* 26 (23): 4837–4846.

530 Wang, X., Horii, A., and Zhang, S. (2008). Designer functionalized self-assembling peptide nanofiber scaffolds for growth, migration, and tubulogenesis of human umbilical vein endothelial cells. *Soft Matter* 4 (12): 2388–2395.
531 Narmoneva, D.A., Vukmirovic, R., Davis, M.E. et al. (2004). Endothelial cells promote cardiac myocyte survival and spatial reorganization. *Circulation* 110 (8): 962–968.
532 Cunha, C., Panseri, S., Villa, O. et al. (2011). 3D culture of adult mouse neural stem cells within functionalized self-assembling peptide scaffolds. *International Journal of Nanomedicine* 6: 943.
533 Cheng, T.-Y., Chen, M.-H., Chang, W.-H. et al. (2013). Neural stem cells encapsulated in a functionalized self-assembling peptide hydrogel for brain tissue engineering. *Biomaterials* 34 (8): 2005–2016.
534 Ellis-Behnke, R.G., Liang, Y.-X., You, S.-W. et al. (2006). Nano neuro knitting: peptide nanofiber scaffold for brain repair and axon regeneration with functional return of vision. *Proceedings of the National Academy of Sciences of the United States of America* 103 (13): 5054–5059.
535 Guo, J., Leung, K.K.G., Su, H. et al. (2009). Self-assembling peptide nanofiber scaffold promotes the reconstruction of acutely injured brain. *Nanomedicine: Nanotechnology, Biology and Medicine* 5 (3): 345–351.
536 Song, H., Zhang, L., and Zhao, X. (2010). Hemostatic efficacy of biological self-assembling peptide nanofibers in a rat kidney model. *Macromolecular Bioscience* 10 (1): 33–39.
537 Gras, S.L., Tickler, A.K., Squires, A.M. et al. (2008). Functionalised amyloid fibrils for roles in cell adhesion. *Biomaterials* 29 (11): 1553–1562.
538 Aggeli, A., Bell, M., Carrick, L.M. et al. (2003). pH as a trigger of peptide β-sheet self-assembly and reversible switching between nematic and isotropic phases. *Journal of the American Chemical Society* 125 (32): 9619–9628.
539 Davies, R.P.W., Aggeli, A., Beevers, A.J. et al. (2006). Self-assembling β-sheet tape forming peptides. *Supramolecular Chemistry* 18 (5): 435–443.
540 Carrick, L.M., Aggeli, A., Boden, N. et al. (2007). Effect of ionic strength on the self-assembly, morphology and gelation of pH responsive β-sheet tape-forming peptides. *Tetrahedron* 63 (31): 7457–7467.
541 Silvertown, J.D., Wong, B.P.Y., Sivagurunathan, K.S. et al. (2017). Remineralization of natural early caries lesions in vitro by P_{11}-4 monitored with photothermal radiometry and luminescence. *Journal of Investigative and Clinical Dentistry* 8 (4): 1–8.
542 Jung, J.P., Jones, J.L., Cronier, S.A. et al. (2008). Modulating the mechanical properties of self-assembled peptide hydrogels via native chemical ligation. *Biomaterials* 29 (13): 2143–2151.
543 Jung, J.P., Nagaraj, A.K., Fox, E.K. et al. (2009). Co-assembling peptides as defined matrices for endothelial cells. *Biomaterials* 30 (12): 2400–2410.
544 Kyle, S., Aggeli, A., Ingham, E., and McPherson, M.J. (2010). Recombinant self-assembling peptides as biomaterials for tissue engineering. *Biomaterials* 31 (36): 9395–9405.

545 Salick, D.A., Kretsinger, J.K., Pochan, D.J., and Schneider, J.P. (2007). Inherent antibacterial activity of a peptide-based beta-hairpin hydrogel. *Journal of the American Chemical Society* 129 (47): 14793–14799.

546 Worthington, P., Pochan, D.J., and Langhans, S.A. (2015). Peptide hydrogels – versatile matrices for 3D cell culture in cancer medicine. *Frontiers in Oncology* 5: 1–10.

547 Schneider, J.P., Pochan, D.J., Ozbas, B. et al. (2002). Responsive hydrogels from the intramolecular folding and self-assembly of a designed peptide. *Journal of the American Chemical Society* 124 (50): 15030–15037.

548 Kumar, V.A., Taylor, N.L., Shi, S. et al. (2015). Highly angiogenic peptide nanofibers. *ACS Nano* 9 (1): 860–868.

549 Galler, K.M., Hartgerink, J.D., Cavendar, A.C. et al. (2012). A customized self-assembling peptide hydrogel for dental pulp tissue engineering. *Tissue Engineering Part A* 18 (1–2): 176–184.

550 Kang, M.K., Colombo, J.S., D'Souza, R.N., and Hartgerink, J.D. (2014). Sequence effects of self-assembling multidomain peptide hydrogels on encapsulated SHED cells. *Biomacromolecules* 15 (6): 2004–2011.

551 Collier, J.H., Rudra, J.S., Gasiorowski, J.Z., and Jung, J.P. (2010). Multi-component extracellular matrices based on peptide self-assembly. *Chemical Society Reviews* 39 (9): 3413–3424.

552 Hauser, C.A.E. and Zhang, S. (2010). Designer self-assembling peptide nanofiber biological materials. *Chemical Society Reviews* 39 (8): 2780–2790.

553 Beniash, E., Hartgerink, J.D., Storrie, H. et al. (2005). Self-assembling peptide amphiphile nanofiber matrices for cell entrapment. *Acta Biomaterialia* 1 (4): 387–397.

554 Storrie, H., Guler, M.O., Abu-Amara, S.N. et al. (2007). Supramolecular crafting of cell adhesion. *Biomaterials* 28 (31): 4608–4618.

555 Hosseinkhani, H., Hosseinkhani, M., and Kobayashi, H. (2006). Proliferation and differentiation of mesenchymal stem cells using self-assembled peptide amphiphile nanofibers. *Biomedical Materials (Bristol, England)* 1 (1): 8–15.

556 Hosseinkhani, H., Hosseinkhani, M., and Kobayashi, H. (2006). Design of tissue engineered nanoscaffold through self-assembly of peptide amphiphile. *Journal of Bioactive and Compatible Polymers* 21 (4): 277–296.

557 Hosseinkhani, H., Hosseinkhani, M., and Kobayashi, H. (2006). Osteogenic differentiation of mesenchymal stem cells in self-assembled peptide-amphiphile nanofibers. *Biomaterials* 27 (22): 4079–4086.

558 Hosseinkhani, H., Hosseinkhani, M., Tian, F. et al. (2006). Ectopic bone formation in collagen sponge self-assembled peptide-amphiphile nanofibers hybrid scaffold in a perfusion culture bioreactor. *Biomaterials* 27 (29): 5089–5098.

559 Tysseling-Mattiace, V.M., Sahni, V., Niece, K.L. et al. (2008). Self-assembling nanofibers inhibit glial scar formation and promote axon elongation after spinal cord injury. *The Journal of Neuroscience* 28 (14): 3814–3823.

560 Shah, R.N., Shah, N.A., Lim, M.M.D.R. et al. (2010). Supramolecular design of self-assembling nanofibers for cartilage regeneration. *Proceedings of the National Academy of Sciences of the United States of America* 107 (8): 3293–3298.

561 Cai, S., Liu, Y., Shu, X.Z., and Prestwich, G.D. (2005). Injectable glycosaminoglycan hydrogels for controlled release of human basic fibroblast growth factor. *Biomaterials* 26 (30): 6054–6067.

562 Hosseinkhani, H., Hosseinkhani, M., Khademhosseini, A., and Kobayashi, H. (2007). Bone regeneration through controlled release of bone morphogenetic protein-2 from 3-D tissue engineered nano-scaffold. *Journal of Controlled Release* 117 (3): 380–386.

563 Hartgerink, J.D., Beniash, E., and Stupp, S.I. (2001). Self-assembly and mineralization of peptide-amphiphile nanofibers. *Science (New York, NY)* 294 (5547): 1684–1688.

564 Morimoto, K., Yamaguchi, H., Iwakura, Y. et al. (1991). Effects of viscous hyaluronate-sodium solutions on the nasal absorption of vasopressin and an analogue. *Pharmaceutical Research* 8 (4): 471–474.

565 Cascone, M.G., Sim, B., and Downes, S. (1995). Blends of synthetic and natural polymers as drug delivery systems for growth hormone. *Biomaterials* 16 (7): 569–574.

566 Moore, A.R. and Willoughby, D.A. (1995). Hyaluronan as a drug delivery system for diclofenac: a hypothesis for mode of action. *International Journal of Tissue Reactions* 17 (4): 153–156.

567 Gowland, G., Moore, A.R., Willis, D., and Willoughby, D.A. (1996). Marked enhanced efficacy of cyclosporin when combined with hyaluronic acid. *Clinical Drug Investigation* 11 (4): 245–250.

568 Miller, J.A., Ferguson, R.L., Powers, D.L. et al. (1997). Efficacy of hyaluronic acid/nonsteroidal anti-inflammatory drug systems in preventing postsurgical tendon adhesions. *Journal of Biomedical Materials Research* 38 (1): 25–33.

569 Luo, Y., Kirker, K.R., and Prestwich, G.D. (2000). Cross-linked hyaluronic acid hydrogel films: new biomaterials for drug delivery. *Journal of Controlled Release* 69 (1): 169–184.

570 Sinha, V.R. and Kumria, R. (2001). Polysaccharides in colon-specific drug delivery. *International Journal of Pharmaceutics* 224 (1): 19–38.

571 Chourasia, M.K. and Jain, S.K. (2004). Polysaccharides for colon targeted drug delivery. *Drug Delivery* 11 (2): 129–148.

572 Goodarzi, N., Varshochian, R., Kamalinia, G. et al. (2013). A review of polysaccharide cytotoxic drug conjugates for cancer therapy. *Carbohydrate Polymers* 92 (2): 1280–1293.

573 Basu, A., Kundru, K.R., Abtew, E., and Domb, A.J. (2015). Polysaccharide-based conjugates for biomedical applications. *Bioconjugate Chemistry* 26 (8): 1396–1412.

574 Yang, J., Han, S., Zheng, H. et al. (2015). Preparation and application of micro/nanoparticles based on natural polysaccharides. *Carbohydrate Polymers* 123: 53–66.

575 Garnett, M.C. (2001). Targeted drug conjugates: principles and progress. *Advanced Drug Delivery Reviews* 53 (2): 171–216.

576 Ojima, I., Geng, X., Wu, X. et al. (2002). Tumor-specific novel taxoid–monoclonal antibody conjugates. *Journal of Medicinal Chemistry* 45 (26): 5620–5623.

577 Haag, R. and Kratz, F. (2006). Polymer therapeutics: concepts and applications. *Angewandte Chemie International Edition* 45 (8): 1198–1215.

578 Khandare, J. and Minko, T. (2006). Polymer–drug conjugates: progress in polymeric prodrugs. *Progress in Polymer Science* 31 (4): 359–397.

579 Yeh, C.-C., Hou, M.-F., Wu, S.-H. et al. (2006). A study of glutathione status in the blood and tissues of patients with breast cancer. *Cell Biochemistry and Function* 24 (6): 555–559.

580 Pasut, G. and Veronese, F.M. (2007). Polymer–drug conjugation, recent achievements and general strategies. *Progress in Polymer Science* 32 (8): 933–961.

581 Warnecke, A., Fichtner, I., Sass, G., and Kratz, F. (2007). Synthesis, cleavage profile, and antitumor efficacy of an albumin-binding prodrug of methotrexate that is cleaved by plasmin and cathepsin B. *Archiv der Pharmazie* 340 (8): 389–395.

582 Homma, A., Sato, H., Tamura, T. et al. (2010). Synthesis and optimization of hyaluronic acid-methotrexate conjugates to maximize benefit in the treatment of osteoarthritis. *Bioorganic and Medicinal Chemistry* 18 (3): 1062–1075.

583 Bildstein, L., Dubernet, C., and Couvreur, P. (2011). Prodrug-based intracellular delivery of anticancer agents. *Advanced Drug Delivery Reviews* 63 (1–2): 3–23.

584 Hou, L., Fan, Y., Yao, J. et al. (2011). Low molecular weight heparin-all-trans-retinoid acid conjugate as a drug carrier for combination cancer chemotherapy of paclitaxel and all-trans-retinoid acid. *Carbohydrate Polymers* 86 (3): 1157–1166.

585 Luo, Q., Wang, P., Miao, Y. et al. (2012). A novel 5-fluorouracil prodrug using hydroxyethyl starch as a macromolecular carrier for sustained release. *Carbohydrate Polymers* 87 (4): 2642–2647.

586 Chau, Y., Tan, F.E., and Langer, R. (2004). Synthesis and characterization of dextran-peptide-methotrexate conjugates for tumor targeting via mediation by matrix metalloproteinase II and matrix metalloproteinase IX. *Bioconjugate Chemistry* 15 (4): 931–941.

587 Sugahara, S.-I., Kajiki, M., Kuriyama, H., and Kobayashi, T.-R. (2008). Carrier effects on antitumor activity and neurotoxicity of AZ10992, a paclitaxel-carboxymethyl dextran conjugate, in a mouse model. *Biological and Pharmaceutical Bulletin* 31 (2): 223–230.

588 Luo, Y., Ziebell, M.R., and Prestwich, G.D. (2000). A hyaluronic acid-taxol antitumor bioconjugate targeted to cancer cells. *Biomacromolecules* 1 (2): 208–218.

589 Dubowchik, G.M., Radia, S., Mastalerz, H. et al. (2002). Doxorubicin immunoconjugates containing bivalent, lysosomally-cleavable dipeptide linkages. *Bioorganic and Medicinal Chemistry Letters* 12 (11): 1529–1532.

590 Chau, Y., Dang, N.M., Tan, F.E., and Langer, R. (2006). Investigation of targeting mechanism of new dextran-peptide-methotrexate conjugates using biodistribution study in matrix-metalloproteinase-overexpressing tumor xenograft model. *Journal of Pharmaceutical Sciences* 95 (3): 542–551.

591 Chau, Y., Padera, R.F., Dang, N.M., and Langer, R. (2006). Antitumor efficacy of a novel polymer-peptide-drug conjugate in human tumor xenograft models. *International Journal of Cancer* 118 (6): 1519–1526.
592 Zhou, W., Wang, Y., Jian, J., and Song, S. (2013). Self-aggregated nanoparticles based on amphiphilic poly(lactic acid)-grafted-chitosan copolymer for ocular delivery of amphotericin B. *International Journal of Nanomedicine* 8: 3715–3728.
593 Lautenschläger, C., Schmidt, C., Lange, K., and Stallmach, A. (2015). Drug delivery strategies for targeted treatment of inflammatory bowel disease. *Zeitschrift Fur Gastroenterologie* 53 (3): 226–234.
594 Lautenschläger, M., Sendker, J., Hüwel, S. et al. (2015). Intestinal formation of trans-crocetin from saffron extract (*Crocus sativus* L.) and in vitro permeation through intestinal and blood brain barrier. *Phytomedicine* 22 (1): 36–44.
595 Pitombeira, N.A.O., Veras Neto, J.G., Silva, D.A. et al. (2015). Self-assembled nanoparticles of acetylated cashew gum: characterization and evaluation as potential drug carrier. *Carbohydrate Polymers* 117: 610–615.
596 Dosio, F., Arpicco, S., Stella, B., and Fattal, E. (2016). Hyaluronic acid for anticancer drug and nucleic acid delivery. *Advanced Drug Delivery Reviews* 97: 204–236.
597 Luo, Y. and Prestwich, G.D. (1999). Synthesis and selective cytotoxicity of a hyaluronic acid-antitumor bioconjugate. *Bioconjugate Chemistry* 10 (5): 755–763.
598 Götte, M. and Yip, G.W. (2006). Heparanase, hyaluronan, and CD44 in cancers: a breast carcinoma perspective. *Cancer Research* 66 (21): 10233–10237.
599 Schätzlein, A.G. (2006). Delivering cancer stem cell therapies - a role for nanomedicines? *European Journal of Cancer (Oxford, England: 1990)* 42 (9): 1309–1315.
600 Auzenne, E., Ghosh, S.C., Khodadadian, M. et al. (2007). Hyaluronic acid-paclitaxel: antitumor efficacy against CD44(+) human ovarian carcinoma xenografts. *Neoplasia (New York, NY)* 9 (6): 479–486.
601 Lee, E., Lee, J., Lee, I.-H. et al. (2008). Conjugated chitosan as a novel platform for oral delivery of paclitaxel. *Journal of Medicinal Chemistry* 51 (20): 6442–6449.
602 Lee, H., Lee, K., and Park, T.G. (2008). Hyaluronic acid-paclitaxel conjugate micelles: synthesis, characterization, and antitumor activity. *Bioconjugate Chemistry* 19 (6): 1319–1325.
603 Oh, E.J., Park, K., Kim, K.S. et al. (2010). Target specific and long-acting delivery of protein, peptide, and nucleotide therapeutics using hyaluronic acid derivatives. *Journal of Controlled Release* 141 (1): 2–12.
604 Pedrosa, S.S., Pereira, P., Correia, A. et al. (2017). Targetability of hyaluronic acid nanogel to cancer cells: in vitro and in vivo studies. *European Journal of Pharmaceutical Sciences* 104: 102–113.
605 Dalerba, P., Cho, R.W., and Clarke, M.F. (2007). Cancer stem cells: models and concepts. *Annual Review of Medicine* 58: 267–284.

606 Zhang, M. and James, S.P. (2005). Synthesis and properties of melt-processable hyaluronan esters. *Journal of Materials Science Materials in Medicine* 16 (6): 587–593.

607 Zhang, M. and James, S.P. (2005). Silylation of hyaluronan to improve hydrophobicity and reactivity for improved processing and derivatization. *Polymer* 46 (11): 3639–3648.

608 Norbedo, S., Dinon, F., Bergamin, M. et al. (2009). Synthesis of 6-amino-6-deoxyhyaluronan as an intermediate for conjugation with carboxylate-containing compounds: application to hyaluronan–camptothecin conjugates. *Carbohydrate Research* 344 (1): 98–104.

609 Leonelli, F., La Bella, A., Francescangeli, A. et al. (2005). A new and simply available class of hydrosoluble bioconjugates by coupling paclitaxel to hyaluronic acid through a 4-hydroxybutanoic acid derived linker. *Helvetica Chimica Acta* 88 (1): 154–159.

610 Rosato, A., Banzato, A., De Luca, G. et al. (2006). HYTAD1-p20: a new paclitaxel-hyaluronic acid hydrosoluble bioconjugate for treatment of superficial bladder cancer. *Urologic Oncology* 24 (3): 207–215.

611 Manju, S. and Sreenivasan, K. (2011). Conjugation of curcumin onto hyaluronic acid enhances its aqueous solubility and stability. *Journal of Colloid and Interface Science* 359 (1): 318–325.

612 Coradini, D., Pellizzaro, C., Abolafio, G. et al. (2004). Hyaluronic-acid butyric esters as promising antineoplastic agents in human lung carcinoma: a preclinical study. *Investigational New Drugs* 22 (3): 207–217.

613 Speranza, A., Pellizzaro, C., and Coradini, D. (2005). Hyaluronic acid butyric esters in cancer therapy. *Anti-Cancer Drugs* 16 (4): 373–379.

614 Coradini, D., Pellizzaro, C., Scarlata, I. et al. (2006). A novel retinoic/butyric hyaluronan ester for the treatment of acute promyelocytic leukemia: preliminary preclinical results. *Leukemia* 20 (5): 785–792.

615 Banzato, A., Bobisse, S., Rondina, M. et al. (2008). A paclitaxel-hyaluronan bioconjugate targeting ovarian cancer affords a potent in vivo therapeutic activity. *Clinical Cancer Research* 14 (11): 3598–3606.

616 De Stefano, I., Battaglia, A., Zannoni, G.F. et al. (2011). Hyaluronic acid-paclitaxel: effects of intraperitoneal administration against CD44(+) human ovarian cancer xenografts. *Cancer Chemotherapy and Pharmacology* 68 (1): 107–116.

617 Galer, C.E., Sano, D., Ghosh, S.C. et al. (2011). Hyaluronic acid–paclitaxel conjugate inhibits growth of human squamous cell carcinomas of the head and neck via a hyaluronic acid-mediated mechanism. *Oral Oncology* 47 (11): 1039–1047.

618 Lee, S.J., Ghosh, S.C., Han, H.D. et al. (2012). Metronomic activity of CD44-targeted hyaluronic acid-paclitaxel in ovarian carcinoma. *Clinical Cancer Research* 18 (15): 4114–4121.

619 Mittapalli, R.K., Liu, X., Adkins, C.E. et al. (2013). Paclitaxel-hyaluronic nanoconjugates prolong overall survival in a preclinical brain metastases of breast cancer model. *Molecular Cancer Therapeutics* 12 (11): 2389–2399.

620 Serafino, A., Zonfrillo, M., Andreola, F. et al. (2011). CD44-targeting for antitumor drug delivery: a new SN-38-hyaluronan bioconjugate for locoregional treatment of peritoneal carcinomatosis. *Current Cancer Drug Targets* 11 (5): 572–585.

621 Montagner, I.M., Merlo, A., Zuccolotto, G. et al. (2014). Peritoneal tumor carcinomatosis: pharmacological targeting with hyaluronan-based bioconjugates overcomes therapeutic indications of current drugs. *PLoS One* 9 (11): e112240.

622 Cai, S., Thati, S., Bagby, T.R. et al. (2010). Localized doxorubicin chemotherapy with a biopolymeric nanocarrier improves survival and reduces toxicity in xenografts of human breast cancer. *Journal of Controlled Release* 146 (2): 212–218.

623 Cai, S., Xie, Y., Bagby, T.R. et al. (2008). Intralymphatic chemotherapy using a hyaluronan–cisplatin conjugate. *The Journal of Surgical Research* 147 (2): 247–252.

624 Cohen, S.M., Mukerji, R., Cai, S. et al. (2011). Subcutaneous delivery of nanoconjugated doxorubicin and cisplatin for locally advanced breast cancer demonstrates improved efficacy and decreased toxicity at lower doses than standard systemic combination therapy in vivo. *American Journal of Surgery* 202 (6): 646–652; discussion 652–653.

625 Venable, R.O., Worley, D.R., Gustafson, D.L. et al. (2012). Effects of intratumoral administration of a hyaluronan–cisplatin nanoconjugate to five dogs with soft tissue sarcomas. *American Journal of Veterinary Research* 73 (12): 1969–1976.

626 Cohen, S.M., Rockefeller, N., Mukerji, R. et al. (2013). Efficacy and toxicity of peritumoral delivery of nanoconjugated cisplatin in an in vivo murine model of head and neck squamous cell carcinoma. *JAMA Otolaryngology–Head & Neck Surgery* 139 (4): 382–387.

627 Yang, Q., Aires, D.J., Cai, S. et al. (2014). In vivo efficacy of nano hyaluronan-conjugated cisplatin for treatment of murine melanoma. *Journal of Drugs in Dermatology : JDD* 13 (3): 283–287.

628 Ling, X., Shen, Y., Sun, R. et al. (2015). Tumor-targeting delivery of hyaluronic acid–platinum(IV) nanoconjugate to reduce toxicity and improve survival. *Polymer Chemistry* 6 (9): 1541–1552.

629 Camacho, K.M., Menegatti, S., and Mitragotri, S. (2016). Low-molecular-weight polymer-drug conjugates for synergistic anticancer activity of camptothecin and doxorubicin combinations. *Nanomedicine (London, England)* 11 (9): 1139–1151.

630 Luo, Y., Bernshaw, N.J., Lu, Z.-R. et al. (2002). Targeted delivery of doxorubicin by HPMA copolymer-hyaluronan bioconjugates. *Pharmaceutical Research* 19 (4): 396–402.

631 Bassi, P.F., Volpe, A., D'Agostino, D. et al. (2011). Paclitaxel-hyaluronic acid for intravesical therapy of *Bacillus* Calmette–Guérin refractory carcinoma in situ of the bladder: results of a phase I study. *The Journal of Urology* 185 (2): 445–449.

632 Montagner, I.M., Banzato, A., Zuccolotto, G. et al. (2013). Paclitaxel-hyaluronan hydrosoluble bioconjugate: mechanism of action in human bladder cancer cell lines. *Urologic Oncology: Seminars and Original Investigations* 31 (7): 1261–1269.

633 Qiu, L., Qiao, M., Chen, Q. et al. (2014). Enhanced effect of pH-sensitive mixed copolymer micelles for overcoming multidrug resistance of doxorubicin. *Biomaterials* 35 (37): 9877–9887.

634 Li, K., Liu, H., Gao, W. et al. (2015). Mulberry-like dual-drug complicated nanocarriers assembled with apogossypolone amphiphilic starch micelles and doxorubicin hyaluronic acid nanoparticles for tumor combination and targeted therapy. *Biomaterials* 39: 131–144.

635 Li, J., Huo, M., Wang, J. et al. (2012). Redox-sensitive micelles self-assembled from amphiphilic hyaluronic acid-deoxycholic acid conjugates for targeted intracellular delivery of paclitaxel. *Biomaterials* 33 (7): 2310–2320.

636 Liu, Y., Sun, J., Lian, H. et al. (2014). Folate and CD44 receptors dual-targeting hydrophobized hyaluronic acid paclitaxel-loaded polymeric micelles for overcoming multidrug resistance and improving tumor distribution. *Journal of Pharmaceutical Sciences* 103 (5): 1538–1547.

637 Thomas, R.G., Moon, M., Lee, S., and Jeong, Y.Y. (2015). Paclitaxel loaded hyaluronic acid nanoparticles for targeted cancer therapy: in vitro and in vivo analysis. *International Journal of Biological Macromolecules* 72: 510–518.

638 Yin, T., Wang, L., Yin, L. et al. (2015). Co-delivery of hydrophobic paclitaxel and hydrophilic AURKA specific siRNA by redox-sensitive micelles for effective treatment of breast cancer. *Biomaterials* 61: 10–25.

639 Peer, D. and Margalit, R. (2004). Loading mitomycin C inside long circulating hyaluronan targeted nano-liposomes increases its antitumor activity in three mice tumor models. *International Journal of Cancer* 108 (5): 780–789.

640 Arpicco, S., Lerda, C., Dalla Pozza, E. et al. (2013). Hyaluronic acid-coated liposomes for active targeting of gemcitabine. *European Journal of Pharmaceutics and Biopharmaceutics* 85 (3 Pt. A): 373–380.

641 Dalla Pozza, E., Lerda, C., Costanzo, C. et al. (2013). Targeting gemcitabine containing liposomes to CD44 expressing pancreatic adenocarcinoma cells causes an increase in the antitumoral activity. *Biochimica et Biophysica Acta (BBA) - Biomembranes* 1828 (5): 1396–1404.

642 Peer, D. and Margalit, R. (2004). Tumor-targeted hyaluronan nanoliposomes increase the antitumor activity of liposomal doxorubicin in syngeneic and human xenograft mouse tumor models. *Neoplasia (New York, NY)* 6 (4): 343–353.

643 Park, J.-H., Cho, H.-J., Yoon, H.Y. et al. (2014). Hyaluronic acid derivative-coated nanohybrid liposomes for cancer imaging and drug delivery. *Journal of Controlled Release* 174: 98–108.

644 Jiang, T., Zhang, Z., Zhang, Y. et al. (2012). Dual-functional liposomes based on pH-responsive cell-penetrating peptide and hyaluronic acid for tumor-targeted anticancer drug delivery. *Biomaterials* 33 (36): 9246–9258.

645 Cho, H.-J., Yoon, H.Y., Koo, H. et al. (2011). Self-assembled nanoparticles based on hyaluronic acid-ceramide (HA-CE) and Pluronic® for tumor-targeted delivery of docetaxel. *Biomaterials* 32 (29): 7181–7190.

646 Huang, J., Zhang, H., Yu, Y. et al. (2014). Biodegradable self-assembled nanoparticles of poly(D,L-lactide-co-glycolide)/hyaluronic acid block copolymers for target delivery of docetaxel to breast cancer. *Biomaterials* 35 (1): 550–566.

647 Song, S., Chen, F., Qi, H. et al. (2014). Multifunctional tumor-targeting nanocarriers based on hyaluronic acid-mediated and pH-sensitive properties for efficient delivery of docetaxel. *Pharmaceutical Research* 31 (4): 1032–1045.

648 Song, S., Qi, H., Xu, J. et al. (2014). Hyaluronan-based nanocarriers with CD44-overexpressed cancer cell targeting. *Pharmaceutical Research* 31 (11): 2988–3005.

649 Liang, D., Wang, A.-T., Yang, Z.-Z. et al. (2015). Enhance cancer cell recognition and overcome drug resistance using hyaluronic acid and α-tocopheryl succinate based multifunctional nanoparticles. *Molecular Pharmaceutics* 12 (6): 2189–2202.

650 Yadav, A.K., Mishra, P., Mishra, A.K. et al. (2007). Development and characterization of hyaluronic acid-anchored PLGA nanoparticulate carriers of doxorubicin. *Nanomedicine: Nanotechnology, Biology and Medicine* 3 (4): 246–257.

651 Yadav, A.K., Mishra, P., Jain, S. et al. (2008). Preparation and characterization of HA-PEG-PCL intelligent core-corona nanoparticles for delivery of doxorubicin. *Journal of Drug Targeting* 16 (6): 464–478.

652 Han, H.S., Thambi, T., Choi, K.Y. et al. (2015). Bioreducible shell-cross-linked hyaluronic acid nanoparticles for tumor-targeted drug delivery. *Biomacromolecules* 16 (2): 447–456.

653 Hu, K., Zhou, H., Liu, Y. et al. (2015). Hyaluronic acid functional amphipathic and redox-responsive polymer particles for the co-delivery of doxorubicin and cyclopamine to eradicate breast cancer cells and cancer stem cells. *Nanoscale* 7 (18): 8607–8618.

654 Park, H.-K., Lee, S.J., Oh, J.-S. et al. (2015). Smart nanoparticles based on hyaluronic acid for redox-responsive and CD44 receptor-mediated targeting of tumor. *Nanoscale Research Letters* 10 (1): 981.

655 He, M., Zhao, Z., Yin, L. et al. (2009). Hyaluronic acid coated poly(butyl cyanoacrylate) nanoparticles as anticancer drug carriers. *International Journal of Pharmaceutics* 373 (1–2): 165–173.

656 Choi, K.Y., Yoon, H.Y., Kim, J.-H. et al. (2011). Smart nanocarrier based on PEGylated hyaluronic acid for cancer therapy. *ACS Nano* 5 (11): 8591–8599.

657 Cho, H.-J., Yoon, I.-S., Yoon, H.Y. et al. (2012). Polyethylene glycol-conjugated hyaluronic acid-ceramide self-assembled nanoparticles for targeted delivery of doxorubicin. *Biomaterials* 33 (4): 1190–1200.

658 Jin, Y.-J., Termsarasab, U., Ko, S.-H. et al. (2012). Hyaluronic acid derivative-based self-assembled nanoparticles for the treatment of melanoma. *Pharmaceutical Research* 29 (12): 3443–3454.

659 Han, H.S., Lee, J., Kim, H.R. et al. (2013). Robust PEGylated hyaluronic acid nanoparticles as the carrier of doxorubicin: mineralization and its effect on tumor targetability in vivo. *Journal of Controlled Release* 168 (2): 105–114.

660 Lee, C.-S. and Na, K. (2014). Photochemically triggered cytosolic drug delivery using pH-responsive hyaluronic acid nanoparticles for light-induced cancer therapy. *Biomacromolecules* 15 (11): 4228–4238.
661 Mezghrani, O., Tang, Y., Ke, X. et al. (2015). Hepatocellular carcinoma dually-targeted nanoparticles for reduction triggered intracellular delivery of doxorubicin. *International Journal of Pharmaceutics* 478 (2): 553–568.
662 Zhong, Y., Zhang, J., Cheng, R. et al. (2015). Reversibly crosslinked hyaluronic acid nanoparticles for active targeting and intelligent delivery of doxorubicin to drug resistant CD44+ human breast tumor xenografts. *Journal of Controlled Release* 205: 144–154.
663 Choi, K.Y., Jeon, E.J., Yoon, H.Y. et al. (2012). Theranostic nanoparticles based on PEGylated hyaluronic acid for the diagnosis, therapy and monitoring of colon cancer. *Biomaterials* 33 (26): 6186–6193.
664 Yao, J., Zhang, L., Zhou, J. et al. (2013). Efficient simultaneous tumor targeting delivery of all-trans retinoid acid and paclitaxel based on hyaluronic acid-based multifunctional nanocarrier. *Molecular Pharmaceutics* 10 (3): 1080–1091.
665 Yoon, H.Y., Koo, H., Choi, K.Y. et al. (2013). Photo-crosslinked hyaluronic acid nanoparticles with improved stability for in vivo tumor-targeted drug delivery. *Biomaterials* 34 (21): 5273–5280.
666 Wang, W., Xi, M., Duan, X. et al. (2015). Delivery of baicalein and paclitaxel using self-assembled nanoparticles: synergistic antitumor effect in vitro and in vivo. *International Journal of Nanomedicine* 10: 3737–3750.
667 Ganesh, S., Iyer, A.K., Gattacceca, F. et al. (2013). In vivo biodistribution of siRNA and cisplatin administered using CD44-targeted hyaluronic acid nanoparticles. *Journal of Controlled Release* 172 (3): 699–706.
668 Nakashima, M., Ichinose, K., Kanematsu, T. et al. (1999). In vitro characteristics and in vivo plasma disposition of cisplatin conjugated with oxidized and dicarboxymethylated dextrans. *Biological and Pharmaceutical Bulletin* 22 (7): 756–761.
669 Tan, M.L., Friedhuber, A.M., Dunstan, D.E. et al. (2010). The performance of doxorubicin encapsulated in chitosan–dextran sulphate microparticles in an osteosarcoma model. *Biomaterials* 31 (3): 541–551.
670 Yousefpour, P., Atyabi, F., Farahani, E.V. et al. (2011). Polyanionic carbohydrate doxorubicin–dextran nanocomplex as a delivery system for anticancer drugs: in vitro analysis and evaluations. *International Journal of Nanomedicine* 6: 1487–1496.
671 Yousefpour, P., Atyabi, F., Farahani, E.V. et al. (2011). Targeted delivery of doxorubicin-utilizing chitosan nanoparticles surface-functionalized with anti-Her2 trastuzumab. *International Journal of Nanomedicine* 6: 1977–1990.
672 Mehvar, R. (2000). Dextrans for targeted and sustained delivery of therapeutic and imaging agents. *Journal of Controlled Release* 69 (1): 1–25.
673 Harada, M., Imai, J., Okuno, S., and Suzuki, T. (2000). Macrophage-mediated activation of camptothecin analogue T-2513-carboxymethyl dextran conjugate (T-0128): possible cellular mechanism for antitumor activity. *Journal of Controlled Release* 69 (3): 389–397.

674 Harada, M., Sakakibara, H., Yano, T. et al. (2000). Determinants for the drug release from T-0128, camptothecin analogue-carboxymethyl dextran conjugate. *Journal of Controlled Release* 69 (3): 399–412.

675 Ochi, Y., Shiose, Y., Kuga, H., and Kumazawa, E. (2005). A possible mechanism for the long-lasting antitumor effect of the macromolecular conjugate DE-310: mediation by cellular uptake and drug release of its active camptothecin analog DX-8951. *Cancer Chemotherapy and Pharmacology* 55 (4): 323–332.

676 Veltkamp, S.A., Witteveen, E.O., Capriati, E. et al. (2008). Clinical and pharmacologic study of the novel prodrug delimotecan (MEN 4901/T-0128) in patients with solid tumors. *Clinical Cancer Research* 14 (22): 7535–7544.

677 Onishi, H., Seno, Y., Pithayanukul, P., and Nagai, T. (1991). Conjugate of N4-(4-carboxybutyryl)-ara-C and ethylenediamine-introduced dextran. Drug release profiles and further in vivo study of its antitumor effects. *Drug Design and Delivery* 7 (2): 139–145.

678 Dang, W., Colvin, O.M., Brem, H., and Saltzman, W.M. (1994). Covalent coupling of methotrexate to dextran enhances the penetration of cytotoxicity into a tissue-like matrix. *Cancer Research* 54 (7): 1729–1735.

679 Nevozhay, D., Budzynska, R., Jagiello, M. et al. (2006). The effect of the substitution level of some dextran-methotrexate conjugates on their antitumor activity in experimental cancer models. *Anticancer Research* 26 (3A): 2179–2186.

680 Nevozhay, D., Budzynska, R., Kanska, U. et al. (2006). Antitumor properties and toxicity of dextran-methotrexate conjugates are dependent on the molecular weight of the carrier. *Anticancer Research* 26 (2A): 1135–1143.

681 Okuno, S., Harada, M., Yano, T. et al. (2000). Complete regression of xenografted human carcinomas by camptothecin analogue-carboxymethyl dextran conjugate (T-0128). *Cancer Research* 60 (11): 2988–2995.

682 Harada, M., Murata, J., Sakamura, Y. et al. (2001). Carrier and dose effects on the pharmacokinetics of T-0128, a camptothecin analogue-carboxymethyl dextran conjugate, in non-tumor- and tumor-bearing rats. *Journal of Controlled Release* 71 (1): 71–86.

683 Sugahara, S.-I., Kajiki, M., Kuriyama, H., and Kobayashi, T.-R. (2007). Complete regression of xenografted human carcinomas by a paclitaxel–carboxymethyl dextran conjugate (AZ10992). *Journal of Controlled Release* 117 (1): 40–50.

684 Inoue, K., Kumazawa, E., Kuga, H. et al. (2003). CM-dextran-polyalcohol-camptothecin conjugate: DE-310 with a novel carrier system and its preclinical data. *Advances in Experimental Medicine and Biology* 519: 145–153.

685 Soepenberg, O., de Jonge, M.J.A., Sparreboom, A. et al. (2005). Phase I and pharmacokinetic study of DE-310 in patients with advanced solid tumors. *Clinical Cancer Research* 11 (2 Pt. 1): 703–711.

686 Munechika, K., Sogame, Y., Kishi, N. et al. (1994). Tissue distribution of macromolecular conjugate, adriamycin linked to oxidized dextran, in rat and mouse bearing tumor cells. *Biological and Pharmaceutical Bulletin* 17 (9): 1193–1198.

687 Danhauser-Riedl, S., Hausmann, E., Schick, H.D. et al. (1993). Phase I clinical and pharmacokinetic trial of dextran conjugated doxorubicin (AD-70, DOX-OXD). *Investigational New Drugs* 11 (2–3): 187–195.
688 Naeye, B., Raemdonck, K., Remaut, K. et al. (2010). PEGylation of biodegradable dextran nanogels for siRNA delivery. *European Journal of Pharmaceutical Sciences* 40 (4): 342–351.
689 Naeye, B., Deschout, H., Röding, M. et al. (2011). Hemocompatibility of siRNA loaded dextran nanogels. *Biomaterials* 32 (34): 9120–9127.
690 Tripathi, S.K., Goyal, R., and Gupta, K.C. (2011). Surface modification of crosslinked dextran nanoparticles influences transfection efficiency of dextran–polyethylenimine nanocomposites. *Soft Matter* 7 (24): 11360–11371.
691 De Backer, L., Naessens, T., De Koker, S. et al. (2015). Hybrid pulmonary surfactant-coated nanogels mediate efficient in vivo delivery of siRNA to murine alveolar macrophages. *Journal of Controlled Release* 217: 53–63.
692 Shatsberg, Z., Zhang, X., Ofek, P. et al. (2016). Functionalized nanogels carrying an anticancer microRNA for glioblastoma therapy. *Journal of Controlled Release* 239: 159–168.
693 Dutta, R. and Mahato, R.I. (2017). Recent advances in hepatocellular carcinoma therapy. *Pharmacology & Therapeutics* 173: 106–117.
694 Raemdonck, K., Naeye, B., Høgset, A. et al. (2010). Prolonged gene silencing by combining siRNA nanogels and photochemical internalization. *Journal of Controlled Release* 145 (3): 281–288.
695 Choi, K.Y., Silvestre, O.F., Huang, X. et al. (2014). Versatile RNA interference nanoplatform for systemic delivery of RNAs. *ACS Nano* 8 (5): 4559–4570.
696 Kang, L., Gao, Z., Huang, W. et al. (2015). Nanocarrier-mediated co-delivery of chemotherapeutic drugs and gene agents for cancer treatment. *Acta Pharmaceutica Sinica B* 5 (3): 169–175.
697 Kim, Y.H., Gihm, S.H., Park, C.R. et al. (2001). Structural characteristics of size-controlled self-aggregates of deoxycholic acid-modified chitosan and their application as a DNA delivery carrier. *Bioconjugate Chemistry* 12 (6): 932–938.
698 Park, J.H., Cho, Y.W., Chung, H. et al. (2003). Synthesis and characterization of sugar-bearing chitosan derivatives: aqueous solubility and biodegradability. *Biomacromolecules* 4 (4): 1087–1091.
699 Park, J.H., Kwon, S., Lee, M. et al. (2006). Self-assembled nanoparticles based on glycol chitosan bearing hydrophobic moieties as carriers for doxorubicin: in vivo biodistribution and anti-tumor activity. *Biomaterials* 27 (1): 119–126.
700 Kato, Y., Onishi, H., and Machida, Y. (2004). N-succinyl-chitosan as a drug carrier: water-insoluble and water-soluble conjugates. *Biomaterials* 25 (5): 907–915.
701 Sato, M., Onishi, H., Kitano, M. et al. (1996). Preparation and drug release characteristics of the conjugates of mitomycin C with glycol-chitosan and N-succinyl-chitosan. *Biological and Pharmaceutical Bulletin* 19 (2): 241–245.
702 Sato, M., Onishi, H., Takahara, J. et al. (1996). In vivo drug release and antitumor characteristics of water-soluble conjugates of mitomycin C with glycol-chitosan and N-succinyl-chitosan. *Biological and Pharmaceutical Bulletin* 19 (9): 1170–1177.

703 Hu, F.-Q., Liu, L.-N., Du, Y.-Z., and Yuan, H. (2009). Synthesis and antitumor activity of doxorubicin conjugated stearic acid-g-chitosan oligosaccharide polymeric micelles. *Biomaterials* 30 (36): 6955–6963.

704 Son, Y.J., Jang, J.-S., Cho, Y.W. et al. (2003). Biodistribution and anti-tumor efficacy of doxorubicin loaded glycol-chitosan nanoaggregates by EPR effect. *Journal of Controlled Release* 91 (1–2): 135–145.

705 Lee, E., Kim, H., Lee, I.-H., and Jon, S. (2009). In vivo antitumor effects of chitosan-conjugated docetaxel after oral administration. *Journal of Controlled Release* 140 (2): 79–85.

706 Lee, E., Lee, J., and Jon, S. (2010). A novel approach to oral delivery of insulin by conjugating with low molecular weight chitosan. *Bioconjugate Chemistry* 21 (10): 1720–1723.

707 Atyabi, F., Moghaddam, F.A., Dinarvand, R. et al. (2008). Thiolated chitosan coated poly hydroxyethyl methacrylate nanoparticles: synthesis and characterization. *Carbohydrate Polymers* 74 (1): 59–67.

708 Anitha, A., Deepa, N., Chennazhi, K.P. et al. (2011). Development of mucoadhesive thiolated chitosan nanoparticles for biomedical applications. *Carbohydrate Polymers* 83 (1): 66–73.

709 Quiñones, J.P., Peniche, H., Peniche, C. et al. (2012). Self-assembled nanoparticles of glycol chitosan – Ergocalciferol succinate conjugate, for controlled release. *Carbohydrate Polymers* 88 (4): 1373–1377.

710 Rajitha, P., Gopinath, D., Biswas, R. et al. (2016). Chitosan nanoparticles in drug therapy of infectious and inflammatory diseases. *Expert Opinion on Drug Delivery* 13 (8): 1177–1194.

711 Jameela, S.R. and Jayakrishnan, A. (1995). Glutaraldehyde cross-linked chitosan microspheres as a long acting biodegradable drug delivery vehicle: studies on the in vitro release of mitoxantrone and in vivo degradation of microspheres in rat muscle. *Biomaterials* 16 (10): 769–775.

712 He, P., Davis, S.S., and Illum, L. (1999). Chitosan microspheres prepared by spray drying. *International Journal of Pharmaceutics* 187 (1): 53–65.

713 Mitra, S., Gaur, U., Ghosh, P.C., and Maitra, A.N. (2001). Tumour targeted delivery of encapsulated dextran-doxorubicin conjugate using chitosan nanoparticles as carrier. *Journal of Controlled Release* 74 (1–3): 317–323.

714 Denkbaş, E.B., Kiliçay, E., Birlikseven, C., and Öztürk, E. (2002). Magnetic chitosan microspheres: preparation and characterization. *Reactive and Functional Polymers* 50 (3): 225–232.

715 Thanou, M., Florea, B.I., Geldof, M. et al. (2002). Quaternized chitosan oligomers as novel gene delivery vectors in epithelial cell lines. *Biomaterials* 23 (1): 153–159.

716 Agnihotri, S.A., Mallikarjuna, N.N., and Aminabhavi, T.M. (2004). Recent advances on chitosan-based micro- and nanoparticles in drug delivery. *Journal of Controlled Release* 100 (1): 5–28.

717 Qi, L., Xu, Z., Jiang, X. et al. (2004). Preparation and antibacterial activity of chitosan nanoparticles. *Carbohydrate Research* 339 (16): 2693–2700.

718 Sinha, V.R., Singla, A.K., Wadhawan, S. et al. (2004). Chitosan microspheres as a potential carrier for drugs. *International Journal of Pharmaceutics* 274 (1–2): 1–33.

719 Kim, K., Kwon, S., Park, J.H. et al. (2005). Physicochemical characterizations of self-assembled nanoparticles of glycol chitosan–deoxycholic acid conjugates. *Biomacromolecules* 6 (2): 1154–1158.
720 Amidi, M., Romeijn, S.G., Borchard, G. et al. (2006). Preparation and characterization of protein-loaded N-trimethyl chitosan nanoparticles as nasal delivery system. *Journal of Controlled Release* 111 (1–2): 107–116.
721 Liu, C., Tan, Y., Liu, C. et al. (2007). Preparations, characterizations and applications of chitosan-based nanoparticles. *Journal of Ocean University of China* 6 (3): 237–243.
722 Park, K., Kim, J.-H., Nam, Y.S. et al. (2007). Effect of polymer molecular weight on the tumor targeting characteristics of self-assembled glycol chitosan nanoparticles. *Journal of Controlled Release* 122 (3): 305–314.
723 Kim, J.-H., Kim, Y.-S., Park, K. et al. (2008). Self-assembled glycol chitosan nanoparticles for the sustained and prolonged delivery of antiangiogenic small peptide drugs in cancer therapy. *Biomaterials* 29 (12): 1920–1930.
724 Kim, J.-H., Kim, Y.-S., Park, K. et al. (2008). Antitumor efficacy of cisplatin-loaded glycol chitosan nanoparticles in tumor-bearing mice. *Journal of Controlled Release* 127 (1): 41–49.
725 Wang, Y.-S., Jiang, Q., Li, R.-S. et al. (2008). Self-assembled nanoparticles of cholesterol-modified O-carboxymethyl chitosan as a novel carrier for paclitaxel. *Nanotechnology* 19 (14): 145101.
726 Ganguly, K., Kulkarni, A.R., and Aminabhavi, T.M. (2015). In vitro cytotoxicity and in vivo efficacy of 5-fluorouracil-loaded enteric-coated PEG-crosslinked chitosan microspheres in colorectal cancer therapy in rats. *Drug Delivery* 23 (8): 1–14.
727 Luque-Alcaraz, A.G., Lizardi-Mendoza, J., Goycoolea, F.M. et al. (2016). Preparation of chitosan nanoparticles by nanoprecipitation and their ability as a drug nanocarrier. *RSC Advances* 6 (64): 59250–59256.
728 Jennings, J.A. and Bumgardner, J.D. (eds.) (2017). *Chitosan Based Biomaterials*, vol. 1. Elsevier.
729 Mao, H.Q., Roy, K., Troung-Le, V.L. et al. (2001). Chitosan-DNA nanoparticles as gene carriers: synthesis, characterization and transfection efficiency. *Journal of Controlled Release* 70 (3): 399–421.
730 Mansouri, S., Lavigne, P., Corsi, K. et al. (2004). Chitosan-DNA nanoparticles as non-viral vectors in gene therapy: strategies to improve transfection efficacy. *European Journal of Pharmaceutics and Biopharmaceutics* 57 (1): 1–8.
731 Xue, W.-J., Feng, Y., Wang, F. et al. (2016). Asialoglycoprotein receptor-magnetic dual targeting nanoparticles for delivery of RASSF1A to hepatocellular carcinoma. *Scientific Reports* 6: 22149.
732 Azzam, T., Eliyahu, H., Makovitzki, E. et al. (2004). Hydrophobized dextran-spermine conjugate as potential vector for in vitro gene transfection. *Journal of Controlled Release* 96 (2): 309–323.
733 Mao, S., Sun, W., and Kissel, T. (2010). Chitosan-based formulations for delivery of DNA and siRNA. *Advanced Drug Delivery Reviews* 62 (1): 12–27.

734 Rudzinski, W.E. and Aminabhavi, T.M. (2010). Chitosan as a carrier for targeted delivery of small interfering RNA. *International Journal of Pharmaceutics* 399 (1–2): 1–11.
735 Jiang, H.-L., Lim, H.-T., Kim, Y.-K. et al. (2011). Chitosan-graft-spermine as a gene carrier in vitro and in vivo. *European Journal of Pharmaceutics and Biopharmaceutics* 77 (1): 36–42.
736 Saboktakin, M.R., Tabar, N.A., Tabatabaie, R.M. et al. (2012). Intelligent drug delivery systems based on modified chitosan nanoparticles. 9 (1): 56–70.
737 Ragelle, H., Vandermeulen, G., and Préat, V. (2013). Chitosan-based siRNA delivery systems. *Journal of Controlled Release* 172 (1): 207–218.
738 Lee, D.W., Yun, K.-S., Ban, H.-S. et al. (2009). Preparation and characterization of chitosan/polyguluronate nanoparticles for siRNA delivery. *Journal of Controlled Release* 139 (2): 146–152.
739 Huh, M.S., Lee, S.-Y., Park, S. et al. (2010). Tumor-homing glycol chitosan/polyethylenimine nanoparticles for the systemic delivery of siRNA in tumor-bearing mice. *Journal of Controlled Release* 144 (2): 134–143.
740 Jeong, E.J., Choi, M., Lee, J. et al. (2015). The spacer arm length in cell-penetrating peptides influences chitosan/siRNA nanoparticle delivery for pulmonary inflammation treatment. *Nanoscale* 7 (47): 20095–20104.
741 Alinejad, V., Hossein Somi, M., Baradaran, B. et al. (2016). Co-delivery of IL17RB siRNA and doxorubicin by chitosan-based nanoparticles for enhanced anticancer efficacy in breast cancer cells. *Biomedicine & Pharmacotherapy* 83: 229–240.
742 Corbet, C., Ragelle, H., Pourcelle, V. et al. (2016). Delivery of siRNA targeting tumor metabolism using non-covalent PEGylated chitosan nanoparticles: identification of an optimal combination of ligand structure, linker and grafting method. *Journal of Controlled Release* 223: 53–63.
743 Van Woensel, M., Wauthoz, N., Rosière, R. et al. (2016). Development of siRNA-loaded chitosan nanoparticles targeting Galectin-1 for the treatment of glioblastoma multiforme via intranasal administration. *Journal of Controlled Release* 227: 71–81.
744 Wang, H., Pan, M., Ni, J. et al. (2016). ClC-7 deficiency impairs tooth development and eruption. *Scientific Reports* 6: 19971.
745 Yoo, H.S., Kim, T.G., and Park, T.G. (2009). Surface-functionalized electrospun nanofibers for tissue engineering and drug delivery. *Advanced Drug Delivery Reviews* 61 (12): 1033–1042.
746 Pillay, V., Dott, C., Choonara, Y.E. et al. (2013). A review of the effect of processing variables on the fabrication of electrospun nanofibers for drug delivery applications. *Journal of Nanomaterials* 2013 (1): 1–22.
747 Hu, X., Liu, S., Zhou, G. et al. (2014). Electrospinning of polymeric nanofibers for drug delivery applications. *Journal of Controlled Release* 185: 12–21.
748 Sridhar, R., Lakshminarayanan, R., Madhaiyan, K. et al. (2015). Electrosprayed nanoparticles and electrospun nanofibers based on natural materials:

applications in tissue regeneration, drug delivery and pharmaceuticals. *Chemical Society Reviews* 44 (3): 790–814.
749 Lim, E.-K. and Chung, B.H. (2016). Preparation of pyrenyl-based multifunctional nanocomposites for biomedical applications. *Nature Protocols* 11 (2): 236–251.
750 Wang, X., Du, Y., Luo, J. et al. (2007). Chitosan/organic rectorite nanocomposite films: structure, characteristic and drug delivery behaviour. *Carbohydrate Polymers* 69 (1): 41–49.
751 Liu, K.-H., Liu, T.-Y., Chen, S.-Y., and Liu, D.-M. (2008). Drug release behavior of chitosan-montmorillonite nanocomposite hydrogels following electrostimulation. *Acta Biomaterialia* 4 (4): 1038–1045.
752 Depan, D., Kumar, A.P., and Singh, R.P. (2009). Cell proliferation and controlled drug release studies of nanohybrids based on chitosan-g-lactic acid and montmorillonite. *Acta Biomaterialia* 5 (1): 93–100.
753 Sun, Y., Chen, Z.-L., Yang, X.-X. et al. (2009). Magnetic chitosan nanoparticles as a drug delivery system for targeting photodynamic therapy. *Nanotechnology* 20 (13): 135102.
754 Aguzzi, C., Capra, P., Bonferoni, C. et al. (2010). Chitosan–silicate biocomposites to be used in modified drug release of 5-aminosalicylic acid (5-ASA). *Applied Clay Science* 50 (1): 106–111.
755 Hua, S., Yang, H., Wang, W., and Wang, A. (2010). Controlled release of ofloxacin from chitosan–montmorillonite hydrogel. *Applied Clay Science* 50 (1): 112–117.
756 Yuan, Q., Shah, J., Hein, S., and Misra, R.D.K. (2010). Controlled and extended drug release behavior of chitosan-based nanoparticle carrier. *Acta Biomaterialia* 6 (3): 1140–1148.
757 Bao, H., Pan, Y., Ping, Y. et al. (2011). Chitosan-functionalized graphene oxide as a nanocarrier for drug and gene delivery. *Small (Weinheim an Der Bergstrasse, Germany)* 7 (11): 1569–1578.
758 Nanda, R., Sasmal, A., and Nayak, P.L. (2011). Preparation and characterization of chitosan–polylactide composites blended with Cloisite 30B for control release of the anticancer drug paclitaxel. *Carbohydrate Polymers* 83 (2): 988–994.
759 Arias, J.L., Reddy, L.H., and Couvreur, P. (2012). Fe_3O_4/chitosan nanocomposite for magnetic drug targeting to cancer. *Journal of Materials Chemistry* 22 (15): 7622–7632.
760 Lim, E.-K., Sajomsang, W., Choi, Y. et al. (2013). Chitosan-based intelligent theragnosis nanocomposites enable pH-sensitive drug release with MR-guided imaging for cancer therapy. *Nanoscale Research Letters* 8 (1): 467.
761 Sun, L., Wang, Y., Jiang, T. et al. (2013). Novel chitosan-functionalized spherical nanosilica matrix as an oral sustained drug delivery system for poorly water-soluble drug carvedilol. *ACS Applied Materials & Interfaces* 5 (1): 103–113.

762 Zhang, D., Sun, P., Li, P. et al. (2013). A magnetic chitosan hydrogel for sustained and prolonged delivery of *Bacillus* Calmette–Guérin in the treatment of bladder cancer. *Biomaterials* 34 (38): 10258–10266.
763 Chandran, P.R. and Sandhyarani, N. (2014). An electric field responsive drug delivery system based on chitosan–gold nanocomposites for site specific and controlled delivery of 5-fluorouracil. *RSC Advances* 4 (85): 44922–44929.
764 Azhar, F.F. and Olad, A. (2014). A study on sustained release formulations for oral delivery of 5-fluorouracil based on alginate–chitosan/montmorillonite nanocomposite systems. *Applied Clay Science* 101: 288–296.
765 Jia, M., Li, Y., Yang, X. et al. (2014). Development of both methotrexate and mitomycin C loaded PEGylated chitosan nanoparticles for targeted drug codelivery and synergistic anticancer effect. *ACS Applied Materials & Interfaces* 6 (14): 11413–11423.
766 Justin, R. and Chen, B. (2014). Characterisation and drug release performance of biodegradable chitosan–graphene oxide nanocomposites. *Carbohydrate Polymers* 103: 70–80.
767 Vasile, B.S., Oprea, O., Voicu, G. et al. (2014). Synthesis and characterization of a novel controlled release zinc oxide/gentamicin–chitosan composite with potential applications in wounds care. *International Journal of Pharmaceutics* 463 (2): 161–169.
768 Taleb, M.F.A., Alkahtani, A., and Mohamed, S.K. (2015). Radiation synthesis and characterization of sodium alginate/chitosan/hydroxyapatite nanocomposite hydrogels: a drug delivery system for liver cancer. *Polymer Bulletin* 72 (4): 725–742.
769 Lin, J., Li, Y., Li, Y. et al. (2015). Drug/dye-loaded, multifunctional PEG–chitosan–iron oxide nanocomposites for methotraxate synergistically self-targeted cancer therapy and dual model imaging. *ACS Applied Materials & Interfaces* 7 (22): 11908–11920.
770 Yadollahi, M., Farhoudian, S., and Namazi, H. (2015). One-pot synthesis of antibacterial chitosan/silver bio-nanocomposite hydrogel beads as drug delivery systems. *International Journal of Biological Macromolecules* 79: 37–43.
771 Prabha, G. and Raj, V. (2016). Preparation and characterization of polymer nanocomposites coated magnetic nanoparticles for drug delivery applications. *Journal of Magnetism and Magnetic Materials* 408: 26–34.
772 Yadollahi, M., Farhoudian, S., Barkhordari, S. et al. (2016). Facile synthesis of chitosan/ZnO bio-nanocomposite hydrogel beads as drug delivery systems. *International Journal of Biological Macromolecules* 82: 273–278.
773 Ding, Y., Yin, H., Shen, S. et al. (2017). Chitosan-based magnetic/fluorescent nanocomposites for cell labelling and controlled drug release. *New Journal of Chemistry* 41 (4): 1736–1743.
774 Shariatinia, Z. and Zahraee, Z. (2017). Controlled release of metformin from chitosan-based nanocomposite films containing mesoporous MCM-41 nanoparticles as novel drug delivery systems. *Journal of Colloid and Interface Science* 501: 60–76.

775 Luo, Y. and Wang, Q. (2014). Recent development of chitosan-based polyelectrolyte complexes with natural polysaccharides for drug delivery. *International Journal of Biological Macromolecules* 64: 353–367.

776 Anal, A.K. and Stevens, W.F. (2005). Chitosan–alginate multilayer beads for controlled release of ampicillin. *International Journal of Pharmaceutics* 290 (1–2): 45–54.

777 Zhao, Q.S., Ji, Q.X., Cheng, X.J. et al. (2010). Preparation of alginate coated chitosan hydrogel beads by thermosensitive internal gelation technique. *Journal of Sol-Gel Science and Technology* 54 (2): 232–237.

778 Tavakol, M., Vasheghani-Farahani, E., and Hashemi-Najafabadi, S. (2013). The effect of polymer and $CaCl_2$ concentrations on the sulfasalazine release from alginate-N,O-carboxymethyl chitosan beads. *Progress in Biomaterials* 2: 1–8.

779 Motwani, S.K., Chopra, S., Talegaonkar, S. et al. (2008). Chitosan–sodium alginate nanoparticles as submicroscopic reservoirs for ocular delivery: formulation, optimisation and in vitro characterisation. *European Journal of Pharmaceutics and Biopharmaceutics* 68 (3): 513–525.

780 Elzatahry, A.A., Mohy Eldin, M.S., Soliman, E.A., and Hassan, E.A. (2009). Evaluation of alginate–chitosan bioadhesive beads as a drug delivery system for the controlled release of theophylline. *Journal of Applied Polymer Science* 111 (5): 2452–2459.

781 Wong, T.W., Chan, L.W., Kho, S.B., and Heng, P.W.S. (2002). Design of controlled-release solid dosage forms of alginate and chitosan using microwave. *Journal of Controlled Release* 84 (3): 99–114.

782 de la Fuente, M., Seijo, B., and Alonso, M.J. (2008). Bioadhesive hyaluronan–chitosan nanoparticles can transport genes across the ocular mucosa and transfect ocular tissue. *Gene Therapy* 15 (9): 668–676.

783 de la Fuente, M., Seijo, B., and Alonso, M.J. (2008). Novel hyaluronic acid-chitosan nanoparticles for ocular gene therapy. *Investigative Ophthalmology & Visual Science* 49 (5): 2016–2024.

784 Contreras-Ruiz, L., de la Fuente, M., García-Vázquez, C. et al. (2010). Ocular tolerance to a topical formulation of hyaluronic acid and chitosan-based nanoparticles. *Cornea* 29 (5): 550–558.

785 Oyarzun-Ampuero, F.A., Brea, J., Loza, M.I. et al. (2009). Chitosan–hyaluronic acid nanoparticles loaded with heparin for the treatment of asthma. *International Journal of Pharmaceutics* 381 (2): 122–129.

786 Lu, H.-D., Zhao, H.-Q., Wang, K., and Lv, L.-L. (2011). Novel hyaluronic acid-chitosan nanoparticles as non-viral gene delivery vectors targeting osteoarthritis. *International Journal of Pharmaceutics* 420 (2): 358–365.

787 Fukuda, M., Peppas, N.A., and McGinity, J.W. (2006). Properties of sustained release hot-melt extruded tablets containing chitosan and xanthan gum. *International Journal of Pharmaceutics* 310 (1–2): 90–100.

788 Phaechamud, T. and Ritthidej, G.C. (2007). Sustained-release from layered matrix system comprising chitosan and xanthan gum. *Drug Development and Industrial Pharmacy* 33 (6): 595–605.

789 Hong, Y., Gong, Y., Gao, C., and Shen, J. (2008). Collagen-coated polylactide microcarriers/chitosan hydrogel composite: injectable scaffold for cartilage regeneration. *Journal of Biomedical Materials Research Part A* 85 (3): 628–637.

790 Wang, F., Li, Z., Khan, M. et al. (2010). Injectable, rapid gelling and highly flexible hydrogel composites as growth factor and cell carriers. *Acta Biomaterialia* 6 (6): 1978–1991.

791 Shelke, N.B., Kadam, R., Tyagi, P. et al. (2011). Intravitreal poly(L-lactide) microparticles sustain retinal and choroidal delivery of TG-0054, a hydrophilic drug intended for neovascular diseases. *Drug Delivery and Translational Research* 1 (1): 76–90.

792 Verma, S., Domb, A.J., and Kumar, N. (2011). Nanomaterials for regenerative medicine. *Nanomedicine (London, England)* 6 (1): 157–181.

793 Verma, S., Garkhal, K., Mittal, A., and Kumar, N. (2011). Biodegradable polymers for emerging clinical use in tissue engineering. In: *Biodegradable Polymers in Clinical Use and Clinical Development* (eds. A.J. Domb, N. Kumar and A. Ezra), 565–629. Wiley.

794 Tian, H., Tang, Z., Zhuang, X. et al. (2012). Biodegradable synthetic polymers: preparation, functionalization and biomedical application. *Progress in Polymer Science* 37 (2): 237–280.

795 Oh, S.H. and Lee, J.H. (2013). Hydrophilization of synthetic biodegradable polymer scaffolds for improved cell/tissue compatibility. *Biomedical Materials (Bristol, England)* 8 (1): 014101.

796 Bidarra, S.J., Barrias, C.C., and Granja, P.L. (2014). Injectable alginate hydrogels for cell delivery in tissue engineering. *Acta Biomaterialia* 10 (4): 1646–1662.

797 Dorsey, S.M., McGarvey, J.R., Wang, H. et al. (2015). MRI evaluation of injectable hyaluronic acid-based hydrogel therapy to limit ventricular remodeling after myocardial infarction. *Biomaterials* 69: 65–75.

798 Shen, Z.-S., Cui, X., Hou, R.-X. et al. (2015). Tough biodegradable chitosan–gelatin hydrogels via in situ precipitation for potential cartilage tissue engineering. *RSC Advances* 5 (69): 55640–55647.

799 Sim, H.J., Thambi, T., and Lee, D.S. (2015). Heparin-based temperature-sensitive injectable hydrogels for protein delivery. *Journal of Materials Chemistry B* 3 (45): 8892–8901.

800 Liu, M., Zeng, X., Ma, C. et al. (2017). Injectable hydrogels for cartilage and bone tissue engineering. *Bone Research* 5: 17014.

801 Sokolsky-Papkov, M., Agashi, K., Olaye, A. et al. (2007). Polymer carriers for drug delivery in tissue engineering. *Advanced Drug Delivery Reviews* 59 (4–5): 187–206.

802 Lee, K.Y., Jeong, L., Kang, Y.O. et al. (2009). Electrospinning of polysaccharides for regenerative medicine. *Advanced Drug Delivery Reviews* 61 (12): 1020–1032.

803 Aravamudhan, A., Ramos, D.M., Nip, J. et al. (2013). Cellulose and collagen derived micro-nano structured scaffolds for bone tissue engineering. *Journal of Biomedical Nanotechnology* 9 (4): 719–731.

804 Wang, Y. and Wang, P.G. (2013). Polysaccharide-based systems in drug and gene delivery. *Advanced Drug Delivery Reviews* 65 (9): 1121–1122.

805 Chadha, R. and Bhandari, S. (2014). Drug–excipient compatibility screening—role of thermoanalytical and spectroscopic techniques. *Journal of Pharmaceutical and Biomedical Analysis* 87: 82–97.

806 Cheng, Y., Ramos, D., Lee, P. et al. (2014). Collagen functionalized bioactive nanofiber matrices for osteogenic differentiation of mesenchymal stem cells: bone tissue engineering. 10 (2): 287–298.

807 Lautenschläger, C., Schmidt, C., Fischer, D., and Stallmach, A. (2014). Drug delivery strategies in the therapy of inflammatory bowel disease. *Advanced Drug Delivery Reviews* 71: 58–76.

808 Rokstad, A.M.A., Lacík, I., de Vos, P., and Strand, B.L. (2014). Advances in biocompatibility and physico-chemical characterization of microspheres for cell encapsulation. *Advanced Drug Delivery Reviews* 67–68: 111–130.

809 Chiu, Y.-L., Chen, S.-C., Su, C.-J. et al. (2009). pH-triggered injectable hydrogels prepared from aqueous N-palmitoyl chitosan: in vitro characteristics and in vivo biocompatibility. *Biomaterials* 30 (28): 4877–4888.

810 Jin, R., Moreira-Teixeira, L.S., Dijkstra, P.J. et al. (2010). Enzymatically-crosslinked injectable hydrogels based on biomimetic dextran–hyaluronic acid conjugates for cartilage tissue engineering. *Biomaterials* 31 (11): 3103–3113.

811 Tan, H. and Marra, K.G. (2010). Injectable, biodegradable hydrogels for tissue engineering applications. *Materials* 3 (3): 1746–1767.

812 Choi, B.G., Park, M.H., Cho, S.-H. et al. (2011). Thermal gelling polyalanine-poloxamine-polyalanine aqueous solution for chondrocytes 3D culture: initial concentration effect. *Soft Matter* 7 (2): 456–462.

813 Li, Y., Rodrigues, J., and Tomás, H. (2012). Injectable and biodegradable hydrogels: gelation, biodegradation and biomedical applications. *Chemical Society Reviews* 41 (6): 2193–2221.

814 Ko, D.Y., Shinde, U.P., and Yeon, B. (2013). Recent progress of in situ formed gels for biomedical applications. *Progress in Polymer Science* 38 (3): 672–701.

815 Yeon, B., Park, M.H., Moon, H.J. et al. (2013). 3D culture of adipose-tissue-derived stem cells mainly leads to chondrogenesis in poly(ethylene glycol)-poly(L-alanine) diblock copolymer thermogel. *Biomacromolecules* 14 (9): 3256–3266.

816 Park, H., Woo, E.K., and Lee, K.Y. (2014). Ionically cross-linkable hyaluronate-based hydrogels for injectable cell delivery. *Journal of Controlled Release* 196: 146–153.

817 Lin, C.-C., Ki, C.S., and Shih, H. (2015). Thiol-norbornene photo-click hydrogels for tissue engineering applications. *Journal of Applied Polymer Science* 132 (8): 1–11.

818 Müller, F.A., Müller, L., Hofmann, I. et al. (2006). Cellulose-based scaffold materials for cartilage tissue engineering. *Biomaterials* 27 (21): 3955–3963.

819 Domingues, R.M.A., Gomes, M.E., and Reis, R.L. (2014). The potential of cellulose nanocrystals in tissue engineering strategies. *Biomacromolecules* 15 (7): 2327–2346.

820 Courtenay, J.C., Johns, M.A., Galembeck, F. et al. (2017). Surface modified cellulose scaffolds for tissue engineering. *Cellulose* 24 (1): 253–267.
821 Kumbar, S.G., Toti, U.S., Deng, M. et al. (2011). Novel mechanically competent polysaccharide scaffolds for bone tissue engineering. *Biomedical Materials (Bristol, England)* 6 (6): 065005.
822 Laurencin, C.T., Kumbar SG, Nukavarapu SP, James R (2010). Mechanically competent natural polymer based porous grafts for bone repair and regeneration. University of Connecticut, University of Virginia. United States patent application US 12/710,637 (25 August 2011).
823 Kumbar, S.G. and Laurencin, C.T. (2016). Natural polymer-based porous orthopedic fixation screw for bone repair and regeneration. University of Connecticut. United States patent application US 13/033,094 (25 August 2011).
824 Filion, T.M., Kutikov, A., and Song, J. (2011). Chemically modified cellulose fibrous meshes for use as tissue engineering scaffolds. *Bioorganic and Medicinal Chemistry Letters* 21 (17): 5067–5070.
825 Czaja, W., Romanovicz, D., and Malcolm Brown, R. (2004). Structural investigations of microbial cellulose produced in stationary and agitated culture. *Cellulose* 11 (3): 403–411.
826 Bodhibukkana, C., Srichana, T., Kaewnopparat, S. et al. (2006). Composite membrane of bacterially-derived cellulose and molecularly imprinted polymer for use as a transdermal enantioselective controlled-release system of racemic propranolol. *Journal of Controlled Release* 113 (1): 43–56.
827 Fu, L., Zhang, Y., Li, C. et al. (2012). Skin tissue repair materials from bacterial cellulose by a multilayer fermentation method. *Journal of Materials Chemistry* 22 (24): 12349–12357.
828 Aravamudhan, A., Ramos, D.M., Ahmed Nada, A.A., and Kumbar, S.G. (2014). Natural polymers: polysaccharides and their derivatives for biomedical applications. In: *Natural and Synthetic Biomedical Polymers*, Chapter 4 (eds. S.G. Kumbar, C.T. Laurencin and M. Deng), 67–89. Oxford: Elsevier.
829 Almeida, I.F., Pereira, T., Silva, N.H.C.S. et al. (2014). Bacterial cellulose membranes as drug delivery systems: an in vivo skin compatibility study. *European Journal of Pharmaceutics and Biopharmaceutics* 86 (3): 332–336.
830 Klemm, D., Heublein, B., Fink, H.-P., and Bohn, A. (2005). Cellulose: fascinating biopolymer and sustainable raw material. *Angewandte Chemie International Edition* 44 (22): 3358–3393.
831 Takada, T., Katagiri, T., Ifuku, M. et al. (2003). Sulfated polysaccharides enhance the biological activities of bone morphogenetic proteins. *The Journal of Biological Chemistry* 278 (44): 43229–43235.
832 Shi, Q., Li, Y., Sun, J. et al. (2012). The osteogenesis of bacterial cellulose scaffold loaded with bone morphogenetic protein-2. *Biomaterials* 33 (28): 6644–6649.
833 Hajiabbas, M., Mashayekhan, S., Nazaripouya, A. et al. (2015). Chitosan-gelatin sheets as scaffolds for muscle tissue engineering. *Artificial Cells, Nanomedicine, and Biotechnology* 43 (2): 124–132.

834 Nair, L.S., Starnes, T., Ko, J.-W.K., and Laurencin, C.T. (2007). Development of injectable thermogelling chitosan–inorganic phosphate solutions for biomedical applications. *Biomacromolecules* 8 (12): 3779–3785.
835 Cheng, Y., Nada, A., Valmikinathan, C.M. et al. (2014). In situ gelling polysaccharide-based hydrogel for cell and drug delivery in tissue engineering. *Journal of Applied Polymer Science* 131 (4): 39934–39943.
836 Funakoshi, T., Majima, T., Iwasaki, N. et al. (2005). Novel chitosan-based hyaluronan hybrid polymer fibers as a scaffold in ligament tissue engineering. *Journal of Biomedical Materials Research Part A* 74 (3): 338–346.
837 Silva, S.S., Caridade, S.G., Mano, J.F., and Reis, R.L. (2013). Effect of crosslinking in chitosan/aloe vera-based membranes for biomedical applications. *Carbohydrate Polymers* 98 (1): 581–588.
838 da Silva, M.L.A., Crawford, A., Mundy, J.M. et al. (2010). Chitosan/polyester-based scaffolds for cartilage tissue engineering: assessment of extracellular matrix formation. *Acta Biomaterialia* 6 (3): 1149–1157.
839 Silva, J.M., Georgi, N., Costa, R. et al. (2013). Nanostructured 3D constructs based on chitosan and chondroitin sulphate multilayers for cartilage tissue engineering. *PLoS One* 8 (2): e55451.
840 Yang, T.-L. and Young, T.-H. (2008). The enhancement of submandibular gland branch formation on chitosan membranes. *Biomaterials* 29 (16): 2501–2508.
841 Yuan, Y., Zhang, P., Yang, Y. et al. (2004). The interaction of Schwann cells with chitosan membranes and fibers in vitro. *Biomaterials* 25 (18): 4273–4278.
842 Simões, M.J., Gärtner, A., Shirosaki, Y. et al. (2011). In vitro and in vivo chitosan membranes testing for peripheral nerve reconstruction. *Acta Médica Portuguesa* 24 (1): 43–52.
843 Poon, Y.F., Cao, Y., Liu, Y. et al. (2010). Hydrogels based on dual curable chitosan-graft-polyethylene glycol-graft-methacrylate: application to layer-by-layer cell encapsulation. *ACS Applied Materials & Interfaces* 2 (7): 2012–2025.
844 Wang, J., Chen, L., Zhao, Y. et al. (2009). Cell adhesion and accelerated detachment on the surface of temperature-sensitive chitosan and poly(N-isopropylacrylamide) hydrogels. *Journal of Materials Science Materials in Medicine* 20 (2): 583–590.
845 Sá-Lima, H., Caridade, S.G., Mano, J.F., and Reis, R.L. (2010). Stimuli-responsive chitosan-starch injectable hydrogels combined with encapsulated adipose-derived stromal cells for articular cartilage regeneration. *Soft Matter* 6 (20): 5184–5195.
846 Yan, L.-P., Wang, Y.-J., Ren, L. et al. (2010). Genipin-cross-linked collagen/chitosan biomimetic scaffolds for articular cartilage tissue engineering applications. *Journal of Biomedical Materials Research Part A* 95 (2): 465–475.
847 Moreira, C.D.F., Carvalho, S.M., Mansur, H.S., and Pereira, M.M. (2016). Thermogelling chitosan–collagen–bioactive glass nanoparticle hybrids as

potential injectable systems for tissue engineering. *Materials Science and Engineering C* 58: 1207–1216.

848 Zhang, L., Ao, Q., Wang, A. et al. (2006). A sandwich tubular scaffold derived from chitosan for blood vessel tissue engineering. *Journal of Biomedical Materials Research Part A* 77 (2): 277–284.

849 Shi, H., Han, C., Mao, Z. et al. (2008). Enhanced angiogenesis in porous collagen–chitosan scaffolds loaded with angiogenin. *Tissue Engineering Part A* 14 (11): 1775–1785.

850 Chen, Y.-H., Wang, I.J., and Young, T.-H. (2009). Formation of keratocyte spheroids on chitosan-coated surface can maintain keratocyte phenotypes. *Tissue Engineering Part A* 15 (8): 2001–2013.

851 Choi, J.S. and Yoo, H.S. (2010). Pluronic/chitosan hydrogels containing epidermal growth factor with wound-adhesive and photo-crosslinkable properties. *Journal of Biomedical Materials Research Part A* 95 (2): 564–573.

852 Han, C.-M., Zhang, L.-P., Sun, J.-Z. et al. (2010). Application of collagen–chitosan/fibrin glue asymmetric scaffolds in skin tissue engineering. *Journal of Zhejiang University Science B* 11 (7): 524–530.

853 Wu, X., Black, L., Santacana-Laffitte, G., and Patrick, C.W. Jr., (2007). Preparation and assessment of glutaraldehyde-crosslinked collagen–chitosan hydrogels for adipose tissue engineering. *Journal of Biomedical Materials Research Part A* 81 (1): 59–65.

854 Mei, L., Hu, D., Ma, J. et al. (2012). Preparation, characterization and evaluation of chitosan macroporous for potential application in skin tissue engineering. *International Journal of Biological Macromolecules* 51 (5): 992–997.

855 Ge, S., Zhao, N., Wang, L. et al. (2012). Bone repair by periodontal ligament stem cellseeded nanohydroxyapatite-chitosan scaffold. *International Journal of Nanomedicine* 7: 5405–5414.

856 Hsueh, Y.-Y., Chang, Y.-J., Huang, T.-C. et al. (2014). Functional recoveries of sciatic nerve regeneration by combining chitosan-coated conduit and neurosphere cells induced from adipose-derived stem cells. *Biomaterials* 35 (7): 2234–2244.

857 Boucard, N., Viton, C., Agay, D. et al. (2007). The use of physical hydrogels of chitosan for skin regeneration following third-degree burns. *Biomaterials* 28 (24): 3478–3488.

858 Gobin, A.S., Butler, C.E., and Mathur, A.B. (2006). Repair and regeneration of the abdominal wall musculofascial defect using silk fibroin-chitosan blend. *Tissue Engineering* 12 (12): 3383–3394.

859 Hao, T., Wen, N., Cao, J.-K. et al. (2010). The support of matrix accumulation and the promotion of sheep articular cartilage defects repair in vivo by chitosan hydrogels. *Osteoarthritis and Cartilage* 18 (2): 257–265.

860 Liang, Y., Liu, W., Han, B. et al. (2011). An in situ formed biodegradable hydrogel for reconstruction of the corneal endothelium. *Colloids and Surfaces B: Biointerfaces* 82 (1): 1–7.

861 Meng, D., Dong, L., Wen, Y., and Xie, Q. (2015). Effects of adding resorbable chitosan microspheres to calcium phosphate cements for bone regeneration. *Materials Science and Engineering C* 47: 266–272.

862 Jayakumar, R., Prabaharan, M., Sudheesh Kumar, P.T. et al. (2011). Biomaterials based on chitin and chitosan in wound dressing applications. *Biotechnology Advances* 29 (3): 322–337.

863 Kozen, B.G., Kircher, S.J., Henao, J. et al. (2008). An alternative hemostatic dressing: comparison of CELOX, HemCon, and QuikClot. *Academic Emergency Medicine* 15 (1): 74–81.

864 Peh, K., Khan, T., and Ch'ng, H. (2000). Mechanical, bioadhesive strength and biological evaluations of chitosan films for wound dressing. *Journal of Pharmacy & Pharmaceutical Sciences* 3 (3): 303–311.

865 Anandan, R., Nair, P.G.V., and Mathew, S. (2004). Anti-ulcerogenic effect of chitin and chitosan on mucosal antioxidant defence system in HCl-ethanol-induced ulcer in rats. *Journal of Pharmacy and Pharmacology* 56 (2): 265–269.

866 Tavaria, F., Jorge, M.P., Ruiz, A.L.T.G. et al. (2016). Anti-proliferative, anti-inflammatory, anti-ulcerogenic and wound healing properties of chitosan. 12: 114–122.

867 Pei, H.N., Chen, X.G., Li, Y., and Zhou, H.Y. (2008). Characterization and ornidazole release in vitro of a novel composite film prepared with chitosan/poly(vinyl alcohol)/alginate. *Journal of Biomedical Materials Research Part A* 85 (2): 566–572.

868 Sung, J.H., Hwang, M.-R., Kim, J.O. et al. (2010). Gel characterisation and in vivo evaluation of minocycline-loaded wound dressing with enhanced wound healing using polyvinyl alcohol and chitosan. *International Journal of Pharmaceutics* 392 (1–2): 232–240.

869 Radhakumary, C., Antonty, M., and Sreenivasan, K. (2011). Drug loaded thermoresponsive and cytocompatible chitosan based hydrogel as a potential wound dressing. *Carbohydrate Polymers* 83 (2): 705–713.

870 Cohen, M.L. (2000). The theory of real materials. *Annual Review of Materials Science* 30 (1): 1–26.

871 Mi, F.-L., Wu, Y.-B., Shyu, S.-S. et al. (2003). Asymmetric chitosan membranes prepared by dry/wet phase separation: a new type of wound dressing for controlled antibacterial release. *Journal of Membrane Science* 212 (1): 237–254.

872 Lu, S., Gao, W., and Gu, H.Y. (2008). Construction, application and biosafety of silver nanocrystalline chitosan wound dressing. *Burns* 34 (5): 623–628.

873 Ong, S.-Y., Wu, J., Moochhala, S.M. et al. (2008). Development of a chitosan-based wound dressing with improved hemostatic and antimicrobial properties. *Biomaterials* 29 (32): 4323–4332.

874 Thomas, V., Yallapu, M.M., Sreedhar, B., and Bajpai, S.K. (2009). Fabrication, characterization of chitosan/nanosilver film and its potential antibacterial application. *Journal of Biomaterials Science Polymer Edition* 20 (14): 2129–2144.

875 Li, L.-H., Deng, J.-C., Deng, H.-R. et al. (2010). Preparation, characterization and antimicrobial activities of chitosan/Ag/ZnO blend films. *Chemical Engineering Journal* 160 (1): 378–382.

876 Vimala, K., Mohan, Y.M., Sivudu, K.S. et al. (2010). Fabrication of porous chitosan films impregnated with silver nanoparticles: a facile approach for superior antibacterial application. *Colloids and Surfaces B: Biointerfaces* 76 (1): 248–258.

877 Nishinari, K. and Takahashi, R. (2003). Interaction in polysaccharide solutions and gels. *Current Opinion in Colloid and Interface Science* 8 (4): 396–400.

878 Liao, Y.-H., Jones, S.A., Forbes, B. et al. (2005). Hyaluronan: pharmaceutical characterization and drug delivery. *Drug Delivery* 12 (6): 327–342.

879 Hunt, N.C. and Grover, L.M. (2010). Cell encapsulation using biopolymer gels for regenerative medicine. *Biotechnology Letters* 32 (6): 733–742.

880 Li, L., He, Z.-Y., Wei, X.-W., and Wei, Y.-Q. (2016). Recent advances of biomaterials in biotherapy. *Regenerative Biomaterials* 3 (2): 99–105.

881 Carney, S.L. and Muir, H. (1988). The structure and function of cartilage proteoglycans. *Physiological Reviews* 68 (3): 858–910.

882 Kim, D.H., Martin, J.T., Elliott, D.M. et al. (2015). Phenotypic stability, matrix elaboration and functional maturation of nucleus pulposus cells encapsulated in photocrosslinkable hyaluronic acid hydrogels. *Acta Biomaterialia* 12: 21–29.

883 Hwang, H.-D., Cho, H.-J., Balakrishnan, P. et al. (2012). Cross-linked hyaluronic acid-based flexible cell delivery system: application for chondrogenic differentiation. *Colloids and Surfaces B: Biointerfaces* 91: 106–113.

884 Toh, W.S., Lim, T.C., Kurisawa, M., and Spector, M. (2012). Modulation of mesenchymal stem cell chondrogenesis in a tunable hyaluronic acid hydrogel microenvironment. *Biomaterials* 33 (15): 3835–3845.

885 Park, H., Choi, B., Hu, J., and Lee, M. (2013). Injectable chitosan hyaluronic acid hydrogels for cartilage tissue engineering. *Acta Biomaterialia* 9 (1): 4779–4786.

886 Yu, Y., Brouillette, M.J., Seol, D. et al. (2015). Use of recombinant human stromal cell-derived factor 1α-loaded fibrin/hyaluronic acid hydrogel networks to achieve functional repair of full-thickness bovine articular cartilage via homing of chondrogenic progenitor cells. *Arthritis & Rheumatology (Hoboken, NJ)* 67 (5): 1274–1285.

887 Amann, E., Wolff, P., Breel, E. et al. (2017). Hyaluronic acid facilitates chondrogenesis and matrix deposition of human adipose derived mesenchymal stem cells and human chondrocytes co-cultures. *Acta Biomaterialia* 52: 130–144.

888 Khademhosseini, A., Eng, G., Yeh, J. et al. (2006). Micromolding of photocrosslinkable hyaluronic acid for cell encapsulation and entrapment. *Journal of Biomedical Materials Research Part A* 79 (3): 522–532.

889 Chung, C. and Burdick, J.A. (2009). Influence of three-dimensional hyaluronic acid microenvironments on mesenchymal stem cell chondrogenesis. *Tissue Engineering Part A* 15 (2): 243–254.

890 Skaalure, S.C., Dimson, S.O., Pennington, A.M., and Bryant, S.J. (2014). Semi-interpenetrating networks of hyaluronic acid in degradable PEG hydrogels for cartilage tissue engineering. *Acta Biomaterialia* 10 (8): 3409–3420.

891 Kayakabe, M., Tsutsumi, S., Watanabe, H. et al. (2006). Transplantation of autologous rabbit BM-derived mesenchymal stromal cells embedded in

hyaluronic acid gel sponge into osteochondral defects of the knee. *Cytotherapy* 8 (4): 343–353.

892 Park, J.Y., Choi, J.-C., Shim, J.-H. et al. (2014). A comparative study on collagen type I and hyaluronic acid dependent cell behavior for osteochondral tissue bioprinting. *Biofabrication* 6 (3): 035004.

893 Kim, J., Kim, I.S., Cho, T.H. et al. (2007). Bone regeneration using hyaluronic acid-based hydrogel with bone morphogenic protein-2 and human mesenchymal stem cells. *Biomaterials* 28 (10): 1830–1837.

894 Chen, W.-H., Lo, W.-C., Hsu, W.-C. et al. (2014). Synergistic anabolic actions of hyaluronic acid and platelet-rich plasma on cartilage regeneration in osteoarthritis therapy. *Biomaterials* 35 (36): 9599–9607.

895 Liu, Y., Shu, X.Z., and Prestwich, G.D. (2006). Osteochondral defect repair with autologous bone marrow-derived mesenchymal stem cells in an injectable, in situ, cross-linked synthetic extracellular matrix. *Tissue Engineering* 12 (12): 3405–3416.

896 Tan, H., Chu, C.R., Payne, K.A., and Marra, K.G. (2009). Injectable in situ forming biodegradable chitosan–hyaluronic acid based hydrogels for cartilage tissue engineering. *Biomaterials* 30 (13): 2499–2506.

897 Zhang, L., Xiao, Y., Jiang, B. et al. (2010). Effect of adipic dihydrazide modification on the performance of collagen/hyaluronic acid scaffold. *Journal of Biomedical Materials Research Part B Applied Biomaterials* 92 (2): 307–316.

898 Erickson, I.E., Kestle, S.R., Zellars, K.H. et al. (2012). High mesenchymal stem cell seeding densities in hyaluronic acid hydrogels produce engineered cartilage with native tissue properties. *Acta Biomaterialia* 8 (8): 3027–3034.

899 Kim, H.J., Kim, K.K., Park, I.K. et al. (2012). Hybrid scaffolds composed of hyaluronic acid and collagen for cartilage regeneration. *Tissue Engineering and Regenerative Medicine* 9 (2): 57–62.

900 Kim, M., S. Garrity, I.E. Erickson et al. (2012). Optimization of macromer density in human MSC-laden hyaluronic acid (HA) hydrogels. *2012 38th Annual Northeast Bioengineering Conference (NEBEC).*

901 Yu, F., Cao, X., Li, Y. et al. (2013). An injectable hyaluronic acid/PEG hydrogel for cartilage tissue engineering formed by integrating enzymatic crosslinking and Diels–Alder "click chemistry". *Polymer Chemistry* 5 (3): 1082–1090.

902 Palumbo, F.S., Fiorica, C., Di Stefano, M. et al. (2015). In situ forming hydrogels of hyaluronic acid and inulin derivatives for cartilage regeneration. *Carbohydrate Polymers* 122: 408–416.

903 Ha, C.-W., Park, Y.-B., Chung, J.-Y., and Park, Y.-G. (2015). Cartilage repair using composites of human umbilical cord blood-derived mesenchymal stem cells and hyaluronic acid hydrogel in a minipig model. *Stem Cells Translational Medicine* 4 (9): 1044–1051.

904 Frith, J.E., Menzies, D.J., Cameron, A.R. et al. (2014). Effects of bound versus soluble pentosan polysulphate in PEG/HA-based hydrogels tailored for intervertebral disc regeneration. *Biomaterials* 35 (4): 1150–1162.

905 Chen, Y.-C., Su, W.-Y., Yang, S.-H. et al. (2013). In situ forming hydrogels composed of oxidized high molecular weight hyaluronic acid and gelatin for nucleus pulposus regeneration. *Acta Biomaterialia* 9 (2): 5181–5193.

906 Freymann, U., Endres, M., Neumann, K. et al. (2012). Expanded human meniscus-derived cells in 3-D polymer-hyaluronan scaffolds for meniscus repair. *Acta Biomaterialia* 8 (2): 677–685.
907 Kim, Y.-M., Oh, S.H., Choi, J.-S. et al. (2014). Adipose-derived stem cell-containing hyaluronic acid/alginate hydrogel improves vocal fold wound healing. *The Laryngoscope* 124 (3): E64–E72.
908 Oberlender, S.A. and Tuan, R.S. (1994). Spatiotemporal profile of N-cadherin expression in the developing limb mesenchyme. *Cell Adhesion and Communication* 2 (6): 521–537.
909 Zou, L., Zou, X., Chen, L. et al. (2008). Effect of hyaluronan on osteogenic differentiation of porcine bone marrow stromal cells in vitro. *Journal of Orthopaedic Research* 26 (5): 713–720.
910 Patterson, J., Siew, R., Herring, S.W. et al. (2010). Hyaluronic acid hydrogels with controlled degradation properties for oriented bone regeneration. *Biomaterials* 31 (26): 6772–6781.
911 Bhakta, G., Rai, B., Lim, Z.X.H. et al. (2012). Hyaluronic acid-based hydrogels functionalized with heparin that support controlled release of bioactive BMP-2. *Biomaterials* 33 (26): 6113–6122.
912 Bhakta, G., Lim, Z.X.H., Rai, B. et al. (2013). The influence of collagen and hyaluronan matrices on the delivery and bioactivity of bone morphogenetic protein-2 and ectopic bone formation. *Acta Biomaterialia* 9 (11): 9098–9106.
913 Park, H.-J., Jin, Y., Shin, J. et al. (2016). Catechol-functionalized hyaluronic acid hydrogels enhance angiogenesis and osteogenesis of human adipose-derived stem cells in critical tissue defects. *Biomacromolecules* 17 (6): 1939–1948.
914 Subramaniam, S., Fang, Y.-H., Sivasubramanian, S. et al. (2016). Hydroxyapatite-calcium sulfate-hyaluronic acid composite encapsulated with collagenase as bone substitute for alveolar bone regeneration. *Biomaterials* 74: 99–108.
915 Zhu, M., Lin, S., Sun, Y. et al. (2016). Hydrogels functionalized with N-cadherin mimetic peptide enhance osteogenesis of hMSCs by emulating the osteogenic niche. *Biomaterials* 77: 44–52.
916 Sahoo, S., Chung, C., Khetan, S., and Burdick, J.A. (2008). Hydrolytically degradable hyaluronic acid hydrogels with controlled temporal structures. *Biomacromolecules* 9 (4): 1088–1092.
917 West, D.C., Hampson, I.N., Arnold, F., and Kumar, S. (1985). Angiogenesis induced by degradation products of hyaluronic acid. *Science* 228 (4705): 1324–1326.
918 Day, A.J. and de la Motte, C.A. (2005). Hyaluronan cross-linking: a protective mechanism in inflammation? *Trends in Immunology* 26 (12): 637–643.
919 Noble, P.W., Lake, F.R., Henson, P.M., and Riches, D.W. (1993). Hyaluronate activation of CD44 induces insulin-like growth factor-1 expression by a tumor necrosis factor-alpha-dependent mechanism in murine macrophages. *Journal of Clinical Investigation* 91 (6): 2368–2377.
920 Park, D., Kim, Y., Kim, H. et al. (2012). Hyaluronic acid promotes angiogenesis by inducing RHAMM-TGFβ receptor interaction via CD44-PKCδ. *Molecules and Cells* 33 (6): 563–574.

921 Rooney, P., Kumar, S., Ponting, J., and Wang, M. (1995). The role of hyaluronan in tumour neovascularization (review). *International Journal of Cancer* 60 (5): 632–636.
922 Perng, C.-K., Wang, Y.-J., Tsi, C.-H., and Ma, H. (2011). In vivo angiogenesis effect of porous collagen scaffold with hyaluronic acid oligosaccharides. *The Journal of Surgical Research* 168 (1): 9–15.
923 Seidlits, S.K., Drinnan, C.T., Petersen, R.R. et al. (2011). Fibronectin–hyaluronic acid composite hydrogels for three-dimensional endothelial cell culture. *Acta Biomaterialia* 7 (6): 2401–2409.
924 Mohandas, A., Anisha, B.S., Chennazhi, K.P., and Jayakumar, R. (2015). Chitosan–hyaluronic acid/VEGF loaded fibrin nanoparticles composite sponges for enhancing angiogenesis in wounds. *Colloids and Surfaces B: Biointerfaces* 127: 105–113.
925 da Silva, L.P., Pirraco, R.P., Santos, T.C. et al. (2016). Neovascularization induced by the hyaluronic acid-based spongy-like hydrogels degradation products. *ACS Applied Materials & Interfaces* 8 (49): 33464–33474.
926 Collier, J.H., Camp, J.P., Hudson, T.W., and Schmidt, C.E. (2000). Synthesis and characterization of polypyrrole–hyaluronic acid composite biomaterials for tissue engineering applications. *Journal of Biomedical Materials Research* 50 (4): 574–584.
927 Wang, X., He, J., Wang, Y., and Cui, F.-Z. (2012). Hyaluronic acid-based scaffold for central neural tissue engineering. *Interface Focus* 2 (3): 278–291.
928 Hemshekhar, M., Thushara, R.M., Chandranayaka, S. et al. (2016). Emerging roles of hyaluronic acid bioscaffolds in tissue engineering and regenerative medicine. *International Journal of Biological Macromolecules* 86: 917–928.
929 Wang, K.K., Nemeth, I.R., Seckel, B.R. et al. (1998). Hyaluronic acid enhances peripheral nerve regeneration in vivo. *Microsurgery* 18 (4): 270–275.
930 Ozgenel, G.Y. (2003). Effects of hyaluronic acid on peripheral nerve scarring and regeneration in rats. *Microsurgery* 23 (6): 575–581.
931 Engler, A.J., Sen, S., Sweeney, H.L., and Discher, D.E. (2006). Matrix elasticity directs stem cell lineage specification. *Cell* 126 (4): 677–689.
932 Hynes, S.R., Rauch, M.F., Bertram, J.P., and Lavik, E.B. (2009). A library of tunable poly(ethylene glycol)/poly(L-lysine) hydrogels to investigate the material cues that influence neural stem cell differentiation. *Journal of Biomedical Materials Research Part A* 89 (2): 499–509.
933 Leipzig, N.D. and Shoichet, M.S. (2009). The effect of substrate stiffness on adult neural stem cell behavior. *Biomaterials* 30 (36): 6867–6878.
934 Tarus, D., Hamard, L., Caraguel, F. et al. (2016). Design of hyaluronic acid hydrogels to promote neurite outgrowth in three dimensions. *ACS Applied Materials & Interfaces* 8 (38): 25051–25059.
935 Cui, F.Z., Tian, W.M., Hou, S.P. et al. (2006). Hyaluronic acid hydrogel immobilized with RGD peptides for brain tissue engineering. *Journal of Materials Science Materials in Medicine* 17 (12): 1393–1401.
936 Zaviskova, K., Tukmachev, D., Dubisova, J. et al. (2018). Injectable hydroxyphenyl derivative of hyaluronic acid hydrogel modified with RGD as

scaffold for spinal cord injury repair. *Journal of Biomedical Materials Research Part A* 106 (4): 1129–1140.
937 Ashton, R.S., Conway, A., Pangarkar, C. et al. (2012). Astrocytes regulate adult hippocampal neurogenesis through ephrin-B signaling. *Nature Neuroscience* 15 (10): 1399–1406.
938 Conway, A., Vazin, T., Spelke, D.P. et al. (2013). Multivalent ligands control stem cell behaviour in vitro and in vivo. *Nature Nanotechnology* 8 (11): 831–838.
939 Conway, A. and Schaffer, D.V. (2014). Biomaterial microenvironments to support the generation of new neurons in the adult brain. *Stem Cells (Dayton, Ohio)* 32 (5): 1220–1229.
940 Krewson, C.E., Klarman, M.L., and Saltzman, W.M. (1995). Distribution of nerve growth factor following direct delivery to brain interstitium. *Brain Research* 680 (1–2): 196–206.
941 Burdick, J.A., Ward, M., Liang, E. et al. (2006). Stimulation of neurite outgrowth by neurotrophins delivered from degradable hydrogels. *Biomaterials* 27 (3): 452–459.
942 Jain, A., Kim, Y.-T., McKeon, R.J., and Bellamkonda, R.V. (2006). In situ gelling hydrogels for conformal repair of spinal cord defects, and local delivery of BDNF after spinal cord injury. *Biomaterials* 27 (3): 497–504.
943 Wang, Y.-C., Wu, Y.-T., Huang, H.-Y. et al. (2008). Sustained intraspinal delivery of neurotrophic factor encapsulated in biodegradable nanoparticles following contusive spinal cord injury. *Biomaterials* 29 (34): 4546–4553.
944 Mo, L., Yang, Z., Zhang, A., and Li, X. (2010). The repair of the injured adult rat hippocampus with NT-3-chitosan carriers. *Biomaterials* 31 (8): 2184–2192.
945 Lam, J., Lowry, W.E., Carmichael, S.T., and Segura, T. (2014). Delivery of iPS-NPCs to the stroke cavity within a hyaluronic acid matrix promotes the differentiation of transplanted cells. *Advanced Functional Materials* 24 (44): 7053–7062.
946 Lim, H.J., Perera, T.H., Wilems, T.S. et al. (2016). Response to di-functionalized hyaluronic acid with orthogonal chemistry grafting at independent modification sites in rodent models of neural differentiation and spinal cord injury. *Journal of Materials Chemistry B* 4 (42): 6865–6875.
947 Nih, L.R., Carmichael, S.T., and Segura, T. (2016). Hydrogels for brain repair after stroke: an emerging treatment option. *Current Opinion in Biotechnology* 40: 155–163.
948 Knowlton, S., Anand, S., Shah, T., and Tasoglu, S. (2018). Bioprinting for neural tissue engineering. *Trends in Neurosciences* 41 (1): 31–46.
949 Alghoul, M., Mendiola, A., Seth, R. et al. (2012). The effect of hyaluronan hydrogel on fat graft survival. *Aesthetic Surgery Journal* 32 (5): 622–633.
950 Cortivo, R., Brun, P., Rastrelli, A., and Abatangelo, G. (1991). In vitro studies on biocompatibility of hyaluronic acid esters. *Biomaterials* 12 (8): 727–730.
951 Benedetti, L., Cortivo, R., Berti, T. et al. (1993). Biocompatibility and biodegradation of different hyaluronan derivatives (Hyaff) implanted in rats. *Biomaterials* 14 (15): 1154–1160.

952 Halbleib, M., Skurk, T., de Luca, C. et al. (2003). Tissue engineering of white adipose tissue using hyaluronic acid-based scaffolds. I: in vitro differentiation of human adipocyte precursor cells on scaffolds. *Biomaterials* 24 (18): 3125–3132.

953 Stillaert, F.B., Di Bartolo, C., Hunt, J.A. et al. (2008). Human clinical experience with adipose precursor cells seeded on hyaluronic acid-based spongy scaffolds. *Biomaterials* 29 (29): 3953–3959.

954 Espandar, L., Bunnell, B., Wang, G.Y. et al. (2012). Adipose-derived stem cells on hyaluronic acid-derived scaffold: a new horizon in bioengineered cornea. *Archives of Ophthalmology (Chicago, Ill.: 1960)* 130 (2): 202–208.

955 Otterlei, M., Ostgaard, K., Skjåk-Braek, G. et al. (1991). Induction of cytokine production from human monocytes stimulated with alginate. *Journal of Immunotherapy* 10 (4): 286–291.

956 Iwamoto, M., Kurachi, M., Nakashima, T. et al. (2005). Structure-activity relationship of alginate oligosaccharides in the induction of cytokine production from RAW264.7 cells. *FEBS Letters* 579 (20): 4423–4429.

957 Park, H., Kang, S.-W., Kim, B.-S. et al. (2009). Shear-reversibly crosslinked alginate hydrogels for tissue engineering. *Macromolecular Bioscience* 9 (9): 895–901.

958 Follin, B., Juhl, M., Cohen, S. et al. (2015). Human adipose-derived stromal cells in a clinically applicable injectable alginate hydrogel: phenotypic and immunomodulatory evaluation. *Cytotherapy* 17 (8): 1104–1118.

959 Venkatesan, J., Bhatnagar, I., Manivasagan, P. et al. (2015). Alginate composites for bone tissue engineering: a review. *International Journal of Biological Macromolecules* 72: 269–281.

960 Ruvinov, E. and Cohen, S. (2016). Alginate biomaterial for the treatment of myocardial infarction: progress, translational strategies, and clinical outlook: from ocean algae to patient bedside. *Advanced Drug Delivery Reviews* 96: 54–76.

961 Balakrishnan, B., Joshi, N., Jayakrishnan, A., and Banerjee, R. (2014). Self-crosslinked oxidized alginate/gelatin hydrogel as injectable, adhesive biomimetic scaffolds for cartilage regeneration. *Acta Biomaterialia* 10 (8): 3650–3663.

962 Park, H. and Lee, K.Y. (2014). Cartilage regeneration using biodegradable oxidized alginate/hyaluronate hydrogels. *Journal of Biomedical Materials Research Part A* 102 (12): 4519–4525.

963 Li, Z., Ramay, H.R., Hauch, K.D. et al. (2005). Chitosan–alginate hybrid scaffolds for bone tissue engineering. *Biomaterials* 26 (18): 3919–3928.

964 Zhao, L., Weir, M.D., and Xu, H.H.K. (2010). An injectable calcium phosphate-alginate hydrogel-umbilical cord mesenchymal stem cell paste for bone tissue engineering. *Biomaterials* 31 (25): 6502–6510.

965 Park, D.-J., Choi, B.-H., Zhu, S.-J. et al. (2005). Injectable bone using chitosan-alginate gel/mesenchymal stem cells/BMP-2 composites. *Journal of Cranio-Maxillofacial Surgery* 33 (1): 50–54.

966 Yu, J., Du, K.T., Fang, Q. et al. (2010). The use of human mesenchymal stem cells encapsulated in RGD modified alginate microspheres in the repair of myocardial infarction in the rat. *Biomaterials* 31 (27): 7012–7020.

967 Wang, Q., Jamal, S., Detamore, M.S., and Berkland, C. (2011). PLGA-chitosan/PLGA-alginate nanoparticle blends as biodegradable colloidal gels for seeding human umbilical cord mesenchymal stem cells. *Journal of Biomedical Materials Research Part A* 96 (3): 520–527.
968 Zhou, H. and Xu, H.H.K. (2011). The fast release of stem cells from alginate-fibrin microbeads in injectable scaffolds for bone tissue engineering. *Biomaterials* 32 (30): 7503–7513.
969 He, X., Dziak, R., Mao, K. et al. (2012). Integration of a novel injectable nano calcium sulfate/alginate scaffold and BMP2 gene-modified mesenchymal stem cells for bone regeneration. *Tissue Engineering Part A* 19 (3–4): 508–518.
970 Lewandowska-Łańcucka, J., Mystek, K., Mignon, A. et al. (2017). Alginate- and gelatin-based bioactive photocross-linkable hybrid materials for bone tissue engineering. *Carbohydrate Polymers* 157: 1714–1722.
971 Chen, G., Kawazoe, N., and Ito, Y. (2018). Photo-crosslinkable hydrogels for tissue engineering applications. In: *Photochemistry for Biomedical Applications: From Device Fabrication to Diagnosis and Therapy* (ed. Y. Ito), 277–300. Singapore: Springer.
972 Pina, S., Oliveira, J.M., and Reis, R.L. (2015). Natural-based nanocomposites for bone tissue engineering and regenerative medicine: a review. *Advanced Materials* 27 (7): 1143–1169.
973 Filardo, G., Perdisa, F., Gelinsky, M. et al. (2018). Novel alginate biphasic scaffold for osteochondral regeneration: an in vivo evaluation in rabbit and sheep models. *Journal of Materials Science Materials in Medicine* 29 (6): 74.
974 Shachar, M., Tsur-Gang, O., Dvir, T. et al. (2011). The effect of immobilized RGD peptide in alginate scaffolds on cardiac tissue engineering. *Acta Biomaterialia* 7 (1): 152–162.
975 Sapir, Y., Kryukov, O., and Cohen, S. (2011). Integration of multiple cell-matrix interactions into alginate scaffolds for promoting cardiac tissue regeneration. *Biomaterials* 32 (7): 1838–1847.
976 Xu, Z. and Lam, M.T. (2018). Alginate application for heart and cardiovascular diseases. In: *Alginates and their Biomedical Applications* (eds. B.H.A. Rehm and M.F. Moradali), 185–212. Singapore: Springer.

10

Biopolymers in Regenerative Medicine: Overview, Current Advances, and Future Trends

Michael R. Behrens and Warren C. Ruder

University of Pittsburgh, Department of Bioengineering, 300 Technology Drive, Pittsburgh, PA 15219, USA

10.1 Introduction

When the integrity of biological tissues is lost due to disease or injury, biopolymers can be used as a therapy to regenerate tissue and restore function. Polymers can be used to create artificial tissues that can be grafted into the damaged region. Restoring lost function with engineered tissue is the goal of regenerative medicine, an emerging medical paradigm that aims to augment the body's natural repair processes [1, 2]. Although many body tissues possess significant regenerative capacity, such as the skin's ability to repair itself when cut, many critical tissues in the body cannot fully recover when damaged [3]. In order to maximize functional regeneration of damaged tissue, engineers can use biopolymers to create artificial tissues that can be implanted at the site of injury. These artificial tissues are designed to mimic the form and function of natural tissue, and biopolymers play a significant role as a material for facilitating this biomimicry [4].

Engineered tissues should be designed to mimic and, ultimately, to replicate natural tissue if regeneration is to be fully realized [5]. Natural biological tissues are composed of cells contained within an extracellular matrix (ECM), a protein-based polymeric scaffold. ECM provides tissue with mechanical structure while also coordinating the biochemical cues that maintain normal cell physiology [6, 7]. Materials that can recapitulate the properties of ECM can be employed in the construction of artificial engineered tissues for regenerative medicine. ECM-mimicking scaffolds can be designed from a variety of materials including metals and ceramics, but biopolymers are often employed as the scaffolding material of choice. Biopolymers are suitable as artificial tissue scaffolds for multiple reasons. First, biopolymer scaffolds can be designed to match the wide range of mechanical properties found in native tissues, from hard tissues such as bone [8] to soft tissues such as liver [9]. Second, biopolymer scaffolds can be designed so that the implanted artificial scaffold degrades at the same rate that local cells replace and remodel it with their own natural ECM. Once the biopolymer scaffold degrades fully, natural ECM ideally is the only remaining scaffold material present within the regenerated tissue,

Biopolymers for Biomedical and Biotechnological Applications, First Edition.
Edited by Bernd H. A. Rehm and M. Fata Moradali.

thus minimizing the long-term persistence of artificial materials *in vivo* [10]. Third, biopolymer scaffolds can be designed to release key growth factors as they dissolve, coordinating cell behaviors as they replicate and form mature, functional tissues [11]. Mechanical properties, degradation rate, and integration with growth factors are all considerations that factor into the design decisions encountered when engineering new biopolymer scaffolds for improved artificial tissues in regenerative medicine.

10.2 Biopolymer Scaffold Assembly

Assembling new biopolymer scaffolds is critical in the field of regenerative medicine. In addition to the desirable properties of biopolymers as a tissue scaffold material, these materials can be assembled with a wide range of techniques to create diverse functioning scaffolds for multiple tissue types. Three predominant manufacturing techniques for biopolymer scaffolds are assembly into bulk hydrogel scaffolds [12], electrospinning of fiber scaffolds [13], and three-dimensional (3D) printing of scaffolds [14]. Each of these techniques provides unique advantages for scaffold assembly, and scaffolds are increasingly designed to incorporate multiple assembly techniques and multiple types of biopolymers in order to better mimic native ECM.

10.2.1 Hydrogel Biopolymer Scaffolds

Hydrogels are composed of multiple crosslinked polymer chains in a 3D matrix that can absorb and retain significant amounts of water. Traditional polymers used for tissue scaffold hydrogels include synthetic polymers such as poly(ethylene oxide) (PEO), poly(vinyl alcohol) (PVA), and poly(propylene fumarate-*co*-ethylene glycol) (P(PF-*co*-EG)) and natural polymers including alginate, chitosan, collagen, and hyaluronic acid (HA) [15]. Hydrogel scaffolds made of these materials can be created with complex shapes through techniques such as emulsification, lithography, and micromolding, which facilitate their application for diverse uses [16]. Hydrogel scaffolds are well suited for use with multiple cell types because this type of scaffold can easily be made porous to promote mass transport and increase surface area for cell adhesion and also can be easily designed to release growth factors as they degrade [15]. Because they can be made from a variety of biopolymer types, hydrogels can be used as scaffolding for multiple tissue types by tuning their material composition to alter biochemical and mechanical properties of the hydrogel to match the tissue of interest. Biopolymer-based hydrogels are particularly suitable for replicating soft tissue types, as well as some hard tissues that are not required to withstand significant mechanical forces during healing, such as cranial defects [17].

There are multiple methods available for creating crosslinked hydrogel scaffolds, including chemical crosslinking schemes such as UV-induced photopolymerization and physical crosslinking schemes such as thermal gelation and crosslinking via ionic interactions [18]. Thermally crosslinked hydrogels can be created from many polymer types such as poly(L-lactic acid) (PLLA), polyethylene glycol (PEG), and chitosan [12]. Some thermally gelling hydrogels can be injected into an injury site as a liquid and are designed to undergo gelation after exposure to 37 °C body temperature, which makes thermal crosslinking a simple and attractive method of gelation [17]. However, thermally crosslinked hydrogels can suffer from problems including syneresis, in which the gel expels a significant fraction of its water content after gelation [19]. A common example of syneresis is the tendency of liquid to accumulate on the surface of yogurt when it is left unstirred for a long period. To address this problem, Mikos and coworkers investigated the use of a composite hydrogel that can gel under both thermal and chemical cues for mesenchymal stem cell (MSC) delivery and bone tissue engineering [17]. The hydrogel is composed of *N*-isopropylacrylamide (NiPAAm) copolymerized with epoxy pendant rings and lactone moieties. NiPAAm forms a thermally gelling hydrogel. The second component of the composite hydrogel is a diamine-functionalized polyamidoamine (PAMAM) crosslinker, which acts as a chemical crosslinker that can hydrolytically degrade over time. Combining thermal and chemical crosslinking gives the dual-gelling hydrogel improved kinetics and stability for injection into tissue defects compared with a single-gelling hydrogel [17]. To make the hydrogel a hospitable environment for cellular attachment, researchers investigated the incorporation of gelatin microparticles into the gel as a porogen that could be enzymatically degraded after gelation, leaving pores in the scaffold to facilitate mass transport and cellular attachment. Testing demonstrated this porogen-laden hydrogel supported MSC survival for up to 28 days and promoted calcium mineralization. This porous hydrogel, laden with MSCs and injected into a rat cranial defect, was able to promote bone bridging across the defect and mineralization of the hydrogel, demonstrating that the use of degradable porogen can be used to improve hard tissue formation in hydrogel scaffolds by recapitulating the natural porous structure of bone.

Photopolymerizable hydrogels are chemically crosslinked when visible or UV light strikes a photoinitiator molecule to release free radicals that interact with the polymer to form crosslinks [20]. Common examples of photopolymerization from other industries include the lithographically patterned photoresists used in the manufacture of electronics and microelectromechanical systems (MEMS) and resins used in dental applications. Advantages of photopolymerization for hydrogel crosslinking include fast curing times, high degrees of spatial and temporal control over crosslinking, including crosslinking in situ, and minimal heat generation [20]. Tuan and coworkers reported the development of a biopolymer hydrogel that employs a photopolymerization crosslinking scheme in a cartilage scaffold [21]. For this hydrogel, methacrylated poly-D,L-lactic acid/polyethylene glycol/poly-D,L-lactic acid (mPDLLA-PEG) was mixed with methacrylated hyaluronic acid (mHA) along with the photoinitiator lithium phenyl-2,4,6-trimethylbenzoylphosphinate (LAP). mPDLLA–PEG was used as the primary structural material, while mHA was incorporated into the scaffold to provide cell-binding ligands and promote a chondrogenic phenotype in cells growing on the scaffold. This hydrogel was photocrosslinked using projection stereolithography to produce 3D hydrogel scaffolds with complex geometries. Prior to crosslinking, the scaffold material was mixed with human adipose-derived stem cells (ASCs). The cells within this hydrogel were able to maintain high viability for up to 28 days and were driven toward a chondrocyte phenotype when cultured with tumor growth factor (TGF)-β3, as indicated by matrix deposition and gene expression analysis. However, material compressive strength was not maintained over the course of the experiment, dropping from 780 kPa at the beginning of the experiment to 240 kPa on day 28, demonstrating that although this material showed promise for development of a chondrocyte phenotype, it was unsuitable as a cartilage engineering material due to the loss of compressive strength over time.

Although there are many potential applications of biopolymer-based hydrogels for tissue engineering scaffold design, these materials often fall short when trying to match the mechanical properties of many tissue types, lacking tensile strength to recapitulate tissues such as tendon and lacking compressive strength to recapitulate the biomechanics of cartilage and bone. Electrospinning, another important method of scaffold fabrication, can be used to create biopolymer scaffolds with improved mechanical properties over hydrogels for applications such as tendon engineering while maintaining an environment that promotes cell attachment and growth.

10.2.2 Electrospinning of Biopolymer Scaffolds

Electrospinning is a method of developing fibrous scaffolds out of polymeric materials by extruding electrically charged liquid polymer into thin fibers, typically several hundred nanometers in diameter [22]. In order to electrospin a polymer, the polymer is placed into a syringe in a liquid solvent. The needle of the syringe, called the spinneret, is attached to the terminal of a DC power supply, and a collector is connected to the other terminal of the power supply, setting up an electric field between the spinneret and the collector. A high voltage (up to 30 kV) is applied across the device, which creates a high charge at the tip of the spinneret. As polymer is extruded at a constant rate from the spinneret, a liquid jet is formed from the polymer, creating a continuous stream of fiber propelled across the gap between the spinneret and the collector (Figure 10.1a) [22]. A solid fiber is formed at the collector because the solvent evaporates while the material is propelled through air toward the collector. A significant degree of control over fiber properties (diameter, surface roughness, and cross-sectional area) can be achieved by controlling variables such as material type, material viscosity, material conductivity, applied voltage, and extruder tip design [13]. The fibers can be randomly aligned at the collector by spinning onto a fixed collector plate or aligned in parallel by collecting the extruded fibers on a rotating mandrel [13]. Common biopolymers used in electrospinning for tissue engineering applications include poly(lactic-*co*-glycolic acid) (PLGA), poly(L-lactic acid-*co*-ε-caprolactone) (PLCL), collagen, silk, fibrinogen, and chitosan [13]. Electrospun fibers offer a biomimetic environment with high porosity within which cells can be successfully cultured in 3D. Recent advances in electrospinning scaffolds for tissue engineering include spinning with new composite materials [24, 25] and inclusion of polymer additives for enhanced

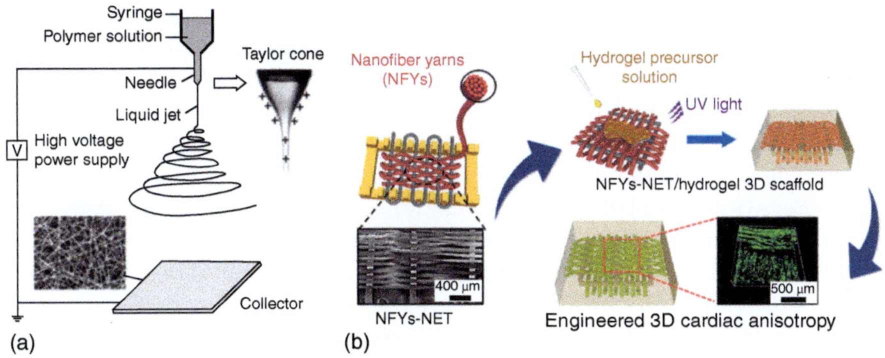

Figure 10.1 Electrospun tissue scaffolds. (a) Typical electrospinning setup. *Source:* Li and Xia 2004 [22]. Reproduced with permission of John Wiley and Sons. (b) A hybrid cardiac scaffold was formed by weaving electrospun nanofiber yarn and surgical sutures into a mat. These woven layers were combined with hydrogel to form a 3D scaffold with controllable layered anisotropy for cardiac engineering. *Source:* Wu et al. 2017 [23]. Reproduced with permission of American Chemical Society.

physical properties such as electrical conductivity, which can be useful for recreating electrically active tissue types such as myocardium [26]. Electrospun fibers can also undergo post-processing such as spinning the fibers into yarn that can act as a macroscopic scaffolding [23] or by braiding sheets of electrospun fibers together for improved tensile strength [27].

One noteworthy variation of the electrospinning technique is coaxial electrospinning. Coaxial electrospun fibers, composed of a core material and an outer shell material, can be formed by electrospinning with a coaxial spinneret setup with an inner needle inside an outer needle [24, 28]. This is a useful technique for creating composite fibers with properties from multiple biopolymers. In particular, the core material can be selected to provide mechanical support, while the shell material can be used to provide biocompatibility. The utility of this method for creating tissue engineering scaffolds has been demonstrated in a study by Ferreira and coworkers. They demonstrated that polycaprolactone (PCL) and gelatin can be co-electrospun to produce new biopolymer scaffolds for use as vascular grafts [24]. Methacrylated gelatin (GelMA) was used to form the outer shell of the electrospun fibers. Gelatin (denatured collagen) is a material that in this study was used to increase the hydrophilicity of the material and helped to reduce thrombosis and hemolysis when interacting with blood. Thrombosis is the formation of blood clots, and hemolysis is the destruction of red blood cells, both of which must be minimized for vascular graft engineering [29]. Inside the gelatin shell, PCL was used as the core of the fiber in order to give strength to the scaffold. Coaxial electrospinning was performed with an inner and outer needle and two syringe pumps. The resultant fibers were photocrosslinked to form a tight mesh. The researchers were able to control the hydrophobicity of the material by changing the ratio of GelMA to PCL, as measured by a water contact angle test. High GelMA concentration (75% GelMA/25% PCL) showed the best wettability. Hemolysis and thrombogenicity induced by this material were both reduced with increased percentage of gelatin in the fibers. Additionally, the biocompatibility of the material was demonstrated by culturing normal human dermal fibroblasts on the scaffold over seven days without any observed drop in cell viability over that time.

Electrospun biopolymer fibers alone can be used as scaffolds, or they can be incorporated as components in more complex scaffold architectures, as required by a particular application. Ma and coworkers have demonstrated a technique for developing hybrid biopolymer scaffolds composed of both electrospun fibers and hydrogel, with complex architecture designed to mimic native myocardium [23]. The base material of the scaffold was a yarn of electrospun nanofibers. The fibers were created from a mix of polylactic acid (PLA) and silk fibroin with carbon nanotube (CNT) additives, which was electrospun onto a water/ethanol solution as the collector. Fibers were then drawn off the liquid surface and onto a rotating mandrel in order to turn the electrospun fibers into a yarn with a precisely controlled diameter. This nanofiber yarn was then woven into a mat with surgical sutures as the cross-woven fibers (Figure 10.1b). This woven mat was seeded with cardiomyocytes in order to promote the alignment of the cardiomyocytes along the axis of the woven fibers to recapitulate native myocardium anisotropy. As an additional biomimetic step, multiple layers of this cardiomyocyte seeded mat were assembled by stacking layers at different fiber angles to mimic the structure of layered myocardium within which each successive layer is rotated several degrees relative to the adjacent layers. This layered structure was encased in a photocrosslinkable GelMA hydrogel and further seeded with endothelial cells in order to promote vascularization of the tissue. All tissues must be vascularized in order to ensure that cells are well perfused (i.e. within sufficient proximity to a capillary bed to receive oxygen and nutrients from the blood while cellular waste products are also removed) [30]. The resultant hybrid tissue construct contained multiple types of biopolymer with complex structure and alignment as well as multiple cell types. It demonstrated improvements in several metrics of myocardium mimicry, including increased actin and connexin 43 expression (connexin 43 is a gap junction protein that helps facilitate electrical signal transmission between myocytes), increased sarcomere length, and increased cell alignment compared with a control tissue consisting of only 3D GelMA hydrogel. This study demonstrates the utility of electrospun fibers as component materials within structurally complex multi-material biopolymer scaffolds. Complex scaffold assembly has been further advanced by a third major method of biopolymer scaffold assembly, 3D bioprinting.

10.2.3 Three-Dimensional Printing of Biopolymer Scaffolds

Tissues exhibit a great deal of spatial complexity, varying in their spatial distribution of ECM and cellular composition, their mechanical properties, their biochemical properties, and their physical shape. Tissues also contain complex structures such as blood vessels that are necessary in order to maintain the health of the tissue by providing access to nutrients, oxygen, and waste removal. Traditional methods for preparing tissue engineering scaffolds, including hydrogels and electrospun scaffolds, often fall short as ideal tissue replacements due to limits in the complexity of scaffolds that can be created with traditional manufacturing methods. Recently, 3D bioprinting has emerged as a cutting-edge technology platform for producing complex biomimetic tissue engineering scaffolds [14]. With 3D bioprinting, cells, biopolymers, and growth factors can be incorporated into bioinks that can be extruded by a three-axis printhead. Once printed, bioinks harden through various mechanisms such as temperature change, photopolymerization, or enzymatic crosslinking. This allows subsequent print layers to be placed above the previous layer, gradually building up a tissue of arbitrary shape in three dimensions [14]. This method can be used to prepare cell-impregnated biopolymer scaffolds based upon digital models of a tissue defect, allowing seamless geometrical integration of the tissue scaffold at the defect site for improved regenerative outcomes [31]. Recent improvements in 3D bioprinting include improvements in multi-material printing [32, 33], development of new bioink formulations [34, 35], alternative crosslinking methods for improved print resolution, and printing of pre-cultured organoids within a polymer matrix [36].

One of the principal advantages of tissue preparation using 3D printing is that complex structures of multiple materials can be created in a single manufacturing step. Zhao and coworkers reported a 3D printed urethra tissue construct composed of multiple biomaterial types [32]. The tissue construct was cylindrical, containing three concentric layers of material. The middle layer contained a PCL/PLCL blended scaffold designed to add strength to the tissue, while the innermost layer contained endothelial cells embedded within a gelatin/fibrin/HA hydrogel to mimic the endothelialized lumen of the urethra, and the outermost layer contained smooth muscle cells within the same type of hydrogel. The researchers were able to match the mechanical properties of a rabbit urethra both by tuning the ratio of PCL to PLCL within the scaffold and by altering the print structure to create complex geometries. By iterating and testing multiple designs, they determined that a spiral-shaped inner scaffold was optimum for providing the necessary mechanical properties to their engineered urethra. This rapid design–build–test cycle is made simple by making alterations to the digital computer-aided design (CAD) model used to generate the scaffold, which demonstrates a major advantage of 3D bioprinting over other scaffold preparation methods. In addition to mechanically mimicking the tissue to be regenerated, a tissue-engineered construct must provide the right microenvironment for cell growth and survival. This was accomplished in this study by co-printing a cell-laden gelatin/fibrin/HA hydrogel within the thermoplastic scaffold using a different extruder head loaded with the hydrogel bioink. The thermoplastic scaffold provided structural integrity, and the hydrogel provided an environment for cell growth. After printing, this construct was incubated with thrombin in order to crosslink the fibrin hydrogel, a necessary step to harden the hydrogel bioink so that it retained its structure. The results of this study showed that the cells within the hydrogel could proliferate and maintain their phenotypes (i.e. endothelial cells on the lumen side, smooth muscle cells on the outside) for up to seven days post-printing (Figure 10.2a).

Many 3D bioprinters such as the printer used in the study by Zhao and coworkers [32] accomplish multi-material printing by employing multiple printheads with separate extruders used for each bioink formulation. Khademhosseini and

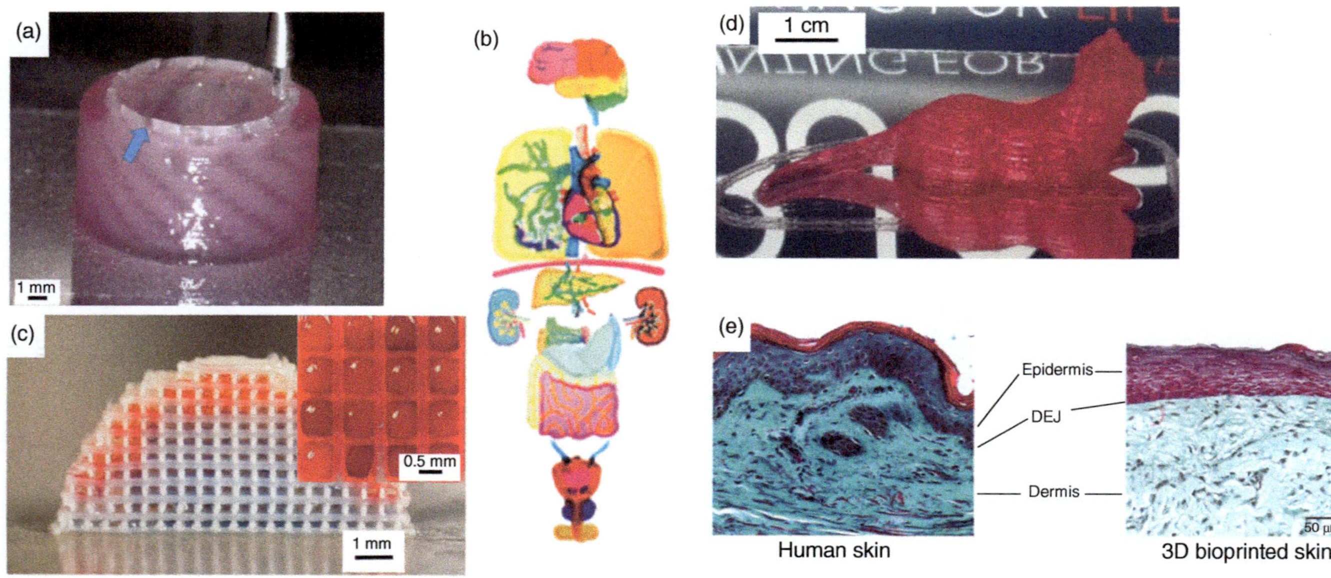

Figure 10.2 Three-dimensional bioprinted constructs. (a) A rabbit urethra tissue construct was prepared by bioprinting a spiral PCL/PLCL frame, with cell-laden fibrin-based hydrogels on the inner and outer surfaces. *Source:* Zhang et al. 2017 [32]. Reproduced with permission of Elsevier. (b) Multiple materials extruded from a single printhead were used to create model tissues resembling human organ systems. The multiple organ systems presented were printed separately and digitally combined into the same image. *Source:* Liu et al. 2017 [33]. Reproduced with permission of John Wiley and Sons. (c) A complex 3D cartilage tissue construct was created by printing a PEGT/PBT frame and filling the interstitial spaces in the scaffold with uniform chondrocyte-laden hydrogel microtissues. *Source:* Mekhileri et al. 2018 [36]. https://iopscience.iop.org/article/10.1088/1758-5090/aa9ef1/meta. Licensed under CC by 3.0. (d) A gelatin-based bioink used for creating the full-thickness human skin tissue can also be used to create arbitrarily shaped, digitally defined shapes. *Source:* Pourchet et al. 2017 [34]. Reproduced with permission of John Wiley and Sons. (e) A full-thickness human dermis was 3D printed from a scaffold-free gelatin-based bioink, and keratinocytes were subsequently cultured on top to create a structure with a dermis and epidermis, creating a distinct dermal-epidermal junction (DEJ). *Source:* Pourchet et al. 2017 [34]. Reproduced with permission of John Wiley and Sons.

coworkers have developed a bioprinter that can extrude up to seven different materials serially from a single printhead, massively decreasing print times by eliminating the need to physically switch between different printheads each time a material change is needed [33]. This was accomplished through development of a printhead receiving inputs from seven reservoirs, each of which contains a different bioink, and pneumatically switching the inputs on and off to control which material was extruded. They demonstrated this capability with nanosilicate-based bioinks to print a variety of complex structures, including a variety of human organ-shaped structures (Figure 10.2b), and structures with spatial gradients of hydroxyapatite. Biological tissues are rarely monolithic and contain spatial gradients of mechanical properties, molecules such as growth factors, and cell types. This makes the ability to print polymer scaffolds with gradient properties highly desirable for tissue engineering research [37].

A different method of assembling complex multi-material tissue constructs has been reported on by Woodfield and coworkers [36]. This study demonstrates a method of creating hybrid tissues consisting of chondrocyte perfused hydrogel microtissues approximately 1 mm in diameter, housed within a polyethylene glycol terephthalate/polybutylene terephthalate (PEGT/PBT) polymer scaffold (Figure 10.2c). A custom printer was developed with multiple printheads, one to lay down layers of thermoplastic polymer for the scaffold, and a second printhead was designed to isolate and deposit sequentially a single microtissue sphere into the interstitial spaces between sections of thermoplastic polymer. The microtissues used in this study were prepared ahead of time by culturing of cells in a photocrosslinkable methacrylated gelatin hydrogel in a high-throughput manner designed to produce many microtissue spheres. To create the full 3D tissue construct, first a layer of scaffold material was printed to create pores sized to house the 1 mm microtissues. Then the second printhead would isolate a single hydrogel microtissue and deposit into an interstitial site in the scaffold layer. This was done without damage to the microtissue and without loss of cell viability. By repeating this process for multiple layers, a 3D tissue construct composed of a stiff biopolymer-based scaffold integrated with modular organoid-like microtissues was assembled. This fabrication method enables modular tissue designs to be realized using prefabricated hydrogel microtissues assembled into custom shapes within a large organ scaffold. This technique could be combined with organoid technology to create tissues out of multiple organoids that closely mimic the physiology and developmental path of native tissue.

The development of multiple types of bioinks developed for 3D bioprinting is another enabling technology for creating more complex tissue structures. Gelatin is commonly used as a bioink for 3D printing because of its favorable cytocompatible properties, including the presence of signaling and binding motifs such as arginylglycylaspartic acid (RGD), its water solubility, and its enzymatic degradability [35]. However, the gelation temperature of many gelatin formulations is below the physiological temperature of 37 °C, so pure gelatin does not form a solid gel in situ. In order to modulate the gelation temperature of gelatin, it is commonly treated to add methacrylate groups to the gelatin, forming GelMA, which enables photopolymerization post-extrusion. Groll and coworkers proposed an alternate technology for crosslinking gelatin bioinks by functionalizing gelatin with thiols that can be dimerized using thiol–ene click chemistry, forming GelAGE (allylated gelatin) [35]. The advantages of GelAGE over GelMA as a bioink include an absence of nondegradable polymeric crosslinks and an ability to be photocrosslinked using visible light-sensitive photoinitiators instead of UV, which can be damaging to cells within the bioink. With this technology, the authors demonstrated viability of chondrocytes within 3D printed constructs for up to three weeks.

Marquette and coworkers leveraged the low phase transition temperature of gelatin to formulate a new bioink for 3D printing structures that maintain complex shapes with minimal support. They reported the development of a novel bioink that was used to print human dermis in a scaffold-free manner [34]. The bioink was formulated by mixing three different biopolymers: gelatin, alginate, and fibrinogen. Gelatin was used as a mechanical agent within the bioink because of its phase transition temperature at 35 °C. The ink was printed onto a cooled substrate, and the gelatin was used to enable the construct to hold its shape throughout the printing process. Alginate, the second biopolymer used, acted as a structural support once crosslinked in the presence of calcium. This was used to retain the shape of the print once the gelatin was solubilized at physiological temperature. Fibrinogen was the last biopolymer used, which both formed a hydrogel and aided in cell maturation due to cell adhesion motifs within the protein such as RGD, which are focal adhesion points for integrin mechanoreceptors on cell surfaces. By mixing this bioink formulation with human fibroblasts, the authors demonstrated printing of complex arbitrary 3D shapes (Figure 10.2d) as well as a 1 cm × 1 cm × 0.5 cm skin patch (Figure 10.2e). To complete the skin patch, a layer of keratinocytes was seeded onto the printed dermal patch. Histological analysis of the 3D printed tissue showed that the synthetic tissue closely approximated normal human skin, including expression of specific markers such as Ki67, cytokeratin 10, and filaggrin. Additionally, this synthetic tissue achieved a natural skin barrier function. Advantages of this method of artificial skin preparation include rapid assembly time and the ability to print skin in complex shapes, as demonstrated by using this bioink mixed with fibroblasts to print a human-ear-shaped structure.

In addition to the composition of bioinks, the method of extrusion and crosslinking in 3D bioprinting can impact the properties of the final 3D tissue. Burdick and coworkers developed a method of in situ crosslinking of 3D printed photocrosslinkable hydrogels that offers improvements in printing resolution and consistency [38]. One of the primary difficulties that must be overcome when printing hydrogel is that the bioink must have a low viscosity during extrusion in order to ensure that forces on the cells embedded within the hydrogel remain low in order to maintain cell viability [38]. However, the viscosity of the hydrogel must also increase rapidly once the material is extruded from a printhead so that the 3D printed construct maintains high spatial resolution. Traditional methods to accomplish this include pre-print crosslinking, in which the hydrogel is crosslinked immediately before it is extruded, often with UV light and a photoinitiator, and post-print crosslinking, in which the crosslinking agent is applied immediately post extrusion. Pre-print crosslinking can be useful to ensure high print resolution, but cells suffer low viability due to the increased

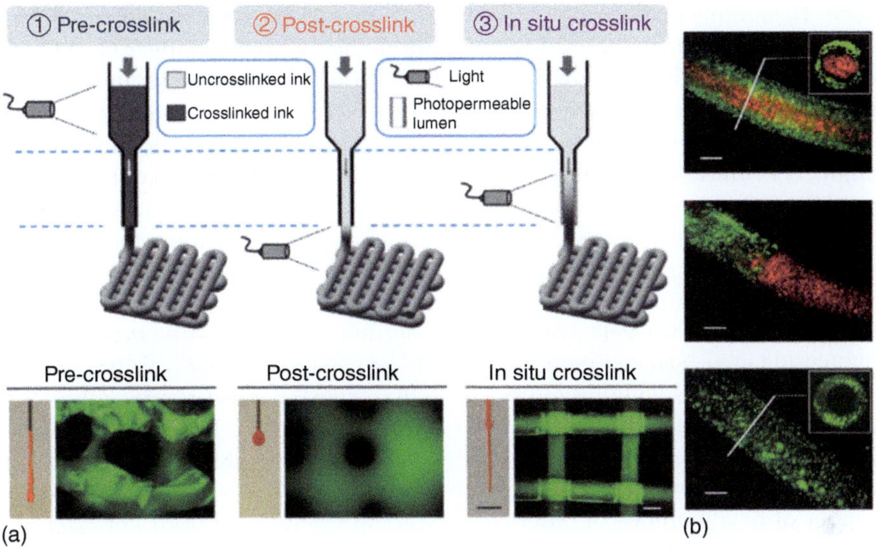

Figure 10.3 In situ crosslinking of photocrosslinkable hydrogels. (a) Printheads capable of photocrosslinking liquid hydrogels while they are in the process of being extruded allow 3D printed constructs to be formed with greater resolution compared with pre- or post-approaches. (b) In situ crosslinking can be used to create cell-laden filaments with complex architecture. *Source:* Ouyang et al. 2017 [38]. Reproduced with permission of John Wiley and Sons.

applied forces during printing of high viscosity hydrogel. Post-print crosslinking allows for high cell viability at the cost of resolution, as the material has time to spread out before it hardens. The Burdick group has developed a method of in situ crosslinking, in which UV light is applied to the bioink while it is being extruded from a glass capillary (Figure 10.3a). This capability was demonstrated using a variety of photocrosslinkable hydrogels including methacrylated hyaluronic acid (MeHA), methacrylated gelatin (GelMA), poly(ethylene glycol) diacrylate (PEGDA), and norbornene-functionalized hyaluronic acid (NorHA). Fibroblasts were mixed into each bioink for this study. In each case, cell viability remained high after seven days, along with high print resolution compared with pre- and post-print crosslinking. This method was used to produce extruded filaments of bioink with complex structures, including two material core–shell filaments, heterogeneous filaments, and hollow filaments (Figure 10.3b). The ability to print these structures at high resolution expands the potential space of scaffold designs that can be employed for tissue engineering.

10.3 Organ System Specific Biopolymer Scaffolds

The human body is a complex system composed of multiple organ systems with widely divergent functions. Accordingly, regeneration of each tissue type requires a tissue-specific approach. Researchers in regenerative medicine are interested in tissue engineering solutions for the entire range of human tissue. One of the key achievements of polymers engineering is the ability to precisely tune mechanical properties. With this in mind, two key fields of tissue engineering can be explored where biomechanical properties are a chief concern of the design engineer: musculoskeletal tissue engineering, consisting engineered bone, muscle, cartilage, and other connective tissue such as ligaments and tendons, as well as cardiovascular tissue engineering, consisting of artificial veins, arteries, and myocardium.

10.3.1 Biopolymers for Musculoskeletal System Regeneration

Designing advanced tissue engineering scaffolds is particularly important for engineering load-bearing tissues and other tissues that will be subjected to high mechanical forces including bone, ligaments, tendons, and cartilage [39]. Not only must the mechanical properties of the native tissue be matched, but the scaffold material should be designed to degrade at a rate coordinated with the regeneration of new tissue [40]. For recapitulating these structural tissues, a number of new methods of biopolymer scaffold design have been investigated. Researchers have created scaffolds using multiple biopolymer types so that the resultant scaffold both acts as a permissive environment for cell growth and as a mechanically robust structure [41]. Additionally, scaffolds have been created using established methods such as electrospinning, with subsequent processing steps such as braiding the resultant fibers into textile-like materials, which had made the scaffolds well suited for ligament and tendon repair [27]. Bone and cartilage tissue scaffolds have also been created from biopolymer microspheres that have been laser sintered to form a bulk scaffold with gradient properties [42].

10.3.1.1 Biopolymers for Bone Regeneration

Bone is a highly regenerative tissue naturally, but in many cases damage to the tissue can be significant enough that natural regeneration falls short, such as in critical size defects. Critical size bone defects are gaps in bone tissue that are large enough to be unbridgeable in the natural healing process [43]. In such cases healing can be improved with an engineered tissue implant that can provide mechanical support during healing and a scaffold on which new bone tissue can grow and propagate [40].

Many methods for manufacturing tissue engineering scaffolds are incompatible with cell survival and therapeutic protein viability due to manufacturing conditions including high temperature, UV light exposure, or presence of organic solvents [41]. In order to maximize cell viability on a bone engineering biopolymer scaffold, Yang and coworkers have developed a method for creating a PLGA/PEG microparticle-based scaffold with mechanical properties similar to cancellous bone, which can be produced under ambient conditions conducive to cell survival and protein viability [41]. PLGA and PEG were melted and blended together at a 93.5 : 6.5 (w/w) ratio, milled into a powder, and sieved to select particles between 50 and 100 μm. These particles were suspended in an aqueous carrier fluid to form a paste and extruded with a 3D bioprinter. To sinter the material after printing, it was incubated at 37 °C for two hours followed by 22 hours submerged in

0.9% sodium chloride. Testing for yield stress and Young's modulus showed that this material displayed properties similar to cancellous bone. To confirm that the manufacturing conditions were biocompatible, lysozyme was mixed into the material prior to printing, and then lysozyme release was measured for up to nine days, indicating that enzymatic activity was retained throughout the manufacturing procedure and subsequent scaffold degradation. Performing tests such as this to confirm biocompatibility through the scaffold manufacturing process is an important concern in tissue engineering.

The musculoskeletal system contains many tissues with gradient properties, such as the osteochondral interface where bone tissue makes a gradual biological and mechanical transition to cartilage tissue. The mechanical complexity of this transitional zone cannot easily be recapitulated with a single scaffold material, and more complex techniques are needed to develop tissue engineering scaffolds that can support regeneration within a damaged transitional zone. While additive manufacturing techniques such as 3D printing are well suited for developing scaffolds with integrated gradients, other techniques such as selective laser sintering of distinct layers of biopolymer-based microparticles have also been explored. For repair of the osteochondral transitional tissue zone, Zhang and coworkers have demonstrated a scaffold based on a gradient of PCL microspheres and HA/PCL microspheres. This scaffold was employed for improvements in osteochondral repair of critical size bone defects in rabbits [42]. The target site for integration of this scaffold was the osteochondral interface in knee joints where bone transitions to cartilage. At this site, the transition from bone to cartilage is gradual, with a biological and mechanical gradient. To recapitulate this gradient, the researchers prepared a layered scaffold by selective laser sintering of the PCL microspheres (Figure 10.4). By layering PCL microspheres on top of HA/PCL microspheres, they built a scaffold designed to sit at the osteochondral interface, with the HA microspheres being used to encourage an osteogenic phenotype and the PCL spheres to encourage a chondral phenotype. This scaffold exhibited

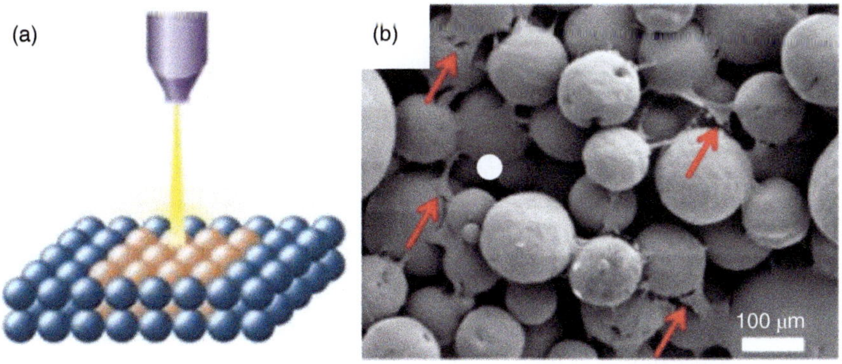

Figure 10.4 Selective laser sintering of PCL microspheres for bone scaffold engineering. (a) Schematic for sintering process. PCL microspheres are arranged and exposed to a laser in order to bond them together. (b) Electron micrograph of PCL microsphere scaffold. Cells cultured on the scaffold are indicated by red arrows. *Source:* Du et al. 2017 [42]. Reproduced with permission of Elsevier.

good mechanical properties and excellent biological properties, supporting the survival and proliferation of cells *in vitro*. This scaffold was implanted in an osteochondral defect in rabbit knees and analyzed at 6 and 12 weeks and compared with an untreated control and a PCL-only scaffold. Histological analysis revealed that the gradient scaffold significantly increased both bone and cartilage repair compared with PCL only and the untreated control. Overall filling of the defect, scaffold degradation, and bone and cartilage morphology all showed significant improvements with the gradient scaffold.

10.3.1.2 Biopolymers for Cartilage Regeneration

Cartilaginous tissue possesses inherently poor regenerative capability [44]. Regenerative medicine strategies that employ biopolymer-based scaffolds that can both promote chondrogenesis and act as mechanical reinforcement are promising treatment options for improving wound healing in cartilage. One of the most common cartilaginous tissues to be injured, and one of the most difficult to treat, is meniscus [45]. Tuan and coworkers have investigated the potential for treating meniscus injury with a bovine meniscus-derived ECM and PEG hydrogel scaffolds [45]. They investigated differences in healing outcomes between scaffolds prepared with the inner and outer portions of the bovine meniscus, which differ subtly in their ECM composition. Bovine meniscus was decellularized, ground into a powder, and mixed with PEG and human bone marrow MSCs. This mixture was photocrosslinked to produce a 3D cell-laden scaffold. They found that the inner ECM scaffold promoted a fibrocartilaginous phenotype in the cells, while outer membrane promoted a fibroblastic phenotype, recapitulating the phenotypic distribution of the cells found in the two distinct meniscus regions. This study demonstrated the potential for decellularized meniscus to act as a scaffold material and also highlight the importance of ECM composition and context in achieving a desired phenotype in the cells grown on the scaffold.

Another biopolymer-based scaffold for treatment of meniscus injury has been developed by the Tuan group as well [46]. In this study, methacrylated gelatin was used as a scaffold for meniscus repair, seeded with ASCs and the growth factor TGF-β3. This photocrosslinkable hydrogel was mixed with TGF-β3 prior to gelation to test the efficacy of scaffold-released growth factor in improving meniscus tear repair outcomes in comparison with treatment with TGF-β3-supplemented media alone. To perform *in vitro* meniscus repair tests, bovine meniscus biopsy explants were harvested, a tear was created, and the hydrogel was injected into the tear site and subsequently polymerized with visible light exposure. Tests of TGF-β3 release indicate that over a 21-day time course, 80% of the available TGF-β3 present in the scaffold was released. Tests were performed with both the presence and absence of TGF-β3 and adipose stem cells in the scaffold. The results of this study showed that methacrylated gelatin mixed with TGF-β3 was a superior scaffold material for supporting the differentiation of ASCs to a chondrogenic phenotype, which resulted in repair of meniscus tear *in vitro* as measured by histological and mechanical testing.

10.3.1.3 Biopolymers for Ligament and Tendon Regeneration

Biomaterials to replace ligaments and tendons must be able to withstand significant tensile loads, as well as support the growth of cells and deposition of ECM into the scaffold as it degrades [47]. To meet these requirements, Yang et al. developed a composite biomaterial with potential as a tendon graft by co-electrospinning of PCL and methacrylated gelatin (gelMA) [48]. This was done by preparing a solution of dissolved PCL and a second solution of dissolved GelMA. Each solution was placed in syringes positioned at opposing sides of a rotating mandrel during electrospinning (Figure 10.5). PCL is a high strength polymer with slow degradation *in vivo* and is used to provide structural support to the graft. However, PCL is hydrophobic and typically displays low wettability

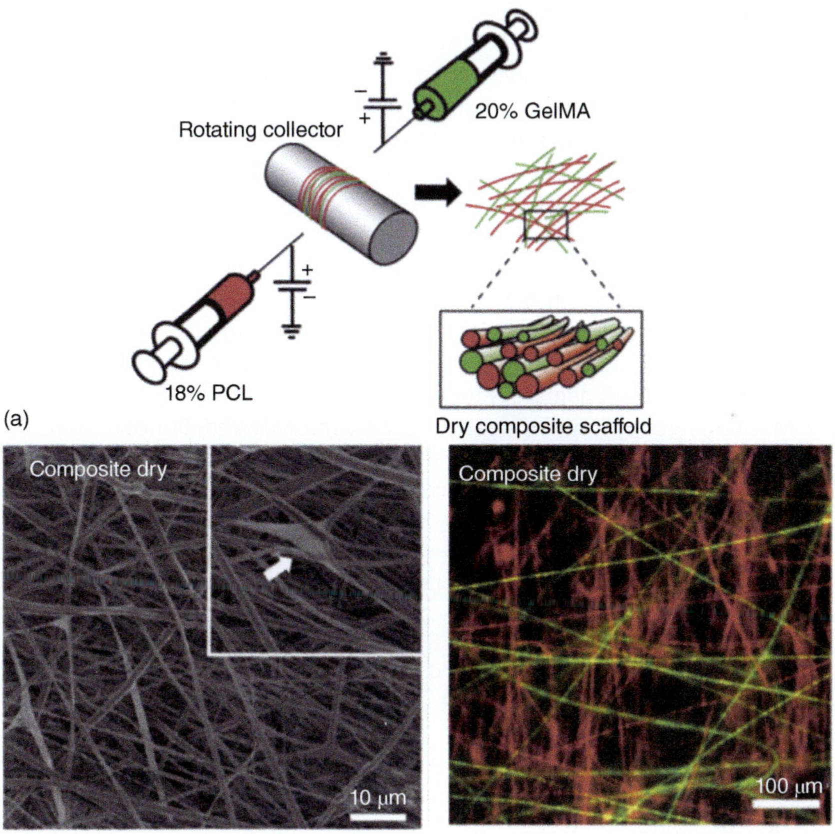

Figure 10.5 Biopolymer scaffolds for ligament and tendon regeneration. (a) Multipolymer electrospun scaffolds for tendon engineering were prepared by dual electrospinning of PCL and methacrylated gelatin (GelMA) onto the same rotating collector. (b) The composite electrospun material, visualized by electron microscopy and fluorescence microscopy. PCL fibers are shown in red, and gelatin fibers are shown in green. *Source:* Yang et al. 2016 [48]. Reproduced with permission of Elsevier.

and poor cell attachment. Methacrylated gelatin is a photocrosslinkable hydrogel that is conducive to cell growth and survival. By co-spinning these two fibers into a mesh, this material takes advantage of the high strength of PCL and the high biocompatibility of gelMA. To form a tendon graft tissue, sheets of the PCL/GelMA graft were placed in a partially overlapping configuration, wetted with a GelMA/photoinitiator solution, and exposed to visible light for crosslinking. Human ASCs cultured on this scaffold developed a chondrogenic phenotype when exposed to TGF-β3, upregulating scleraxis and tenascin C, two markers of a chondrocyte phenotype. Although this scaffold was able to promote a chondrogenic phenotype in the cells cultured in it, it lacked adequate mechanical strength for tendon tissue replacement, particularly tensile strength. The authors suggested that this material could be further improved by assembling the scaffolds with textile patterns, such as braids and weaves to improve tensile strength.

Researchers have also investigated textile patterning in electrospun scaffolds for improving scaffold tensile strength in tendon engineering applications. To this end, researchers conducted a study to test the mechanical and biological properties of two different biopolymer scaffolds with different textile patterns [27]. Nanofibrous sheets of PCL and PLLA were electrospun onto a rotating mandrel in order to align the fibers. Then, three sheets of those fibers were combined either by stacking in layers or by braiding layers together. Human bone marrow-derived stem cells were seeded onto these scaffolds for analysis. The researchers concluded that stacked sheets of electrospun fibers created scaffolds that supported greater cell proliferation and migration, while braided scaffolds showed improved tensile strength. This result demonstrates the feasibility of using textile patterns to improve mechanical properties of electrospun nanofiber scaffolds; however more work is needed to improve the suitability of these scaffolds as an environment for cell culture.

10.3.2 Biopolymers for Cardiovascular System Regeneration

The 2010 Global Burden of Disease Study by Lozano et al. estimated that one in four deaths worldwide is attributable to cardiovascular disease (CVD), up from one in five in 1990 [49]. CVD remains the leading cause of death in the United States, leading to over 660 000 deaths in 2016 [50]. For multiple types of CVD culminating in end-stage heart failure, the primary treatment option is organ transplantation. The use of heart transplantation as a treatment option has been rising, with over 5000 procedures performed globally in 2015 [51]. Survival rates from heart transplantation have also been increasing overtime as procedures are refined. However, organ transplantation as a treatment option for CVD is limited due to factors such as access to donor hearts, chronic immunosuppression treatment, and complications arising from transplant rejection [51]. Other treatment options are necessary to augment or replace transplantation in many cases. Mechanical circulatory support devices such as left ventricular assist devices are increasing in prevalence and are often employed for circulatory support while the patient awaits a transplant. Mechanical support devices are not without their own set of complications, however, which can include infection, device failure, perioperative hemorrhage, thromboembolism, and hemolysis [52]. New tissue engineering strategies employing biopolymers to repair, replace, and regenerate damaged cardiovascular tissue show great promise for closing the gaps in current treatment methodologies.

10.3.2.1 Biopolymers for Vascular Regeneration

Blood vessels are load-bearing tissue designed to withstand normal blood pressure and maintain their structural integrity [53]. Replacement tissue for blood vessel regeneration must perform this task without inducing thrombosis or hemolysis, so excellent biocompatibility of the biopolymer used in blood vessel assembly is a requirement of blood vessel scaffold design [53]. Vascular grafts made of nondegradable polymers including polyethylene terephthalate (PET) (commonly called Dacron) and expanded polytetrafluoroethylene (ePTFE) (commonly called Gore-Tex) are used for small diameter vascular grafts but suffer from low long-term patency rates [29]. Patency rate is defined as the likelihood that a vessel will remain open over a given length of time. New generations of vascular grafts are being developed with new kinds of biopolymers, which are designed to act as traditional tissue engineering scaffolds by degrading at the rate that cells can remodel the tissue in order to produce a natural tissue at the end of the healing process [54]. In order to minimize the chances of reactions with substances in the blood that lead to thrombosis, the luminal surface of these grafts must present an endothelial-like surface. Additional mechanical and mass transport concerns arise in cases where the graft is expected to exhibit vasoactivity, dilating and constricting in response to physiological signals to regulate blood flow [53].

In addition to synthetic vascular scaffolds composed of biopolymers such as the coaxially electrospun PCL/GelMA fiber scaffolds developed by Ferreira and coworkers that were previously discussed [24], new sources of natural biopolymers are being assessed as materials for vascular tissue engineering. Researchers have studied whether amniotic membrane (AM) could be a suitable material for vascular tissue engineering by testing its hemolytic and thrombogenic activity [55]. AM is the innermost layer of the placenta and is a potential source of natural biomaterial. The tests showed that both the epithelial and mesenchymal sides of the AM display low hemolysis and thrombogenesis, comparable with ePTFE, a current clinically employed synthetic vascular graft material. Platelet aggregation was lower on both sides of the AM than ePTFE. Analysis of platelet morphology attached to AM showed that platelets were more round and less dendritic and spread out, compared with ePTFE and glass. In this study, the AM was not decellularized before these tests. The authors conclude that AM may be a suitable material for vascular grafts.

While developing grafts for repair of large and small diameter blood vessels is an important goal in regenerative medicine, vascularization of all engineered tissues is a central challenge underlying all endeavors in regenerative medicine [30]. All cells within a tissue must be within close proximity (100–200 μm) of capillaries supplying blood in order to survive long term [56]. Engineered tissues that are grafted in place without vasculature often suffer from loss of viability and cell death shortly after engraftment. New strategies for inducing angiogenesis in grafted tissue, the sprouting of new blood vessels from existing vasculature, are of great interest. A common method to encourage angiogenesis within an engineered tissue graft is to seed the graft with pro-angiogenetic growth factors such as vascular endothelial growth factor (VEGF) [56]. However, in many cases, finer control over the patterning of new vasculature is required in order to form efficient branching fluidic channels with low tortuosity. A study by Segura and coworkers looked at the impact of integrin-specific activation on angiogenesis morphology in hydrogel scaffolds, *in vivo* and *in vitro* [57]. Integrin-specific activation was accomplished in this study by engineering fibronectin to present new cell-binding ligands and mixing the recombinant fibronectin into tissue scaffolds. Fibronectin is a protein component of the ECM and is responsible for mediating many of the interactions between cells and the ECM by presenting cell-binding ligands such as RGD. The principal class of cell membrane proteins that bind to RGD domains is integrins. Integrins are heterodimeric proteins composed of an α subunit and a β subunit (e.g. $\alpha2\beta1$). In addition to mediating cell attachment to ECM, integrins are signaling molecules that relay information from the environment into the cell, dictating cellular phenotype in response to specific binding cues [58]. The Segura lab engineered recombinant fibronectin fragments to specifically interact with $\alpha3/\alpha5\beta1$ integrins and $\alpha v\beta1$ integrins, which are integrins present in endothelial cells that have been implicated in control of vascular patterning [57]. The engineered fibronectin fragments were integrated into a HA hydrogel matrix along with nanocapsules for secreting VEGF. Vascular patterning in the skin of mice implanted with the HA hydrogel showed that αv-specific hydrogel increased the tortuosity of new blood vessels compared with $\alpha3/\alpha5$. In an ischemic stroke model in mice, $\alpha3/\alpha5\beta1$-specific hydrogel implanted at the stroke site showed improvements in vessel integrity and reduced tortuosity compared with $\alpha v\beta1$-specific hydrogel. These results showed that incorporating engineered ECM components into tissue engineering scaffolds was an effective strategy for improving vascularization, which is important for regeneration of all tissue types.

10.3.2.2 Biopolymers for Cardiac Regeneration

Extending the useful lifetime of a damaged heart is of great interest as an alternative to transplantation and mechanical device support after a heart attack. In post-myocardial infarction (MI, commonly known as "heart attack"), the infarcted tissue undergoes rapid remodeling, wall thinning, and degradation of contractile function, which can ultimately lead to congestive heart failure. One potential regenerative medicine treatment for MI is the immediate application of a tissue-engineered cardiac patch to the site of injury. These constructs can support and regenerate areas of damaged myocardium after MI and may be able to extend the useful lifetime of a damaged heart [59]. Cardiac patches must be able to act as either a replacement for the damaged tissue or act to repair the damage and regenerate a healthy functional tissue. There are multiple hurdles that must be overcome in order to develop safe, clinically efficacious cardiac patches. One challenge to developing cardiac patches is correct sourcing of cells necessary to seed the cardiac patch, and equally important is the development of polymer scaffolds that can act as mechanical support for the cells while providing these cells with the correct queues for growth, survival, and function. Cells directly applied to the damaged area without a scaffold often have difficulty achieving a foothold on which to grow and successfully integrating into the native tissue [60]. Without support from a scaffold, cells alone are often washed away or suffer low rates of viability [60]. Accordingly, design of biopolymer-based scaffold material that fulfills these requirements is a major research goal.

Cardiac tissue is composed of laminated sheets of cardiomyocytes that are oriented to align the direction of their contractions. Successive layers of the myocardium are slightly rotated relative to the adjacent layers, creating a helical alignment pattern throughout the thickness of the myocardium [61]. This architecture causes myocyte contractions propagating through the heart to gives rise to the large organ-level contractions necessary for driving blood flow. Cardiomyocytes are active cells that are electrically coupled to their neighbor cells in order to synchronize their beating. This coordinated beating throughout the myocardium is what provides sufficient motive force for pumping [61]. Much of the activity in the cardiac regenerative medicine field is focused on creating new biopolymer tissue scaffolds that can recapitulate the structure of native myocardium. Recently, new scaffolds for cardiac tissue engineering that promote coordinated electrical activity [62], anisotropic cell alignment [63], and tissue-matched biomechanics [25] have been reported.

Tuning the electrical properties of a 3D printed PCL biopolymer scaffold was the goal of a study by Yoon and coworkers. To do this, the researchers characterized the physical properties of 3D printed PCL scaffolds infused with multiwall CNTs [62]. CNTs were added to the polymer bioink in order to increase the conductivity of the scaffold to improve its suitability as a cardiac patch material. Electrical conductivity in the scaffold helped the cardiomyocytes to synchronize their beating. Researchers tested PCL, and PCL mixed with 1%, 3%, and 5% CNTs, and measured the thermal, mechanical, and biodegradation properties of the resultant material. They also assayed these materials for H9C2 myoblast growth. The study demonstrated that it is possible to tune the melting temperature, elastic modulus, hardness, electrical conductivity, and enzymatic degradation rate of PCL–CNT scaffolds by changing the CNT fraction. Conductivity of the material was increased up to 9 orders of magnitude for 5% CNT composition compared with PCL without CNT. The authors concluded that PCL mixed with 1% CNT demonstrated optimal properties for use as a scaffold for myoblasts.

The effect of fiber alignment within biopolymer cardiac patch scaffold has also been investigated. The native ECM in the left ventricle, where tissue is often damaged during MI, is aligned along the circumferential direction of the heart. This ECM alignment serves to align myotubes and promote pumping action in the heart. Accordingly, controlling the orientation of electrospun fibers in a cardiac patch scaffold may improve the effectiveness of the cardiac patch by influencing the orientation of newly formed myotubes within the patch. This hypothesis was tested by Wagner and coworkers [63]. These investigators aimed to determine the role fiber alignment played in the regenerative effectiveness of a biodegradable electrospun poly(ester carbonate urethane)urea (PECUU) cardiac patch. This patch was tested in an *in vivo* rat model of chronic infarction.

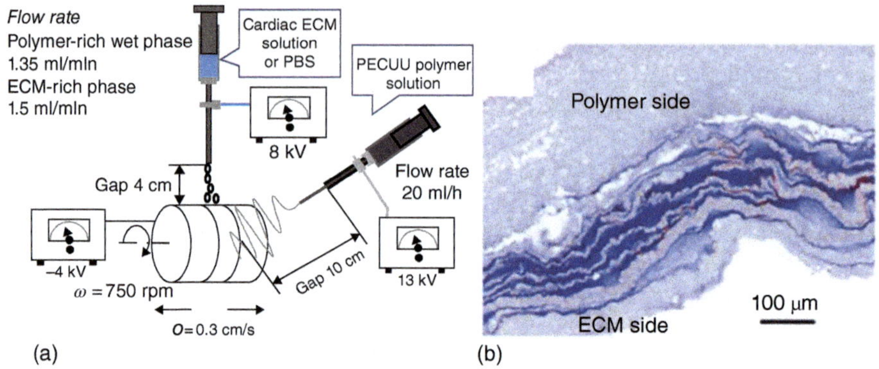

Figure 10.6 Cardiac biopolymer scaffolds. (a) A hybrid cardiac patch was formed with electrospun PECUU polymer, and cardiac ECM on the epicardial side of the patch. (b) Distinct layers were created with PECUU as the polymer side, and cardiac ECM on the ECM side of the patch. *Source:* D'Amore et al. 2016 [63]. Reproduced with permission of Elsevier.

The PECUU cardiac patches were prepared by electrospinning onto a rotating mandrel in order to align the fibers parallel to each other. When cardiac patches prepared from these electrospun fibers were grafted onto the infarct site in the rat heart, they were aligned either circumferentially or longitudinally with the native myocardium's direction. This study delivered a negative result, and the researchers found no significant difference in cardiac function or regeneration due to fiber alignment. Wagner et al. then in this same study reported on the creation of a third scaffold by integrating decellularized porcine ECM into the cardiac patch. This patch was bilayered, with an ECM-rich side and a polymer-only side (Figure 10.6a). With this ECM-integrated patch, a new experiment was performed in order to determine the effectiveness of decellularized ECM as a scaffold material. The patch with the epicardial facing ECM-rich layer demonstrated improvements in mechanical support, prevention of LV wall degeneration, improvements in echocardiogram measured functionality, decreased scar formation, and improvement in angiogenesis compared with the PECUU-only scaffolds, suggesting that material composition may be a more important determinant of scaffold success than scaffold alignment.

10.4 Summary and Outlook

Looking toward the future of tissue engineering, new biopolymer technologies will be required to further improve scaffold design and create engineered tissues that can be widely adopted in the clinic. Successful scaffolds must balance biocompatibility, biodegradability, mechanical properties, and manufacturability while promoting cell attachment of the correct cell types, preventing attachment of incorrect cell types (as in platelet adhesion during thrombosis), and promoting proliferation, differentiation, and ECM deposition. The field of synthetic biology, which employs the tools of genetic engineering to create new materials, fuels, and therapies, will be valuable for producing new biopolymers with unique properties advantageous for tissue engineering. The beginnings of this paradigm can be seen in the Segura group's creation of recombinant fibronectin for hydrogels with engineered RGD domains to promote angiogenesis [57]. Developing new biopolymers and new methods of producing biopolymer scaffolds for tissue engineering is a critical step in the advancement of regenerative medicine.

References

1 Freed, L.E., Vunjak-Novakovic, G., Biron, R.J. et al. (1994). Biodegradable polymer scaffolds for tissue engineering. *Bio/Technology* 12: 689–693.

2 Atala, A., Kasper, F.K., and Mikos, A.G. (2012). Engineering complex tissues. *Science Translational Medicine* 4: 160rv112.

3 Baddour, J.A., Sousounis, K., and Tsonis, P.A. (2012). Organ repair and regeneration: an overview. *Birth Defects Research Part C: Embryo Today: Reviews* 96: 1–29.

4 Zhang, Z., Ortiz, O., Goyal, R., and Kohn, J. (2014). Biodegradable polymers. In: *Principles of Tissue Engineering (Fourth Edition)* (eds. R. Lanza, R. Langer and J. Vacanti), 441–473. Boston: Academic Press.

5 Ma, P.X. (2008). Biomimetic materials for tissue engineering. *Advanced Drug Delivery Reviews* 60: 184–198.

6 Hynes, R.O. (2009). The extracellular matrix: not just pretty fibrils. *Science* 326: 1216.

7 Badylak, S.F., Freytes, D.O., and Gilbert, T.W. (2009). Extracellular matrix as a biological scaffold material: Structure and function. *Acta Biomaterialia* 5: 1–13.

8 Hutmacher, D.W. (2000). Scaffolds in tissue engineering one and cartilage. In: *The Biomaterials: Silver Jubilee Compendium* (ed. D.F. Williams), 175–189. Oxford: Elsevier Science.

9 Kumbar, S.G., James, R., Nukavarapu, S.P., and Laurencin, C.T. (2008). Electrospun nanofiber scaffolds: engineering soft tissues. *Biomedical Materials* 3: 034002.

10 Zhu, Y. and Wagner, W.R. (2019). Design principles in biomaterials and scaffolds. In: *Principles of Regenerative Medicine (Third Edition)* (eds. A. Atala, R. Lanza, A.G. Mikos and R. Nerem), 505–522. Boston: Academic Press.

11 Lee, K., Silva Eduardo, A., and Mooney David, J. (2011). Growth factor delivery-based tissue engineering: general approaches and a review of recent developments. *Journal of the Royal Society Interface* 8: 153–170.

12 Lee, K.Y. and Mooney, D.J. (2001). Hydrogels for tissue engineering. *Chemical Reviews* 101: 1869–1880.

13 Pham, Q.P., Sharma, U., and Mikos, A.G. (2006). Electrospinning of polymeric nanofibers for tissue engineering applications: a review. *Tissue Engineering* 12: 1197–1211.

14 Murphy, S.V. and Atala, A. (2014). 3D bioprinting of tissues and organs. *Nature Biotechnology* 32: 773.

15 Drury, J.L. and Mooney, D.J. (2003). Hydrogels for tissue engineering: scaffold design variables and applications. *Biomaterials* 24: 4337–4351.

16 Khademhosseini, A. and Langer, R. (2007). Microengineered hydrogels for tissue engineering. *Biomaterials* 28: 5087–5092.
17 Vo, T.N., Shah, S.R., Lu, S. et al. (2016). Injectable dual-gelling cell-laden composite hydrogels for bone tissue engineering. *Biomaterials* 83: 1–11.
18 Hennink, W.E. and van Nostrum, C.F. (2012). Novel crosslinking methods to design hydrogels. *Advanced Drug Delivery Reviews* 64: 223–236.
19 Ekenseair, A.K., Boere, K.W.M., Tzouanas, S.N. et al. (2012). Synthesis and characterization of thermally and chemically gelling injectable hydrogels for tissue engineering. *Biomacromolecules* 13: 1908–1915.
20 Nguyen, K.T. and West, J.L. (2002). Photopolymerizable hydrogels for tissue engineering applications. *Biomaterials* 23: 4307–4314.
21 Sun, A.X., Lin, H., Beck, A.M. et al. (2015). Projection stereolithographic fabrication of human adipose stem cell-incorporated biodegradable scaffolds for cartilage tissue engineering. *Frontiers in Bioengineering and Biotechnology* 3: 115.
22 Li, D. and Xia, Y. (2004). Electrospinning of nanofibers: reinventing the wheel? *Advanced Materials* 16: 1151–1170.
23 Wu, Y., Wang, L., Guo, B., and Ma, P.X. (2017). Interwoven aligned conductive nanofiber yarn/hydrogel composite scaffolds for engineered 3D cardiac anisotropy. *ACS Nano* 11: 5646–5659.
24 Coimbra, P., Santos, P., Alves, P. et al. (2017). Coaxial electrospun PCL/Gelatin-MA fibers as scaffolds for vascular tissue engineering. *Colloids and Surfaces B: Biointerfaces* 159: 7–15.
25 Liu, Y., Wang, S., and Zhang, R. (2017). Composite poly(lactic acid)/chitosan nanofibrous scaffolds for cardiac tissue engineering. *International Journal of Biological Macromolecules* 103: 1130–1137.
26 Wang, L., Wu, Y., Hu, T. et al. (2017). Electrospun conductive nanofibrous scaffolds for engineering cardiac tissue and 3D bioactuators. *Acta Biomaterialia* 59: 68–81.
27 Rothrauff, B.B., Lauro, B.B., Yang, G. et al. (2017). Braided and stacked electrospun nanofibrous scaffolds for tendon and ligament tissue engineering. *Tissue Engineering Part A* 23: 378–389.
28 Yarin, A.L. (2011). Coaxial electrospinning and emulsion electrospinning of core–shell fibers. *Polymers for Advanced Technologies* 22: 310–317.
29 Xue, L. and Greisler, H.P. (2003). Biomaterials in the development and future of vascular grafts. *Journal of Vascular Surgery* 37: 472–480.
30 Novosel, E.C., Kleinhans, C., and Kluger, P.J. (2011). Vascularization is the key challenge in tissue engineering. *Advanced Drug Delivery Reviews* 63: 300–311.
31 Li, L., Yu, F., Shi, J. et al. (2017). In situ repair of bone and cartilage defects using 3D scanning and 3D printing. *Scientific Reports* 7: 9416.
32 Zhang, K., Fu, Q., Yoo, J. et al. (2017). 3D bioprinting of urethra with PCL/PLCL blend and dual autologous cells in fibrin hydrogel: an in vitro evaluation of biomimetic mechanical property and cell growth environment. *Acta Biomaterialia* 50: 154–164.
33 Liu, W., Zhang, Y.S., Heinrich, M.A. et al. (2017). Rapid continuous multimaterial extrusion bioprinting. *Advanced Materials* 29: 1604630.

34 Pourchet, L.J., Thepot, A., Albouy, M. et al. (2017). Human skin 3D bioprinting using scaffold-free approach. *Advanced Healthcare Materials* 6: 1601101.

35 Bertlein, S., Brown, G., Lim, K.S. et al. (2017). Thiol–ene clickable gelatin: a platform bioink for multiple 3D biofabrication technologies. *Advanced Materials* 29: 1703404.

36 Mekhileri, N.V., Lim, K.S., Brown, G.C.J. et al. (2018). Automated 3D bioassembly of micro-tissues for biofabrication of hybrid tissue engineered constructs. *Biofabrication* 10: 024103.

37 Bracaglia, L.G., Smith, B.T., Watson, E. et al. (2017). 3D printing for the design and fabrication of polymer-based gradient scaffolds. *Acta Biomaterialia* 56: 3–13.

38 Ouyang, L., Highley, C.B., Sun, W., and Burdick, J.A. (2017). A generalizable strategy for the 3D bioprinting of hydrogels from nonviscous photo-crosslinkable Inks. *Advanced Materials* 29: 1604983.

39 Vacanti, C.A. and Vacanti, J.P. (2014). Engineering cartilage and other structural tissues: principles of bone and cartilage regeneration. In: *Principles of Tissue Engineering (Fourth Edition)* (eds. R. Lanza, R. Langer and J. Vacanti), 1267–1274. Boston: Academic Press.

40 Shrivats, A.R., Alvarez, P., Schutte, L., and Hollinger, J.O. (2014). Bone regeneration, Principles of Tissue Engineering (Fourth Edition). In: (eds. R. Lanza, R. Langer and J. Vacanti), 1201–1221. Boston: Academic Press.

41 Sawkins, M.J., Mistry, P., Brown, B.N. et al. (2015). Cell and protein compatible 3D bioprinting of mechanically strong constructs for bone repair. *Biofabrication* 7: 035004.

42 Du, Y., Liu, H., Yang, Q. et al. (2017). Selective laser sintering scaffold with hierarchical architecture and gradient composition for osteochondral repair in rabbits. *Biomaterials* 137: 37–48.

43 Bosch, C., Melsen, B., and Vargervik, K. (1998). Importance of the critical-size bone defect in testing bone-regenerating materials. *Journal of Craniofacial Surgery* 9: 310–316.

44 Simon, T.M. and Jackson, D.W. (2018). Articular cartilage: injury pathways and treatment options. *Sports Medicine and Arthroscopy Review* 26: 31–39.

45 Shimomura, K., Rothrauff, B.B., and Tuan, R.S. (2017). Region-specific effect of the decellularized meniscus extracellular matrix on mesenchymal stem cell-based meniscus tissue engineering. *The American Journal of Sports Medicine* 45: 604–611.

46 Sasaki, H., Rothrauff, B.B., Alexander, P.G. et al. (2018). In vitro repair of meniscal radial tear with hydrogels seeded with adipose stem cells and TGF-beta3. *The American Journal of Sports Medicine* 46: 2402–2413.

47 Goulet, F., Auger, F.A., Cloutier, R. et al. (2014). Tendon and ligament tissue engineering. In: *Principles of Tissue Engineering (Fourth Edition)* (eds. R. Lanza, R. Langer and J. Vacanti), 1275–1287. Boston: Academic Press.

48 Yang, G., Lin, H., Rothrauff, B.B. et al. (2016). Multilayered polycaprolactone/gelatin fiber-hydrogel composite for tendon tissue engineering. *Acta Biomaterialia* 35: 68–76.

49 Lozano, R., Naghavi, M., Foreman, K. et al. (2012). Global and regional mortality from 235 causes of death for 20 age groups in 1990 and 2010: a systematic analysis for the Global Burden of Disease Study 2010. *The Lancet* 380: 2095–2128.

50 Underlying Cause of Death 1999-2016 on CDC WONDER Online Database (2017). *Data are from the Multiple Cause of Death Files, 1999–2016, as compiled from data provided by the 57 vital statistics jurisdictions through the Vital Statistics Cooperative Program*. Centers for Disease Control and Prevention, National Center for Health Statistics.

51 Lund, L.H., Khush, K.K., Cherikh, W.S. et al. (2017). The registry of the International Society for Heart and Lung Transplantation: thirty-fourth adult heart transplantation report – 2017; focus theme: allograft ischemic time. *The Journal of Heart and Lung Transplantation* 36: 1037–1046.

52 Birks, E.J. (2010). Left ventricular assist devices. *Heart* 96: 63.

53 Nerem, R.M. (2003). Role of mechanics in vascular tissue engineering. *Biorheology* 40: 281–287.

54 Soletti, L., Hong, Y., Guan, J. et al. (2010). A bilayered elastomeric scaffold for tissue engineering of small diameter vascular grafts. *Acta Biomaterialia* 6: 110–122.

55 Kakavand, M., Yazdanpanah, G., Ahmadiani, A., and Niknejad, H. (2015). Blood compatibility of human amniotic membrane compared with heparin-coated ePTFE for vascular tissue engineering. *Journal of Tissue Engineering and Regenerative Medicine* 11: 1701–1709.

56 Rouwkema, J., Rivron, N.C., and van Blitterswijk, C.A. (2008). Vascularization in tissue engineering. *Trends in Biotechnology* 26: 434–441.

57 Li, S., Nih, L.R., Bachman, H. et al. (2017). Hydrogels with precisely controlled integrin activation dictate vascular patterning and permeability. *Nature Materials* 16: 953–961.

58 Hynes, R.O. (2002). Integrins: bidirectional, allosteric signaling machines. *Cell* 110: 673–687.

59 Atala, A., Lanza, R., Mikos, T., and Nerem, R. (2018). *Principles of Regenerative Medicine*. Elsevier Science.

60 Kitsara, M., Agbulut, O., Kontziampasis, D. et al. (2017). Fibers for hearts: a critical review on electrospinning for cardiac tissue engineering. *Acta Biomaterialia* 48: 20–40.

61 Li, Z. and Guan, J. (2011). Hydrogels for cardiac tissue engineering. *Polymers* 3: 740–761.

62 Ho, C.M.B., Mishra, A., Lin, P.T. et al. (2016). 3D printed polycaprolactone carbon nanotube composite scaffolds for cardiac tissue engineering. *Macromolecular Bioscience* 17: 1600250.

63 D'Amore, A., Yoshizumi, T., Luketich, S.K. et al. (2016). Bi-layered polyurethane – extracellular matrix cardiac patch improves ischemic ventricular wall remodeling in a rat model. *Biomaterials* 107: 1–14.

Index

Biopolymers for Biomedical and Biotechnological Applications, First Edition.
Edited by Bernd H. A. Rehm and M. Fata Moradali.

d

n